Dynamics of Molecular Biology

Dynamics of Molecular Biology

Edited by Erik Pierre

New York

Published by Syrawood Publishing House,
750 Third Avenue, 9th Floor,
New York, NY 10017, USA
www.syrawoodpublishinghouse.com

Dynamics of Molecular Biology
Edited by Erik Pierre

International Standard Book Number: 978-1-68286-651-1 (Hardback)

Cataloging-in-Publication Data

Dynamics of molecular biology / edited by Erik Pierre.
p. cm.
Includes bibliographical references and index.
ISBN 978-1-68286-651-1
1. Molecular biology. 2. Biomolecules. I. Pierre, Erik.
QH506 .D96 2019
572.8--dc23

TABLE OF CONTENTS

PREFACE

Molecular biology is the study of biomolecules, their activities and interactions with each other in various systems of the cell. This area of study has relevance across other scientific fields of research such as genetics, bioinformatics, immunology, biochemistry, etc. Molecular cloning, gel electrophoresis and polymerase chain reaction are some of the important techniques of molecular biology. Different approaches, evaluations, methodologies and advanced studies in this field have been included in this book. While understanding the long-term perspectives of the topics, the book also makes an effort in highlighting their impact as a modern tool for the growth of this discipline. Scientists and students actively engaged in this field will find this book full of crucial and unexplored concepts.

This book is a result of research of several months to collate the most relevant data in the field.

When I was approached with the idea of this book and the proposal to edit it, I was overwhelmed. It gave me an opportunity to reach out to all those who share a common interest with me in this field. I had 3 main parameters for editing this text:

1. Accuracy – The data and information provided in this book should be up-to-date and valuable to the readers.
2. Structure – The data must be presented in a structured format for easy understanding and better grasping of the readers
3. Universal Approach – This book not only targets students but also experts and innovators in the field, thus my aim was to present topics which are of use to all

Thus, it took me a couple of months to finish the editing of this book.

I would like to make a special mention of my publisher who considered me worthy of this opportunity and also supported me throughout the editing process. I would also like to thank the editing team at the back-end who extended their help whenever required.

Editor

1

How to perform RT-qPCR accurately in plant species? A case study on flower colour gene expression in an azalea (*Rhododendron simsii* hybrids) mapping population

Ellen De Keyser[1*], Laurence Desmet[1], Erik Van Bockstaele[1,2] and Jan De Riek[1]

Abstract

Background: Flower colour variation is one of the most crucial selection criteria in the breeding of a flowering pot plant, as is also the case for azalea (*Rhododendron simsii* hybrids*)*. Flavonoid biosynthesis was studied intensively in several species. In azalea, flower colour can be described by means of a 3-gene model. However, this model does not clarify pink-coloration. The last decade gene expression studies have been implemented widely for studying flower colour. However, the methods used were often only semi-quantitative or quantification was not done according to the MIQE-guidelines. We aimed to develop an accurate protocol for RT-qPCR and to validate the protocol to study flower colour in an azalea mapping population.

Results: An accurate RT-qPCR protocol had to be established. RNA quality was evaluated in a combined approach by means of different techniques e.g. SPUD-assay and Experion-analysis. We demonstrated the importance of testing noRT-samples for all genes under study to detect contaminating DNA. In spite of the limited sequence information available, we prepared a set of 11 reference genes which was validated in flower petals; a combination of three reference genes was most optimal. Finally we also used plasmids for the construction of standard curves. This allowed us to calculate gene-specific PCR efficiencies for every gene to assure an accurate quantification. The validity of the protocol was demonstrated by means of the study of six genes of the flavonoid biosynthesis pathway. No correlations were found between flower colour and the individual expression profiles. However, the combination of early pathway genes (*CHS, F3H, F3'H* and *FLS*) is clearly related to co-pigmentation with flavonols. The late pathway genes *DFR* and *ANS* are to a minor extent involved in differentiating between coloured and white flowers. Concerning pink coloration, we could demonstrate that the lower intensity in this type of flowers is correlated to the expression of *F3'H*.

Conclusions: Currently in plant research, validated and qualitative RT-qPCR protocols are still rare. The protocol in this study can be implemented on all plant species to assure accurate quantification of gene expression. We have been able to correlate flower colour to the combined regulation of structural genes, both in the early and late branch of the pathway. This allowed us to differentiate between flower colours in a broader genetic background as was done so far in flower colour studies. These data will now be used for eQTL mapping to comprehend even more the regulation of this pathway.

Keywords: RT-qPCR, Flower colour, RNA quality, noRT, Standard curves, Reference genes, Gene expression, Pink

* Correspondence: Ellen.dekeyser@ilvo.vlaanderen.be
[1]Institute for Agricultural and Fisheries Research (ILVO)-Plant Sciences Unit, Caritasstraat 21, 9090, Melle, Belgium
Full list of author information is available at the end of the article

Background

As for all flowering plants, flower characteristics and especially flower colour are among the most important features for pot azalea (*Rhododendron simsii* hybrids) breeding. Flavonoids account for this pigmentation in azalea [1,2]. The flavonoid biosynthesis pathway is one of the best studied biochemical pathways in plants, especially in petunia and snapdragon [3-7]. Flavonoids are synthesized by a branched pathway that yields both coloured pigments (anthocyanins) and colourless co-pigments (flavonols). In De Cooman et al. [8], it was observed that the azalea co-pigment formation follows a slightly aberrant pathway compared to anthocyanin production (Figure 1). Anthocyanins tend to occur mainly as cyanidins, azaleatin is the most common flavonol in azalea [2]. Azalea flower colour ranges from purple through carmine red, red, pink and white. Furthermore, azalea flowers can also be picotee type, with a different-coloured centre and margin, or flecked. The latter is expected to be caused by transposon activities [9]. Flower colour segregation in azalea can be predicted by a Mendelian model encompassing 3 major genes (P, W & Q; [10]). Purple flower colour is dominant over all other colours and is encoded by P. In the absence of the allele for P, W differentiates between (red) coloured (W-) and white flowers (ww). Q encodes for co-pigmentation by means of flavonols; in combination with the allele for W it results in carmine red flowers. Red flowers are recessive for Q (qq). This model does not clarify the presence of pink flowers, but the authors suggested pink to be a gradation in pigment. Also Sasaki et al. [11] state that flower colour intensity is determined by the amount of anthocyanin present. By means of image analysis, De Keyser et al. [12] recently confirmed in azalea that pink can be seen as (carmine) red at a lower intensity level. Studying the gene expression levels of the flavonoid biosynthetic genes could be informative to shed a light on this pink mystery as well. By means of the transgenic approach, Nakamura et al. [13] created pink torenia plants by down regulation of *flavonoid 3′-hydroxylase* (*F3′H*) and *flavonoid 3′,5′-hydroxylase* (*F3′5′H*) genes and also Boase et al. [14] reported that the suppression of the latter gene resulted in reduced colour intensity. The past decade, genetic engineering is explored widely for the modification of floricultural plants (reviewed in [15]). Expression levels of the targeted genes were always determined in order to identify their correlation to the flower colour phenotype [13,16-18]. The exploration of natural flower colour differences by means of gene expression studies is only done between a limited number of genotypes, e.g. in cyclamen [19], *Ipomoea* [20], *Freesia hybrida* [21], azalea [22,23] or *Oncidium* [24]. No data are currently available on the consistent effect of the studied genes in other genotypes with the same flower colour. Moreover, the quantification methods used in the aforementioned studies are not the most accurate. Some studies still describe the use of Northern blots [18,24] or semi-quantitative RT-PCR (reversed transcription PCR) [16,19,21,23,24], others do use quantitative RT-PCR (RT-qPCR) but limit themselves to the comparative Cq (quantification cycle) method [25] in combination with the use of only a single non-validated reference gene. However, multiple, assay-validated reference genes are considered to be an essential component of a consistent RT-qPCR assay [26-30]. mRNA quantification can potentially be a very powerful and reliable technique for investigating gene expression, but only if handled thoughtfully [26,31]. Due to the sensitivity and in order to increase accuracy, the technique was optimised intensively the past decades at all crucial steps from RNA isolation up to the final quantification (reviewed in [31,32]).

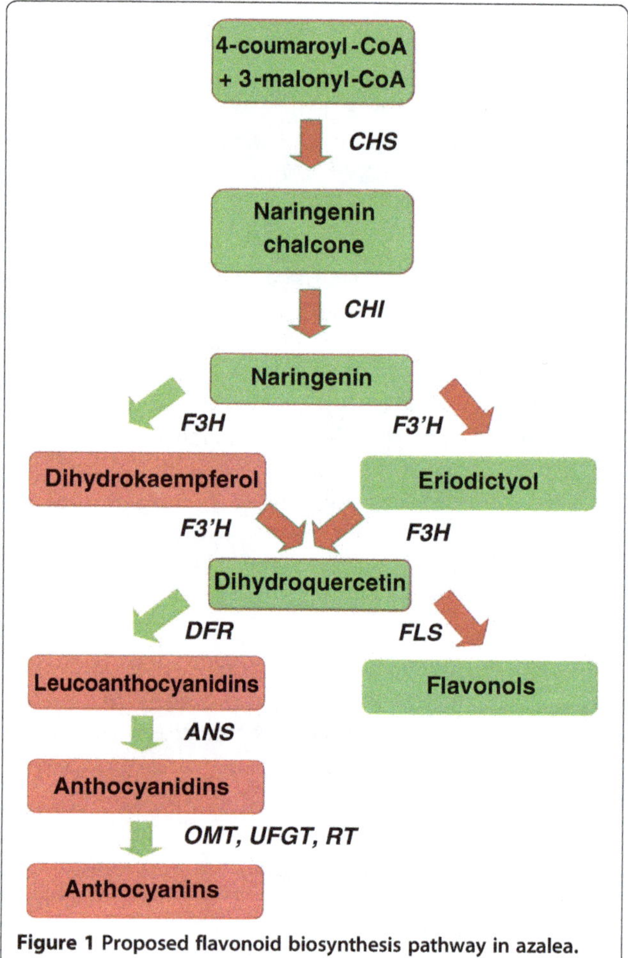

Figure 1 Proposed flavonoid biosynthesis pathway in azalea. The pathway only leads to the production of cyanidin pigments and is redrafted after [8,34]. *CHS: chalcone synthase; CHI: chalcone isomerase; F3H: flavanone 3-hydroxylase; F3′H: flavonoid 3′-hydroxylase; DFR: dihydroflavonol 4-reductase; ANS: anthocyanidin synthase; OMT: O-methyltransferase; UFGT: UDP-glucose:flavonoid 3-O-glucosyltransferase; RT: rhamnosyl transferase; FLS: flavonol synthase.*

MIQE-guidelines (*M*inimum *I*nformation for Publication of *Q*uantitative Real-time PCR *E*xperiments; [26]) were set in order to stimulate the scientific community to quantify in an accurate manner and also to provide all essential data when publishing gene expression studies. However, in plant science, still too many papers on gene expression are published with inaccurate quantification [27-29,33], as was also illustrated for flower colour.

Hence, the aim of this paper is dual.

1. The establishment of a reliable RT-qPCR protocol for transcriptional profiling that can be applied in all plant species, even when only limited transcriptomic data are available. Optimisation at crucial steps is described into detail, with a focus on RNA quality, reference gene validation, the use of noRT (no Reversed Transcriptase) samples and the implementation of plasmid-derived standard curves for PCR efficiency correction.
2. Study of gene expression in relation to flower colour in an azalea mapping population to identify correlations that are not limited to specific genotypes but are consistent over the whole azalea gene pool. Ultimately, the idea is to use these gene expression data to study flower colour in a genetical genomics approach.

Results

Sampling

In azalea flowering, generally four developmental stages are considered: closed buds (stage 1), buds showing colour at the top but with the scales still present (stage 2), candle stage without any scales left (stage 3) and the opened flower (stage 4) [23]. Expression of both the early gene *CHS* (*chalcone synthase*) and the late gene *DFR* (*dihydroflavonol 4-reductase*) appeared to be highest in stage 3 [23], hence this stage was selected for the evaluation of flower colour gene expression. Nakatsuka et al. [34] report a higher expression in azalea for some of the early flavonoid biosynthesis genes in stage 2, but these are only 2-fold differences. We therefore preferred to quantify the expression profile of all genes on the same sample, which would allow us to correlate expression profiles of the different genes in our analysis.

RNA quality control

Azalea RNA concentration varied tremendously between samples and was for some samples too low (Additional file 1) to test all genes in one RT-qPCR experiment. Hence we decided to extract RNA in duplicate from each sample. These technical replicates were then pooled after DNase treatment and purified together as one sample. RNA purity was measured spectrophotometrically. Contaminating proteins are displayed at an absorbance optimum of 280 nm, an $A_{260/280}$ ratio above 1.8 is considered of an acceptable RNA purity although 2 would be optimal [35]. Concerning polysaccharide and polyphenol contamination, $A_{260/230}$ is measured. A value of 2.5 means free of contamination [36], 2 is acceptable. However, the absorbance ratio's only reflect RNA purity [26,37] but not RNA integrity [37]. Absorption ratios were satisfying, except for low-concentrated samples (<15 ng/μl) where both $A_{260/230}$ and $A_{260/280}$ were clearly decreased. The low absorption ratios could indicate the presence of potential inhibitors. However, the reliability of the measurement can also be questioned in case of low RNA concentrations.

Performing a SPUD assay is considered to be the method of choice to evaluate the influence of inhibitory components on the RT-qPCR performance [32,38]. Therefore a subset of 14 randomly selected samples was used for a SPUD analysis. The difference in mean Cq-value between the SPUD control and RNA/cDNA samples did not exceed the variation within the SPUD control group (Figure 2) and remained below the proposed cut-off value of 1 Cq [39]. This confirmed that no PCR inhibitors were present in spite of the low absorption ratios in 3 samples (Additional file 1).

Finally, RNA integrity was checked on the same subset of samples. In order to see how degradation evolved in our own material, we constructed a degradation series. A decrease of the ribosomal peaks and a shift in the electropherogram towards the so-called fast region [40] is clearly noticed (Figure 3). A visible degradation was also spotted on the gel-view (Figure 3). For low-concentrated samples, gel views were even the only reliable indicator for quality since the signal was too weak to verify on the electropherogram. Based on the degradation series, RNA was considered to be degraded when the 25S/18S rRNA ratio was below 1; degradation also becomes very well noticeable in the virtual gel view at this point (lane 4 and 5, Figure 3). According to these settings, all tested RNA samples were graded as good quality. Consequently, the robustness of our RNA isolation procedure from flower petals was demonstrated; RNA samples could even be placed for 15 hours at room temperature, without any visible degradation (data not shown). Hence, RNA quality results were extrapolated to all cDNA samples isolated from azalea flower buds in this study.

Amplification specificity

Amplification of DNA in cDNA samples could result in an overestimation of the actual gene expression level of a gene or, even worse, in the false detection of expression. Developing primers spanning an intron or targeting exon-exon junctions can prevent co-amplification of DNA during RT-qPCR. Alignments with homologous sequences were made for all target genes (Table 1).

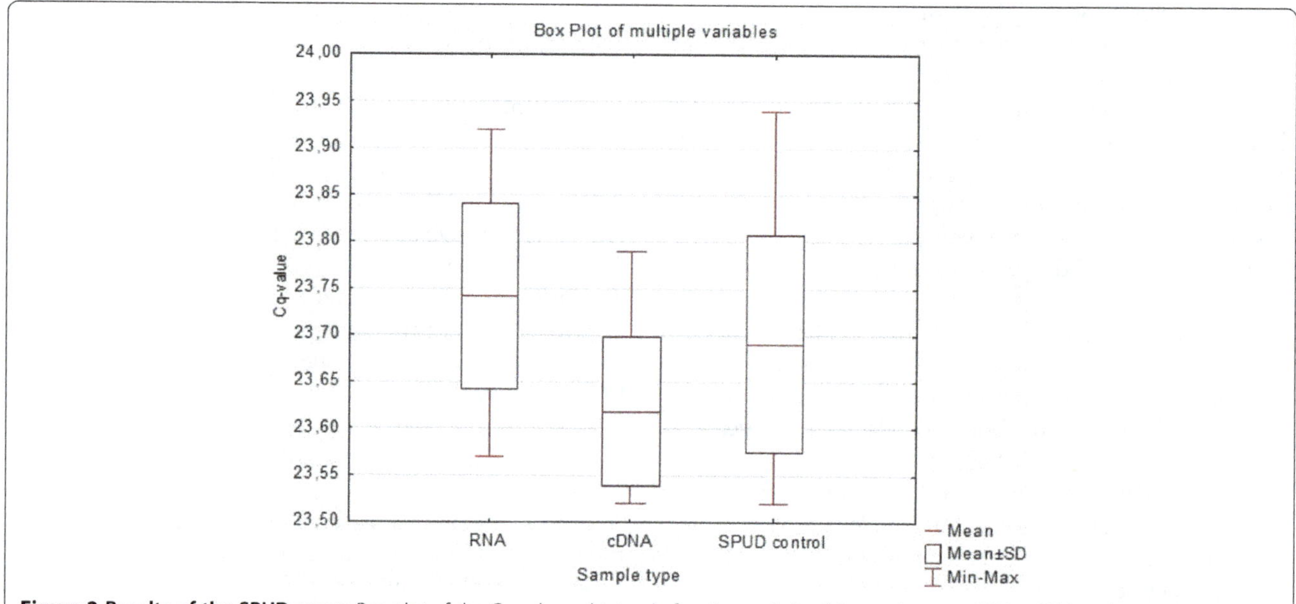

Figure 2 Results of the SPUD assay. Box plot of the Cq values obtained after the analysis of 3 sample types (RNA, cDNA and control) in a SPUD assay. For RNA/cDNA 14 different samples were measured in duplicate, 14 replicates were used for the SPUD control.

No introns were present in *CHS*; intron-spanning primers were developed in *ANS* (*anthocyanidin synthase)* and *DFR*. In *FLS (flavonol synthase)* and *F3′H (flavonoid 3′-hydroxylase)* primers amplified a single exon but were located at the 3′ end of the sequence to reduce the influence of RNA degradation. The azalea *F3H* (*flavanone 3-hydroxylase)* fragment was too short and covered only a single exon. EST (Expressed Sequence Tags) sequences of the reference genes (Table 2) could not be evaluated for the presence of introns since their functional annotation was not specific enough. Hence, not all primers were intron-spanning and some introns were too small to prevent co-amplification of DNA [32]. Therefore DNA contamination had to be checked for after all. NoRTs were included for all samples and amplification was performed on these noRTs with all primer sets (both reference and target genes). In case of amplification of noRTs, contamination was considered to be negligible when the difference in Cq between the noRT and the sample was above seven cycles. In that case, at

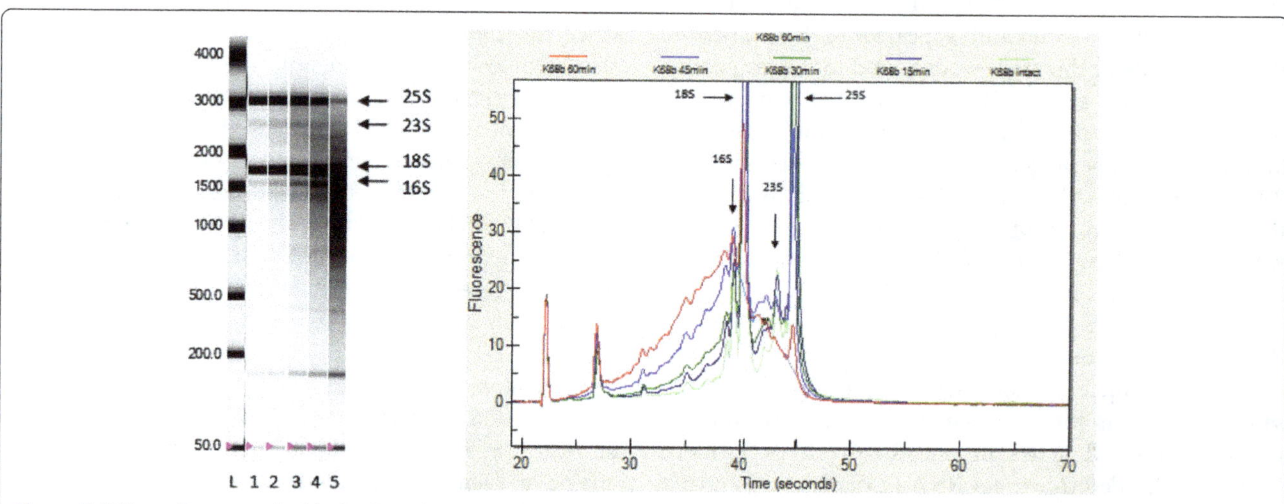

Figure 3 RNA quality control with the Experion (Bio-Rad). Electropherogram (right) and virtual gel-view (left) of an RNA degradation series that was constructed by heating an RNA sample for 0, 15, 30, 45 and 60 min at 80°C. The loading marker and small RNA band and cytoplasmic 18S and 25S as well as 16S and 23S chloroplast and mitochondrial ribosomal bands are indicated with arrows. Lanes: (L) size standard, (1) intact RNA, (2) 15 min, (3) 30 min, (4) 45 min, (5) 60 min. Intensity settings can vary between lanes.

Table 1 Target genes

Code	Gene	Acc. No.	Primer (5'-3')	Ampl.	Position
ANS	*anthocyanidin synthase*	AB289596	CCAAGAATCCGTCCGACTACA GGTTAGGCCTCTCAGGTGCTT	65 bp	Exon1/2
CHS	*chalcone synthase*	AJ413277	TGGGAATCAACGGTTTTGGAA CTCGGGCTTAAGGCTCAACTT	151 bp	Exon1
DFR	*dihydroflavonol 4-reductase*	AJ413278	CGTCATGAGGCTGCTTGAAC AAAGCTCCCTTCCTCGTTGAG	151 bp	Exon1/2
F3H	*flavanone 3-hydroxylase*	AB289594	GGGCTCCAGGCCACTAGAG ATGGTCGCCCAAATTGACAA	87 bp	Exon2
F3'H	*flavonoid 3'-hydroxylase*	AB289597	AAGAGCTGGACTCAATTGTTGGA CCTTGATGATGGCTTGGAGGTA	87 bp	Exon3
FLS	*flavonol synthase*	AB289599	CAAGGATGTCATGGGCTGTGT CGTTAATGAGCTCCGGAATAGG	75 bp	Exon3

Primer pairs for target genes were developed using Primer Express 2.0 (Applied Biosystems). EMBL accession numbers and the length of the amplicons are indicated. The position of the amplicons at the genomic DNA level is marked.

Table 2 Reference genes

Gene	Acc. No.	Functional annotation	Primer (5'-3')	Ampl.
GAPDH	FN552706	*Glyceraldehyde 3-phosphate dehydrogenase*	TCGGAATCAACGGTTTTGGAA CACTTGACCGTGAACACTGT	151 bp
HK5	AM932886	*Histone H3*	GAAACTCCCATTCCAGAGGCT GCATGGATGGCACAGAGGTT	153 bp
HK47	AM932894	*Nucleosome assembly protein*	GGTATAGGATTGACAATCCCAAGG CATTCAATCTCCGTCCCTATCG	151 bp
HK65	AM932901	*Protein kinase regulatory subunit γ*	CGGCAGTTAGGAGCTACCTCG CCCTCACCGTCCACAACATAG	151 bp
HK92	FN552699	*Heterotrimeric G-protein,a subunit*	ATCACAGTCATCCATGCCAATG CGCCGCCAATTTCTGATAGT	151 bp
HK96	AM932905	*Expansin*	AGGTTCACAATCAATGGCCAC TGTTGCTCTGCCAATTCTGC	151 bp
HK112	FN552700	*3-deoxy-D-arabino-heptulo-sonate 7-phosphate synthase*	CTCCTCCCTTCCTCCCAATC GTAACCGTTGTGCTCCCTACAGTC	152 bp
HK129	AM932909	*Protein phosphatase*	TGCAAAGATCGAATGCACGA CCTGCAAACGGAACTCGAGA	165 bp
HK134	FN552701	*Chlorophyll a/b binding protein CP24 precursor*	CGGTTGCTCCCAAAAAGTCTT CTCCGCTTCTCGGTACCACT	158 bp
HK156	FN552702	*Cytochrome P450 mRNA*	AGCCATGACCATCTTCGCTT GGCGATGATGCAAACGAGTT	156 bp
HK164	FN552703	*Chlorophyll a/b binding protein*	AAAACCTCTTCTCTTGCAAACCAT CTTGCCGACAGACTTCCTCAT	151 bp
HK173	FN552704	*Pyruvate dehydrogenase*	GGTGCGAGATTGGTATTTGGA TTGAACTCCCAAAGCCATTGT	151 bp
HK190	FN552705	*Protein disulphide isomerase*	CGTATCGATCATCGGCTCGT CACACCACGGAGCGTAGAACT	152 bp

Primer pairs for candidate reference genes were developed using Primer Express 2.0 (Applied Biosystems) based on EST-fragments (described by their EMBL accession number). The length of the amplicons and the putative function annotated to the sequences is indicated.

least 128-fold less contaminating DNA was present compared to cDNA. This is even above the five cycles that are the default setting for the same feature in qBase$^+$ (Biogazelle), the software module that was developed by Hellemans et al. [41] for RT-qPCR data analysis. Only three samples amplified using the *DFR primers* and one sample using the *F3′H* primers were considered to be contaminated. Hence, these particular data were discarded from the dataset and only a single biological replicate was used instead for further calculations.

Reference genes

The possible conservation of gene expression stability across different plant species [27] was an opportunity to select conventionally used reference genes in azalea. However, in a crop with only little sequence information available, this required degenerate PCR, with a low success-rate. Only *GAPDH* (*glyceraldehyde 3-phosphate dehydrogenase*) could be isolated as such. Hence, 13 fragments were selected based on putative functions from an azalea EST database [42] as candidate reference genes (Table 2). Amplification patterns of two of these genes (HK134 and HK190) did not satisfy in flower petals (data not shown). The expression of the 11 remaining reference genes was determined in petals of eight azalea cultivars and standard-curve derived quantities were imported into geNorm [30]. With a pair wise variation $V_{2/3}$ of 0.145, the use of two reference genes seems sufficient (see Additional file 2). However, this value is nearby the proposed cut-off value of 0.15 and with $V_{3/4}$ being only 0.108, three reference genes appeared to be most favourable for normalisation of gene expression in azalea flower buds. These validated reference genes have an optimal M-value (for homogeneous tissues) below 0.5 (M = 0.368 [41]) and belong to different functional classes. Hence they are not likely to be co-regulated, what enforces their trustworthiness for combination into a normalisation factor [30]. Unfortunately, when analysing the second assay, quite some noRTs amplified with one of the selected reference genes (HK173). Therefore this gene had to be eliminated as a reference gene for the final analysis. Hence, normalisation was done with a normalisation factor based on two reference genes (HK5 and HK129). The normalisation factor had a less optimal M-value of 0.524 over the three assays, still this solution was preferred over using unreliable expression data for normalisation.

Standard curves

Plasmids containing the fragments of interest were used for the construction of a relative dilution series. Initially, reproducibility and stability of these dilution series was a major problem. However, this problem could be circumvented by linearization of the plasmids [43] and by diluting the linear fragments in a yeast tRNA solution. The addition of a carrier such as yeast tRNA prevents the loss of very little quantities in the smallest dilution steps [44]. In this way, the error on the linear regression of the dilution series was not worth mentioning. The SD(E) values (Additional file 3) were always below 0.01.

It is possible to analyze a standard curve only once for each gene and to apply the derived PCR efficiency in all further analysis. However, we preferred to work with run-specific amplification efficiencies to avoid the introduction of confounding technical variation. This was the best option, since amplification efficiencies of the individual standard curves clearly differ in time (Additional file 3), The PCR efficiency of e.g. HK129 varied between 0.94 and 0.81. The efficiencies for *F3′H* and certainly for *DFR* were far below the optimum, but by using the run-specific amplification efficiency, this difference in efficiency was accounted for and calculation errors were significantly reduced between assays.

Flower colour gene expression

We aimed at finding gene expression differences for six key genes of the flavonoid biosynthesis pathway between four flower colour groups: white, red, carmine red and pink in an azalea mapping population. Initially we selected five seedlings from each flower colour group in combination with the (pink-coloured) parents of the crossing population (assay 1; see Additional file 1). No significant correlations were found between the colour grouping and the gene expression levels of the individual genes (data not shown). Since these data were in due course to be used for eQTL (expression Quantitative Trait Locus) mapping, we gradually expanded the dataset in order to determine the minimal sample size with sufficient power in eQTL mapping. First 29 samples were added to the dataset (assay 2; see Additional file 1) and Kruskal-Wallis analysis was performed to determine the power of eQTL mapping. This yielded only highly significant ($p < 0.001$) correlations for *CHS*. Eventually, we needed a total of 70 siblings to obtain enough power to detect (preliminary) eQTLs for 50% of the genes (Figure 4). We therefore considered 70 samples (+2 parents) to be sufficient for our gene expression study.

The results of all three assays were hence combined in a single dataset with 23 white flowers, 22 red, 19 carmine red and 8 pink ones. Due to the spread of the analysis over 3 different time points, inter-run calibration (IRC) was required to correct for potential run-to-run variation. Using (multiple) IRCs as advised by [41] was not feasible since these were not implemented consequently in every assay. Instead, the overall gene expression level per plate (and per gene) was used for inter-run calibration. The geometric mean was preferred over the arithmetic mean for calculating this IRC factor, as the former controls better for possible outlying values [30]. To verify

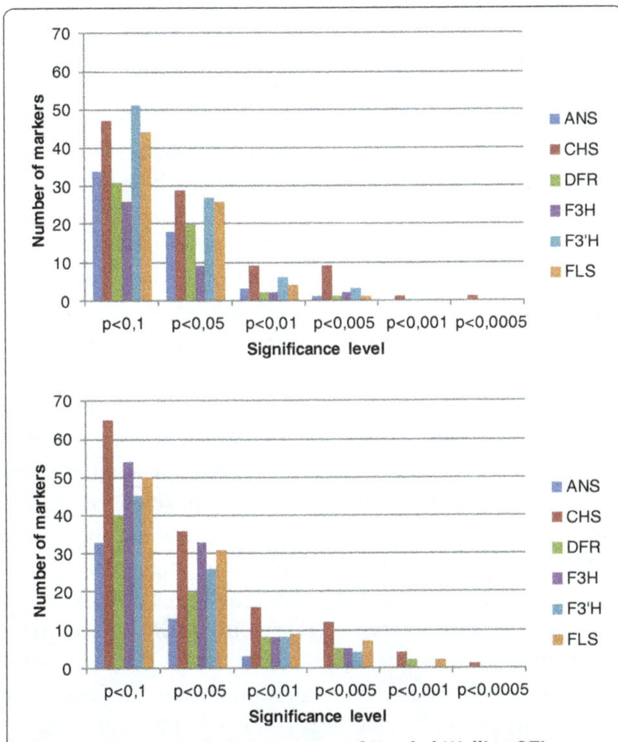

Figure 4 Power analysis by means of Kruskal-Wallis eQTL mapping. Preliminary eQTL mapping by means of Kruskal-Wallis analysis was performed in MapQTL®5 [91]. Two population sizes were compared: 49 siblings (upper panel) and 70 siblings (lower panel). For each gene (*ANS, CHS, DFR, F3H, F3'H* and *FLS*) the number of markers (vertical axis) that correlated at a certain significance level (horizontal axis) is given.

whether our methodology did not introduce bias in the dataset, we decided to compare the outcome of both calculation methods. For this purpose, the samples of the total dataset were split up again after averaging the calibrated normalised relative quantities (CNRQ) of the biological replicates. All gene expression results, both CNRQ and NRQ (normalised relative quantities) per assay, are shown in Additional file 4. Mantel-analysis confirmed the consistency of the inter-run calibration method applied. The (C)NRQ values in both matrices were significantly correlated at the level of $p = 0.001$ for assay 2 and 3 and $p = 0.004$ for assay 1.

The mean difference in Cq values between technical replicates varied between 0.07 and 0.27 cycles. However, the variation in the technical replicates was considered negligible compared to biological variation. The fold differences of CNRQ values of some biological replicates varied noticeably (see Additional file 5). This was most pronounced for *F3′H* with a substantial higher mean and maximum fold difference. The latter is due to sample 234, which shows a lot of variation for the other genes as well. The biological variation in *DFR* expression is less pronounced, but with a mean/median of 1.76/1.38 still rather high.

No correlation could be found between the flower colour groups and gene expression levels (Table 3). Nevertheless, the expression of some genes appeared to be correlated to others, for *CHS* and *FLS* there was even a significant correlation with all other genes (Table 3). The flavonoid biosynthesis pathway can be partitioned among early and late pathway genes, but the breaking point differs between species [45,46]. In azalea, *F3H* and *F3′H* are considered as early pathway genes together with *CHS* and *FLS*; *ANS* and *DFR* are some of the late pathway genes [8]. Taking different combinations of early or late pathway genes as an input for discriminant analysis, some of these combinations appeared to be able to distinguish to a minor extent between flower colour groups (Table 4). Combining the expression of all 4 early pathway genes could classify 51.4% of the samples in the correct colour group. Co-pigmentation of flavonols cannot be visualised in white flowers and therefore the interpretation of the expression profiles in this group can be misleading, certainly for *FLS*. When white flowers were omitted from the dataset, already 65.3% of the samples could be assigned to the correct flower colour group based on the same combination of early pathway genes. In case we classified samples according to flower colour intensity (pink versus (carmine) red), the expression levels of the early pathway genes could assign over 85% of the samples correctly. Even the combination of all genes performed very well for this purpose. Interestingly, when we compared the *F3′H* gene expression levels between both groups (Mann–Whitney U-test), a significant difference ($p = 0.0425$) was found. When [13] down regulated this gene in torenia, flower colour turned to pink as well. These results confirm that *F3′H* gene expression is an important factor for the establishment of flower colour intensity in azalea as well.

When samples were classified according to their co-pigmentation pattern (Q/q [12]), again the combined information of the early pathway genes could discriminate best between both classes (68.1% correct classifications, Table 4). Also the combination of all six genes scores quite well in grouping the samples (63.8%). The difference

Table 3 Spearman correlation analysis

	ANS	*CHS*	*DFR*	*F3H*	*F3'H*	*FLS*
Colour	0.123	−0.170	0.067	0.091	0.126	−0.152
ANS	1.000	0.329*	0.352*	0.549*	0.171	0.509*
CHS		1.000	0.309**	0.740*	0.617*	0.630*
DFR			1.000	0.078	0.214	0.307*
F3H				1.000	0.496*	0.582*
F3'H					1.000	0.418*

Non-parametric correlation was calculated between the log-transformed CNRQ values (geometric mean of biological replicates) of six genes and flower colour (white, pink, red and carmine red). *: significant at $p < 0.01$.

Table 4 Results of the assignments after discriminant analysis

Genes included	Grouping variable				
	Colour	Colour (no white)	Intensity	W	Q
CHS/F3H/F3'H	40.3%	59.7%	81.6%	59.7%	51.1%
CHS/F3H/F3'H/FLS	51.4%	65.3%	85.7%	58.3%	68.1%
DFR/ANS	27.8%	32.7%	57.1%	55.6%	51.1%
All genes	52.8%	55.1%	81.6%	55.6%	63.8%

Log-transformed CNRQ values of a combination of genes was used to calculate a discriminant function to predict classification according to 5 classes: colour (white, red, carmine red or pink); colour (no white; only red, carmine red and pink); intensity (pink versus (carmine) red); W (coloured versus white) and Q (co-pigmentation versus no co-pigmentation). The percentage of correctly assigned samples is presented.

between coloured and white flowers (W/w [12]) can be evaluated most reliable based on the expression of *CHS, F3H* and *F3'H*. The addition of *FLS* gene expression slightly reduces the information content (58.3% versus 59.7%), most likely due to the fact that flavonols have no impact on the phenotypic classification of W. However, when we look at the effect of the late pathway genes *ANS* and *DFR*, we can conclude that the expression of these genes is mainly involved in differentiating between white and coloured flowers as well.

Discussion

Optimisation of the RT-qPCR protocol

A good RT-qPCR experiment should always be based on a well-thought sampling protocol. Gene expression experiments essentially reflect a snapshot of RNA at the moment of extraction. Therefore, biological replicates are a prerequisite [26]. In this study, biological replicates were gathered on different flowers of a single plant. Indeed, sampling on two independent plants would have been a better approach since any influence of the physiological condition of the plant onto the overall gene expression would have been taken into account. However, when evaluating gene expression in a crossing population with only one plant per genotype, this is not an option. Growing all plants together at optimal conditions and sampling in a standardized way was therefore expected to be sufficient to fade out this effect as much as possible.

RT-qPCR has become the method of choice for gene expression analysis, but it suffers from considerable pitfalls, e.g. when it comes to evaluation of the RNA quality. Reporting on RNA quality assessment is one of the key-elements of the MIQE-guidelines [26] but is currently not done in 3 out of 4 published gene expression studies in plants [33]. Moreover, the results of the quality assessments are often not shown in the other 25%, although this information is crucial for the significance of the published results. Nevertheless, this parameter has a major impact on RT-qPCR performance [33,39,47,48], but there is no gold standard to define RNA quality and every method can have a different appreciation [39]. Absorption ratio's only reflect RNA purity [37], whereas a SPUD-assay can evaluate for the inhibitory effect of these impurities [32,38]. Our results demonstrate that only looking at the absorption ratios can lead to wrong assumptions concerning the RNA quality. In spite of the low absorption ratios of several samples, no PCR inhibition was seen in the SPUD assay, indicating the acceptable quality of our samples. Assessing PCR efficiency in a test sample by serial dilution of the sample can be an alternative method to identify inhibition [32] but is not so obvious in case of low concentrated samples. D'haene and Hellemans [49] demonstrate that inhibitors can be derived from the shape of the amplification curve, but this is not an objective method. Hence, we advise to perform a SPUD assay on a representative subset of the samples every time a new sample type, treatment and/or extraction protocol is used.

To assess RNA integrity as well, microfluidic capillary electrophoresis was implemented. This technology recently gained interest in the plant RNA community (reviewed in [33]), but is partly based on the ribosomal peak ratio (28S/18S). Since the relationship between this ratio and mRNA integrity appears to be unclear [40,48,50,51], RIN (RNA Integrity Number [36]) and RQI (RNA Quality Indicator [52]) values that take into account the complete electropherogram were introduced as a more solid measure for RNA integrity. However, these values were initially assigned by using electropherograms of various mammalian tissues to train the software in an adaptive learning approach. In plants, no 28S rRNA is present, instead there is a 25S rRNA peak. In addition, total RNA in chloroplast-containing plant tissues also consists of 16S and 23S rRNA [53], adding 2 extra peaks. These rRNA peaks will be recognized as degradation peaks by the software, leading to a miscalculation of the RIN/RQI value and an underestimation of the true integrity of the material in plants. This is clearly seen in the result of Pico de Coana et al. [54]. Moreover, an optimal 28S/18S rRNA ratio of 2 is without any evidence extrapolated to plant 25S/18S rRNA [55]. These researchers rely on the software outputs, but they omit to look at the raw data to decide on the true quality of the RNA. Microfluidic capillary electrophoresis in plant science can be of great value (when the technology is available) but should always be restricted to a visual evaluation of the electropherograms and virtual gel views. The construction of a degradation series can then help to decide on the level of RNA integrity of specific samples.

Co-purification of traces of DNA during RNA extraction is inevitable, therefore noRT samples have been analysed in all cases. As is also asked for in the MIQE-guidelines [26], noRT results should always be given when gene expression data are published. However, far too often papers are published in which qPCR data are lacking results of the noRTs. How these authors (and the readers) can be sure that the so-called gene expression differences are not false positive signals? In the case the use of noRTs is described, it is not always clear what these noRTs exactly consist of. Some researchers just add RNA as a control in the RT-qPCR (e.g. [56,57]). However, to control in addition for DNA contamination during the cDNA synthesis step, we handled the RNA for noRT samples in exactly the same way as the normal samples. The same compounds were added, except off course the RT enzyme, as advised by Nolan et al. [32]. Suppliers of reversed transcriptase enzymes should provide special kits with additional buffers and primers for this purpose and this is unfortunately not always feasible. As an alternative, one could indeed use diluted RNA as a noRT sample and add the RT-reaction mixture as an additional sample in the analysis to control for potential contamination in this mixture. Even more crucial, in our opinion, is the analysis of noRTs with all primers. Often only a single gene is used to control for genomic DNA contamination [22,56-58]. The fact that in our dataset an individual sample was suffering from contamination when one specific gene was amplified, but not when the other genes were amplified, strengthens the need to test all primer sets on all noRT samples. Also Laurell et al. [59] state that the sensitivity towards genomic DNA contamination differs greatly between assays. These authors developed ValidPrime as an efficient alternative for the use of noRT controls, but currently no such assays are available for plant studies yet.

For normalisation of gene expression data, reference genes are indispensable [30]. The use of reference genes controls for variations in extraction yield, reverse-transcription and efficiency of amplification. It is without question that multiple, assay-validated reference genes are considered to be an essential component of a consistent qPCR assay [26], also in plant science [27-29]. In azalea, we aimed at developing a basic set of reference genes for application in all azalea gene expression studies. Czechowski et al. [60] demonstrated that the commonly used reference genes were not always the best candidates. Also GAPDH was not withdrawn as a reliable reference gene in our analysis. Therefore alternatives were looked for. Microarray data can be an ideal source of reference genes [61], but are lacking in azalea. Coker and Davies [62] took advantage of EST data for reference gene selection in tomato. Since a limited set of 62 ESTs was available in azalea [42], candidate reference genes were selected from this dataset. The proposed set of 11 azalea reference genes is a valuable toolbox for future qPCR research in azalea. However, each experimental condition demands a specific set of reference genes [63,64] and even different lab protocols seem to have an influence on reference gene selection [65]. Therefore, validation of this set in the desired tissues and conditions will be essential to select the appropriate assay-specific reference genes.

Several quantification strategies with altered normalisation methods are available, all depending on the PCR efficiency (E) for their calculations [25,41]. The quantification approach can have a serious impact on the final results [66]. Assuming an optimal PCR efficiency is not recommended [26,41]. The use of sample-specific amplification efficiencies [67-70] has become more common in RT-qPCR studies [71] since it allows quantification without standard curves. However, the outcome of using sample-specific amplification efficiencies can vary drastically depending on the settings and is reported to increase the random error [72]. Recently, Regier and Frey [66] demonstrated that using the average target specific efficiency (based on sample specific efficiency estimations) can be an alternative to the standard curve method in case a reliable algorithm is used (e.g. LinReg). Nevertheless, the use of standard curves remains the most precise method [73,74]. Based on the equation of a standard curve, the qPCR efficiency can be calculated. In our study, plasmid DNA was used for standard curve construction. Hellemans et al. [41] advise to make the dilution series with a sample that mimics as much as possible the samples to be analysed in qPCR [41], most often this is a mixture of representative cDNA samples [57,75]. Plasmid DNA consists of a different sample matrix, what can result in altered efficiencies due to the presence of different kinds of inhibitory components [76]. However, the absence of PCR inhibitors was controlled for by means of the SPUD assay. Moreover, in absolute quantification studies the use of plasmid DNA to construct a dilution series is even preferred [77]. Especially in case of the limited availability of cDNA, plasmid DNA also has the advantage of being available plentiful and is therefore a valuable alternative for the construction of standard curves.

Flower colour gene expression

Optimisation at all stages of the RT-qPCR has resulted in a reliable protocol for quantification of gene expression in azalea. We also aimed at studying the correlation between flower colour and the expression of candidate genes of the flavonoid biosynthesis pathway in a broader genetic background in contrast with what is currently reported in other ornamentals [19-22,24]. Moreover, we ultimately wanted to use flower colour as a model system for genetical genomics [78] in azalea. Most crucial was therefore the minimal required population size with sufficient power for eQTL mapping [79]. With 4 different flower colour groups, conventional power analysis [80] was not an option. But according to Shi et al. [81] even in small populations the power should already be sufficient to detect eQTLs. Therefore we started with a small subpopulation of 20 plants and gradually expanded to a final population of 70 siblings. This stepwise approach forced us to use an alternative method for inter-run calibration. The performance of a Mantel-test validated the approach for our assay. However, this method of inter-run calibration cannot automatically be considered to be trustworthy in other experiments. We believe that the rather small expression differences between our samples and genes had a significant impact here. Experiments in which large expression differences are measured are more likely to suffer from using the average gene expression as an inter-run calibrator and we therefore want to encourage the use of inter-run calibration as described in Hellemans et al. [41]. However, after validation with a Mantel-test, one could use the described methodology when lacking proper inter-run calibrators. The use of 3 biological replicates could have allowed to identify outlier values in some samples with high biological variation. However, these values do reflect the true variation present in the flower buds and can therefore not be neglected. These data clearly reinforce the substantial interest of using biological (rather than technical) replicates in every qPCR experiment.

The individual expression profiles were not discriminative enough to differentiate between colour groups. Also in other species, no such correlations have been reported since most studies limit themselves to the comparison of gene expression between few cultivars with different flower colours [19-22,24]. The use of multiple genotypes in each flower colour group certainly complicates the analysis. When the biological variation within a genotype is already substantial, detecting differences between genotypes is even harder. Only when the expression of *F3′H* was compared between pink and (carmine) red flowers, a significant expression difference was found. This implicates that there clearly is a link between the flower colour intensity and the *F3′H* expression. Similar conclusions can be drawn from the combined effect of early pathway genes (so including *F3′H*) on flower colour intensity, with very high percentages of correctly assigned genotypes. With a transgenic approach in torenia, Nakamura et al. [13] also demonstrated that the regulation of *F3′H* is crucial to manipulate flower colour intensity. Also *F3′5′H* is reported to be involved in pink [13,14] but this gene is only of interest for the production of dephinidin derivatives [82]. Delphidin pigments can be present in purple azalea flowers, but this colour was not present in the studied population. Therefore the expression of this gene was not determined. Besides these two flavonoid biosynthetic genes, pale-anthocyanin coloration can also be the result of a mutation in a putative *glutathione S-tranferase* gene that is responsible for the transport of pigments to the vacuole [83]. Therefore it would certainly be interesting to determine the expression of such transporter genes as well. HPLC measurements of the pigment types and concentrations could add even more to the elucidation of pink in azalea.

Also for the other genes, the combination of expression profiles was highly informative, since flower colour regulation is known to occur mainly via a coordinated transcriptional control of structural genes [5,7]). Especially the early pathway genes *CHS, F3H, F3′H* and *FLS* can discriminate rather well between the colour groups when white flowers are omitted from the analysis and these genes are most suited to differentiate for co-pigmentation as well. This makes sense, since the early pathway is indeed responsible for the production of the flavonols as co-pigments. To be able to include white flowers in the analysis, HPLC data would be needed to score for the presence of flavonols. The late pathway genes *ANS* and *DFR* are less informative but are still helpful for the classification of coloration. This could implicate that the difference between white and coloured flowers is situated rather at the regulation of the late pathway gene expression. Also in potato, *DFR* is known to be involved in the difference between white and coloured tubers [84] and Jung et al. [85] reported that the regulation of white pigmentation in potato is situated at the transcriptional level.

Due to the actual presence of gene expression differences that are related to the transcriptional regulation of the flavonoid biosynthetic pathway, these data are well-suited for eQTL mapping. For this purpose, not only the expression profiles of the individual genes but also the discriminant functions will be used as a first step towards *a priori* eQTL mapping [86] on the genetic map of the population under study [87]. As such, the gene expression information will be used in a genetical genomics approach [78] to evaluate the impact of the entire pathway on the flower colour. This can confirm the existence of a co-regulation network and will help to understand more the observed variation in flower colour. Moreover, the presence of markers for *myb*-functional genes on the genetic map can be valuable candidate genes potentially co-localising with flower colour eQTLs.

Conclusions

To conclude, we are convinced that optimisation at crucial steps resulted in the development of a reliable protocol for gene expression analysis that is not only applicable to azalea, but can easily be used on other plant material as well. Currently in plant research, validated and qualitative RT-qPCR protocols are still rare. A pool of azalea reference genes was constructed, three of them are sufficient for normalisation of gene expression in flower petals, but the remaining genes can in the future also be used for normalisation in other azalea tissues, e.g. leaves and shoots. We also stressed on the importance of a multi-level RNA quality control, to evaluate both RNA purity and RNA integrity, with special attention for the bottlenecks for automated procedures on plant RNA. Furthermore, the co-amplification of contaminating DNA in few samples showed the importance of analysing noRT samples with all genes under study. Finally the advantages of using plasmid-derived standard curves in every analysis was demonstrated as well.

The accurate protocol resulted in the quantification of several flavonoid biosynthesis genes in a subset of 70 siblings of an azalea mapping population. The expression of *F3′H* could differentiate between pink and (carmine) red flower colour groups. The combined regulation of the early pathway genes clearly has an impact on the co-pigmentation and the late pathway genes *ANS* and *DFR* are to a minor extent involved in differentiating between white and coloured flower phenotypes. These gene expression profiles will now be used as eQTLs to study flower colour in a genetical genomics approach. This might help us to point-out the actual genes that are encompassed in W and Q. Providing more detailed data on pigment composition (HPLC) in the petals of the different genotypes could even add an additional level (mQTLs or metabolite QTLs) of information to this map-based approach.

Methods

RNA isolation

RNA was isolated from flower buds in the candle stage (25–30 mm) [23] of 70 siblings of the 'GxH' crossing population [87] and both parents ('98-13-4' and 'Sima'). From each plant, two individual buds were sampled (a and b) as biological replicates. For reference gene selection, candle stage flower buds of eight azalea cultivars ('Hellmut Vogel' and seven of its flower colour sports: 'Paloma', 'Hector', 'Mw. Troch', 'Nordlicht', 'Terra Nova', 'Zalm Vogel' and 'Super Nova') displaying a range of colours were used. Approximately 70 mg of petal tissue (other bud organs were carefully removed) was weighed per sample in duplicate in pre-cooled 2 ml safe-lock tubes (Eppendorf). Three zirconium beads were added to the tubes and the plant material was crushed in a pre-cooled block of the Retsch Tissuelyser (Qiagen) for 2 times 30 s at 30 Hz. After a short centrifugation (30 s, 4°C, full speed), the tubes were placed on ice and RNA was isolated according to the protocol of the RNAqueous kit® (Ambion) in combination with the Plant RNA Isolation Aid (Ambion). Elution was done in three steps (40/25/25 µl) and eluents were pooled. DNase treatment occurred on 80 µl of RNA with the DNA-*free* kit (Ambion). 10 µl DNaseI buffer and 1.5 µl rDNaseI were added, followed by an incubation step of 30 min at 37°C. DNase Inactivation Reagent (10 µl) was added and samples were incubated for 2 min at room temperature. After centrifugation (90 s, 10000 g) the supernatant was transferred to a new tube. Duplicate samples were finally pooled and purified [88] using 0.3 M Sodium Acetate pH5.5 (Ambion). Two and a half volumes of 100% EtOH was added and samples were incubated for at least 15 min at −80°C or overnight at −20°C. Supernatant was removed after 25 min centrifugation (14000 rpm, 4°C) and 1 ml 70% EtOH was added. Again tubes were centrifuged for 20 min at the same conditions and supernatant was discarded. The RNA pellet was dried in a vacuum-desiccator and resolved in 25 µl of RNase-free water. Samples were stored at −80°C until cDNA synthesis.

RNA quantity/quality

RNA was quantified by means of the NanoDrop spectrophotometer (Isogen). The presence of inhibitory components was evaluated (on a subset of 14 samples, Additional file 1) by means of the SPUD-assay developed by [30,32]. A stock solution of 5 µM of the 101 bp SPUD amplicon (Sigma) was diluted $1/10^8$ in yeast tRNA (50 ng/µl; Invitrogen). 0.5 µl of the diluted amplicon, 0.48 µM of both forward and reverse SPUD primers (Invitrogen), 0.1 µM of the dual-labelled (Fam-Tamra) SPUD probe (MWG-Biotech) and 1× LightCycler480 Probes Master Mix (Roche) was combined in a total volume of 10 µl in a white 384-well plate (Roche). For each sample, 1 µl of RNA or 2 µl of cDNA was added and all samples were analysed in duplicate. In the SPUD control samples, no RNA or cDNA was added; NTCs (No Template Control) were included as well. Plates were sealed with an adhesive film. Cycling conditions in the LightCycler480 (Roche) were 10 min at 95°C, followed by 45 cycles of 10s 95°C, 30 s 60°C and 1s 72°C. Fluorescence data were recorded every cycle at the end of the annealing/elongation step at 60°C. Data were analysed using the LightCycler480 software version 1.5 (Roche). Cq-values were exported to Microsoft Excel for further calculations. Finally, RNA quality and quantity was also determined on the same subset of samples using the Experion microfluidic capillary electrophoresis system (Bio-Rad) in combination with the RNA StdSens Chips (Bio-Rad). A degradation series was prepared by heating an RNA sample for 15, 30, 45 and 60 min at 80°C in a PCR machine.

Reverse transcription

First strand cDNA synthesis was performed with the SuperScript III First-Strand Synthesis SuperMix (Invitrogen) according to the manufacturers protocol and starting from 100 ng of RNA or 6 µl for low-concentrated samples (< 17 ng/µl). Oligo$(dT)_{20}$ was used for priming and all incubations occurred in a Perkin Elmer 2720 (Applied Biosystems). As a control for DNA contamination, noRTs were created in the same way as samples, except for the SuperScript III/RnaseOUT Enzyme Mix that was omitted in these cases. Both cDNA and noRT samples were diluted 1/3 and stored at −20°C.

Reference genes

Homolog's of commonly used reference genes (*ubiquitin, GAPDH, β-actin, α-6-tubulin, TATA-box binding protein, elongation factor α*) were searched for in azalea with degenerate primers; gene-isolation was only doing well for *GAPDH*. The fragment was cloned using the TOPO TA Cloning Kit (Invitrogen) and sequenced in order to develop specific RT-qPCR primers (Table 2). Twelve candidate reference genes were selected out of 62 annotated genes from a *Rhododendron simsii* hybrid 'Flamenco' EST library [42] and qPCR primers were developed with melting temperatures 58-60°C, primer lengths 20–24 bp and amplicon lengths 151–165 bp. (Primer Express 2.0, Table 2). Primers were at first tested on the EST containing plasmids. Primer pairs that amplified the proper fragment were, together with *GAPDH* primers, tested in duplo in a RT-qPCR assay on cDNA from flower petals of 8 azalea cultivars. PCR analysis was carried out in an ABI7000 thermocycler (Applied Biosystems). Amplification mixture consisted of 12.5 µl of SYBR Green I Master Mix (Applied Biosystems), 7.5 pmol of both primers and 2 µl cDNA in a total volume of 25 µl. Cycling conditions were 2 min 50°C, 10 min 95°C and 40 cycles of 15 s 95°C and 1 min 60°C. For melting curve analysis, cycling conditions were 15 s 95°C, 15 s 60°C followed by ramping from 60°C to 95°C with a ramp speed of 2% and a final step of 15 s 95°C. Cq-values were averaged and transformed to quantities using standard curves. These data were used for reference gene selection using geNorm software [30].

Standard curves

Amplified fragments of both reference and target genes were cloned using the TOPO TA Cloning Kit (Invitrogen) containing TOP10F' chemically competent cells and the pCR2.1-TOPO cloning vector. For *CHS* and *DFR*, full length cDNA sequences were previously cloned [23]. Plasmid DNA was purified (GFX *Micro* Plasmid Prep Kit, Amersham) and linearised using 10 U of *Hind*III (Invitrogen) for 2 h at 37°C, followed by an enzyme inactivation step for 10 min at 70°C. The stock concentration of plasmids was diluted to a working solution of 1 ng/µl in 50 ng/µl yeast tRNA (Invitrogen). Standard curves were constructed as six log10 dilutions of this working solution in yeast tRNA (50 ng/µl). To prevent extrapolation, the range of the standard curve was set to cover Cq values of the cDNA samples. It must also be strengthened that the diluted aliquots were never stored longer as 24 h at 4°C to preserve quality [89] were and prepared newly from the same stock of plasmid DNA stored at −20°C if needed again later. Standard curves were used for calculation of PCR efficiencies ($E = 10^{(-1/slope)} - 1$).

Quantification

Six RT-qPCR primer sets were developed in azalea for genes coding for key enzymes in the flavonoid biosynthesis pathway: *chalcone synthase* (*CHS*), *flavanone 3-hydroxylase* (*F3H*), *flavonoid 3′-hydroxylase* (*F3′H*), *anthocyanidin synthase* (*ANS*), *dihydroflavonol 4-reductase* (*DFR*) and *flavonol synthase* (*FLS*) (Table 1). *CHS* and *DFR* were *R. simsii* hybrid sequences [9], the others from *R. Xpulchrum* [34]. Primers were designed using Primer Express 2.0 (Applied Biosystems). Primers were targeted to the 3′ end and preferably spanning an intron. Intron/exon positions were predicted based on homologies with poplar or *Arabidopsis* sequences. Small amplicon sizes were preferred because this gives more consistent results [48]. All samples, noRTs, NTCs and standard curves were measured in duplicate in a LightCycler480 (Roche). In a white 384-well plate (Roche), 375 nM of each primer and 5 µl of LightCycler480 SYBR Green I Master (Roche) was used with 2 µl of sample in a total volume of 10 µl. Plates were sealed with an adhesive film. Cycling conditions were 5 min at 95°C, followed by 40 cycles of 10 s 95°C, 12 s 60°C and 10 s 72°C. Data acquisition was done at the end of every cycle. Melting curve analysis was performed as follows: 5 s 95°C, 1 min 65°C and heating to 97°C with a ramp rate of 0.06°C/s. Data acquisition occurred 10 times for every °C. Data were analysed using the LightCycler480 software version 1.5 (Roche). We started with gene expression analysis on 20 siblings and both parent plants. In a second phase, 29 new siblings were analysed and finally a third assay was run with 21 seedlings for gene expression analysis (See Additional file 1). Within an assay, the sample-maximisation method was preferred and samples were analysed in a single plate per gene. The 2nd derivative method of Luu-The et al. [90] was selected for Cq determination in every run. Cq-values were exported to Microsoft Excel; technical replicates were averaged geometrically. For combining the 3 assays, the overall gene expression level per plate and per gene (geometric mean) was used for inter-run calibration. Gene specific amplification efficiencies derived from standard curves and a normalisation factor [30] based on two validated reference genes (HK5 and HK129) was used for calculation of (calibrated) normalised relative quantities ((C)NRQ). Biological replicates were averaged geometrically as well.

Data analysis

Log-transformed data were used as an input for statistics. SPSS Statistics 19 software package was used for all statistical data analysis. Kruskal-Wallis (in MapQTL®5 [91]) was used as an alternative for power analysis to determine the required population size. Power was sufficient when at least half of the genes correlated with markers at the level of $p < 0.001$. To verify the inter-run calibration method, two calculation methods were compared for each assay: standard quantification in the individual assay (NRQ-values) and the same subset of samples calculated within the global dataset of 72 samples (CNRQ-values). Bivariate spearman correlation coefficients were calculated between log-transformed values of all samples for every gene, resulting in assay-specific correlation matrices. Correlation matrices of comparable datasets were used as an input for Mantel analysis [92] by means of the Mantel nonparametric test calculator [93].

Additional files

Additional file 1: RNA concentration and purity. Description: RNA quantity and purity was measured of each biological replicate per sample using a NanoDrop spectrophotometer. For each sample, the assay is indicated in which the sample was analysed. Flower colour is indicated as well (0 = white, 1 = red, 2 = carmine red, 3 = pink). Samples used for analysis in the SPUD-assay and the Experion are indicated with an *.

Additional file 2: Evaluation of the optimal number of reference genes for normalization. Description: A cut-off value of 0.15 is proposed (top panel). Average expression stability (M) of the reference genes tested in azalea. M is calculated at each step during stepwise exclusion of the least stable reference gene. Genes are ranked from the least (left) to the most stable (right). Only genes with an M-value < 0.5 are valid in homogeneous samples (lower panel). Both graphs are generated in GeNorm [30].

Additional file 3: PCR efficiencies of the standard curves. Description: Summary of slopes and derived PCR efficiencies (E) of the standard curves of dilution series analysed on different plates in 3 independent assays. E and the standard deviation on E (SD(E)) were calculated according to the formulas described in Hellemans et al. [41].

Additional file 4: Description: Gene expression results. In the left part of the table, gene expression values were calculated on samples of a single assay (assay 1, 2 or 3). On the right, results are presented per assay but calculations occurred on the entire dataset of 72 samples. For each sample the geometric mean of the biological replicates is presented and (C)NRQ values have been log-transformed.

Additional file 5: Description: Fold differences between 2 CNRQ values of biological replicates. Samples are grouped according to flower colour (0 = white, 1 = red, 2 = carmine red, 3 = pink). Empty cells indicate one of the biological replicates was discarded after noRT analysis.

Competing interests

The authors declare that they have no competing interests.

Authors' contributions

EDK was responsible for design of the study, data analysis and statistics and drafted the manuscript; LD was involved in the acquisition of data and assisted in data analysis; EVB participated in the study's design and contributed to editing the manuscript; JDR conceived the study, participated in data analysis and helped to draft the manuscript. All authors read and approved the final manuscript.

Acknowledgements

The authors are grateful to Romain Uytterhaegen for his outstanding technical assistance in plant cultivation. The whole staff of the biotech lab of ILVO-Plant Sciences Unit was very much appreciated for their support. We want to express our gratitude to Bio-Rad laboratories for giving us the opportunity to test the Experion on our plant material. This research was funded by the Ministry of Agriculture of Flanders.

Author details

[1]Institute for Agricultural and Fisheries Research (ILVO)-Plant Sciences Unit, Caritasstraat 21, 9090, Melle, Belgium. [2]Department for Plant Production, Ghent University, Coupure links 653, 9000, Ghent, Belgium.

References

1. De Loose R: **Flavonoid glycosides in the petals of some *Rhododendron* species and hybrids.** *Phytochem* 1968, **9:**875–879.
2. De Loose R: **The flower pigments of the Belgian hybrids of *Rhododendron simsii* and other species and varieties from *Rhododendron* subseries obtusum.** *Phytochem* 1969, **88:**253–259.
3. Gerats AGM, Martin C: **Flavonoid synthesis in Petunia hybrida: genetics and molecular biology of flower colour.** In *Recent advances in phytochemistry.* Edited by Stafford HA, Ibrahim RK. New York: Plenum Press; 1992:165–200.
4. Holton TA, Cornish EC: **Genetics and biochemistry of anthocyanin biosynthesis.** *Plant Cell* 1995, **7:**1071–1083.
5. Mol J, Grotewold R, Koes RE: **How genes paint flowers and seeds.** *Trends Plant Sci* 1998, **3:**212–217.
6. Winkel-Shirley B: **Biosynthesis of flavonoids and effects of stress.** *Curr Opin Plant Biol* 2002, **5:**218–223.
7. Schwinn K, Venail J, Shang Y, Mackay S, Alm V, Butelli E, Oyama R, Bailey P, Davies K, Martin C: **A small family of MYB-regulatory genes controls floral pigmentation intensity and patterning in the genus *Antirrhinum*.** *Plant Cell* 2006, **18:**831–851.
8. De Cooman L, Everaert ESW, Faché P, Vande Casteele K, Van Sumere CF: **Flavonoid biosynthesis in petals of *Rhododendron simsii*.** *Phytochem* 1993, **33:**1419–1426.
9. De Schepper S, Debergh P, Van Bockstaele E, De Loose M, Gerats A, Depicker A: **Genetic and epigenetic aspects of somaclonal variation: flower colour bud sports in azalea, a case study.** *South African J Bot* 2003, **69:**117–128.
10. Heursel J, Horn W: **A hypothesis on the inheritance of flower colours and flavonoids in *Rhododendron simsii* Planch.** *Zeitschrift für Pflanzenzüchtung* 1977, **79:**238–249.
11. Sasaki N, Nishizaki Y, Uchida Y, Wakamatsu E, Umemoto N, Momose M, Okamura M, Yoshida H, Yamaguchi M, Nakayama M, Ozeki Y, Itoh Y: **Identification of the *glutathione S-transferase* gene responsible for flower color intensity in carnations.** *Plant Biotechnol* 2012, **29:**223–227.
12. De Keyser E, Lootens P, Van Bockstaele E, De Riek J: **Image analysis for QTL mapping of flower colour and leaf characteristics in pot azalea (*Rhododendron simsii* hybrids).** *Euphytica* 2013, **189:**445–460.
13. Nakamura N, Fukuchi-Mizutani M, Fukui Y, Ishiguro K, Suzuki K, Suzuki H, Okazaki K, Shibata D, Tanaka Y: **Generation of pink flower varieties from blue *Torenia hybrida* by redirecting the flavonoid biosynthetic pathway from delphinidin to pelargonidin.** *Plant Biotechnol* 2010, **27:**375–383.
14. Boase MR, Lewis DH, Davies KM, Marshall GB, Patel D, Schwinn KE, Deroles SC: **Isolation and antisense suppression of *flavonoid 3',5'-hydroxylase* modifies flower pigments and colour in cyclamen.** *BMC Plant Biol* 2010, **10:**107.
15. Nishihara M, Nakatsuka T: **Genetic engineering of flavonoid pigments to modify flower color in floricultural plants.** *Biotechnol Lett* 2011, **33:**433–441.
16. Chen W-H, Hsu C-Y, Cheng H-Y, Chang H, Chen H-H, Ger M-J: **Downregulation of putative UDP-glucose: flavonoid 3-O-glucosyltransferase gene alters flower coloring in *Phalaenopsis*.** *Plant Cell Rep* 2011, **30:**1007–1017.
17. Kamiishi Y, Otani M, Takagi H, Han D-S, Mori S, Tatsuzawa F, Okuhara H, Kobayashi H, Nakano M: **Flower color alteration in the liliaceous ornamental Tricyrtis sp. By RNA interference-mediated suppression of the chalcone synthase gene.** *Mol Breeding* 2012, **30:**671–680.

18. Nakatsuka T, Mishiba K, Kubota A, Abe Y, Yamamura S, Nakamura N, Tanaka Y, Nishihara M: **Genetic engineering of novel flower colour by suppression of anthocyanin modification genes in gentian.** *J Plant Physiol* 2010, **167**:231–237.
19. Akita Y, Kitamura S, Hase Y, Narumi I, Ishizaka H, Kondo E, Kameari N, Nakayama M, Tanikawa N, Morita Y, Tanaka A: **Isolation and characterization of the fragrant cyclamen *O-methyltransferase* involved in flower coloration.** *Planta* 2011, **234**:1127–1136.
20. Yamamizo C, Noda N, Ohmiya A: **Anthocyanin and carotenoid pigmentation in flowers of section *Mina*, subgenus *Quamoclit*, genus *Ipomoea*.** *Euphytica* 2012, **184**:429–440.
21. Sui X, Gao X, Ao M, Wang Q, Yang D, Wang M, Fu Y, Wang L: **cDNA cloning and characterization of UDP-glucose: anthocyanidin 3-*O*-glucosyltransferase in *Freesia hybrida*.** *Plant Cell Rep* 2011, **30**:1209–1218.
22. Mizuta D, Nakatsuka A, Miyajima I, Ban T, Kobayashi N: **Pigment composition patterns and expression analysis of flavonoid biosynthesis genes in the petals of evergreen azalea 'Oomurasaki' and its red flower sport.** *Plant Breeding* 2010, **129**:558–562.
23. De Schepper S, Debergh P, Van Bockstaele E, De Loose M: **Molecular characterisation of flower colour genes in azalea sports (*Rhododendron simsii* hybrids).** *Acta Hort* 2001, **552**:143–150.
24. Liu X-J, Chuang Y-N, Chiou C-Y, Chin D-C, Shen F-Q, Yeh K-W: **Methylation effect on chalcone synthase gene expression determines anthocyanin pigmentation in floral tissues of two *Oncidium* orchid cultivars.** *Planta* 2012, **236**:401–409.
25. Livak KJ, Schmittgen TD: **Analysis of relative gene expression data using real-time quantitative PCR and the 2(–Delta Delta C(T)) Method.** *Methods* 2001, **25**:402–408.
26. Bustin SA, Benes V, Garson JA, Hellemans J, Huggett J, Kubista M, Mueller R, Nolan T, Pfaffl MW, Shipley GL, Vandesompele J, Wittwer CT: **The MIQE guidelines: *M*inimum *I*nformation for publication of *Q*uantitative real-time PCR *E*xperiments.** *Clin Chemistry* 2009, **55**:611–622.
27. Guttierrez L, Mauriat M, Guénin S, Pelloux J, Lefebvre J, Louvet R, Rusterrucci C, Moritz T, Guerineau F, Bellini C, Van Wuytswinkel O: **The lack of a systematic validation of reference genes: a serious pitfall undervalued in reverse transcription – polymerase chain reaction (RT-PCR) analysis in plants.** *Plant Biotechnol J* 2008, **6**:609–618.
28. Guttierrez L, Mauriat M, Pelloux J, Bellini C, Van Wuytswinkel O: **Towards a systematic validation of references in real-time RT-PCR.** *Plant Cell* 2008, **20**:1734–1735.
29. Guénin S, Mauriat M, Pelloux J, Van Wuytswinkel O, Bellini C, Guttierrez L: **Normalization of qRT-PCR data: the necessity of adopting a systematic, experimental conditions-specific, validation of references.** *J Exp Bot* 2009, **60**:487–493.
30. Vandesompele J, De Preter K, Pattyn F, Poppe B, Van Roy N, De Paepe A, Speleman F: **Accurate normalization of real-time quantitative RT-PCR data by geometric averaging of multiple internal control genes.** *Genome Biol* 2002, **3**:1–11.
31. Thellin O, Elmoualij B, Heinen E, Zorzi W: **A decade of improvements in quantification of gene expression and internal standard selection.** *Biotechnol Adv* 2009, **4**:323–333.
32. Nolan T, Hands RE, Bustin SA: **Quantification of mRNA using real-time RT-PCR.** *Nat Protoc* 2006, **1**:1559–1582.
33. Die JV, Roman B: **RNA quality assessment: a view from plant qPCR studies.** *J Exp Bot* 2012, **63**:6069–6077.
34. Nakatsuka A, Mizuta D, Kii Y, Myajima I, Kobayashi N: **Isolation and expression analysis of flavonoid biosynthesis genes in evergreen azalea.** *Sci Hort* 2008, **118**:314–320.
35. Baelde HJ, Cleton-Jansen AM, Van Beerendonck H, Namba K: **High quality RNA isolation from tumors with low cellularity and high extra-cellular matrix component for cDNA microarrays. Application to chondrosarcoma.** *J Clin Pathol* 2001, **54**:778–782.
36. Shultz DJ, Craig R, Cox-Foster DL, Mumma RO, Medford JI: **RNA isolation from recalcitrant plant tissue.** *Plant Mol Biol Rep* 1994, **12**:310–316.
37. Mueller O, Hahnenberger K, Dittmann M, Yee H, Dubrow R, Nagle R, Ilsley D: **A microfluidic system for high-speed reproducible DNA sizing and quantitation.** *Electrophoresis* 2000, **21**:128–134.
38. Nolan T, Hands RE, Ogunkolade W, Bustin SA: **SPUD: A quantitative PCR assay for the detection of inhibitors in nucleic acid preparations.** *Anal Biochem* 2006, **351**:308–310.
39. Vermeulen J, De Preter K, Lefever S, Nuytens J, De Vloed F, Derveaux S, Hellemans J, Speleman F, Vandesompele J: **Measurable impact of RNA quality on gene expression results from quantitative PCR.** *Nucl Acids Res* 2011, **39**:1–12.
40. Schroeder A, Mueller O, Stocker S, Salowsky R, Leiber M, Gassmann M, Lightfoot S, Menzel W, Granzow M, Ragg T: **The RIN: and RNA integrity number for assigning integrity values to RNA measurements.** *BMC Mol Biol* 2006, **7**:3.
41. Hellemans J, Mortier G, De Paepe A, Speleman F, Vandesompele J: **qBase relative quantification framework and software for management and automated analysis of real-time quantitative PCR data.** *Genome Biol* 2007, **8**:. doi:10.1186/gb-2007-8-2-r19.
42. De Keyser E, De Riek J, Van Bockstaele E: **Discovery of species-wide EST-derived markers in *Rhododendron* by intron-flanking primer design.** *Mol Breeding* 2009, **23**:171–178.
43. Ovstebo R, Bente Foss Haug K, Lande K, Kierulf P: **PCR-based calibration curves for studies of quantitative gene expression in human monocytes: development and evaluation.** *Clinical Biochem* 2003, **49**:425–432.
44. Wang QT, Xiao W, Mindrinos M, Davis RW: **Yeast tRNA as carrier in the isolation of microscale RNA for global amplification and expression profiling.** *Biotechniques* 2002, **33**:788–796.
45. Quattrocchio F, Wing JF, van der Woude K, Souer E, de Vetten N, Mol J, Koes R: **Molecular analysis of the *anthocyanin2* gene of petunia and its role in the evolution of flower color.** *Plant Cell* 1999, **11**:1433–1444.
46. Streisfeld MA, Rausher MD: **Altered *trans*-regulatory control of gene expression in multiple anthocyanin genes contributes to adaptive flower color evolution in *Mimulus aurantiacus*.** *Mol Biol Evol* 2009, **26**:433–444.
47. Bustin SA, Nolan T: **Pitfalls of quantitative real-time reverse-transcription polymerase chain reaction.** *J Biomol Tech* 2004, **15**:155–166.
48. Fleige S, Walf V, Huch S, Prgomet C, Sehm J, Pfaffl M: **Comparison of relative mRNA quantification models and the impact of RNA integrity in quantitative real-time RT-PCR.** *Biotechnol letters* 2006, **28**:1601–1613.
49. D'haene B, Hellemans J: **The importance of quality control during qPCR data analysis.** *Int Drug Disc* 2010:18–24.
50. Imbeaud S, Graudens E, Boulanger V, Barlet X, Zaborski P, Eveno E, Mueller O, Schroeder A, Auffray C: **Towards standardization of RNA quality assessment using user-independent classifiers of microcapillary electrophoresis traces.** *Nucl Acids Res* 2005, **33**:e56. doi:10.1093/nar/gni054.
51. Pfaffl MW, Fleige S, Riedmaier I: **Validation of lab-on-chip capillary electrophoresis systems for total RNA quality and quantity control.** *Biotechnol Biotechnol Equip* 2008, **22**:839–843.
52. Denisov V, Strong W, Walder M, Gringich J, Wintz H: **Development and validation of RQI: an RNA quality indicator for the ExperionTM automated gel electrophoresis system.** *Tech Note Bio-Rad* 2008, **5761**.
53. Krupp G: *Stringent RNA quality control using the Agilent 2100 bioanalyzer*, Agilent Technologies application note. 2005.
54. Pico de Coana Y, Parody N, Fernandez-Caldas E, Alonso C: **A modified protocol for RNA isolation from high polysaccharide containing *Cupressus arizonica* pollen. Applications for RT-PCR and phage display library construction.** *Mol Biotechnol* 2009. doi:10.1007/s12033-009-9219-z.
55. Philips MA, D'auria JC, Luck K, Gershenzon J: **Evaluation of candidate reference genes for real-time quantitative PCR of plant samples using purified cDNA as template.** *Plant Mol Biol Rep* 2009, **27**:407–416.
56. Onate-Sanchez L, Vicente-Carbajosa J: **DNA-free RNA isolation protocols for Arabidopsis thaliana, including seeds and siliques.** *BMC Reseach Notes* 2008, **1**:93.
57. De Santis C, Smith-Keune C, Jerry DR: **Normalizing RT-qPCR data: are we getting the right answers? An appraisal of normalization approaches and internal reference genes from a case-study in the Finfish *Lates Calcarifer*.** *Mar Biotechnol* 2011, **13**:170–180.
58. Grunwald U, Guo W, Fischer K, Isayenkov S, Ludwig-Muller J, Hause B, Yan X, Kuster H, Franken P: **Overlapping expression patterns and differential transcript levels of phosphate transporter genes in arbuscular mycorrhizal, P_i-fertilised and phytohormone treated *Medicago truncatula* roots.** *Planta* 2009, **229**:1023–1034.
59. Laurell H, Iacovoni JS, Abot A, Svec D, Maoret J-J, Arnal J-F, Kubista M: **Correction of RT-qPCR data for genomic DNA-derived signals with ValidPrime.** *Nucl Acids Res* 2012, **40**:e51.
60. Czechowski T, Stitt M, Altmann T, Udvardi MK, Scheible WR: **Genome-wide identification and testing of superior reference genes for transcript normalization in Arabidopsis.** *Plant Physiol* 2005, **139**:5–17.

61. Popovici V, Goldstein DR, Antonov J, Jaggi R, Delorenzi M, Wirapati P: **Selecting control genes for RT-qPCR using public microarray data.** *BMC Bioinforma* 2009, **10**: . doi:1186/1471-2105-10-42.
62. Coker JS, Davies E: **Selection of candidate housekeeping controls in tomato plants using EST data.** *Biotechniques* 2003, **35**:740–748.
63. Cruz F, Kalaoun S, Nobile P, Colombo C, Almeida J, Barros LMG, Romano E, Grossi-de-Sa MF, Vaslin M, Alves-Ferreira M: **Evaluation of coffee reference genes for relative expression studies by quantitative real-time RT-PCR.** *Mol Breeding* 2009, **23**:607–616.
64. Klie M, Debener T: **Identification of superior reference genes for data normalisation of expression studies via quantitative PCR in hybrid roses (*Rosa hybrida*).** *BMC Reserach Notes* 2011, **4**:518.
65. Mallona I, Lischewski S, Weiss J, Hause B, Egea-Cortines M: **Validation of reference genes for quantitative real-time PCR during leaf and flower development in *Petunia hybrida*.** *BMC Plant Biol* 2010, **10**. doi:10.1186/1471-2229-10-4.
66. Regier N, Frey B: **Experimental comparison of relative RT-qPCR quantification approaches for gene expression studies in poplar.** *BMC Mol Biol* 2010, **11**:57.
67. Ramakers C, Ruijter JM, Lekanne-Deprez RH, Moorman AFM: **Assumption-free analysis of quantitative real-time PCR data.** *Neuroscience Lett* 2002, **339**:62–66.
68. Liu W, Saint DA: **A new quantitative method of real time reverse transcription polymerase chain reaction assay based on simulation of polymerase chain reaction kinetics.** *Anal Biochem* 2002, **302**:52–59.
69. Liu W, Saint DA: **Validation of a quantitative method for real time PCR kinetics.** *Biochem Biophys Res Comm* 2002, **294**:347–353.
70. Lalam N: **Estimation of the reaction efficiency in polymerase chain reaction.** *J Theor Biol* 2006, **242**:947–953.
71. Rieu I, Powers SJ: **Real-time quantitative RT-PCR: design, calculations and statistics.** *Plant Cell* 2009, **21**:1031–1033.
72. Nordgard O, Kvaloy JT, Farmen RK, Heikkila R: **Error propagation in relative real-time reverse transcription polymerase chain reaction quantification models: the balance between accuracy and precision.** *Anal Biochem* 2006, **356**:182–193.
73. Marino JH, Cook P, Miller KS: **Accurate and statistically verified quantification of relative mRNA abundances using SYBR Green I and real-time RT-PCR.** *J Immunol Meth* 2003, **283**:291–306.
74. Cook P, Fu C, Hickey M, Han E-S, Miller K: **SAS programs for real-time RT-PCR having multiple independent samples.** *Bioinformatics* 2004, **37**:990–995.
75. Derveaux S, Vandesompele J, Hellemans J: **How to do successful gene expression analysis using real-time PCR.** *Methods* 2010, **50**:227–230.
76. Kunert R, Gach JS, Vorauer-Uhl K, Engel E, Katinger H: **Validated method for quantification of genetically modified organisms in samples of maize flour.** *J Agricul Food Chem* 2006, **54**:678–681.
77. Taverniers I, Van Bockstaele E, De Loose M: **Cloned plasmid DNA fragments as calibrators for controlling GMOs: diffeent real-time duplex quantitative PCR methods.** *Anal Bioanal Chem* 2004, **378**:1198–1207.
78. Jansen RC, Nap JP: **Genetical genomics: the added value from segregation.** *Trends Genet* 2001, **17**:388–391.
79. Joosen RVL, Ligterink W, Hilhorst HWM, Keurentjes JJB: **Advances in genetical genomics of plants.** *Curr Genomics* 2009, **10**:540–549.
80. Motulsky H: **Choosing an appropriate sample size.** In *Intuitive Biostatistics*. New York: Oxford University Press; 1995:195–204.
81. Shi C, Uzarowska A, Ouzunova M, Landbeck M, Wenzel G, Lübberstedt T: **Identification of candidate genes associated with cell wall digestibility and eQTL (expression quantitative trait loci) analysis in a Flint x Flint maize recombinant inbred line population.** *BMC Genomics* 2007, **8**:22.
82. Schwinn KE, Davies KM: **Flavonoids.** In *Plant Pigments and their manipulation*, Annual Plant Reviews, Volume 14. Edited by Davies K. Oxford: Blackwell Publishing Ltd; 2004:92–149.
83. Larsen ES, Alfenito MR, Briggs WR, Walbot V: **A carnation anthocyanin mutant is complemented by the *glutathione S-transferases* encoded by maize *Bz2* and petunia *An9*.** *Plant Cell Rep* 2003, **21**:900–904.
84. De Jong WS, Eannetta NT, De Jong DM, Bodis M: **Candidate gene analysis of anthocyanin pigmentation loci in the *Solanaceae*.** *Theor Appl Genet* 2004, **108**:423–432.
85. Jung CS, Griffiths HM, De Jong DM, Cheng S, Bodis M, Kim TS, De Jong WS: **The potato developer (D) locus encodes an R2R3 MYB transcription factor that regulates expression of multiple anthocyanin structural genes in tuber skin.** *Theor Appl Genet* 2009, **120**:45–57.
86. Kliebenstein DJ, West MA, van Leeuwen H, Loudet O: **Dioerge RW and St Clair DA: Identification of QTLs controlling gene expression networks defined a priori.** *BMC Bioinforma* 2006, **7**:308.
87. De Keyser E, Shu QI, Van Bockstaele E, De Riek J: **Multipoint-likelihood maximization mapping on 4 segregating populations to achieve an integrated framework map for QTL analysis in pot azalea (*Rhododendron simsii* hybrids).** *BMC Mol Biol* 2010, **11**:1.
88. Moore D, Dowhan D: **Manipulation of DNA.** *Current Protocols Mol Biol* 2002, **59**:2.1.1–2.1.10.
89. Dhanasekaran S, Doherty TM, Kenneth J: **Comparison of different standards for real-time PCR-based absolute quantification.** *J Immunol Methods* 2010, **354**:34–39.
90. Luu-The V, Paquet N, Calvo E, Cumps J: **Improved real-time RT-PCR method for high-throughput measurements using second derivative calculation and double correction.** *Biotechniques* 2005, **38**:287–293.
91. Van Ooijen JW: *MapQTL®5, software for the mapping of quantitative trait loci in experimental populations.* Wageningen, The Netherlands: Kyazma BV; 2004.
92. Mantel N: **The detection of disease clustering and a generalized regression approach.** *Cancer Res* 1976, **27**:209–220.
93. Liedloff AC: *Mantel Nonparametric Test Calculator. Version 2.0.* Australia: School of Natural Resource Sciences, Queensland University of Technology; 1999.

2

Genetic transformation of *Fusarium avenaceum* by *Agrobacterium tumefaciens* mediated transformation and the development of a USER-Brick vector construction system

Lisette Quaade Sørensen[1], Erik Lysøe[3], Jesper Erup Larsen[1], Paiman Khorsand-Jamal[1], Kristian Fog Nielsen[2] and Rasmus John Normand Frandsen[1*]

Abstract

Background: The plant pathogenic and saprophytic fungus *Fusarium avenaceum* causes considerable in-field and post-field losses worldwide due to its infections of a wide range of different crops. Despite its significant impact on the profitability of agriculture production and a desire to characterize the infection process at the molecular biological level, no genetic transformation protocol has yet been established for *F. avenaceum*. In the current study, it is shown that *F. avenaceum* can be efficiently transformed by *Agrobacterium tumefaciens* mediated transformation. In addition, an efficient and versatile single step vector construction strategy relying on Uracil Specific Excision Reagent (USER) Fusion cloning, is developed.

Results: The new vector construction system, termed USER-Brick, is based on a limited number of PCR amplified vector fragments (core USER-Bricks) which are combined with PCR generated fragments from the gene of interest. The system was found to have an assembly efficiency of 97% with up to six DNA fragments, based on the construction of 55 vectors targeting different polyketide synthase (PKS) and PKS associated transcription factor encoding genes in *F. avenaceum*. Subsequently, the Δ*FaPKS3* vector was used for optimizing *A. tumefaciens* mediated transformation (ATMT) of *F. avenaceum* with respect to six variables. Acetosyringone concentration, co-culturing time, co-culturing temperature and fungal inoculum were found to significantly impact the transformation frequency. Following optimization, an average of 140 transformants per 10^6 macroconidia was obtained in experiments aimed at introducing targeted genome modifications. Targeted deletion of *FaPKS6* (FA08709.2) in *F. avenaceum* showed that this gene is essential for biosynthesis of the polyketide/nonribosomal compound fusaristatin A.

Conclusion: The new USER-Brick system is highly versatile by allowing for the reuse of a common set of building blocks to accommodate seven different types of genome modifications. New USER-Bricks with additional functionality can easily be added to the system by future users. The optimized protocol for ATMT of *F. avenaceum* represents the first reported targeted genome modification by double homologous recombination of this plant pathogen and will allow for future characterization of this fungus. Functional linkage of *FaPKS6* to the production of the mycotoxin fusaristatin A serves as a first testimony to this.

Keywords: Single step cloning, ATMT, *Agrobacterium tumefaciens* mediated transformation, *Fusarium avenaceum*, USER-Brick, Genome modification, Transformation, Fusaristatin, *FaPKS6*, Mycotoxin, LC-MS, MS-MS, Polyketide, Nonribosomal peptide

* Correspondence: rasf@bio.dtu.dk
[1]Eukaryotic Molecular Cell Biology Group, Department of Systems Biology, The Technical University of Denmark, Søltofts Plads building 223, DK-2800 Kgs., Lyngby, Denmark
Full list of author information is available at the end of the article

Background

The plant pathogenic fungus *Fusarium avenaceum* displays a wide host range and causes diseases such as root rot and ear blight of cereals [1]. The financial losses are mainly due to crown rot and head blight of wheat and the accompanying contamination of the harvested grains with mycotoxins [2]. In addition, to the direct in field effects on harvest yields, this species is also responsible for post-harvest rot of many crops, such as swede turnip [3], apple [4], broccoli [5] and potato tubers [6]. In temperate regions of the world, such as Scandinavia, Russia and Canada, *F. avenaceum* has in several reports been found to be the dominant species in connection with head blight. However, in recent years an increasing prevalence in warmer regions, throughout the world, has also been reported [2,7]. During infections and in post-harvest situations *F. avenaceum* has been reported to produce various bioactive secondary metabolites, including antibiotic Y, chlamydosporol, aurofusarin, enniatins, and the mycotoxins fusarin C, 2-amino-14,16-dimethyloctadecan-3-ol, and moniliformin [2,8].

In spite of its significant impact on agricultural production only a few studies have aimed at elucidating the molecular genetic basis of this species' broad host range, large geographical distribution and potential for biosynthesis of secondary metabolites [9]. One of the major hurdles for such studies in this species, and many others is the lack of reliable genetic transformation protocols and a basic toolbox for performing targeted genome modifications. The *Agrobacterium tumefaciens* mediated transformation (ATMT) technique has proven capable of transforming a wide range of different fungal species [10].

The current study concerns the establishment of an efficient transformation protocol for *F. avenaceum*, via ATMT, and the development of a new strategy for single step construction of binary vectors compatible with ATMT of filamentous fungi. The new USER-Brick vector system is compatible with two different popular vector backbone series, pAg1 and ppPK2/pPZP-201BK, and allows for high throughput vector construction. The system provides a versatile toolbox for the construction of plasmids that can be used to address typically posed biological questions, such as 1) what effect will deletion or 2) overexpression of a given gene have on the phenotype, 3) when, and where in the mycelium is the gene in question expressed, and 4) the subcellular localization of a given protein. The system was tested by performing targeted modification (deletion and overexpression) of fourteen different loci in the *F. avenaceum* genome, and by overexpression of 30 PKS associated transcription factor encoding genes from random loci in the *F. avenaceum* genome.

Methods

Organisms

E. coli DH5α was used in connection with USER cloning experiments. Chemically competent *E. coli* cells were produced as described in [11]. *Agrobacterium tumefaciens* LBA4404 was used for ATMT experiments. Transformation and cultivation of *E. coli* and *A. tumefaciens* was performed as described in [12]. *F. avenaceum* IBT41708 and *Fusarium graminearum* PH-1 were obtained from the fungal culture collection at The Technical University of Denmark. Macroconidia for both fungal species were produced using sporulation medium as described in [13]. For the selective steps of ATMT the fungi were cultured at 26°C in darkness on Defined Fusarium Medium (DFM): 12.5 g/L glucose, 1.32 g/L L-asparagine, 0.517 g/L $MgSO_4$, 1.524 g/L KH_2PO_4, 0.746 g/L KCl, 0.04 mg/L $Na_2B_4O_7$, 0.4 mg/L $CuSO_4$, 1.2 mg/L $FeSO_4$, 0.7 mg/L $MnSO_4$, 0.8 mg/L $NaMoO_2$, 10 mg/L $ZnSO_4$.

Amplification of USER-Bricks

The current core USER-Bricks includes vector backbones, selection markers, promoters and fluorescent markers as shown in Figure 1 top panel. PCR reactions were performed using X7 [14] or Pfu Turbo Cx Hotstart DNA polymerase (Agilent Technologies) in 50 µl reactions with Phusion HF buffer (New England Biolabs). Core USER-Bricks were amplified from purified plasmid DNA, 1 ng/50 µl reaction, using 2-deoxyuridine containing primers from Integrated DNA Technologies (Leuven, Belgium) as specified in Figure 2. Genomic inserts (Figure 1, centre panel) unique for the individual target were amplified from purified *F. avenaceum* genomic DNA, 10 ng/50 µl reactions, using the gene specific primers listed in Additional file 1: Tables S1 and S2, with the relevant overhangs specified in Figure 3. For targeted integration 1500 bp long homologous recombination sequences were used, and in the case of expression from random loci the coding sequence of the gene plus 500 bp downstream were amplified. A G-Storm GS2 thermal cycler (G-Storm, Somerton, UK) was used for PCR, with the following program: 98°C for 10 min, 35 cycles (98°C for 20 sec, 60°C for 20 sec, 72°C for 2 min), 72°C for 5 min. The PCR amplicons were gel purified using the illustra GFX PCR DNA and Gel Band Purification Kit (GE Healthcare), to eliminate the plasmid and genomic DNA that had served as the templates during the PCR reactions.

Construction of plasmids from USER-Bricks

The USER-Bricks and 'gene specific fragments' needed for the different types of experiments are described in Figure 1. For USER cloning reactions 1 µl of the needed purified USER-Bricks and 2 µl of the required gene specific inserts

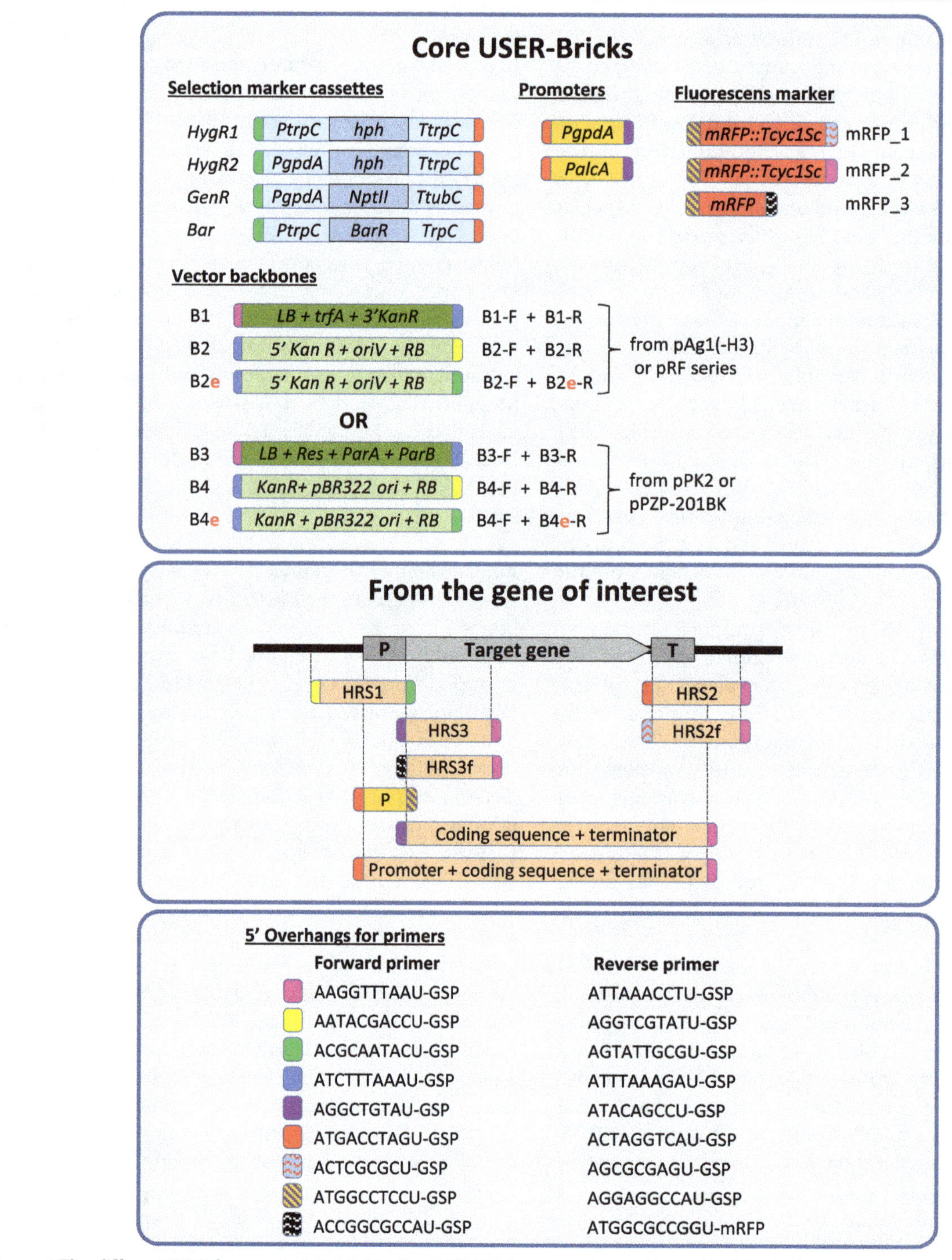

Figure 1 The different DNA fragments (=bricks) in the USER-Brick vector system. The ends of the Bricks are colour coded based on which overhangs that are compatible for fusion. ***Top panel***: The core USER-Brick includes backbones, selection markers, promoters and fluorescent marker fragments. ***Centre panel***: The placement of the different types of PCR amplicons in relation to the gene of interest. ***Bottom panel***: Sequences of the 5′ overhang found on the primers for amplifying the different USER-Bricks in the two panels above.

were mixed with 1.2 units of 'USER enzyme mix' (New England Biolabs) and 10xTaq DNA polymerase buffer (Sigma-Aldrich) to a final concentration of 1x, in a total volume of 12 μl. The reactions were incubated for 25 min at 37°C and then for 25 min at 25°C using a thermal cycler. The entire reaction volume was used for transformation of 50 μl chemically competent *E. coli* DH5α cells, as described in [12]. Ten of the resulting transformants were

Name:	Sequence (5' to 3')	Purpose	Tem-plate	PCR product	Compatible overhangs
Backbone USER-Bricks based on pAg1-H3/pRF-HU2 plasmids					
B1-F	AAGGTTTAAUTCACTGGCCGTCGTTTTA	Generic	A	2165 bp	HRS2-R, CDS-R
B1-R	ATTTAAAGAUCCGCGCGAGC	Generic	A		B2-F
B2-F	ATCTTTAAAUGGAGTGTCTTCTTCCCA	Generic	A	1658 bp	B1-R
B2-R	AATACGACCUTCGTGACTCCCTTAATTCT	Generic	A		HRS1-F
B2e-R	AGTATTGCGUTCGTGACTCCCTTAATTCT	Ectopic express.	A	1658 bp	Marker-F
Backbone USER-Bricks based on ppPK2 or pPZP-201BK					
B3-F	AAGGTTTAAUCGTTTCCCGCCTTCAGTTTAAA	Generic	B	3096 bp	HRS2-R
B3-R	AGCGGCTAAUCAAGGCTTCACCC	Generic	B		B2-F
B4-F	ATTAGCCGCUACAAGATCGTAAAGA	Generic	B	2937 bp	B1-R
B4-R	AATACGACCUTGTGTTATTAAGTTGTCTAAGCGTCA	Generic	B		HRS1-F
B4e-R	AGTATTGCGUTGTGTTATTAAGTTGTCTAAGCGTCA	Ectopic express.	B	2937 bp	Marker-F
Dominant antibiotic markers for fungal selection:					
HygR1-F	ACGCAATACUAGTCGGGGGATCCTCTAG	Hygromycin marker 1	A, C	2480 bp	B2-R, B4-R
HygR1-R	ACTAGGTCAUGGGCCCATCGATGATCAG	*PtrpC:hph:TtrpC*			HRS2-F, CDS-R, promoter-F
HygR2-F	ACGCAATACUATCTTTCGACACTGAAATACGT	Hygromycin marker 2	D	3989 bp	B2-R, B4-R
HygR2-R	ACTAGGTCAUTCGAGTGGAGATGTGGAGTGGG	*PpgdA:hph:TtrpC*			HRS2-F, CDS-R, promoter-F
GenR-F	ACGCAATACUCATGCAACATGCATGTACTGTCTGAT	Geneticin marker	E	1440 bp	B2-R, B4-R
GenR-R	ACTAGGTCAUCGCGGCTTCGAATCGTGG	*PgpdA:NPTII:NcTtubC*			HRS2-F, CDS-R, promoter-F
Bar-F	ACGCAATACUAAGAAGGATTACCTCTAAACAAGTG	Bar resistance	F	1677 bp	B2-R, B4-R
Bar-R	ACTAGGTCAUTCGACAGAAGATGATATTGAAGGAGC	*PtrpC:barR:TtrpC*			HRS2-F, CDS-R, promoter-F
Promoters:					
PgpdA-F	ATGACCTAGUGCCAGCCCGAATTCCCTTGTATC	Constitutive	G	2320 bp	Marker-R
PgpdA-R	ATACAGCCUGGGTGATGTCTGCTCAAGCGGGG				CDS-F
PalcA-F	ATGACCTAGUCTCCCCGATGACATACAGGAGG	Inducible	G	853 bp	Marker-R
PalcA -R	ATACAGCCUTTGAGGCGAGGTGATAGGATTG				CDS-F
Transcriptional reporter:					
mRFP-F	ATGGCCTCCUCCGAGGACGTCATCA	mRFP::*Tcyc1Sc*	H	891 bp	promoter-R
mRFP-R	ATTAAACCTUCTTCGAGCGTCCCAAAACCTTC				B1-F
mRFPf-R	AGCGCGAGUCTTCGAGCGTCCCAAAACCTTC		H	891 bp	HRS2f-F
mRFPft-R	ATGGCGCCGGUGGAGTGGCGG	mRFP	H	675	HRS3f-F

Figure 2 Primer sequences for amplifying the different USER-Bricks, specifying template, product size and compatible USER-Brick fragments. *PCR templates:* A = pAg1-H3, pRF-HU2 [15], B = pPK2 or pPZP-201BK [16], C = pANT-hyg(R) [17] or pCSN43 (Fungal Genetics Stock Center), D = pAN7-1 [18], E = pSM334 [19], F = pBARKS1 [20], G = *A. nidulans genomic DNA* or pRF-HUE, pRF-HU2E [15], H = pWJ1350 [21] or plasmids derived from the original *Discosoma* sp. study [22]. In primer sequences: U = 2-deoxyuridine.

analysed by colony-PCR, using the gene specific primers, and restriction enzyme digestion to verify correct assembly. In cases where the plasmids did not yield the expected results, the presence of the core USER-Bricks was tested by PCR.

The seven different types of vector constructs that the USER–Brick system currently allows for are shown in Figures 4 and 5, including information on which fragments should be combined in the individual case. Additional file 1: Table S1 summarizes the USER Bricks needed for the different constructions. These can either be constructed using the pAg1-H3/pRF-HU2 (B1 + B2) or the ppPK2/pPZP-201BK (B3 + B4) vector backbones, but note that the two vector backbone systems are not mixable. For the construction of 'targeted deletion' vectors, exemplified with the B1 + B2 backbone, the following was combined: B1 + B2 + HygR1 + upstream homologous recombination sequence (HRS) + downstream HRS, where the two HRS surround the target (Figure 4C and Additional file 1: Table S2). For '*in locus* overexpression' constructions: B1 + B2 + HygR1 + PgpdA + upstream HRS + downstream HRS, where the downstream HRS includes the start of the gene to be overexpressed (Figure 4D and Additional file 1: Table S2). For overexpression of genes from a random locus in the recipient fungus genome, B1 + B2e + HygR1 + PgpdA + CDS + promoter element were combined (Figure 4B and Additional file 1: Table S3). Figure 4 also describes the construction of vectors for random ectopic expression using the gene's natural promoter (Figure 4A). Figure 5, shows the strategies

Purpose	Primer name	Sequence (5′ to 3′)	Compatible overhang
Deletion / *in locus* overexpression	HRS1-F [A]	AGGTCGTATU-GSP	B2-R
Deletion / *in locus* overexpression	HRS1-R [A]	AGTATTGCGU-GSP	marker F
Deletion	HRS2-F [B]	ATGACCTAGU-GSP	marker R
Deletion	HRS2-R [B]	ATTAAACCTU-GSP	B1-F
In locus and ectopic expression*	HRS3 or CDS-F [C]	AGGCTGTAU-GSP	promoter-R
In locus and ectopic expression	HRS3 or CDS-R [C]	ATTAAACCTU-GSP	B1-F
Transcription reporter	PromoterX-F	ATGACCTAGU-GSP	marker R
Transcription reporter	PromoterX-R	AGGAGGCCAU-GSP	mRFP-F
Expression from fixed locus	CDSf-R	AGCGCGAGU-GSP	HRS2f-F
Expression from fixed locus	HRS2f-F	ACTCGCGCU-GSP	CDSf-R
N'terminal mRFP tagging*	HRS3f-F	ACCGGCGCCAU-GSP	mRFPtf-R

Figure 3 Primer overhangs that should be added to the 5′ end of the gene specific primers to allow for construction of the specified types of vector constructs. A = natural promoter regions of the target gene; B = the terminator region of the gene and C: for *in locus* overexpression experiments the first 1500 bp of the gene so that the AU in the forward primer (HRS3, CDS-F and HRS3f) is part of the start codon, and for ectopic expression experiments so that the entire coding sequence and terminator are amplified.

for expression from a predefined (fixed) locus in the genome (Figure 5A), transcriptional reporters from random (Figure 5C) and fixed loci (Figure 5B) and N'terminal mRFP tagging (Figure 5D). For experiments comparing the effect of the vector backbone a pPK2::Δ*FaPGL1/PKS3* plasmid was constructed (USER-Bricks: B3 + B4 + HygR1 + FaPKS3-U1/U2 + FaPKS3-U3/U4).

Optimizing ATMT of *F. avenaceum*

The effects of hygromycin B (Invivogen), geneticin/G418 (Invivogen) and DL-phosphinothricin/Basta (Life Science) on germination of *F. avenaceum* was tested with the purpose of identifying the minimal concentration that allowed efficient inhibition during ATMT experiments. Five different concentrations of the three antibiotics were tested, hygromycin B: 75, 100, 150, 200, 250 μg/ml; geneticin: 150, 200, 250, 300, 350 μg/ml and for DL-phosphinothricin (Basta™) 200, 400, 600, 800, 1000 μg/ml. Macroconidia were initially plated onto black filter papers on IMAS medium, incubated for 48 hours and then transferred to DFM medium with the specified selection regimes. Following 10 days of incubation at 26°C in darkness, the amount of mycelium on and beneath the filters was evaluated.

ATMT was optimized using the pAg1-H3/pRF-HU2 vector series in the *A. tumefaciens* LBA4404 strain background with respect to six parameters. In these experiments, unless otherwise specified, the *A. tumefaciens* strain was pre-induced with 200 μM acetosyringone until OD_{600} 0.7, mixed 1:1 with $5*10^5$ macroconidia per agar plate, co-cultured for 2 days at 28°C. The following parameters were the subject of optimization: pre-induction of *A. tumefaciens* (+/− acetosyringone), acetosyringone concentration during co-culturing (0, 200 and 500 μM), co-cultivation time (24, 48, 72 and 96 hours), co-cultivation temperature (24°C, 26°C and 28°C), *F. avenaceum* inoculum (8×10^4, 2×10^5, 5×10^5 and 1×10^6 macroconidia/agar plate). The optimization was performed with three biological replicates, each with ten technical replicates (plates). The average number of obtained transformants per 10^6 used spores was compared using a two tailed Student's *T*-Test assuming unequal variances, performed in Microsoft Excel 2010. The pRF-HU2::Δ*FaPGL1/PKS3* plasmid, targeting the *PKS3/PGL1* locus, was used for all experiments, if not otherwise stated. For testing geneticin and DL-phosphinothricin based selection markers pRF-GU2::Δ*FaPGL1/PKS3* and pRF-BU2::Δ*FaPGL1/PKS3* plasmids were constructed.

Following optimization the system was used for performing targeted modification of thirteen different loci spread across the *F. avenaceum* genome and for random integration of 30 different overexpression constructs from random loci. In these experiments *A. tumefaciens* was grown to OD_{600} of 0.7 in LB media supplemented with kanamycin, co-cultivation was performed for 72 hours at 26°C, using 2×10^5 macroconidia per plate, and selection with 150 μg/ml hygromycin B for six days.

Targeted gene replacement in *F. graminearum*

The gene targeting efficiency in *F. graminearum* was assessed by targeted integration into the *PKS3/PGL1* (FGSG_17168) locus using a pRF-HU2::Δ*FgPGL1/PKS3* plasmid containing the hygromycin resistance marker (HygR1) flanked by two 1500 bp homologous recombination sequences amplified from the *F. graminearum* genome. The ATMT was performed as described in [23].

PCR based genotyping of transformants from targeted genome modification experiments

In the experiments aimed at comparing gene targeting efficiency at the *PGL1/PKS3* locus in *F. avenaceum* and *F. graminearum*, 100 *F. avenaceum* and 104 *F. graminearum* transformants were randomly selected for PCR genotyping using four different primer pairs (described below). The targeting efficiency for the thirteen other

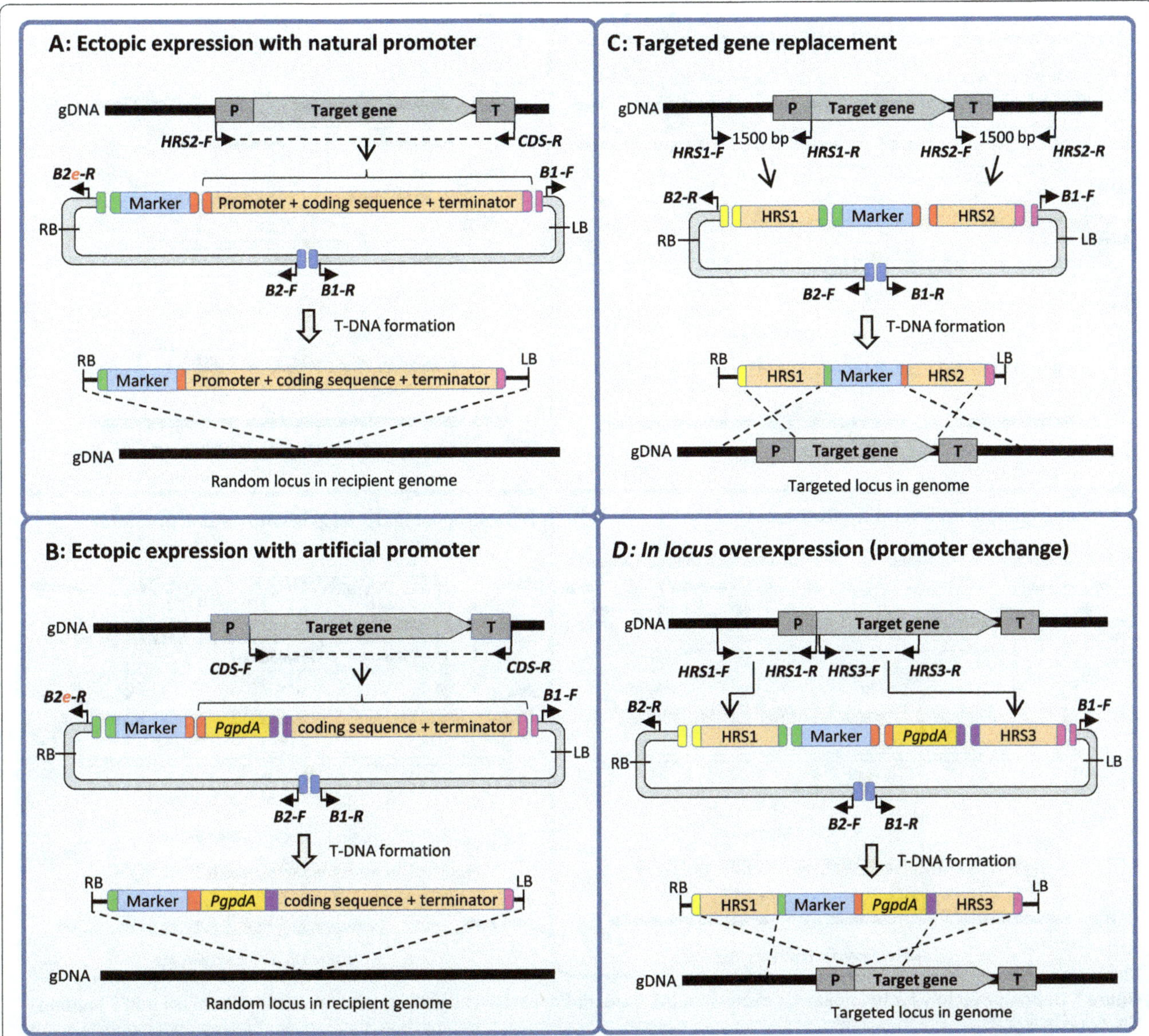

Figure 4 Design of vectors for random heterologous expression with the gene's natural promoter (A), with an alternative promoter (B), for targeted gene replacement (C) and *in locus* overexpression (D). A) Expression of the gene of interest from a random locus in the genome, driven by the gene's natural promoter. Note the use of the B2e USER-Brick to allow for direct fusion of the selection marker cassette with the B2 vector backbone. **B)** Overexpression of the gene of interest from a random genomic locus, with the expression driven by a heterologous promoter, in this case the *gpdA* promoter from *Aspergillus nidulans*. Note the use of the B2e USER-Brick to allow for direct fusion of the selection marker cassette with the B2 vector backbone. **C)** Replacement of the gene of interest. Note that the HRS1 fragment can also be reused for in locus overexpression experiments. **D)** *In locus* overexpression of the gene of interest by targeted integration of a strong constitutive promoter. Note that the HRS1 fragment can be reused for deletion experiments. Primers are represented by solid black arrows. Aberrations: gDNA = genomic DNA; P = promoter; CDS = coding sequence; T = terminator; RB & LB = right & left borders defining the T-DNA region; T-DNA = transfer DNA.

targeted loci (deletions and *in locus* overexpression) in the *F. avenaceum* genome was determined by PCR genotyping of ten randomly selected transformants.

Genomic DNA for the study was obtained by transferring a small scrape of mycelium to a 1.5 ml Eppendorf tube with 100 ml 10:1 TE buffer, which was cooked for 10 minutes in a 750 Watt microwave oven at full effect. The supernatant was then diluted 100 times with MilliQ water, and 1 μl was used as a template in 15 μl PCR reactions with the different test primers (Additional file 1: Table S4 and S5). The different transformants were initially tested using a primer pair targeting the introduced dominant selection marker, with the purpose of verifying that the diluted DNA was of PCR grade. For deletion constructs; a primer pair (T1/T2) amplifying an internal fragment of the gene that was targeted for replacement

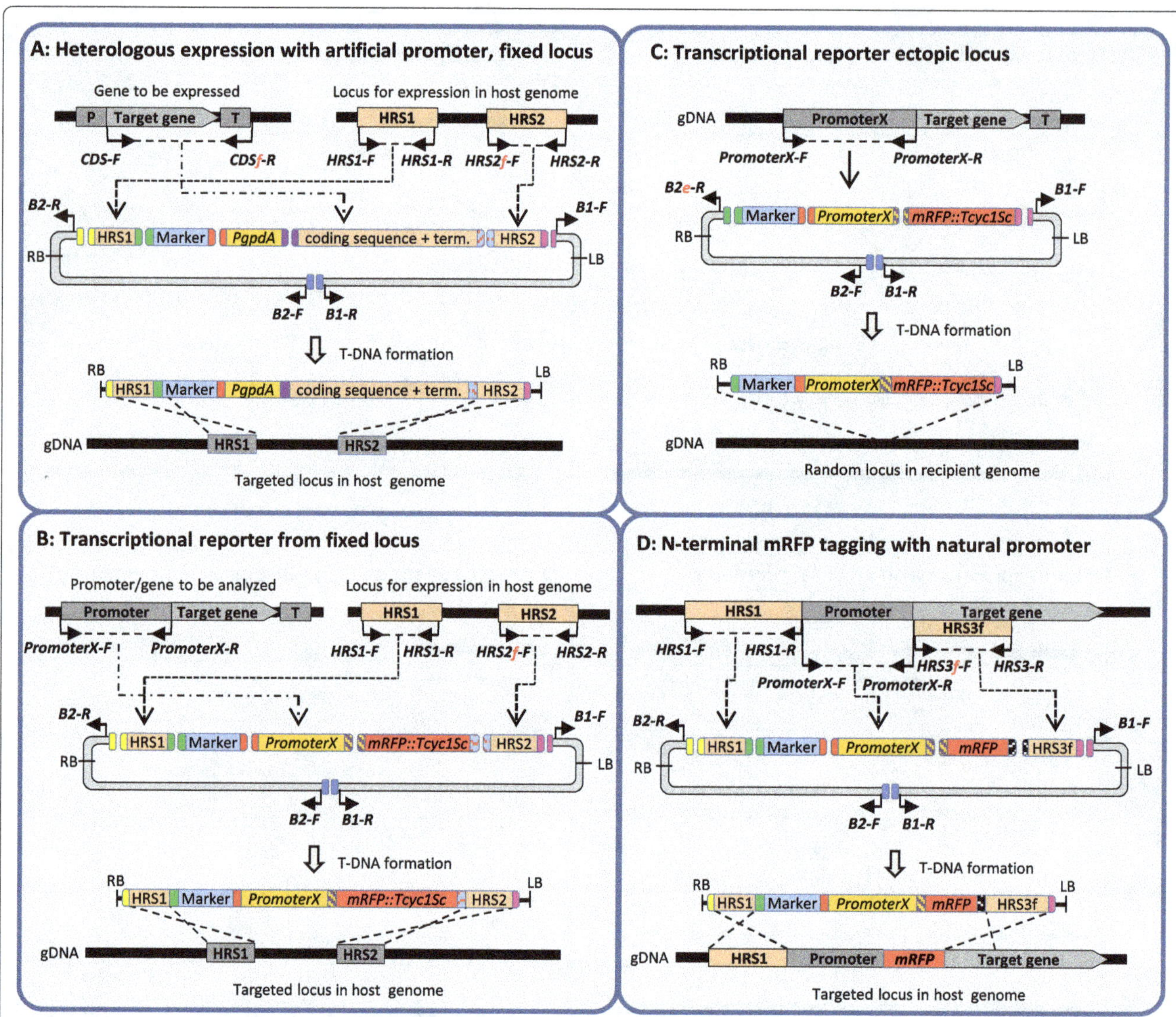

Figure 5 Design of vectors for heterologous expression (A), transcriptional report constructs (B and C) and N'terminal mRFP tagging (D). A) Heterologous expression of a gene of interest from a predefined locus in the genome. Note the unique overhangs on the HRS2f (targeted locus) and CDSf-R (gene to be expressed) fragments that allow for fusion of the two. **B)** Targeted integration of transcriptional reporter construct to monitor the expression of the gene of interest. **C)** Transcriptional reporter construct for random integration. **D)** N'terminal mRFP tagging of gene of interest. The promoter element in the setup can either be the gene's natural promoter, which should be as short as possible to limit the change of recombination, or one of the heterologous promoters (*PgpdA* or *PalcA*). Primers are represented by solid black arrows. Aberrations: gDNA = genomic DNA; P = promoter; CDS = coding sequence; T = terminator; RB & LB = right & left borders defining the T-DNA region; T-DNA = transfer DNA.

was then used, with the purpose of testing for gene replacement. For deletion and *in locus* constructs, two different primer pairs, each bridging one of the used targeting sequences (homologous recombination), were used to test whether homologous recombination had occurred. For deletions the RF-1/T3 and RF-2/T4 primers were used and for *in locus* the RF-3/T1 and RF-2/T2 primers were used (Additional file 1: Table S4). For ectopic expression constructs; two primer pairs (RF-3/T1 and T2/'PKS12-A4/A4-T1') were used to test for the presence of the start and end of the expression cassette (Additional file 1: Table S5). The PCR results were analysed by agarose gel electrophoresis (RunOne system, Embi Tec) or automated capillary electrophoresis (LabChip GX, PerkinElmer) (Table 1).

Chemical analysis of the Δ*FaPKS6* strains

The wild type and three Δ*FaPKS6* transformants, a class 2 and two class 3 (replacement), were grown in darkness for 9 days on YES medium at 28°C. Following incubation the metabolites were extracted by means of the micro-scale method [24], using 3:2:1 (ethyl acetate/dichlormethane/methanol 3/2/1 with 1% formic acid). The samples were extracted for 1 h in an ultrasonic bath (Branson™ Bransonic

Table 1 Standard test primers for the analysis of the obtained transformants

Primer name	Sequence (5′ to 3′)	Amplicon	Target
Primers for testing of the selection markers			
HygR-T-F	AGCTGCGCCGATGGTTTCTACAA	588 bp	Test for marker gene
HygR-T-R	GCGCGTCTGCTGCTCCATACAA		
GenR-T-F	AGCCCATTCGCCGCCAAGTTCT	480 bp	
GenR-T-R	GCAGCTGTGCTCGACGTTGTCA		
BAR-T-F	TCAGATCTCGGTGACGGGCA	552 bp	
BAR-T-R	ATGAGCCCAGAACGACGCC		
Generic primers for testing targeted integration			
RF-1 (HygR constructs)	5′-AAATTTTGTGCTCACCGCCTGGAC	*	T-DNA
RF-2 (HygR constructs)	5′-TCTCCTTGCATGCACCATTCCTTG	*	T-DNA
RF-3 (PgpdA promoter)	5′-TTGCGTCAGTCCAACATTTGTTGCCA	*	T-DNA
RF-7 (GenR constructs)	5′-CTTTGCGCCCTCCCACACAT	*	T-DNA
RF-6 (GenR constructs)	5′-TCAGACACTCTAGTTGTTGACCCCT	*	T-DNA
RF-8 (BarR constructs)	5′-CTGCACTTTTATGCGGTCACACA	*	T-DNA
RF-9 (BarR constructs)	5′-CCTAGGCCACACCTCACCTTATTCT	*	T-DNA

The RF-1 to RF-9 primers anneal inside the different marker genes or introduced promoter and points out into the surrounding genome. These primers should be combined with primers annealing in the genome sequence surrounding the integration site. * The size will depend on the primer design.

Ultrasonic Cleaner Model 2510), the organic phase was moved to new vials, and evaporated to dryness using N_2. The samples were resuspended in 500 μl HPLC grade methanol and ultrasonicated for 20 min., filtered through a PFTE filter and analysed by Ultra high performance liquid chromatography-quadruple Time of Flight mass spectrometry (UHPLC-qTofMS).

UHPLC-qTOFMS analysis, of 0.3-2 μL extracts, was conducted on an Agilent 1290 UHPLC equipped with a photo diode array detector scanning 200–640 nm, and coupled to an Agilent 6550 qTOF (Santa Clara, CA, USA) equipped with an electrospray source (ESI). Separation was performed at 60°C and at a flow of 0.35 ml/min on a 2.1 mm ID, 250 mm, 2.7 μm Agilent Poroshell phenyl hexyl column using a water-acetonitrile gradient solvent system containing 20 mM formic acid. The gradient started at 10% acetonitrile and was increased to 100% acetonitrile within 15 min, keeping this for 4 min, returning to 10% acetonitrile in 1 min, and equilibrating for the next sample in 4 min. Samples were analyzed in both ESI^+ and ESI^- scanning m/z 50 to 1700, and with automated data-dependent MS/MS on all major detected peaks, using collision energies of 10, 20 and 40 eV for each MS/MS experiment. A MS/MS exclusion time of 0.04 min was used to get MS/MS of less abounded ions.

Data files were analysed in Masshunter 6.0 (Agilent technologies) in three different ways: i) Aggressive dereplication using lists of elemental composition and the Search by Formula (10 ppm mass accuracy) of all described *Fusarium* metabolites as well as restricted lists of only *F. avenaceum* and closely related species [25]; ii) Searching the acquired MS/MS spectra in an in-house database of approx. 1200 MS/MS spectra of fungal secondary metabolites acquired at 10, 20 and 40 eV [26]; iii) all major UV/Vis and peaks in the BPC chromatograms not assigned to compounds (and not present in the media blank samples) were then also registered. For absolute verification authentic reference standards were available from 130 Fusarium compounds and furthermore 100 have been tentatively identified based on original producing strains, UV/Vis, LogD and MS/MS [25,27].

Results and discussion

The USER-Brick vector construction system

The past decade has seen an overwhelming blossom of new cloning systems that allow for easy vector construction, including systems based on Gateway [28], In-Fusion/CloneEZ [29], LIC [30], Gibson assembly [31] and USER cloning. The Uracil Specific Excision Reagent (USER) cloning method dates back to 1991 [32], however, it was initially largely ignored due to the costs associated with synthesising 2-deoxyuridine containing primers and the lack of a proofreading DNA polymerase which would not stall when encountering uracil-containing DNA segments. The technique was revitalized by Nour-Eldin and co-workers [33] based on the identification of the 2-deoxyuridine tolerant Pfu Cx polymerase, and further developed into the restriction enzyme digestion free USER Fusion cloning strategy by Geu-Flores [34]. The great advantage of the USER Fusion strategy compared to classical USER cloning and old-fashioned type II restriction enzyme or nicking enzyme based cloning techniques, is that it allows for scarless fusion of multiple fragments in a single cloning step, and only relies on the presence of an A-

N_{8-15}-T motif in the junctions (Geu-Flores et al. 2007). Complementary overhangs which allow for directional assembly of multiple fragments is obtained via unique 5′ overhangs on the used primers. The overhangs include a 2-deoxyuridine base which later can be excised and allow for the formation of sticky ends [34,35].

In the past, several USER based vector systems that allow for the construction of vectors compatible with fungal transformations have been presented [15,36]. Although these systems are superior to standard cloning strategies, with respect to ease of experimental design and cloning efficiency, their Achilles heel has always been the dependency on restriction enzyme digestion and nicking of the recipient plasmid. This is due to the dependency of complete digestion of the recipient vector to eliminate false positives which will otherwise shroud the desired correct transformants.

The new USER-Brick system relies entirely on PCR based amplification of all vector elements, which eliminates the time usage, costs and problems associated with the enzymatic digestion of the recipient plasmid. The unique 5′ overhangs introduced into the ends of the individual USER-Bricks during the amplification step ensures directional assembly of all fragments based on standard base pairing rules. The efficacy of the system was tested by constructing 27 different vectors targeting 14 different loci in the genome of *F. avenaceum* and 30 vectors for overexpression of genes from random loci. PCR based testing of 9–10 transformants from each of the 57 constructed vectors, showed that the system displays an assembly fidelity of 97% +/− 0.3 when fusing up to six DNA fragments in a single reaction (Additional file 1: Table S6). In the 16 cases, app. 3%, where the resulting targeted genome modification plasmids did not test positive for the two expected gene specific fragments subsequent PCR tests for the Core-USER bricks in all cases confirmed that they were present.

In the tested setup the gene specific fragments, amplified from genomic DNA, were gel purified to eliminate unspecific products and primer dimers. However, in experiments with up to five fragments unpurified DNA was found to be as efficient as purified DNA, if no unspecific products were detectable by gel electrophoresis. This was not the case for the core USER-Bricks where purification was found to be essential to eliminate the vector DNA that had served as templates for the PCR reactions, which otherwise produced a high rate of false positives.

The developed system provides a versatile toolbox that can easily be expanded with additional USER-Bricks, such as alternative selection marker cassettes, promoters and fluorescent reporters, if need be. Either by adding the 5′ overhangs specified in Figure 2 to new bricks with similar functionality as those already used in the system or by designing compatible overhangs for new bricks with novel functionalities.

Initial ATMT of *F. avenaceum*

The efficiency of ATMT is highly dependent on abiotic factors, such as media composition, temperature and incubation time, and the optimal transformation conditions are typically unique for the individual fungal species [37]. In the case of *F. avenaceum* no protocol for transformation or targeted genome modification had been established, why we set out to develop and optimize a protocol that would allow future analysis of this species via a reverse genetic approach.

To determine the minimum concentration of hygromycin, geneticin and DL-phosphinothricin that could be used in ATMT experiments with *F. avenaceum*, the ATMT process was mimicked without bacterial cells and the growth of the fungus was recorded. The lowest usable hygromycin B concentration was found to be between 75 and 100 μg/ml, why the highest of these were used in subsequent transformation experiments. Similarly for geneticin and DL-phosphinothricin the minimum useful concentrations were found to be 300 μg/ml and 600 μg/ml, respectively.

For the initial transformation experiment random integration of the T-DNA region with the hygromycin resistance cassette, from pRF-HU2, was attempted using ATMT conditions that have previously worked well for *F. graminearum* [12]. The experiments yielded an average transformation frequency of 14 +/−5.1 transformants per 10^6 *F. avenaceum* spores based on an experiment with five technical replicates. Encouraging results which showed that ATMT based transformation of *F. avenaceum* was possible but would require optimization to yield a useful genetic engineering tool.

Optimization of ATMT for *F. avenaceum*

Pre-induction of A. tumefaciens

The transformation frequency in ATMT of fungi has previously been shown to be influenced by pre-culturing of the *A. tumefaciens* cells with acetosyringone to induce the virulence response of the bacterium. Examples include *Coccidioides immitis* [38] and *Aspergillus awamori* where as little as 6 h of pre-induction yielded a 10 times increase in the number of transformants [39]. However, the opposite negative effect has also been reported for *Beauveria bassiana* [40]. In the case of *F. avenaceum* pre-induction of the used *A. tumefaciens* strain, for 16 hours, did not result in a statistically significant higher number of colonies (Figure 6). Similar results have been reported for *Aspergillus carbonarius* [41] and *Magnaporthe grisea* [42]. Mullins and co-workers found that pre-induction was not essential for the success of ATMT in the case of *Fusarium oxysporum*, but that the transformation process was delayed compared to experiments including pre-induction [43].

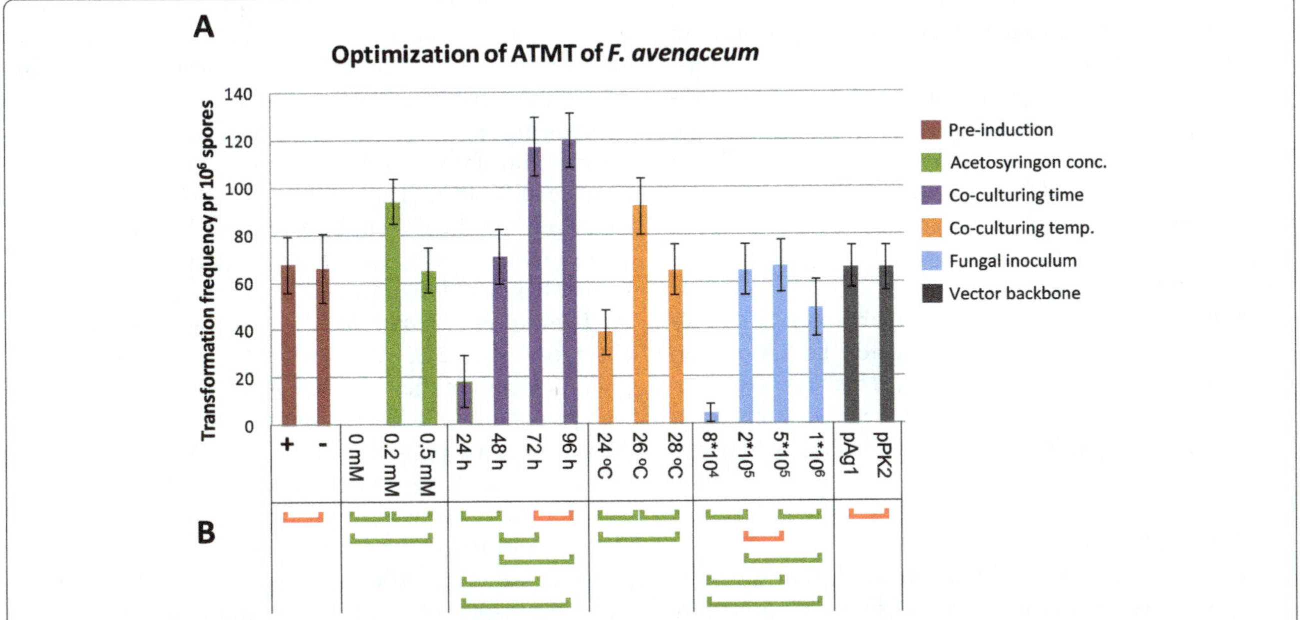

Figure 6 Results from optimization of the *A. tumefaciens* mediated transformation of *F. avenaceum*. A) The average number of transformants per 10^6 spores obtained from three replicates which each included ten agar plates. **B)** Result from Student's *T*-Test comparing the different incubation conditions: green bars represent statistically significant different results for the two compared conditions, and red bars represent not statistically significant different results (further details can be found in the additional file).

Acetosyringone concentration

The presence of a bacterial virulence inducer, such as acetosyringone, during the co-cultivation period, was found to be essential for the successful transformation of *F. avenaceum*. An acetosyringone concentration of 200 μM was found to give the highest number of transformants, while 500 μM resulted in a significant reduction in the number of obtained transformants (Figure 6). This is in contrast to results for *B. bassiana* where an increase in the concentration in the range from 100 μM to 800 μM resulted in increasing numbers of transformants [44]. The observed reduction, for *F. avenaceum*, in the average number of obtained transformants with the higher inducer concentration has also been observed for *Colletotrichum lagenarium* where acetosyringone concentrations above 50 μM resulted in a reduction in transformation frequency [45].

Co-culturing time

The co-culturing time was also found to impact the transformation frequency, and a significant increase was observed when the co-culturing time was prolonged from 24 h to 48 h and further to 72 h (Figure 6). Increasing it from 72 h to 96 h did not result in a statistically significant higher number of transformants, suggesting that the *F. avenaceum* mycelium is more susceptible to transformation within the first 72 hours or that the effects of the applied inducer was lost after this time point. A similar effect has been reported for *Cryptococcus neoformans* [46], while Michielse et al. 2004 found a decrease in the transformation frequency for *A. awamori* when extending the co-culturing time from 48 h to 72 h [47]. Compared to *F. graminearum* where the optimal co-culturing time is 48 h, the maximum transformation efficiency for *F. avenaceum* is delayed which is possibly due to a slower germination and growth rate of the latter species, as described by Summerell and co-workers [48].

Co-culturing temperature

The temperature at which the co-cultivation was performed also significantly affected the transformation frequency. Incubation at 26°C gave the highest number of transformants while lower, and higher temperatures yielded significantly lower numbers of transformants (Figure 6). This effect has also been observed for other fungal species, such as the ascomycete *Botrytis cinerea* [49], the zygomycete *Mortierella alpine* [50] and the basidiomycete *Hebeloma cylindrosporum* [51]. In all cases, the authors have noted that the effect is linked to the optimal germination/growth temperature of the fungus that is being transformed.

Fungal inoculum

Quite interestingly the number of used macroconidia also impacted the number of obtained transformants significantly. The two medium concentrations tested yielded the same number of transformants, while an increase from $5*10^5$ to $1*10^6$ macroconidia per plate lead to a significant reduction (Figure 6). A similar reduction has been reported for *Paecilomyces fumosoroseus* [52]. The biological explanation for the reduction is possibly self-inhibition of germination at higher macroconidia concentrations, as described for other fungal species such as *Colletotrichum fragariae* [53]. The effect of the bacterial inoculum was not tested in the current study, and the concentration was kept constant at an OD_{600} of 0.7 at the stage of co-culturing for all experiments.

Binary vector backbone

In addition to the tested abiotic factors, several studies have also shown that the used vector backbone can affect the transformation efficiency. This was not seen for *F. avenaceum* when the two tested vector backbones, pAg1 and pPK2, yielded similar transformation frequencies (Figure 6).

ATMT with optimized conditions

The identified optimal conditions differed from the optimal conditions for transformation of *F. graminearum* by requiring a longer co-culturing time (increase from 48 h to 72 h), lower co-culturing temperature (reduction from 28°C to 26°C) and no pre-induction of *A. tumefaciens*. The identified optimal conditions were combined in a new ATMT protocol for *F. avenaceum* and used in a series of new transformation experiments, targeted modification of thirteen different loci, which yielded an average of 140 transformants per 10^6 transformed spores. This shows that the contribution of the individual parameters to some extent are additive. However part of the expected gain was lost when the optimal conditions were combined, as one would expect the average transformation frequency to reach ~160 per 10^6 spores based on the experiments where the individual parameters were manipulated. Targeted genome modification of the *FaPKS3* locus using vectors with the geneticin and DL-phosphinothricin selection markers yielded similar transformation frequencies.

Targeted genome modification in *F. avenaceum* versus *F. graminearum*

To test the difference in targeted integration efficiency for *F. avenaceum* and *F. graminearum*, the orthologous *PKS3/PGL1* locus was chosen as the target. The *PGL1* gene encodes a polyketide synthase that previously has been linked to the formation of the purple-black perithecial pigment of *F. graminearum* [54].

Following transformation, approximately 100 randomly selected transformants from each of the two species were genotyped by PCR. The analysis revealed a targeting efficiency of 85% in *F. graminearum* and 74% in *F. avenaceum* (Table 2). The targeting efficiency in other fungal species has been found to vary greatly, ranging from 0.04% for *Blastomyces dermatitidis* [55] to, the more frequently observed level, 29% observed for *Aspergillus awamori* [39]. Members of the *Fusarium* genus in general display a high targeting efficiency, both when transformed by ATMT and protoplast based protocols [56].

Genotyping of the transformants from the experiments aimed at targeted deletion or overexpression of thirteen other loci in the *F. avenaceum* genome resulted in comparative targeting efficiencies (deletion: 72.1% and *in locus* overexpression: 81.9%) as found for the *PGL1/PKS3* locus (Table 2 and Additional file 1: Table S7). However, surprisingly, in the twelve experiments aimed at *in locus* overexpression by insertion of a marker gene and promoter between the target gene's natural promoter and its start site, only ectopic (class 3) and double crossover (class 1) transformants were observed (Table 2 and Additional file 1: Table S7).

Random integration of T-DNA via non-homologous recombination (NHEJ) was more frequent in *F. avenaceum* (10%) than in *F. graminearum* (~1%) (Table 2). The low frequency of random integration in *F. graminearum* concurs well with previous reports by Malz et al. 2006, who found that ATMT based random mutagenesis of *F. graminearum* resulted in 50 times fewer transformants than observed for *F. pseudograminearum* [57]. In the thirty transformations (ectopic overexpression of TFs), which depended on integration via the NHEJ pathways in *F. avenaceum*, significantly fewer transformants were obtained (25 +/−3 per 10^6 spores) compared to the number of transformants obtained in the above discussed experiments (140 per 10^6), which depended on HR integrations. This difference is likely

Table 2 Results from PCR based genotyping of targeted integration into *F. graminearum* and *F. avenaceum*

	No. tested	Class 1 (NHEJ/NHEJ)	Class 2 (NHEJ/HR)	Class 3 (HR/HR)	Gene targeting efficiency (%)
F. graminearum PKS3	104	1	15	88	84.6%
F. avenaceum PKS3	100	10	16	74	74.0%
Other loci in *F. avenaceum*					
Deletion of 13 PKSs	129	13	23	93	72.1%
***In locus* overexpression 12 PKSs**	116	20	1	95	81.9%

NHEJ = non-homologous end joining; HR = homologous recombination. Class 1 transformants represent random integration events; Class 2 represents transformants where the T-DNA has integration by HR at one end and via an unknown mechanism at the opposite end (likely NHEJ); Class 3 transformants represent the desired gene targeting event where both ends have integrated by HR. The number for 'other loci' in *F. avenaceum* represents the average for all 13 deletions and 12 in locus overexpression experiments; see Additional file 1: Table S7 for results for the individual locus.

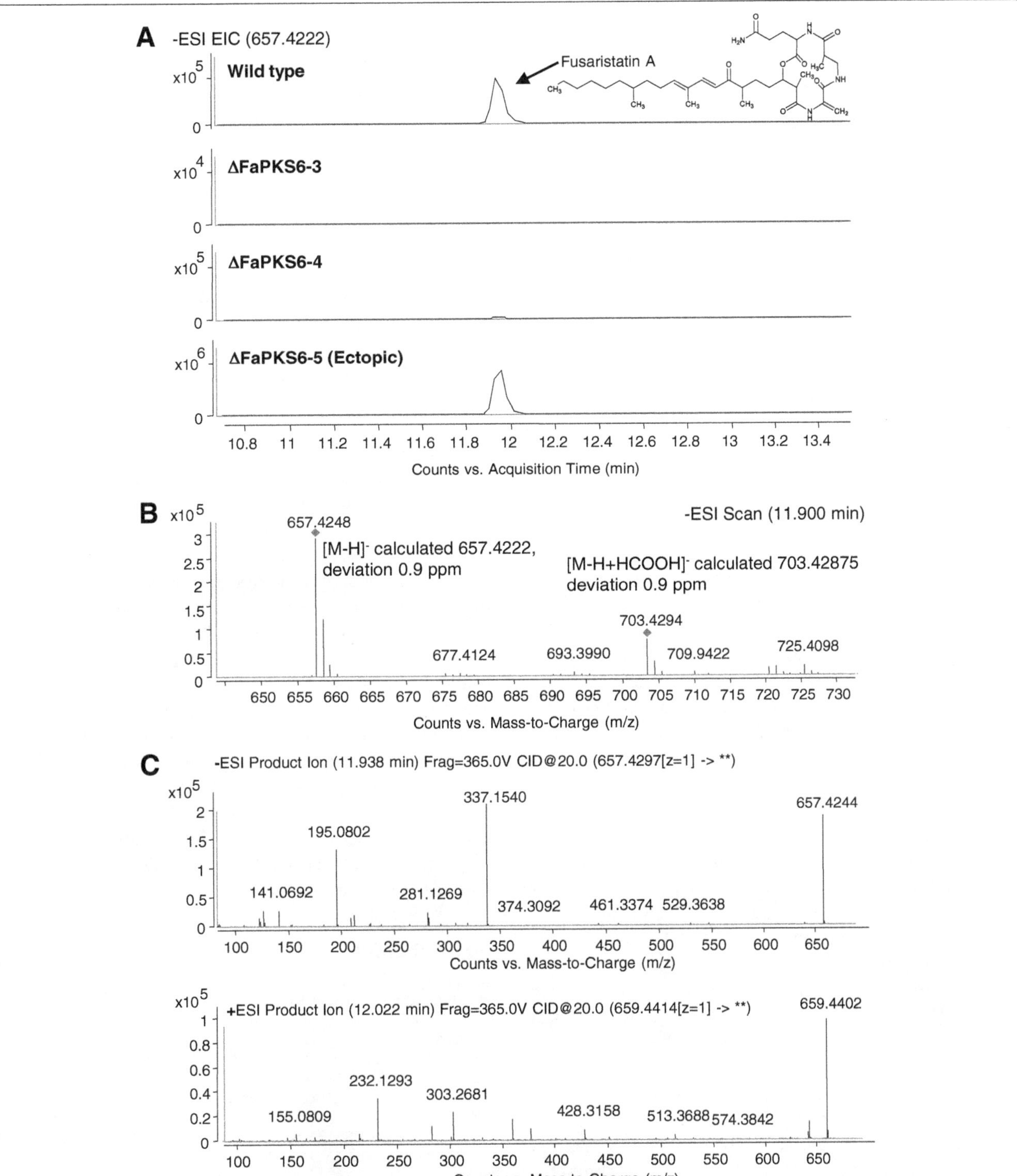

Figure 7 Linking of *FaPKS6* to the formation of fusaristatin A. A) Extracted ion chromatogram, at 657.4222 Scan Frag = 365.0 V, for the wild type and three Δ*FaPKS6* transformants grown on YES medium. The two confirmed *PKS6* replacement strains T3 and T4 have lost the ability to produce fusaristatin A whereas the Δ*FaPKS6*-T5 strain (failed replacement) retains the ability. **B)** Mass spectrum (m/z plot) for the peak at 11.949 minutes and identification of the molecular ion $[M\text{-}H]^-$ 657.4248 and $[M\text{-}H + HCOOH]^-$ 703.4294 adduct. **C)** MS/MS based fragmentation in negative and positive mode, using a CID of 20.0 eV: Top panel shows fragmentation of the $[M\text{-}H]^-$ mass ion and the bottom panel shows fragmentation of the $[M + H]^+$ ion.

due to the fact that integration via the homologous recombination (HR) pathway is not an option due to the lack of homologous recombination sequences in the introduced T-DNA. This supports the initial findings that experimental strategies relying on random mutagenesis by ATMT is feasible in *F. avenaceum*, which is not the case in *F. graminearum*. PCR based genotyping of the obtained transformants showed that 85% (average for all thirty experiments) tested positive for both ends of the TF expression cassette, while 15% showed various levels of truncation (Additional file 1: Table S8).

Linking of FaPKS6 to production of fusaristatin A

To show that the developed molecular toolkit, USER-Brick system and ATMT of *F. avenaceum*, can be used for functional characterization of genes, we analysed the impact on production of secondary metabolites in the constructed *FaPKS6* deletion strain.

Chemical profiling of the Δ*FaPKS6* strains compared to the wild type, by UHPLC-qTOFMS, showed that the deletion strain had lost the ability to produce a compound with a retention time of ~11.9 min compared to the wild type (Figure 7A, A). Based on the identification of the $[M-H]^-$ and $[M + HCOO]^-$ adducts from ESI^- and the $[M + H]^+$ and $[M + Na]^+$ from ESI^+ elemental compositions could be unambiguously assigned as $C_{36}H_{58}N_4O_7$. The elemental composition was searched in Antibase2012 [27] and an in-house database [25] showing only one compound with this elemental composition, fusaristatin A (Figure 7A). With the UV/Vis spectrum not providing any structural information, the dereplication was, besides the elemental composition, based on: i) retention time, where fusaristatin A elutes very close to the enniatins (available as reference standards); ii) and MS/HRMS where ammonia was lost in ESI^+, which is a distinct feature of primary amides (Figure 7C) also the ESI^- MS/MS spectrum showed m/z 529.3638 which is consistent with loss of glutamine and m/z 337.1540 with loss of the β-aminoisobutyric acid-glutamine and breakage partial loss of the fatty acid chain at the carbonyl bond. From ESI^+ MS/MS, the fragments m/z 513.3688 was consistent with loss of glutamine including the oxygen from the fatty acid moiety and m/z 428.3158 with loss of β-aminoisobutyric acid-glutamine including the oxygen from the fatty acid moiety. Finally, the compound was recently identified in the very closely related *F. tricinctum* [58]. Fusaristatin A was first isolated and characterized from the endophytic *Fusarium* sp. YG-45 strain [59], and has later also been identified in the endophyte *Phomopsis longicolla* [60].

The identification is further supported by the domain structure of FaPKS6, and other *Fusarium* PKS6 orthologs, which classifies it as a potential highly reducing PKS with a C-methyltransferase domain. FaPKS6 likely catalyses the formation of the partially reduced decaketide portion of fusaristatin A, which is decorated with three C-bound methyl groups. Following PKS synthesis of the decaketide, it is subsequently elongated by the addition of dehydroalanine, 3-aminoisobutyric acid and glutamine to the carboxylic acid end of the polyketide, likely catalysed by a nonribosomal peptide synthetase (NRPS). A possible candidate for this activity is FaNRPS7 (FA08708) which has three modules and is encoded by a gene located next to the *FaPKS6* gene (FA08709).

This is the first time fusaristatin A has been reported in *F. avenaceum* and the first time its biosynthesis has been linked to a specific polyketide synthase.

Collectively, the linking of *FaPKS6* to fusaristatin A in *F. avenaceum* shows that the developed USER-Brick system and ATMT protocol can be used for efficient functional characterization of genes in the *F. avenaceum* genome.

Conclusion

The USER-Brick strategy offers reduced experimental costs (no restriction enzymes), easier experimental design, quicker experimental flow, higher cloning efficiency, higher reproducibility between experiments and most importantly, flexibility in the design possibilities. Optimization of the ATMT protocol for *F. avenaceum* yielded an average transformation frequency of 140 transformants per 10^6 spores in experiments aimed at targeted genome modification. In addition, the optimization process revealed that acetosyringone concentration, co-culturing time and temperature significantly impact the transformation frequency. Combined, the developed USER-Brick system and ATMT protocol offers a generous molecular toolbox that allows for efficient genome engineering in *F. avenaceum* and other filamentous fungi. In the current study, the system was used to successfully link the *FaPKS6* gene to the biosynthesis of fusaristatin A.

Additional file

Additional file 1: Figure S1. Pre-induction of *A. tumefaciens* has no effect on the transformation frequency. **Figure S2.** The effect of three different acetosyringone concentrations in the co-culturing medium. **Figure S3.** Increasing the co-culturing time to 72 h result in increased numbers of transformants. **Figure S4.** The optimal co-culturing temperature when transforming *F. avenaceum* is 26°C. **Figure S5.** The optimal number of macroconidia per transformation plate was found to be between 2*10^5 and 5*10^5. **Figure S6.** The vector backbone was not found to affect the transformation frequency. **Table S1.** List of USER-Bricks needed for different types of vector constructs. **Table S2.** Gene specific primers for vector construction: deletion and *in locus* overexpression. **Table S3.** Gene specific primers for vector construction: Ectopic overexpression of PKS associated transcription factors. **Table S4.** Primers for validation of genomic modifications in *F. avenaceum* and *F. graminearum*. **Table S5.** Primers for validation of genomic modifications in *F. avenaceum*. **Table S6.** Vector construction efficiency with the USER-Brick approach. **Table S7.** Results from PCR based genotyping of the constructed *F. avenaceum* PKS deletion and *in locus* overexpression strains. **Table S8.** Results from PCR based genotyping of the constructed *F. avenaceum* strains expressing PKS associated transcription factors from random loci in the genome.

Abbreviations

ATMT: *Agrobacterium tumefaciens* mediated transformation; HR: Homologous recombination; NHEJ: Non-homologous end-joining; PCR: Polymerase chain reaction; PKS: Polyketide synthase encoding gene; T-DNA: Transfer DNA; USER: Uracil specific excision reaction.

Competing interests

The authors declare that they have no competing interests.

Authors' contributions

LQS and PKJ performed the transformation experiments aimed at targeted genome modification of FaPKSs encoding genes. JEL contributed with experiments aimed at ectopic overexpression of TF encoding genes. The metabolomics profiling and mass spectrometry based identification of Fusaristatin A was performed by KFN. RJNF conceived the study, developed the USER-Brick platform, optimized ATMT of *F. avenaceum*, wrote and edited the manuscript. All authors read and approved the final manuscript.

Acknowledgements

The research was funded by the 'Danish Research Council for Independent Research — Technology and Production' grant no. 09–069707 and a Young Elite Researcher grant no. 09–076147. We thank Agilent Technologies for the Thought Leader Donation of the 6550 QTOF instrument and UHPLC.

Author details

[1]Eukaryotic Molecular Cell Biology Group, Department of Systems Biology, The Technical University of Denmark, Søltofts Plads building 223, DK-2800 Kgs., Lyngby, Denmark. [2]Metabolic Signaling and Regulation group, Department of Systems Biology, The Technical University of Denmark, Søltofts Plads building 221, DK-2800 Kgs., Lyngby, Denmark. [3]Bioforsk–Norwegian Institute of Agricultural and Environmental Research, Høgskoleveien 7, Ås 1430, Norway.

References

1. Desjardins AE: ***Gibberella* from *A* (*venaceae*) to *Z* (*eae*).** *Annu Rev Phytopathol* 2003, **41:**177–198.
2. Uhlig S, Jestoi M, Parikka P: ***Fusarium avenaceum* - the North European situation.** *Int J Food Microbiol* 2007, **119:**17–24.
3. Peters Rick D, Barasubiye Tharcisse DJ: **Dry rot of rutabaga caused by *Fusarium avenaceum*.** *Hortoscience* 2007, **42:**737–739.
4. Sørensen JL, Phipps RK, Nielsen KF, Schroers H-J, Frank J, Thrane U: **Analysis of *Fusarium avenaceum* metabolites produced during wet apple core rot.** *J Agric Food Chem* 2009, **57:**1632–1639.
5. Mercier J, Markhlouf JMR: ***Fusarium avenaceum*, a pathogen of stored broccoli.** *Can Plant Dis Surv* 1991, **71:**161–162.
6. Aprasad KS, Bateman GL, Read PJ: **Variation in pathogenicity on potato tubers and sensitivity to thiabendazole of the dry rot fungus *Fusarium avenaceum*.** *Potato Res* 1997, **40**(4):357–365.
7. Kulik T, Pszczółkowska A, Łojko M: **Multilocus Phylogenetics show high intraspecific variability within *Fusarium avenaceum*.** *Int J Mol Sci* 2011, **12**(9):5626–5640.
8. Sørensen JL, Giese H: **Influence of carbohydrates on secondary metabolism in *Fusarium avenaceum*.** *Toxins (Basel)* 2013, **5:**1655–1663.
9. Herrmann M, Zocher R, Haese A: **Effect of disruption of the enniatin synthetase gene on the virulence of *Fusarium avenaceum*.** *Mol Plant Microbe Interact* 1996, **9:**226–232.
10. Frandsen RJN: **A guide to binary vectors and strategies for targeted genome modification in fungi using *Agrobacterium tumefaciens*-mediated transformation.** *J Microbiol Methods* 2011, **87**(3):247–262.
11. Sambrook J, Fritsch EF, Maniatis T: *Molecular Cloning: A Laboratory Manual.* Cold Spring Harbor, NY: Cold Spring Harbor Laboratory Press; 1989:931–957.
12. Frandsen RJN, Frandsen M, Giese H: **Targeted gene replacement in Fungal Pathogens via *Agrobacterium tumefaciens*-mediated transformation.** *Methods Mol Biol* 2012, **835:**17–45.
13. Yoder WT, Christianson LM: **Species-specific primers resolve members of *Fusarium* section *Fusarium*. Taxonomic status of the edible "Quorn" fungus reevaluated.** *Fungal Genet Biol* 1998, **23:**68–80.
14. Nørholm MHH: **A mutant Pfu DNA polymerase designed for advanced uracil-excision DNA engineering.** *BMC Biotechnol* 2010, **10:**21.
15. Frandsen RJN, Andersson JA, Kristensen MB, Giese H: **Efficient four fragment cloning for the construction of vectors for targeted gene replacement in filamentous fungi.** *BMC Mol Biol* 2008, **9:**70.
16. Covert SF, Kapoor P, Lee M, Briley A, Nairn CJ: ***Agrobacterium tumefaciens*-mediated transformation of *Fusarium circinatum*.** *Mycol Res* 2001, **105**(3):259–264.
17. Fulton TR, Ibrahim N, Losada MC, Grzegorski D, Tkacz JS: **A melanin polyketide synthase (PKS) gene from *Nodulisporium* sp. that shows homology to the pks1 gene of *Colletotrichum lagenarium*.** *Mol Gen Genet* 1999, **262:**714–720.
18. Punt PJ, Oliver RP, Dingemanse MA, Pouwels PH, van den Hondel CA: **Transformation of *Aspergillus* based on the hygromycin B resistance marker from *Escherichia coli*.** *Gene* 1987, **56:**117–124.
19. Flaherty JE, Pirttilä AM, Bluhm BH, Woloshuk CP: **PAC1, a pH-regulatory gene from *Fusarium verticillioides*.** *Appl Environ Microbiol* 2003, **69:**5222–5227.
20. Pall ML, Brunelli JP: **A series of six compact fungal transformation vectors containing polylinkers with multiple unique restriction sites.** *Fungal Genet Newsl* 1993, **40:**59–62.
21. Lisby M, Mortensen UH, Rothstein R: **Colocalization of multiple DNA double-strand breaks at a single Rad52 repair centre.** *Nat Cell Biol* 2003, **5:**572–577.
22. Campbell RE, Tour O, Palmer AE, Steinbach PA, Baird GS, Zacharias DA, Tsien RY: **A monomeric red fluorescent protein.** *Proc Natl Acad Sci U S A* 2002, **99:**7877–7882.
23. Frandsen RJN, Nielsen NJ, Maolanon N, Sørensen JC, Olsson S, Nielsen J, Giese H: **The biosynthetic pathway for aurofusarin in *Fusarium graminearum* reveals a close link between the naphthoquinones and naphthopyrones.** *Mol Microbiol* 2006, **61:**1069–1080.
24. Smedsgaard J: **Micro-scale extraction procedure for standardization screening of fungal metabolite production in cultures.** *J Chromatogr A* 1997, **760:**264–270.
25. Klitgaard A, Iversen A, Andersen MR, Larsen TO, Frisvad JC, Nielsen KF: **Aggressive dereplication using UHPLC-DAD-QTOF: screening extracts for up to 3000 fungal secondary metabolites.** *Anal Bioanal Chem* 2014, in press.
26. Broecker S, Herre S, Wüst B, Zweigenbaum J, Pragst F: **Development and practical application of a library of CID accurate mass spectra of more than 2,500 toxic compounds for systematic toxicological analysis by LC-QTOF-MS with data-dependent acquisition.** *Anal Bioanal Chem* 2011, **400:**101–117.
27. Nielsen KF, Månsson M, Rank C, Frisvad JC, Larsen TO: **Dereplication of microbial natural products by LC-DAD-TOFMS.** *J Nat Prod* 2011, **74:**2338–2348.
28. Gardiner DM, Jarvis RS, Howlett BJ: **The ABC transporter gene in the sirodesmin biosynthetic gene cluster of *Leptosphaeria maculans* is not essential for sirodesmin production but facilitates self-protection.** *Fungal Genet Biol* 2005, **42:**257–263.
29. Sleight SC, Bartley BA, Lieviant JA, Sauro HM: **In-Fusion BioBrick assembly and re-engineering.** *Nucleic Acids Res* 2010, **38:**2624–2636.
30. Aslanidis C, de Jong PJ: **Ligation-independent cloning of PCR products (LIC-PCR).** *Nucleic Acids Res* 1990, **18:**6069–6074.
31. Gibson DG, Young L, Chuang R-Y, Venter JC, Hutchison CA, Smith HO: **Enzymatic assembly of DNA molecules up to several hundred kilobases.** *Nat Methods* 2009, **6:**343–345.
32. Nisson PE, Rashtchian A, Watkins PC: **Rapid and efficient cloning of Alu-PCR products using uracil DNA glycosylase.** *PCR Methods Appl* 1991, **1:**120–123.
33. Nour-Eldin HH, Hansen BG, Nørholm MHH, Jensen JK, Halkier BA: **Advancing uracil-excision based cloning towards an ideal technique for cloning PCR fragments.** *Nucleic Acids Res* 2006, **34:**e122.
34. Geu-Flores F, Nour-Eldin HH, Nielsen MT, Halkier BA: **USER fusion: a rapid and efficient method for simultaneous fusion and cloning of multiple PCR products.** *Nucleic Acids Res* 2007, **35:**e55.
35. Olsen LR, Hansen NB, Bonde MT, Genee HJ, Holm DK, Carlsen S, Hansen BG, Patil KR, Mortensen UH, Wernersson R: **PHUSER (Primer Help for USER): a novel tool for USER fusion primer design.** *Nucleic Acids Res* 2011, **39**(Suppl 2):W61–W67.
36. Hansen BG, Salomonsen B, Nielsen MT, Nielsen JB, Hansen NB, Nielsen KF, Regueira TB, Nielsen J, Patil KR, Mortensen UH: **Versatile enzyme expression and characterization system for *Aspergillus nidulans*, with the *Penicillium brevicompactum* polyketide synthase gene from the mycophenolic acid gene cluster as a test case.** *Appl Environ Microbiol* 2011, **77:**3044–3051.

37. Michielse CB, Hooykaas PJJ, van den Hondel CAMJJ, Ram AFJ: *Agrobacterium*-mediated transformation as a tool for functional genomics in fungi. *Curr Genet* 2005, **48:**1–17.
38. Abuodeh RO, Orbach MJ, Mandel MA, Das A, Galgiani JN: **Genetic transformation of *Coccidioides immitis* facilitated by *Agrobacterium tumefaciens*.** *J Infect Dis* 2000, **181:**2106–2110.
39. Michielse CB, Ram AFJ, van den Hondel CAMJJ: **The *Aspergillus nidulans amdS* gene as a marker for the identification of multicopy T-DNA integration events in *Agrobacterium*-mediated transformation of *Aspergillus awamori*.** *Curr Genet* 2004, **45:**399–403.
40. Leclerque A, Wan H, Abschütz A, Chen S, Mitina GV, Zimmermann G, Schairer HU: ***Agrobacterium*-mediated insertional mutagenesis (AIM) of the entomopathogenic fungus *Beauveria bassiana*.** *Curr Genet* 2004, **45:**111–119.
41. Morioka LRI, Furlaneto MC, Bogas AC, Pompermayer P, Duarte RTD, Vieira MLC, Watanabe MAE, Fungaro MHP: **Efficient genetic transformation system for the ochratoxigenic fungus *Aspergillus carbonarius*.** *Curr Microbiol* 2006, **52:**469–472.
42. Rho HS, Kang S, Lee YH: ***Agrobacterium tumefaciens*-mediated transformation of the plant pathogenic fungus, *Magnaporthe grisea*.** *Mol Cells* 2001, **12:**407–411.
43. Mullins ED, Chen X, Romaine P, Raina R, Geiser DM, Kang S: ***Agrobacterium*-mediated transformation of *Fusarium oxysporum*: An Efficient Tool for Insertional Mutagenesis and Gene Transfer.** *Phytopathology* 2001, **91:**173–180.
44. Fang W, Zhang Y, Yang X, Zheng X, Duan H, Li Y, Pei Y: ***Agrobacterium tumefaciens*-mediated transformation of *Beauveria bassiana* using an herbicide resistance gene as a selection marker.** *J Invertebr Pathol* 2004, **85:**18–24.
45. Tsuji G, Fujii S, Fujihara N, Hirose C, Tsuge S, Shiraishi T, Kubo Y: ***Agrobacterium tumefaciens* -mediated transformation for random insertional mutagenesis in *Colletotrichum lagenarium*.** *J Gen Plant Pathol* 2003, **69:**230–239.
46. McClelland CM, Chang YC, Kwon-Chung KJ: **High frequency transformation of *Cryptococcus neoformans* and *Cryptococcus gattii* by *Agrobacterium tumefaciens*.** *Fungal Genet Biol* 2005, **42:**904–913.
47. Michielse CB, Ram AFJ, Hooykaas PJJ, van den Hondel CAMJJ: **Role of bacterial virulence proteins in *Agrobacterium*-mediated transformation of *Aspergillus awamori*.** *Fungal Genet Biol* 2004, **41:**571–578.
48. Leslie JF, Summerell BA: *The Fusarium Laboratory Manual.* Ames IA, US: Blackwell Publishing; 2006. 132 and 176.
49. Rolland S, Jobic C, Fèvre M, Bruel C: ***Agrobacterium*-mediated transformation of *Botrytis cinerea*, simple purification of monokaryotic transformants and rapid conidia-based identification of the transfer-DNA host genomic DNA flanking sequences.** *Curr Genet* 2003, **44:**164–171.
50. Ando A, Sumida Y, Negoro H, Suroto DA, Ogawa J, Sakuradani E, Shimizu S: **Establishment of *Agrobacterium tumefaciens*-mediated transformation of an oleaginous fungus, *Mortierella alpina* 1S-4, and its application for eicosapentaenoic acid producer breeding.** *Appl Environ Microbiol* 2009, **75:**5529–5535.
51. Combier J-P, Melayah D, Raffier C, Gay G, Marmeisse R: ***Agrobacterium tumefaciens*-mediated transformation as a tool for insertional mutagenesis in the symbiotic ectomycorrhizal fungus *Hebeloma cylindrosporum*.** *FEMS Microbiol Lett* 2003, **220:**141–148.
52. Lima IGP, Duarte RTD, Furlaneto L, Baroni CH, Fungaro MHP, Furlaneto MC: **Transformation of the entomopathogenic fungus *Paecilomyces fumosoroseus* with *Agrobacterium tumefaciens*.** *Lett Appl Microbiol* 2006, **42:**631–636.
53. Miyagawa H, Inoue M, Yamanaka H, Tsurushima Tetsu UT: **Chemistry of Spore Germination Self-Inhibitors from the Plant Pathogenic Fungus *Colletotrichum fragariae*.** In *Agrochem Discov Chapter 6*, Volume 774. Edited by Baker DR, Umetsu NK. Washington, DC: American Chemical Society; 2000:62–71 [ACS Symposium Series (Series editor)].
54. Gaffoor I, Brown DW, Plattner R, Proctor RH, Qi W, Trail F: **Functional analysis of the polyketide synthase genes in the filamentous fungus *Gibberella zeae* (anamorph *Fusarium graminearum*).** *Eukaryot Cell* 2005, **4:**1926–1933.
55. Gauthier GM, Sullivan TD, Gallardo SS, Brandhorst TT, Vanden Wymelenberg AJ, Cuomo CA, Suen G, Currie CR, Klein BS: **SREB, a GATA transcription factor that directs disparate fates in *Blastomyces dermatitidis* including morphogenesis and siderophore biosynthesis.** *PLoS Pathog* 2010, **6:**e1000846.
56. Duyvesteijn RGE, van Wijk R, Boer Y, Rep M, Cornelissen BJC, Haring MA: **Frp1 is a *Fusarium oxysporum* F-box protein required for pathogenicity on tomato.** *Mol Microbiol* 2005, **57:**1051–1063.
57. Malz S, Grell MN, Thrane C, Maier FJ, Rosager P, Felk A, Albertsen KS, Salomon S, Bohn L, Schäfer W, Giese H: **Identification of a gene cluster responsible for the biosynthesis of aurofusarin in the *Fusarium graminearum* species complex.** *Fungal Genet Biol* 2005, **42:**420–433.
58. Ola ARB, Thomy D, Lai D, Brötz-Oesterhelt H, Proksch P: **Inducing secondary metabolite production by the endophytic fungus *Fusarium tricinctum* through coculture with *Bacillus subtilis*.** *J Nat Prod* 2013, **76:**2094–2099.
59. Shiono Y, Tsuchinari M, Shimanuki K, Miyajima T, Murayama T, Koseki T, Laatsch H, Funakoshi T, Takanami K, Suzuki K: **Fusaristatins A and B, two new cyclic lipopeptides from an endophytic *Fusarium* sp.** *J Antibiot (Tokyo)* 2007, **60:**309–316.
60. Lim C, Kim J, Choi JN, Ponnusamy K, Jeon Y, Kim S-U, Kim JG, Lee C: **Identification, fermentation, and bioactivity against *Xanthomonas oryzae* of antimicrobial metabolites isolated from *Phomopsis longicolla* S1B4.** *J Microbiol Biotechnol* 2010, **20:**494–500.

3

VprBP (DCAF1): a promiscuous substrate recognition subunit that incorporates into both RING-family CRL4 and HECT-family EDD/UBR5 E3 ubiquitin ligases

Tadashi Nakagawa[1*], Koushik Mondal[2] and Patrick C Swanson[2*]

Abstract

The terminal step in the ubiquitin modification system relies on an E3 ubiquitin ligase to facilitate transfer of ubiquitin to a protein substrate. The substrate recognition and ubiquitin transfer activities of the E3 ligase may be mediated by a single polypeptide or may rely on separate subunits. The latter organization is particularly prevalent among members of largest class of E3 ligases, the RING family, although examples of this type of arrangement have also been reported among members of the smaller HECT family of E3 ligases. This review describes recent discoveries that reveal the surprising and distinctive ability of VprBP (DCAF1) to serve as a substrate recognition subunit for a member of both major classes of E3 ligase, the RING-type CRL4 ligase and the HECT-type EDD/UBR5 ligase. The cellular processes normally regulated by VprBP-associated E3 ligases, and their targeting and subversion by viral accessory proteins are also discussed. Taken together, these studies provide important insights and raise interesting new questions regarding the mechanisms that regulate or subvert VprBP function in the context of both the CRL4 and EDD/UBR5 E3 ligases.

Keywords: VprBP, DCAF1, DDB1, Cul4, CRL4, EDD, UBR5, Dyrk2, Merlin, Katanin, UNG2, LGL2, Mcm10, Histone H3, RORα, Methyl degron, p53, TERT, telomerase, RAG1, V(D)J recombination, HIV, Vpr, Vpx, UL35, Ubiquitin, E3 ubiquitin ligase, RING, HECT, WD40 repeat

Introduction

Virtually all cellular processes are subjected to some level of regulation by the ubiquitin modification system which mediates the attachment of ubiquitin or ubiquitin-like molecules to proteins in the involved pathways (for review, see [1]). It is also now appreciated that this system can be subverted by pathogens to disable host responses and alter the cellular environment to benefit the microorganism (for reviews, see [2-4]). Ubiquitin is a 76 amino acid protein that is covalently attached to a target protein through a series of enzymatic steps in which free ubiquitin is initially coupled to an activating enzyme (E1) in an ATP-dependent reaction, and then transferred to the catalytic cysteine of a conjugating enzyme (E2). In the last step, the ubiquitin-linked E2 associates with an ubiquitin ligase (E3), which catalyzes the transfer of ubiquitin to an ε-amino group of a lysine residue in the targeted protein. Additional ubiquitin molecules may be appended to lysine residues in ubiquitin to form polyubiquitin chains. The number of attached ubiquitin molecules and the specific lysine residue(s) used to link them together dictate whether the outcome of ubiquitination mainly serves to alter the function of the target protein, or triggers its degradation through the proteasome pathway.

There are two major classes of E3 ubiquitin ligases (termed E3 ligase henceforth) that differ in how they mediate ubiquitin transfer (for reviews, see [1,5]). Those that contain a RING (*r*eally *i*nteresting *n*ew *g*ene) domain, or the related U-box domain, facilitate the direct transfer of ubiquitin from the ubiquitin-E2 conjugate to

* Correspondence: tnakagaw@med.tohoku.ac.jp; pswanson@creighton.edu
[1]Department of Cell Proliferation, United Center for Advanced Research and Translational Medicine, Graduate School of Medicine, Tohoku University, Sendai 900-8575, Japan
[2]Department of Medical Microbiology and Immunology, Creighton University, 2500 California Plaza, Omaha, NE 68178, USA

the target protein without forming a covalent intermediate. By contrast, those with a HECT (*h*omologous to *E*6-AP *c*arboxy *t*erminus) domain first undergo a transthioesterification reaction which transfers ubiquitin from the E2 enzyme to an active site cysteine residue in the HECT domain before the ubiquitin is ultimately coupled to the target protein. Most eukaryotic organisms have only a single E1 activating enzyme, but express tens of E2 conjugating enzymes and several hundred or more E3 ligases. Substrate specificity is largely determined by the E3 ligase; however, the substrate binding and catalytic activity of a given E3 ligase may or may not be found within the same molecule. The cullin RING ligases (CRLs) are a large group of E3 ligases with separable substrate binding and catalytic activities (for review, see [6,7]). These E3 ligases have a modular organization in which one of the cullin family members of scaffold proteins binds both a small RING-containing catalytic subunit (Roc1 [*r*egulator *o*f *c*ullins 1] or Roc2; also called Rbx1 [*R*ING *box* protein] or Rbx2), and a cullin-specific adaptor protein. In most cases, the adaptor protein, in turn, binds a substrate recognition subunit that recruits and positions the substrate in proximity to the catalytic subunit for ubiquitination. The CRL is defined by the cullin scaffold protein, of which there are seven members in humans and mice (i.e. CRL1 contains Cul1). Because the CRL adaptor protein is generally cullin-specific, it is often not included in the designation, but the substrate recognition subunit is included after the CRL in superscript. For example, Damaged DNA *b*inding protein 1 (DDB1) is the adaptor protein for the CRL4 E3 ligase. The *DDB*1-*Cul*4 *a*ssociating *f*actors (DCAFs) that comprise the substrate recognition subunits for the CRL4 E3 ligase are indicated as $CRL4^{DCAF}$. This convention will be followed here.

The HECT-domain E3 ligases are characterized by the presence of a C-terminal HECT domain (for review, see [8]). This family has been further divided into three broad subgroups based on the presence or absence of additional WW or RCC1 (*r*egulator of *c*hromatin *c*ondensation 1)-like domains (RLDs) in the amino-terminal region of the protein. These include the Nedd4/Nedd4-like subgroup which contains WW domains (identified by a signature pair of tryptophan residues), the HERC (*HE*CT and *RC*C1-like domain) subgroup which contain RLDs, and a subgroup which harbors neither WW domains nor RLDs (non-WW/non-RLD). In contrast to the CRL family of E3 ligases, HECT-domain E3 ligases more commonly function as a single subunit E3 ligase, mediating both substrate recognition and ubiquitination. However, for some HECT E3 ligases, such as Nedd4, adaptor proteins may be engaged to mediate substrate recruitment or influence the subcellular distribution of the E3 ligase to direct ubiquitination of localized substrates [9]. The degree of flexibility of adaptor proteins and substrate recognition subunits to service multiple E3 ligases is important for understanding how the substrates they recruit are regulated in a variety of spatial and temporal contexts.

Here we review and discuss the discovery and characterization of *V*iral *p*rotein *R b*inding *p*rotein (VprBP, also called DCAF1), and its emerging role as a dual-purpose substrate recognition subunit for two distinct E3 ligases: the RING-family member CRL4 and the "non-WW/non-RLD" HECT-family member EDD/UBR5 (*E*3 ligase identified by *d*ifferential *d*isplay/*ub*iquitin protein ligase E3 component n-*r*ecognin 5) (Figure 1). Regulatory roles of the CRL4 and EDD/UBR5 E3 ligases for which there is no known involvement of VprBP as a substrate recognition molecule will not be discussed in depth here.

VprBP: discovery and early association with the CRL4 E3 ligase

In addition to the main Gag, Pol, and Env open reading frames (ORFs), primate lentiviruses, including human immunodeficiency virus 1 (HIV-1), have additional ORFs encoding a variety of other regulatory and accessory factors, one of which is named Viral protein *R* (Vpr). By the early 1990s, Vpr was known to play a key role in promoting viral replication but its function remained enigmatic. Hypothesizing that Vpr targets a host protein to mediate its function, Zhao *et al.* used purified, bacterially expressed Vpr as bait to isolate Vpr-interacting proteins from HeLa cell nuclear extracts by co-immunoprecipitation (co-IP) [10]. This screen yielded a protein initially called Vp*r i*nteracting *p*rotein (RIP), which was renamed VprBP in their subsequent study in which the authors cloned the VprBP cDNA, mapped the Vpr-interacting region to residues 636–1507 of VprBP [11], and showed that VprBP binding to Vpr promotes cytoplasmic retention of Vpr. However, the normal function of VprBP and the significance of VprBP-mediated cytoplasmic sequestration of Vpr for HIV replication remained unclear.

A major breakthrough in understanding the normal role of VprBP came from proteomic screens designed to identify DDB1-associated substrate recognition subunits for the CRL4 E3 ligase [12-15]. VprBP was identified in three of these studies [12-14] (called DCAF1 by Jin *et al.* [12] and Angers *et al.* [13]). Most identified DDB1-interacting proteins share a common structural motif called a WD40 domain (for reviews, see [16,17]). The WD40 domain contains multiple WD40 repeats (ranging from 4–16), each spanning 40–60 residues and bearing a signature Trp-Asp (WD) dipeptide at its C-terminus, although the Trp-Asp dipeptide is sometimes substituted by Tyr-Asp, Ile-Asp, or Trp-Cys in the DCAFs. In most DCAFs, including VprBP, the Trp-Asp dipeptide is also

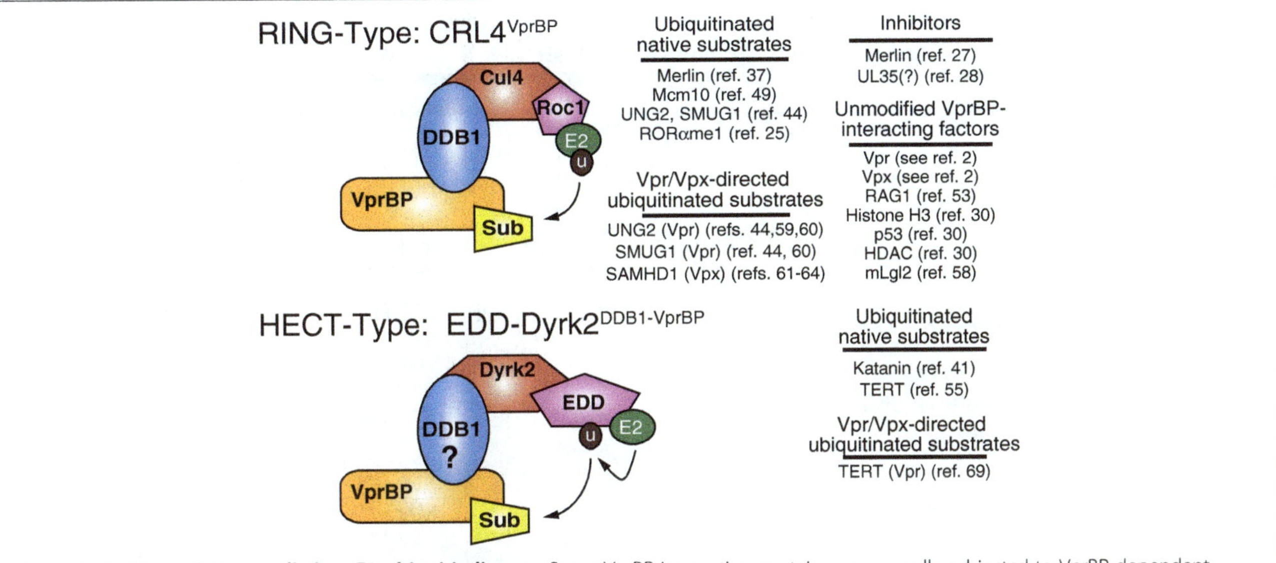

Figure 1 VprBP services two distinct E3 ubiquitin ligases. Some VprBP-interacting proteins are normally subjected to VprBP-dependent ubiquitination in unperturbed cells (native), whereas others are native or novel substrates that undergo accelerated Vpr- or Vpx-dependent degradation in the context of CRL4^{VprBP} or EDD-Dyrk2$^{DDB1\text{-}VprBP}$ complexes. Merlin and UL35 may act to inhibit the CRL4 ligase. For several VprBP-interacting proteins, no evidence of ubiquitin modification has been reported. In some of these examples (e.g. mLgl2), the identity of the VprBP-associated E3 ligase has not been formally established. DDB1 may or may not physically link the EDD-Dyrk2 E3 ligase to substrates through VprBP in all cases. For additional details, see text.

followed by an X-Arg (or occasionally X-Lys) dipeptide. This sequence, designated a WDXR motif, is often evolutionarily conserved between orthologous DCAFs, and mutational analysis suggests that the basic residue within the WDXR motif is required for stable association of DCAF proteins to DDB1. Each WD40 repeat is composed of four anti-parallel beta-strands that form a beta-propeller; consecutive beta-propellers are connected by a peptide linker.

DDB1 itself contains three WD40 beta-propeller domains (BPA, BPB, and BPC) and a C-terminal helical domain [13]. Crystal structures of DDB1 bound to different substrate recognition subunits, including DDB2 [18-20] and Cockayne Syndrome A (CSA) [18], reveal a common mode of binding in which the WD40 beta-propeller domain in DDB2 and CSA stabilizes association with DDB1 by mediating hydrophobic contacts with the BPA domain of DDB1 and by anchoring an amino-terminal helix-loop-helix motif that is inserted into a cavity at the interface between the BPA and BPC domains of DDB1. Notably, some viral proteins, such as hepatitis B virus X protein, are able to subvert the CRL4 E3 ligase by using an alpha-helical motif to bind DDB1 at the BPA-BPC interface in a manner analogous to cellular CRL4 substrate receptors [21]. Whether VprBP shows a mode of DDB1 binding similar to DDB2 and CSA remains unclear, as no readily identifiable counterpart to the helix-loop-helix motif in DDB2 and CSA has been detected in VprBP [18,21]. VprBP preferentially binds DDB1 associated with a form of Cul4 that has been post-translationally modified, most likely by NEDD8 although this was not experimentally confirmed [22]. In the next sections, we will review current knowledge regarding VprBP domain structure and organization, discuss the different cellular processes regulated by VprBP through its association with the CLR4 E3 ligase and its more recently uncovered affiliation with the EDD/UBR5 E3 ligase, and highlight recent insights on the effects of VprBP targeting by viral accessory proteins.

VprBP: domain organization and structural features of VprBP-DDB1 and CRL4^{VprBP} complexes

The human and murine VprBP genes express two or three predicted spliced transcript variants encoding distinct isoforms. The most similar and best characterized human and murine isoforms of VprBP are 1507 and 1506 amino acids, respectively, and share 98% identity. A predicted Armadillo-type fold domain encompasses most of the amino-terminal half of the protein (residues 80–796) (see Figure 2). This region consists of tandemly arrayed Armadillo repeat motifs, each about 40 amino acids in length, which are predicted to form layers of alpha helices that assemble into a right-handed superhelix [23,24]. These structures present a large solvent-accessible surface area well-suited for mediating contact with interacting proteins. Within this region lies a small chromo-like domain (residues 562–593), similar to those found in chromatin remodeling proteins, that has recently been implicated in mediating interactions with monomethylated proteins ([25], see below). A central

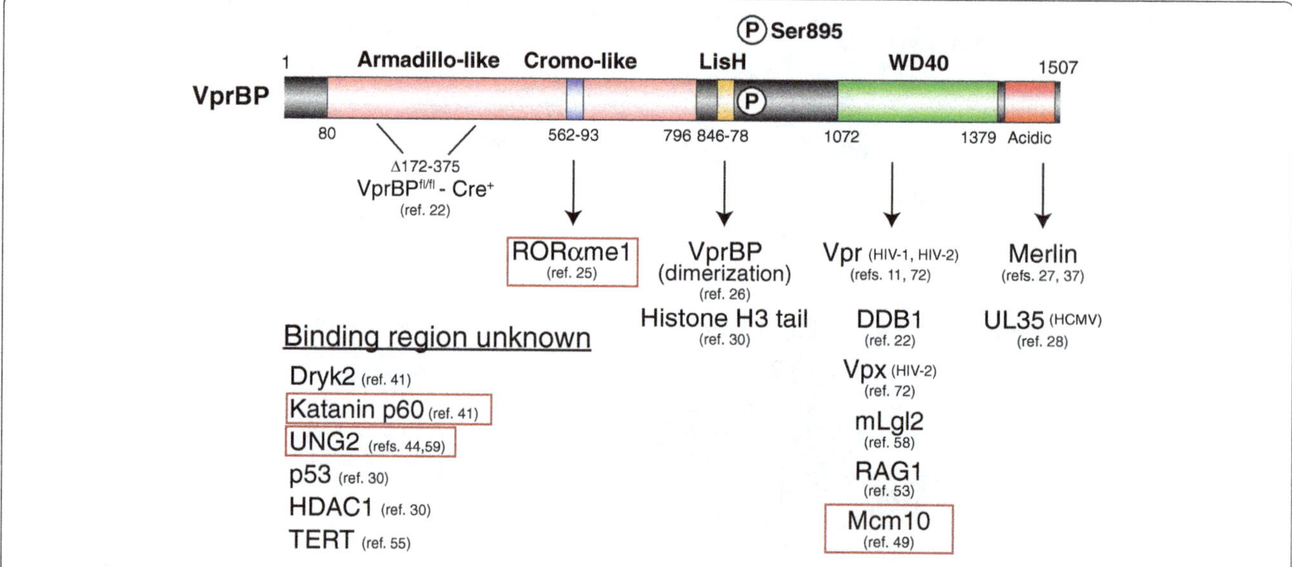

Figure 2 VprBP structural motifs and interacting proteins. The 1507 amino acid human VprBP isoform encoded by transcript variant ENST00000563997 in the Ensembl database (release 71) [29] is shown with the domain features for this isoform as described in the database or in the text. VprBP-interacting proteins are identified below the diagram; those for which the binding site has been mapped are shown below the region mediating the association. Proteins targeted for VprBP-dependent ubiquitination are boxed. The region removed after Cre-mediated deletion of a conditional allele [22], and the location of a DNA-PK-dependent phosphorylation site [30] are also indicated.

region, designated by its homology to Lis1 (LisH; residues 846–878), contains a L-X_2-L-$X_{3\text{-}5}$-L-$X_{3\text{-}5}$-L sequence motif that adopts an alpha-helical conformation and is required to mediate oligomerization of VprBP [26]. The carboxyl-terminal half of the protein contains a WD40 domain consisting of four WD40 repeat motifs, which, as discussed above, not only mediate interactions with DDB1, but also support contacts with other substrate and interacting proteins (see below). This region is followed by a stretch of ~100 amino acids rich in acidic residues that may interact with factors that disrupt or subvert the activity of the associated E3 ligase [27,28]. Notably, the acidic region is absent from a shorter murine isoform of VprBP.

To gain insight into the structural and conformational features of VprBP-containing DDB1 and CRL4 complexes, Ahn *et al.* used size-exclusion chromatography coupled to in-line *m*ulti-*a*ngle *l*ight *s*cattering (SEC-MALS) and SDS-PAGE to analyze fully assembled and purified DDB1-VprBP and CRL4^{VprBP} complexes prepared with a form of VprBP lacking the Arm-like domain (VprBP$_{817\text{-}1507}$) [26]. The authors elegantly showed that both complexes contained equimolar amounts of each of the individual proteins, and had an apparent mass that was approximately twice that predicted based on the molecular weight of each subunit, leading them to conclude that both complexes assemble dimeric structures. Further examination of CRL4^{VprBP} complexes by electron microscopy revealed evidence of two-fold rotational symmetry in the structures. A comparison of the *in vitro* ubiquitination activity of CRL4$^{VprBP(817-1507)}$ and CRL4$^{VprBP(1005-1507)}$ (which lacks the LisH motif) showed that CRL4$^{VprBP(817-1507)}$ was at least 2-fold more efficient, suggesting that VprBP-mediated dimerization promotes CRL4^{VprBP} ubiquitin transfer activity.

VprBP may also be subjected to post-translational modification. Kim *et al.* reported that DNA-PK phosphorylates VprBP *in vitro* at Ser895, and showed that phospho-Ser895-specific polyclonal antibodies detect phosphorylated VprBP in U2OS cells after etoposide-induced DNA damage [30]. As discussed below, phosphorylation at this site is reported to alleviate VprBP-mediated repression of p53-dependent transcription.

Physiological roles and binding partners of VprBP

VprBP has been implicated in regulating a variety of normal cellular processes, including proliferation, DNA replication, cell cycle progression, telomere maintenance, DNA damage responses, and competition between neighboring cells. The evidence supporting a role for VprBP in these processes, and the context and targets of the associated E3 ligase machinery, where known, are discussed in the following sections.

Proliferation, DNA replication, and cell cycle

Proliferating cells undergo repeated cycles of DNA replication (DNA synthesis or S phase) and cell division (mitosis or M phase), which are temporally separated by gap phases (G1 or G2 phases) (for review, see [31]). Potential replication initiation sites, called replication origins, are marked, or "licensed", by the formation of a pre-replication complex (pre-RC) that includes the ORC1-6, CDT1, CDC6, and MCM2-7, beginning late in M phase and proceeding through the G1 phase. During S phase, pre-RCs (30,000-50,000 in mammals) are then activated, or "fired", following recruitment of DNA replication machinery such as DNA polymerases. DNA replication propagates bidirectionally from the origins until the whole genome is precisely duplicated. Importantly, not all origins are used at the very onset of S phase, but origins are fired in a temporal order by which DNA replication is regulated during S phase [32,33]. Precise DNA replication is of utmost importance to transmit genetic information intact to daughter cells. High fidelity DNA duplication is ensured by S phase checkpoint activation, which inhibits late origin firing in a transient manner to provide time for DNA repair when cells encounter DNA damage during S phase. If damaged DNA is not repaired, cells exit S phase and undergo G2 arrest [34,35].

Recent studies by McCall *et al.* provide several lines of evidence suggesting VprBP is involved in regulating DNA replication [22]. First, silencing VprBP expression was shown to suppress proliferation in U2OS and Rb-inactivated (E7 transduced) WI38 cells and increase the percentage of cells in the S and G2 phases in HeLa cells. Second, VprBP and Cul4A were found to exhibit cell cycle-dependent binding to chromatin in HeLa cells, with the association being primarily restricted to the early S and G2 phases. Third, also in HeLa cells, BrdU labeling studies showed that VprBP silencing markedly reduced DNA replication in middle to late S phase. These cells were not responsive to S-phase inhibitors, indicating that DNA synthesis was completely blocked after VprBP silencing. Fourth, DNA fiber-labeling experiments provided evidence that VprBP silencing in HeLa cells increases the frequency of firing of new replication origins, but does not change the frequency of replication termination or dramatically alter the rate of DNA synthesis. To reconcile the seemingly opposing effects of VprBP silencing on BrdU incorporation during S phase and replication origin firing and DNA synthesis rates, McCall *et al.* suggested that VprBP functions to either stabilize the replication fork or regulates the temporal order of early and late origin firing during S phase [22]. The replication defects observed by BrdU labeling experiments in HeLa cells were further confirmed in primary mouse embryonic fibroblasts (MEFs) in which VprBP expression was conditionally disrupted using a Cre-loxP approach. However, in this case, MEFs lacking VprBP did not accumulate in S phase, but were instead reduced in frequency. This outcome was attributed to an increase in apoptosis observed in VprBP-deficient MEFs as assessed by Annexin V staining. The precise mechanism for how VprBP facilitates DNA replication and S phase transit in cells remains unclear, as neither the E3 ligase nor the putative substrate(s) that VprBP recruits to it for ubiquitination were established in this study.

Merlin (also called schwannomin) is a tumor suppressor encoded by the *n*eurofibromatosis type 2 (NF2) gene. Inactivating NF2 mutations have been identified in variety of nervous system tumors including schwannomas, ependymomas, meningiomas and mesotheliomas [36]. To determine the mechanism by which Merlin exerts its anti-proliferative activity, two different groups employed *t*andem *a*ffinity *p*urification and *m*ass spectrometry (TAP-MS) to identify novel binding partners of Merlin, with both groups identifying DDB1 and VprBP as Merlin-interacting proteins [27,37]. In both studies, pull-down experiments established that Merlin associates with the $CRL4^{VprBP}$ E3 ligase complex, and were used to map Merlin interactions to the far C-terminal acidic region of VprBP (Figure 2). Li *et al.* went further by showing that Merlin specifically co-IP's with the unique components of the $CRL4^{VprBP}$ E3 ligase, but not the alternative $EDD\text{-}Dyrk2^{DDB1\text{-}VprBP}$ E3 ligase, from nuclear-soluble, but not cytoplasmic/membrane, cell fractions, leading the authors to conclude Merlin translocates to the nucleus to engage $CRL4^{VprBP}$ [27]. Furthermore, this group found that Merlin association with VprBP depended on maintaining Merlin in a "closed" conformation. This closed and active form of Merlin can be converted to an "open", inactive form by Ser518 phosphorylation, which disrupts an intramolecular interaction between the N- and C-terminal regions of Merlin [38,39].

Using a transient overexpression system, Huang and Chen showed that VprBP downregulates Merlin in a dose-dependent manner by triggering its ubiquitination, and that this effect could be partially overcome by proteasome inhibition [37], leading this group to conclude Merlin is targeted for degradation by $CRL4^{VprBP}$ E3 ligase. However, this conclusion has been challenged by Li *et al.*, who provided evidence that Merlin actually inhibits ubiquitination mediated by the $CRL4^{VprBP}$ E3 ligase [27]. Specifically, levels of VprBP-associated ubiquitination products were shown to diminish as a function of Merlin expression using a Cos7 cell transfection system. Conversely, levels of these products were elevated when Merlin expression was silenced, or the Merlin binding site at the C-terminus of VprBP was removed ($VprBP_{1\text{-}1417}$). These authors also showed that silencing Merlin accelerates proliferation in a normal human Schwann cell line or human umbilical vein endothelial cells, but this effect is largely reversed by concomitant VprBP silencing. Finally, genetic evidence for Merlin's function as a $CRL4^{VprBP}$ inhibitor comes from the author's characterization of a panel of tumor-derived mutations, which fell into three catagories: those that impair Merlin's ability to translocate into the nucleus, those that fail to bind VprBP, and those that bind VprBP, but do not inhibit CRL4^{VprBP} activity.

Another cell cycle-related protein regulated by VprBP is the microtubule-severing enzyme katanin, which is composed of a regulatory p80 subunit and a catalytic p60 subunit [40]. The association between VprBP and the catalytic subunit of p60 katanin was discovered by Maddika and Chen through a TAP-MS screen for potential substrates of the *d*ual specificity *y*rosine-phosphorylation-*r*egulated *k*inase 2 (Dyrk2), which was of interest for its involvement in various signaling pathways activated during cellular, developmental, and oncogenic processes [41]. This screen identified the HECT E3 ligase EDD (UBR5), DDB1, and VprBP as predominant Dyrk2-associated proteins. Co-IP experiments confirmed these interactions and further excluded Cul4 and Roc1 as binding partners, and also established that Dyrk2 kinase activity is dispensable for binding VprBP. Dyrk2 silencing abolished VprBP association with EDD, but not DDB1, suggesting that Dyrk2 functions as an adaptor bridging EDD to the DDB1-VprBP complex. Moreover, Dyrk2 protein levels were not altered by DDB1, VprBP, or EDD silencing, nor did they fluctuate during cell cycle, suggesting that Dyrk2 did not undergo ubiquitin-dependent degradation mediated by EDD. Alternatively, katanin p60 was considered a potential substrate of the EDD-Dyrk2$^{DDB1\text{-}VprBP}$ E3 ligase because its homologue in *Caenorhabditis elegans*, MEI-1, was previously shown to undergo phosphorylation-dependent ubiquitin-mediated degradation by the Dyrk2 homologue MBK2 during meiotic maturation in *C. elegans* [42]. In support of this possibility, Maddika and Chen demonstrated that katanin p60 levels in HeLa cells were increased by silencing of EDD, Dyrk2 or VprBP, and were decreased when these molecules were overexpressed [41]. However, overexpressing a kinase-inactive form of Dyrk2 had no effect on katanin p60 levels, suggesting its degradation depends on Dyrk2-mediated phosphorylation. Consistent with these results, the EDD-Dyrk2$^{DDB1\text{-}VprBP}$ complex was shown to support *in vitro* phosphorylation and ubiquitination of katanin p60. Interestingly, silencing of EDD or Dyrk2 in HeLa cells caused G2/M arrest that was alleviated by concomitant silencing of katanin p60. Overexpressing katanin p60 had a similar effect that was overcome by co-expressing wild-type, but not kinase-inactive Dyrk2.

DNA damage responses

Mammalian cells have at least five uracil DNA glycosylases, among which UNG2 plays a critical role in repairing misincorporated or spontaneously generated uracil in genomic DNA [43]. Previously, the HIV protein Vpr was reported to load UNG2 onto VprBP for degradation (see below). More recently, however, Wen *et al.* found that UNG2 is directly targeted by CRL4^{VprBP} independently of Vpr, but Vpr can enhance VprBP-dependent UNG2 degradation [44]. Thus, silencing DDB1 or VprBP substantially increased UNG2 levels in transiently transfected 293T cells. Interestingly, when overexpressed, UNG2 causes DNA damage [45], colocalizes with the DNA damage marker γH2AX, and is toxic to cells [46], suggesting UNG2 is a potentially very important substrate of VprBP.

Mcm10 is a DNA replication initiation factor which recruits DNA polymerase α to the pre-RC to initiate DNA replication after CDK2 is activated [47]. UV radiation triggers Mcm10 proteolysis and causes stalling of DNA replication to facilitate DNA repair and avoid erroneous DNA synthesis [48]. Given the known role that CRL1^{Skp2} and CRL4^{Cdt2} E3 ligases play in irradiation-induced DNA damage responses, Kaur *et al.* suspected that one of these two E3 ligases might be responsible for Mcm10 degradation after UV-induced DNA damage [49]. Using a knock-down approach in HeLa cells, the authors excluded the involvement of CRL1^{Skp2}, as well as Cdt2, in UV-induced Mcm10 degradation, but showed that silencing DDB1, Cul4, or Roc1 prevented the loss of Mcm10 after UV treatment. Systematic silencing of known CRL4 substrate recognition subunits identified VprBP as a likely candidate that links Mcm10 to the CRL4 ligase. Pull-down experiments confirmed Mcm10 association with the CRL4^{VprBP} complex, and delimited the interacting region to VprBP$_{864\text{-}1507}$ (Figure 2). The CRL4^{VprBP} E3 ligase was also found to support Mcm10 ubiquitination *in vitro* and *in vivo.*

Hrecka *et al.* provided early evidence suggesting that silencing VprBP expression induced p53 target genes, such as p21, suggesting VprBP directly regulates p53-dependent transcription [50]. To test this possibility, Kim *et al.* analyzed transcription of p53-dependent genes induced after DNA damage, including p21 and Noxa, in MEFs in which VprBP expression was conditionally disrupted [30]. The authors showed that loss of VprBP expression increased p21 and Noxa expression levels induced by etoposide treatment. Using a cell-free assay to evaluate p53-dependent transcription from reconstituted nucleosomal substrates in reactions containing p300 HAT and Acetyl Co-A, the authors next showed that intact, but not denatured, VprBP suppressed p53-dependent transcription on nucleosomal, but not histone-free, DNA templates, as long as VprBP was added before p300, suggesting VprBP may block p300-dependent acetylation. This outcome was traced to VprBP's ability to bind to unmodified histone H3 tails and was minimally mediated by VprBP$_{(751-1507)}$, which encompasses the LisH and WD40 domains as well as the acidic C-terminus (Figure 2). Consistent with this finding, ChIP experiments demonstrated that knockdown of histone deacetylase (HDAC1) or VprBP, or etoposide treatment led to increased histone H3 acetylation and p53 occupancy at p53-dependent promoters; conversely, treatment with etoposide led to loss of HDAC1 and VprBP occupancy of the same promoters. These data suggested that VprBP physically associates with HDAC1 and p53. Evidence supporting this possibility was obtained using co-IP experiments, but the interacting regions in VprBP were not mapped. Because VprBP levels remained constant after DNA damage, the authors speculated that VprBP is subjected to post-translational modification after DNA damage to release VprBP bound to unmodified histone H3 tails. By screening a panel of kinases, acetyltransferases, and methyltransferases implicated in DNA damage responses, the authors identified DNA-PK as able to post-translationally modify VprBP. The phosphorylation site was mapped to Ser895, and an alanine substitution at this residue was demonstrated to attenuate p53-dependent transcription, and impair histone acetylation and release of VprBP from p53-dependent promoters in response to DNA damage. These data, taken together, led the authors to propose a model in which VprBP, in cooperation with HDAC1, removes and blocks acetylation of histone H3 tails to repress p53-dependent transcription. In response to DNA damage, DNA-PK phosphorylates VprBP at Ser895, which releases VprBP binding to histone H3, thereby allowing p53-dependent promoter activation.

The products of the recombination activating genes 1 and 2 (RAG1/2) are required to initiate V(D)J recombination, a form of site-specific DNA rearrangement that is responsible for assembling the immunoglobulin (Ig) and T cell receptor genes in B and T lymphocytes, respectively [51]. The N-terminal region of RAG1 (residues 1–383 of 1040), though not absolutely required for V(D)J recombination activity, is evolutionarily conserved and is necessary for efficient and high-fidelity rearrangement of the endogenous antigen receptor loci. The N-terminal region of RAG1 contains a RING domain that functions *in vitro* as an E3 ligase, but the *in vivo* functionality and physiological targets of RAG1 E3 ligase activity, if any, remains unclear. Moreover, since many RING-type E3 ligases form multi-subunit assemblies [52], RAG1 could plausibly associate with other accessory proteins to support its putative function as an E3 ligase. In support of this possibility, Kassmeier *et al.* showed that under mild purification conditions, novel proteins co-purified with full-length RAG1, one of which was identified as VprBP using mass spectrometry [53]. Subsequent experiments provided evidence that DDB1, Cul4A, and Roc1, but not EDD and Dyrk2, co-purify full-length RAG1, and that RAG1 association with the CRL4^{VprBP} complex is mediated primarily by RAG1 interactions with the WD40 domain of VprBP (Figure 2). This co-purified complex supported E3 ligase activity *in vitro*: this activity was not directed at the RAG proteins themselves, and was not obviously impaired by cysteine mutations in the RING domain of RAG1, leading the authors to speculate that RAG1 serves as a scaffold to recruit the CRL4^{VprBP} to ubiquitinate one or more as-yet unidentified substrates. Evidence for the physiological significance of VprBP function in lymphocytes was obtained by conditionally disrupting VprBP expression in the B cell lineage. These animals showed an arrest in B cell development at the pro-B-to-pre-B cell transition, which was associated with an increase in S phase and apoptotic cells. Evidence for abnormal repair of V(D)J recombination intermediates was also detected. Recently, we have found that B cell development can be partially rescued by enforced expression of Bcl2, but most mature B cells emerging in this case express the lambda light chain, rather than the more commonly expressed kappa light chain (Palmer and Swanson, unpublished data). This observation suggests that VprBP is not absolutely required for V(D)J recombination, although its involvement in the DNA repair phase of this process cannot yet be excluded. An intriguing alternative possibility is that VprBP regulates the accessibility or positioning of the kappa locus to permit RAG-mediated DNA rearrangements over the ~3 Mb span of the kappa locus. Because the lambda locus is only about 1/10th the size of the kappa locus and contains many fewer gene segments, it may not require significant architectural changes to support V(D)J recombination, and therefore may undergo RAG-mediated rearrangement in a largely VprBP-independent manner.

Telomerase regulation

Telomerase is an enzyme responsible for adding DNA sequence repeats to chromosomal ends after DNA replication, which is necessary to maintain genomic stability and prevent gradual loss of genetic information with each cell division. The telomerase holoenzyme contains the *t*elomerase *r*everse *t*ranscriptase (TERT) subunit, telomerase RNA (TERC), and several accessory protein subunits [54]. Based on the observation that telomerase activity is not well correlated with TERT transcript levels among different cell types, and that TERT protein exhibits a short half-life in HeLa cells, Jung *et al.* speculated that TERT is regulated at the post-translational level by phosphorylation and/or ubiquitin-dependent degradation [55]. These authors identified Dyrk2 as a candidate TERT regulatory kinase by systematically overexpressing a panel of protein kinases and evaluating their ability to down-regulate TERT protein in HeLa cells. The authors subsequently showed that Dyrk2 interacts with and phosphorylates TERT at Ser457, and promotes its *in vitro* ubiquitination. Notably, however, kinase-inactive Dyrk2 binds but does not phosphorylate

TERT, and does not trigger its degradation in cells. Consistent with these findings, depletion of Dyrk2 was shown to upregulate TERT expression and increased the half-life of the protein. Interestingly, although knockdown of Dyrk2 disrupted TERT association with VprBP and EDD, it had little effect on TERT-DDB1 interactions. This result is puzzling and remains unexplained. However, it raises the possibility that DDB1, rather than VprBP, is responsible for bringing TERT into proximity to EDD. In this regard, it is noteworthy that the region of VprBP interacting with TERT was not mapped, and the Dyrk2 and TERT truncation mutants used to map Dyrk2-TERT interactions were not evaluated for the presence of co-purifying DDB1 or VprBP, which could conceivably bridge Dyrk2 and TERT in the assay. Finally, the authors showed that TERT-Dyrk2 protein interactions were primarily limited to the G2/M phase, and were correlated with a decrease in TERT protein levels. In contrast, TERT-Dyrk2 association was not apparent during S phase, when telomerase activity is expected to be highest. Taken together, these data led the authors to suggest a model in which the TERT protein synthesizes a telomere repeat sequence at the early S phase of the cell cycle, and as the cell transitions to the G2/M phase, the EDD-Dyrk2 E3 ligase targets TERT protein for degradation to suppress telomerase activity.

Turnover of methylated proteins

*E*nhancer of *z*este *h*omolog 2 (EZH2) is a methyltransferase that specifically mediates histone H3K27 methylation and has been found to be deregulated in a variety of cancer types [56]. Given its likely role in tumorigenesis, Lee *et al.* were interested in identifying potential non-histone targets of EZH2 activity [25]. A computational screen of proteins carrying the amino acid sequence R-K-S in histone H3 that is targeted by EZH2 identified the orphan nuclear receptor RORα as carrying a potential acceptor site for methylation by EZH2. Subsequent studies demonstrated that EZH2 catalyzes monomethylation of RORα at K38 *in vitro*, and that loss of EZH2 activity or a K38A RORα mutation increased cellular RORα protein levels and extended its half-life, but did not alter RORα transcript levels, suggesting RORα methylation triggered its degradation. This hypothesis was further supported by the observation that wild-type, but not K38A, RORα protein ubiquitination products were increased in cells treated with the proteasome inhibitor MG132. To determine the putative ubiquitin ligase responsible for mediating RORα degradation, FLAG-RORα was purified from stably transfected HEK293 cells treated with MG132, and RORα–associating proteins were identified by mass spectrometry, yielding VprBP and DDB1 as RORα-interacting proteins. Subsequent studies established that Cul4 associated with RORα in a VprBP, DDB1, and EZH2-dependent manner, and showed that the VprBP-RORα interaction required active EZH2, suggesting the interaction was methylation dependent. The VprBP interacting region was mapped to a putative chromo-like domain in the N-terminal region (residues 562–593, Figure 2). The authors established the functional significance of RORα methylation-induced, ubiquitin-dependent degradation, by showing that transcription from RORα-dependent promoters and target genes was enhanced by a K38A RORα mutation or by silencing VprBP or EZH2 expression. Consistent with data suggesting EZH2 regulates RORα levels, the authors showed that breast cancer cells and tumors which overexpress EZH2 have lower RORα levels compared to normal controls. In addition, tumor cell growth in soft agar could be suppressed in the breast cancer cell line MCF7 by ectopically expressing RORα, inhibiting EZH2 activity, or silencing VprBP.

Neighboring cell competition

In the *Drosophila* model system, faster growing cells were discovered to induce apoptosis of surrounding slower growing cells in mixed cell culture by a process called "cell competition". This process was found to depend on a tumor suppressor gene called *L*ethal *g*iant *l*arvae (Lgl) [57]. Using the mammalian Lgl homologue (mLgl2) as bait to search for factors that mediate Lgl-dependent cell competition, Tamori *et al.* identified VprBP as a Lgl-binding protein by mass spectrometry and showed that a C-terminal VprBP fragment containing the WD40 domains and the acidic region is minimally required to associate with mLgl2 [58] (Figure 2). In *Drosophila*, disruption of the VprBP homolog (called Mahjong) caused growth retardation and lethality at the late pupal stage. In both *Drosophila* and mammalian cells, loss of Mahjong/VprBP expression rendered cells susceptible to apoptosis initiated by neighboring wild-type cells, possibly through the activation of the JNK pathway. In this study, the authors did not report whether components of the CRL4 or EDD-Dyrk2 E3 ubiquitin ligases associate with the VprBP-mLgl2 complex, nor whether Lgl (or mLgl2) undergoes VprBP-dependent ubiquitination. Thus, the mechanism by which VprBP regulates Lgl to influence cell competition remains unclear.

VprBP as a target of viral proteins

VprBP is targeted by the Vpr and Vpx accessory factors of primate lentiviruses, where they are thought to highjack the function of the $CRL4^{VprBP}$ E3 ligase. The background and significance of these discoveries have been recently reviewed [2], and so will not be covered in depth here. What is notable for this review is that engagement of $CRL4^{VprBP}$ by Vpr (but not Vpx) causes cell cycle arrest at the G2 phase. By contrast, Vpx association with the $CRL4^{VprBP}$ E3 ligase is required to overcome host restriction and enable viral infectivity of myeloid cells. Vpr and Vpx appear to facilitate recruitment of different substrates to the $CRL4^{VprBP}$ ligase: Vpr promotes loading of the uracil DNA glycosylases UNG2 and SMUG1 (*s*ingle-strand-selective *m*onofunctional *u*racil-DNA *g*lycosylase 1) [44,59,60], whereas Vpx recruits SAMHD1 (*s*terile *a*lpha *m*otif and *HD* domain-containing protein-1), a deoxyribonucleoside triphosphate (dNTP) triphosphohydrolase thought to inhibit HIV/SIV infection by maintaining cellular dNTP levels below the threshold required for robust viral reverse transcriptase activity [61-66]. As discussed above, Vpr may enhance the normal turnover of UNG2 by $CRL4^{VprBP}$, whereas Vpx is thought to recruit SAMHD1 as a novel target of $CRL4^{VprBP}$ to inhibit SAMHD1 enzymatic activity [67] and promote its degradation [61]. Notably, the Vpr-mediated G2 arrest phenotype does not depend on Vpr binding to UNG2 [68].

Interestingly, Wang *et al.* recently reported that Vpr also enhances TERT degradation mediated by VprBP [69], which may provide a mechanism to explain the downregulation of telomerase activity following HIV-1 infection [70,71]. These authors showed that knockdown of VprBP, but not Cul4A, alleviated Vpr-induced TERT degradation; a Vpr Q65R mutation that abolishes binding to VprBP also prevented Vpr-induced TERT degradation. The involvement of EDD-Dyrk2 E3 ligase was implied by the finding that EDD and DDB1 associate with Vpr in co-IP experiments, but their requirement for Vpr-induced TERT degradation was not formally established. While Dyrk2 was found to promote TERT ubiquitylation *in vitro*, its presence in Vpr immunoprecipitates and its requirement for Vpr-mediated TERT degradation in cells was not tested.

Based on an alignment of multiple VprBP sequences obtained from a yeast two-hybrid screen for Vpr-interacting proteins, Le Rouzic *et al.* concluded that Vpr interacts with the WD40 domain of VprBP [72]. This finding contrasts somewhat with data published by Zhang *et al.* showing that a VprBP fragment encompassing residues 636–1507, but not 1100–1507 (which contains most of the WD40 domain), co-IP's Vpr from insect cells expressing both proteins [11].

Another viral protein that targets VprBP is human cytomegalovirus (HCMV) UL35 protein [28]. Like Merlin, UL35 binds to the far C-terminal acidic region of VprBP (Figure 2), and, like Vpr [73,74], UL35 overexpression triggers G2 arrest associated with cdc2 phosphorylation [28]. Silencing VprBP expression was shown to block UL35-mediated cdc2 phosphorylation, but its effect on cell cycle was not reported [28]. Although DDB1 was found to co-IP with UL35, interactions with other components of the CRL4 or EDD-Dyrk2 E3 ligases were not formally established. Thus, whether UL35 functions to inhibit or hijack one or the other E3 ligase currently remains unclear.

Remaining questions

A number of very interesting questions regarding the normal function of VprBP and its targeting by viral accessory proteins remain to be answered. One important question centers on whether VprBP levels are limiting in the cell in comparison to the CRL4 and EDD-Dyrk2 E3 ligases, and if so, what conditions are necessary to enable VprBP to service these two E3 ligases if both are expressed in the same cell. Since VprBP is a demonstrated substrate of DNA-PK [30], one might imagine that phosphorylation, or an alternative post-translational modification, could provide a mechanism to influence which E3 ligase VprBP associates with at a given time within the cell. Of course, similar mechanisms may act on the core E3 ligase components themselves to prevent VprBP association, but this would presumably affect interactions with other substrate recognition subunits as well, and may therefore be less likely to occur. The dual specificity of VprBP for both the CRL4 and EDD-Dyrk2 E3 ligases also raises some concerns about whether effects observed in VprBP knock-down experiments can be uniquely attributed to inhibiting one or the other E3 ligase, particularly in situations where the CRL4 and EDD/UBR E3 ligases may have overlapping regulatory roles, such as during DNA damage responses [75-77].

Another central unanswered question pertains to the molecular mechanism by which Merlin binding to VprBP inhibits $CRL4^{VprBP}$ activity. Among possible scenarios that could be imagined include competitive inhibition of substrate binding or blocking of conformational changes that facilitate ubiquitin transfer. In this respect, it is intriguing that Merlin binds to the same site on VprBP as UL35, but for reasons that remain unclear, does not trigger G2 arrest like UL35 [28]. The observation that Vpr also induces G2 arrest, but binds VprBP in a different location than UL35 (Figure 2), raises the possibility that Vpr and UL35 share a common mechanism for inducing G2 arrest that results from inhibiting, rather than redirecting, E3 ligase activity.

A third major question is the relationship between VprBP oligomerization and VprBP function and its implication for experimental interpretation. For example, Kim *et al.* reported that VprBP-mediated histone H3 binding and suppression of p53 target gene transcription was supported by $VprBP_{751-1507}$, which contains the LisH and WD40 domains, but not by $VprBP_{910-1507}$, which lacks the LisH region, leading the authors to conclude that the LisH region is responsible for histone H3 binding [30]. However, since the LisH region has been shown to mediate VprBP oligomerization [26], an alternative interpretation is that enhanced activity of $VprBP_{751-1507}$ relative to $VprBP_{910-1507}$ is attributed to the ability of $VprBP_{751-1507}$ to oligomerize when binding histone H3 on nucleosomal DNA substrates, which could enhance otherwise weak intrinsic protein-protein interactions by avidity.

Given that DDB1 as well as several putative substrate and other interacting proteins associate with VprBP through the WD40 domain (Figure 2), another key question is the following: how can the WD40 domain accommodate both DDB1 and one or more of these other interacting proteins in the same macromolecular complex? One explanation may be that DDB1 interacts with only a portion of the WD40 domain in VprBP (which permits other proteins to access this region), and may also rely on contacts outside of the WD40 domain to maintain stable association with VprBP. In support of these possibilities, Kassmeier *et al.* showed that a $VprBP_{1-1000}$ fragment lacking the WD40 domain associates with DDB1 [53], and residual DDB1 binding was detected with a $VprBP_{1-750}$ fragment by others [22]. Structural studies in several laboratories suggest many DCAFs interact with DDB1 through two different interfaces: one mediated by part of the WD40 domain which makes hydrophobic contacts with the BPA domain of DDB1, and the other mediated by an amino-terminal alpha-helical motif that is inserted into a cleft between the BPA and BPC domains [18-21]. Although a comparable alpha-helical motif was not identified in VprBP using multiple sequence alignment approaches [18,21], the sequence motif is quite short (13–20 residues) and not highly conserved, so its presence may yet remain unrecognized somewhere within the first 750–1000 residues of VprBP. Furthermore, the WD40 domain in DDB1-associated substrate recognition subunits may contact other ligands using the face opposite the one used to bind DDB1. For example, DDB2 uses one side of its WD40 beta-propeller to contact DDB1 and the other side to bind substrate DNA [18,20]. The WD40 domain in VprBP may similarly be used to associate with both DDB1 and substrate molecules.

Another interesting question is whether VprBP functions solely to promote polyubiquitination and degradation of targeted substrates. For example, Mcm10 was shown to undergo VprBP-dependent polyubiquitination and degradation in response to DNA damage [49]. However, VprBP silencing alone does not affect Mcm10 protein levels, suggesting VprBP does not normally control Mcm10 turnover. Interestingly, Mcm10 is mono- to diubiquitinated during the G1-to-S phase transition, which promotes its function during DNA replication in budding yeast [78], and Mcm10 silencing causes cell cycle arrest at the G2 phase by checkpoint activation in mammalian cells [79]. These data raise the possibility that VprBP positively regulates Mcm10 function by mediating mono to diubiquitination in mammalian cells.

In summary, VprBP has been implicated in regulating a wide variety of normal cellular processes and is targeted by host and viral accessory proteins to inhibit or subvert its affiliated E3 ligases. At present, our understanding of whether or how VprBP can switch its allegiance between the CRL4 and EDD-Dyrk2 E3 ligases, especially in pathways where both E3 ligases may play a regulatory role, remains limited. Clarifying the existence and nature of this interplay will greatly improve our understanding of how VprBP regulates cellular processes through the CRL4 and EDD-Dyrk2 E3 ligases, and allow us to recognize the full implications of their inhibition or hijacking by host and pathogen-associated factors.

Competing interests

The authors declare that they have no competing interests.

Authors' contributions

TN, KM, and PCS all contributed in drafting and editing the manuscript, and all authors read and approved the final manuscript.

Acknowledgments

PCS gratefully acknowledges funding from the National Institutes of Health (5R01 GM102487). TN thanks Yue Xiong (University of North Carolina at Chapel Hill) for continuous encouragement and support.

References

1. Komander D, Rape M: **The ubiquitin code.** *Annu Rev Biochem* 2012, **81**:203–229.
2. Romani B, Cohen EA: **Lentivirus Vpr and Vpx accessory proteins usurp the cullin4-DDB1 (DCAF1) E3 ubiquitin ligase.** *Curr Opin Virol* 2012, **2**(6):755–763.
3. Viswanathan K, Fruh K, DeFilippis V: **Viral hijacking of the host ubiquitin system to evade interferon responses.** *Curr Opin Microbiol* 2010, **13**(4):517–523.
4. Hicks SW, Galan JE: **Hijacking the host ubiquitin pathway: structural strategies of bacterial E3 ubiquitin ligases.** *Curr Opin Microbiol* 2010, **13**(1):41–46.
5. Metzger MB, Hristova VA, Weissman AM: **HECT and RING finger families of E3 ubiquitin ligases at a glance.** *J Cell Sci* 2012, **125**(Pt 3):531–537.
6. Sarikas A, Hartmann T, Pan ZQ: **The cullin protein family.** *Genome Biol* 2011, **12**(4):220.
7. Duda DM, Scott DC, Calabrese MF, Zimmerman ES, Zheng N, Schulman BA: **Structural regulation of cullin-RING ubiquitin ligase complexes.** *Curr Opin Struct Biol* 2011, **21**(2):257–264.
8. Scheffner M, Kumar S: **Mammalian HECT ubiquitin-protein ligases: Biological and pathophysiological aspects.** *Biochim Biophys Acta* 2013. Epub ahead of print.
9. Shearwin-Whyatt L, Dalton HE, Foot N, Kumar S: **Regulation of functional diversity within the Nedd4 family by accessory and adaptor proteins.** *Bioessays* 2006, **28**(6):617–628.
10. Zhao LJ, Mukherjee S, Narayan O: **Biochemical mechanism of HIV-I Vpr function. Specific interaction with a cellular protein.** *J Biol Chem* 1994, **269**(22):15577–15582.
11. Zhang S, Feng Y, Narayan O, Zhao LJ: **Cytoplasmic retention of HIV-1 regulatory protein Vpr by protein-protein interaction with a novel human cytoplasmic protein VprBP.** *Gene* 2001, **263**(1–2):131–140.

12. Jin J, Arias EE, Chen J, Harper JW, Walter JC: **A family of diverse Cul4-Ddb1-interacting proteins includes Cdt2, which is required for S phase destruction of the replication factor Cdt1.** *Mol Cell* 2006, **23**(5):709–721.
13. Angers S, Li T, Yi X, MacCoss MJ, Moon RT, Zheng N: **Molecular architecture and assembly of the DDB1-CUL4A ubiquitin ligase machinery.** *Nature* 2006, **443**(7111):590–593.
14. He YJ, McCall CM, Hu J, Zeng Y, Xiong Y: **DDB1 functions as a linker to recruit receptor WD40 proteins to CUL4-ROC1 ubiquitin ligases.** *Genes Dev* 2006, **20**(21):2949–2954.
15. Higa LA, Wu M, Ye T, Kobayashi R, Sun H, Zhang H: **CUL4-DDB1 ubiquitin ligase interacts with multiple WD40-repeat proteins and regulates histone methylation.** *Nat Cell Biol* 2006, **8**(11):1277–1283.
16. Higa LA, Zhang H: **Stealing the spotlight: CUL4-DDB1 ubiquitin ligase docks WD40-repeat proteins to destroy.** *Cell Div* 2007, **2**:5.
17. Lee J, Zhou P: **DCAFs, the missing link of the CUL4-DDB1 ubiquitin ligase.** *Mol Cell* 2007, **26**(6):775–780.
18. Fischer ES, Scrima A, Bohm K, Matsumoto S, Lingaraju GM, Faty M, Yasuda T, Cavadini S, Wakasugi M, Hanaoka F, *et al*: **The molecular basis of CRL4DDB2/CSA ubiquitin ligase architecture, targeting, and activation.** *Cell* 2011, **147**(5):1024–1039.
19. Yeh JI, Levine AS, Du S, Chinte U, Ghodke H, Wang H, Shi H, Hsieh CL, Conway JF, Van Houten B, *et al*: **Damaged DNA induced UV-damaged DNA-binding protein (UV-DDB) dimerization and its roles in chromatinized DNA repair.** *Proc Natl Acad Sci USA* 2012, **109**(41):E2737–E2746.
20. Scrima A, Konickova R, Czyzewski BK, Kawasaki Y, Jeffrey PD, Groisman R, Nakatani Y, Iwai S, Pavletich NP, Thoma NH: **Structural basis of UV DNA-damage recognition by the DDB1-DDB2 complex.** *Cell* 2008, **135**(7):1213–1223.
21. Li T, Robert EI, van Breugel PC, Strubin M, Zheng N: **A promiscuous alpha-helical motif anchors viral hijackers and substrate receptors to the CUL4-DDB1 ubiquitin ligase machinery.** *Nat Struct Mol Biol* 2010, **17**(1):105–111.
22. McCall CM, de Marval PL M, Chastain PD, Jackson SC 2nd, He YJ, Kotake Y, Cook JG, Xiong Y: **Human immunodeficiency virus type 1 Vpr-binding protein VprBP, a WD40 protein associated with the DDB1-CUL4 E3 ubiquitin ligase, is essential for DNA replication and embryonic development.** *Mol Cell Biol* 2008, **28**(18):5621–5633.
23. Peifer M, Berg S, Reynolds AB: **A repeating amino acid motif shared by proteins with diverse cellular roles.** *Cell* 1994, **76**(5):789–791.
24. Huber AH, Nelson WJ, Weis WI: **Three-dimensional structure of the armadillo repeat region of beta-catenin.** *Cell* 1997, **90**(5):871–882.
25. Lee JM, Lee JS, Kim H, Kim K, Park H, Kim JY, Lee SH, Kim IS, Kim J, Lee M, *et al*: **EZH2 generates a methyl degron that is recognized by the DCAF1/DDB1/CUL4 E3 ubiquitin ligase complex.** *Mol Cell* 2012, **48**(4):572–586.
26. Ahn J, Novince Z, Concel J, Byeon CH, Makhov AM, Byeon IJ, Zhang P, Gronenborn AM: **The Cullin-RING E3 Ubiquitin Ligase CRL4-DCAF1 Complex Dimerizes via a Short Helical Region in DCAF1.** *Biochemistry* 2011, **50**(8):1359–1367.
27. Li W, You L, Cooper J, Schiavon G, Pepe-Caprio A, Zhou L, Ishii R, Giovannini M, Hanemann CO, Long SB, *et al*: **Merlin/NF2 suppresses tumorigenesis by inhibiting the E3 ubiquitin ligase CRL4(DCAF1) in the nucleus.** *Cell* 2010, **140**(4):477–490.
28. Salsman J, Jagannathan M, Paladino P, Chan PK, Dellaire G, Raught B, Frappier L: **Proteomic profiling of the human cytomegalovirus UL35 gene products reveals a role for UL35 in the DNA repair response.** *J Virol* 2012, **86**(2):806–820.
29. Flicek P, Amode MR, Barrell D, Beal K, Brent S, Carvalho-Silva D, Clapham P, Coates G, Fairley S, Fitzgerald S, *et al*: **Ensembl 2012.** *Nucleic Acids Res* 2012, **40**(Database issue):D84–D90.
30. Kim K, Heo K, Choi J, Jackson S, Kim H, Xiong Y, An W: **Vpr-binding protein antagonizes p53-mediated transcription via direct interaction with H3 tail.** *Mol Cell Biol* 2012, **32**(4):783–796.
31. Sclafani RA, Holzen TM: **Cell cycle regulation of DNA replication.** *Annu Rev Genet* 2007, **41:**237–280.
32. Symeonidou IE, Taraviras S, Lygerou Z: **Control over DNA replication in time and space.** *FEBS Lett* 2012, **586**(18):2803–2812.
33. Mechali M: **Eukaryotic DNA replication origins: many choices for appropriate answers.** *Nat Rev Mol Cell Biol* 2010, **11**(10):728–738.
34. Bartek J, Lukas C, Lukas J: **Checking on DNA damage in S phase.** *Nat Rev Mol Cell Biol* 2004, **5**(10):792–804.
35. Ciccia A, Elledge SJ: **The DNA damage response: making it safe to play with knives.** *Mol Cell* 2010, **40**(2):179–204.
36. Ammoun S, Hanemann CO: **Emerging therapeutic targets in schwannomas and other merlin-deficient tumors.** *Nat Rev Neurol* 2011, **7**(7):392–399.
37. Huang J, Chen J: **VprBP targets Merlin to the Roc1-Cul4A-DDB1 E3 ligase complex for degradation.** *Oncogene* 2008, **27**(29):4056–4064.
38. Kissil JL, Johnson KC, Eckman MS, Jacks T: **Merlin phosphorylation by p21-activated kinase 2 and effects of phosphorylation on merlin localization.** *J Biol Chem* 2002, **277**(12):10394–10399.
39. Xiao GH, Beeser A, Chernoff J, Testa JR: **p21-activated kinase links Rac/Cdc42 signaling to merlin.** *J Biol Chem* 2002, **277**(2):883–886.
40. Roll-Mecak A, McNally FJ: **Microtubule-severing enzymes.** *Curr Opin Cell Biol* 2010, **22**(1):96–103.
41. Maddika S, Chen J: **Protein kinase DYRK2 is a scaffold that facilitates assembly of an E3 ligase.** *Nat Cell Biol* 2009, **11**(4):409–419.
42. Lu C, Mains PE: **The C. elegans anaphase promoting complex and MBK-2 /DYRK kinase act redundantly with CUL-3/MEL-26 ubiquitin ligase to degrade MEI-1 microtubule-severing activity after meiosis.** *Dev Biol* 2007, **302**(2):438–447.
43. Visnes T, Doseth B, Pettersen HS, Hagen L, Sousa MM, Akbari M, Otterlei M, Kavli B, Slupphaug G, Krokan HE: **Uracil in DNA and its processing by different DNA glycosylases.** *Philos Trans R Soc Lond B Biol Sci* 2009, **364**(1517):563–568.
44. Wen X, Casey Klockow L, Nekorchuk M, Sharifi HJ, de Noronha CM: **The HIV1 protein Vpr acts to enhance constitutive DCAF1-dependent UNG2 turnover.** *PLoS One* 2012, **7**(1):e30939.
45. Elder RT, Zhu X, Priet S, Chen M, Yu M, Navarro JM, Sire J, Zhao Y: **A fission yeast homologue of the human uracil-DNA-glycosylase and their roles in causing DNA damage after overexpression.** *Biochem Biophys Res Commun* 2003, **306**(3):693–700.
46. Zeitlin SG, Chapados BR, Baker NM, Tai C, Slupphaug G, Wang JY: **Uracil DNA N-glycosylase promotes assembly of human centromere protein A.** *PLoS One* 2011, **6**(3):e17151.
47. Zhu W, Ukomadu C, Jha S, Senga T, Dhar SK, Wohlschlegel JA, Nutt LK, Kornbluth S, Dutta A: **Mcm10 and And-1/CTF4 recruit DNA polymerase alpha to chromatin for initiation of DNA replication.** *Genes Dev* 2007, **21**(18):2288–2299.
48. Sharma A, Kaur M, Kar A, Ranade SM, Saxena S: **Ultraviolet radiation stress triggers the down-regulation of essential replication factor Mcm10.** *J Biol Chem* 2010, **285**(11):8352–8362.
49. Kaur M, Khan MM, Kar A, Sharma A, Saxena S: **CRL4-DDB1-VPRBP ubiquitin ligase mediates the stress triggered proteolysis of Mcm10.** *Nucleic Acids Res* 2012, **40**(15):7332–7346.
50. Hrecka K, Gierszewska M, Srivastava S, Kozaczkiewicz L, Swanson SK, Florens L, Washburn MP, Skowronski J: **Lentiviral Vpr usurps Cul4-DDB1 [VprBP] E3 ubiquitin ligase to modulate cell cycle.** *Proc Natl Acad Sci USA* 2007, **104**(28):11778–11783.
51. Schatz DG, Swanson PC: **V(D)J Recombination: Mechanisms of Initiation.** *Annu Rev Genet* 2011, **14**:167–202.
52. Deshaies RJ, Joazeiro CA: **RING domain E3 ubiquitin ligases.** *Annu Rev Biochem* 2009, **78:**399–434.
53. Kassmeier MD, Mondal K, Palmer VL, Raval P, Kumar S, Perry GA, Anderson DK, Ciborowski P, Jackson S, Xiong Y, *et al*: **VprBP binds full-length RAG1 and is required for B-cell development and V(D) J recombination fidelity.** *EMBO J* 2011, **31**(4):945–958.
54. Gomez DE, Armando RG, Farina HG, Menna PL, Cerrudo CS, Ghiringhelli PD, Alonso DF: **Telomere structure and telomerase in health and disease (review).** *Int J Oncol* 2012, **41**(5):1561–1569.
55. Jung HY, Wang X, Jun S, Park JI: **Dyrk2-associated EDD-DDB1-VprBP E3 ligase inhibits telomerase by TERT degradation.** *J Biol Chem* 2013, **288**(10):7252–7262.
56. Chang CJ, Hung MC: **The role of EZH2 in tumour progression.** *Br J Cancer* 2012, **106**(2):243–247.
57. Grzeschik NA, Amin N, Secombe J, Brumby AM, Richardson HE: **Abnormalities in cell proliferation and apico-basal cell polarity are separable in Drosophila lgl mutant clones in the developing eye.** *Dev Biol* 2007, **311**(1):106–123.
58. Tamori Y, Bialucha CU, Tian AG, Kajita M, Huang YC, Norman M, Harrison N, Poulton J, Ivanovitch K, Disch L, *et al*: **Involvement of Lgl and Mahjong/VprBP in cell competition.** *PLoS Biol* 2010, **8**(7):e1000422.

59. Ahn J, Vu T, Novince Z, Guerrero-Santoro J, Rapic-Otrin V, Gronenborn AM: **HIV-1 Vpr loads uracil DNA glycosylase-2 onto DCAF1, a substrate recognition subunit of a cullin 4A-ring E3 ubiquitin ligase for proteasome-dependent degradation.** *J Biol Chem* 2010, **285**(48):37333–37341.
60. Schrofelbauer B, Yu Q, Zeitlin SG, Landau NR: **Human immunodeficiency virus type 1 Vpr induces the degradation of the UNG and SMUG uracil-DNA glycosylases.** *Journal of virology* 2005, **79**(17):10978–10987.
61. Hrecka K, Hao C, Gierszewska M, Swanson SK, Kesik-Brodacka M, Srivastava S, Florens L, Washburn MP, Skowronski J: **Vpx relieves inhibition of HIV-1 infection of macrophages mediated by the SAMHD1 protein.** *Nature* 2011, **474**(7353):658–661.
62. Ahn J, Hao C, Yan J, DeLucia M, Mehrens J, Wang C, Gronenborn AM, Skowronski J: **HIV/simian immunodeficiency virus (SIV) accessory virulence factor Vpx loads the host cell restriction factor SAMHD1 onto the E3 ubiquitin ligase complex CRL4DCAF1.** *J Biol Chem* 2012, **287**(15):12550–12558.
63. Laguette N, Sobhian B, Casartelli N, Ringeard M, Chable-Bessia C, Segeral E, Yatim A, Emiliani S, Schwartz O, Benkirane M: **SAMHD1 is the dendritic- and myeloid-cell-specific HIV-1 restriction factor counteracted by Vpx.** *Nature* 2011, **474**(7353):654–657.
64. Berger A, Sommer AFR, Zwarg J, Hamdorf M, Welzel K, Esly N, Panitz S, Reuter A, Ramos I, Jatiani A, *et al*: **SAMHD1-deficient CD14+ cells from individuals with Aicardi-Goutieres syndrome are highly susceptible to HIV-1 infection.** *PLoS pathogens* 2011, **7**(12):e1002425.
65. Goldstone DC, Ennis-Adeniran V, Hedden JJ, Groom HCT, Rice GI, Christodoulou E, Walker PA, Kelly G, Haire LF, Yap MW, *et al*: **HIV-1 restriction factor SAMHD1 is a deoxynucleoside triphosphate triphosphohydrolase.** *Nature* 2011, **480**(7377):379–382.
66. Powell RD, Holland PJ, Hollis T, Perrino FW: **Aicardi-Goutieres syndrome gene and HIV-1 restriction factor SAMHD1 is a dGTP-regulated deoxynucleotide triphosphohydrolase.** *The J Biol Chem* 2011, **286**(51):43596–43600.
67. Delucia M, Mehrens J, Wu Y, Ahn J: **HIV-2 and SIVmac accessory virulence factor Vpx down-regulates SAMHD1 catalysis prior to proteasome-dependent degradation.** *J Biol Chem* 2013, **288**(26):19116–19126.
68. Selig L, Benichou S, Rogel ME, Wu LI, Vodicka MA, Sire J, Benarous R, Emerman M: **Uracil DNA glycosylase specifically interacts with Vpr of both human immunodeficiency virus type 1 and simian immunodeficiency virus of sooty mangabeys, but binding does not correlate with cell cycle arrest.** *J Virol* 1997, **71**(6):4842–4846.
69. Wang X, Singh S, Jung H-Y, Yang G, Jun S, Sastry KJ, Park J-I: **HIV-1 Vpr Protein Inhibits Telomerase Activity via the EDD-DDB1-VPRBP E3 Ligase Complex.** *The J Biol Chem* 2013, **288**(22):15474–15480.
70. Ballon G, Ometto L, Righetti E, Cattelan AM, Masiero S, Zanchetta M, Chieco-Bianchi L, De Rossi A: **Human immunodeficiency virus type 1 modulates telomerase activity in peripheral blood lymphocytes.** *J Infect Dis* 2001, **183**(3):417–424.
71. Franzese O, Adamo R, Pollicita M, Comandini A, Laudisi A, Perno CF, Aquaro S, Bonmassar E: **Telomerase activity, hTERT expression, and phosphorylation are downregulated in CD4(+) T lymphocytes infected with human immunodeficiency virus type 1 (HIV-1).** *J Med Virol* 2007, **79**(5):639–646.
72. Le Rouzic E, Belaidouni N, Estrabaud E, Morel M, Rain JC, Transy C, Margottin-Goguet F: **HIV1 Vpr arrests the cell cycle by recruiting DCAF1/VprBP, a receptor of the Cul4-DDB1 ubiquitin ligase.** *Cell Cycle* 2007, **6**(2):182–188.
73. Re F, Braaten D, Franke EK, Luban J: **Human immunodeficiency virus type 1 Vpr arrests the cell cycle in G2 by inhibiting the activation of p34cdc2-cyclin B.** *J Virol* 1995, **69**(11):6859–6864.
74. He J, Choe S, Walker R, Di Marzio P, Morgan DO, Landau NR: **Human immunodeficiency virus type 1 viral protein R (Vpr) arrests cells in the G2 phase of the cell cycle by inhibiting p34cdc2 activity.** *J Virol* 1995, **69**(11):6705–6711.
75. Munoz MA, Saunders DN, Henderson MJ, Clancy JL, Russell AJ, Lehrbach G, Musgrove EA, Watts CK, Sutherland RL: **The E3 ubiquitin ligase EDD regulates S-phase and G(2)/M DNA damage checkpoints.** *Cell Cycle* 2007, **6**(24):3070–3077.
76. Li JM, Jin J: **CRL Ubiquitin Ligases and DNA Damage Response.** *Front Oncol* 2012, **2**:29.
77. Ling S, Lin WC: **EDD inhibits ATM-mediated phosphorylation of p53.** *J Biol Chem* 2011, **286**(17):14972–14982.
78. Das-Bradoo S, Ricke RM, Bielinsky AK: **Interaction between PCNA and diubiquitinated Mcm10 is essential for cell growth in budding yeast.** *Mol Cell Biol* 2006, **26**(13):4806–4817.
79. Park JH, Bang SW, Kim SH, Hwang DS: **Knockdown of human MCM10 activates G2 checkpoint pathway.** *Biochem Biophys Res Commun* 2008, **365**(3):490–495.

4

Dissecting domains necessary for activation and repression of splicing by muscleblind-like protein 1

Christopher Edge, Clare Gooding and Christopher WJ Smith*

Abstract

Background: Alternative splicing contributes to the diversity of the proteome, and provides the cell with an important additional layer of regulation of gene expression. Among the many RNA binding proteins that regulate alternative splicing pathways are the Muscleblind-like (MBNL) proteins. MBNL proteins bind YGCY motifs in RNA via four CCCH zinc fingers arranged in two tandem arrays, and play a crucial role in the transition from embryonic to adult muscle splicing patterns, deregulation of which leads to Myotonic Dystrophy. Like many other RNA binding proteins, MBNL proteins can act as both activators or repressors of different splicing events.

Results: We used targeted point mutations to interfere with the RNA binding of MBNL1 zinc fingers individually and in combination. The effects of the mutations were tested in assays for splicing repression and activation, including overexpression, complementation of siRNA-mediated knockdown, and artificial tethering using MS2 coat protein. Mutations were tested in the context of both full length MBNL1 as well as a series of truncation mutants. Individual mutations within full length MBNL1 had little effect, but mutations in ZF1 and 2 combined were more detrimental than those in ZF 3 and 4, upon splicing activation, repression and RNA binding. Activation and repression both required linker sequences between ZF2 and 3, but activation was more sensitive to loss of linker sequences.

Conclusions: Our results highlight the importance of RNA binding by MBNL ZF domains 1 and 2 for splicing regulatory activity, even when the protein is artificially recruited to its regulatory location on target RNAs. However, RNA binding is not sufficient for activity; additional regions between ZF 2 and 3 are also essential. Activation and repression show differential sensitivity to truncation of this linker region, suggesting interactions with different sets of cofactors for the two types of activity.

Background

Pre-mRNA splicing is a critical part of mRNA maturation, and alternative splicing is a well established method of generating diversity and exerting control over the proteome. It is now recognised that the vast majority of transcripts are alternatively spliced, allowing production of many protein isoforms from a single gene (for review see [1]). The process is controlled so that certain isoforms are restricted to specific cell types, developmental stages, or conditions [2,3]. Alternative splicing is controlled in large part by a variety of a protein factors which can positively or negatively influence splicing at adjacent splice sites. Early investigations suggested that proteins of the SR family generally act as splicing activators, while proteins of the hnRNP family typically act as repressors. More recent global analyses of the activities of RNA binding proteins has indicated that many of them show both activator or repressor activity, depending on the site at which they bind to the target pre-mRNA [4].

Loss of regulation of alternative splicing can lead to a variety of diseases, including Myotonic Dystrophy (DM1), which is caused by expansions of CUG repeats, which bind and sequester muscleblind like (MBNL) proteins [5]. MBNL proteins normally control the transition from embryonic to adult isoforms of a sub-set of muscle-specific proteins in heart and skeletal muscle cells [6-8]. In DM1, embryonic isoforms of important muscle proteins are expressed, which causes the various clinical symptoms [9,10]. For example, myotonia is casued by deregulation of a MBNL-controlled splicing event in the skeletal muscle chloride channel (CLCN1) [11].

MBNL is a four zinc-finger (ZF) containing protein (of the type $CX_7CX_{4-6}CX_3H$). The ZF domains are arranged

* Correspondence: cwjs1@cam.ac.uk
Department of Biochemistry, University of Cambridge, Tennis Court Road, Cambridge CB2 1QW, UK

in two tandem arrays in the N-terminal part of the protein (Figure 1A). The RNA binding faces in each didomain are arranged back-to-back, creating a predicted anti-parallel alignment of RNA binding to adjacent ZFs [12,13]. SELEX experiments have determined the optimal MBNL binding sequence to consist of multiple YGCY motifs [14], explaining the binding to CUG expansions. By using U-tracts with two GC steps and manipulating the spacing between them, it has been shown that MBNL can bind the two sites with as little as a 1 nt spacer separating them, or in a second binding conformation with a spacer of around 17 nt [15], suggesting multiple modes of RNA-protein interaction. The published crystal structures of MBNL1 ZF domains [13] shows how the two domains in the ZF34 tandem array interact with the RNA. Key aromatic residues in ZF3 and 4 (F202 and Y236) intercalate between the bases of the GC step, while specific hydrogen bonds from the GC bases to side chains in the protein partly explain the binding specificity of MBNL-1.

The MBNL1 gene is comprised of 12 exons, 10 of them protein coding, with the ZFs encoded by exons 2–6. Extensive alternative splicing of exons encoding the linker between ZFs 2 and 3, and the C-terminal end of the protein leads to multiple functionally distinct protein isoforms [16]. Structure-function analyses of MBNL1 and 3 have been performed by generating N- and C–terminal truncations and analysing the effect on splicing regulation. In this analysis, the regions of MBNL required for splicing repression and activation differed. Activation required the entire linker sequence between ZF 2 and 3, while repression required only a small N-terminal portion of the linker [17]. A second structure-function analysis involved targeted mutations to impair RNA binding by the different ZF domains, and analysis of the consequences upon MBNL-repressed and MBNL-activated events [18]. Although activity is usually linked to RNA binding, there is a subset of events where the affinity of MBNL for the RNA is not correlated with activity.

Artificial recruitment systems have been used to great effect to analyze the function of splicing factors and other RNA binding proteins. This method involves expressing the protein of interest as a fusion with a heterologous RNA binding protein, such as MS2 coat-protein, and replacing the normal binding site on the target RNA with an MS2 binding site. This circumvents the normal mode of RNA binding and allows the dissection of splicing activator or repressor domains. This approach has been used to investigate SR proteins [19] hnRNP and other RNA binding proteins including hnRNP A1 [20], PTB [21], MBNL1 [22], RbFOX [23] and hnRNPL [24].

Here we use targeted mutations to disrupt RNA binding by individual ZF domains of MBNL1 combined with larger deletions to analyse the splicing activation and repression function of MBNL1 in both MS2-tethered and non-tethered splicing assays. We find that full length MBNL1 is remarkably tolerant of mutation to individual ZFs or pairs of ZFs in a simple cotransfection assay. However, in MS2 tethering assays of the N-terminal part of MBNL1 containing the four ZF domains, mutation of ZF3 and 4 has no effect on splicing repression, but mutation of ZF1 and 2 is highly deleterious. In contrast, for activation no mutations or pairs of mutations drastically reduce activity. When artificially recruited, only the first two zinc fingers plus a small N-terminal portion of the linker sequence between ZF 2 and 3 is required for repression, whereas for activation the whole linker sequence is needed even though this region plays no part in RNA binding. For both activation or repression, disruption of RNA binding by ZFs 1 and 2 is highly deleterious for activity. Our results further highlight the distinct requirements of different regions of MBNL1 for splicing repression and activation.

Results

Effect of MBNL RNA binding mutations on MBNL-regulated splicing events

Based on high resolution structures of the TIS11d [25] and MBNL proteins [12,13] we designed point mutations in each MBNL zinc finger that would disrupt RNA binding, without severely altering the overall fold and structure of the domain. We targeted conserved aromatic residues F36, Y68, F202 and Y236 in ZF 1–4 respectively, and mutated them to alanine (Figure 1A). The mutations were introduced individually, in combinations in the two di-domains (MBNL-FL-M12 and -M34) and into all four ZF domains simultaneously. Similar mutations have since been reported by others [18,26]. In order to confirm that the mutations disrupt RNA binding, recombinant MBNL1 aa 2–253 was produced with all four zinc fingers mutated and compared to wildtype protein in UV crosslinking assays. While the wild-type crosslinked to the RNA the mutant did not (Figure 1B, lower panel).

We next tested the effects of the MBNL ZF mutants in assays for splicing repression and splicing activation by MBNL1 in HeLa cells. To test splicing repressor activity we used a *Tpm1* minigene with a point mutation of the branch point of exon 3, which increases exon 3 skipping in HeLa cells [22,27]. This minigene responds modestly to simple overexpression of MBNL1. However, upon knockdown of MBNL1 (Figure 1E) exon skipping is reduced substantially (from 35 to 13%, Figure 1C, lanes 1, 2); complementation with overexpressed MBNL1 restores exon skipping to 53% (Figure 1C, lane 10). As a model MBNL-activated exon we used a minigene construct containing a Vldlr exon flanked by globin exons [10], which responds to MBNL1 overexpression by increasing

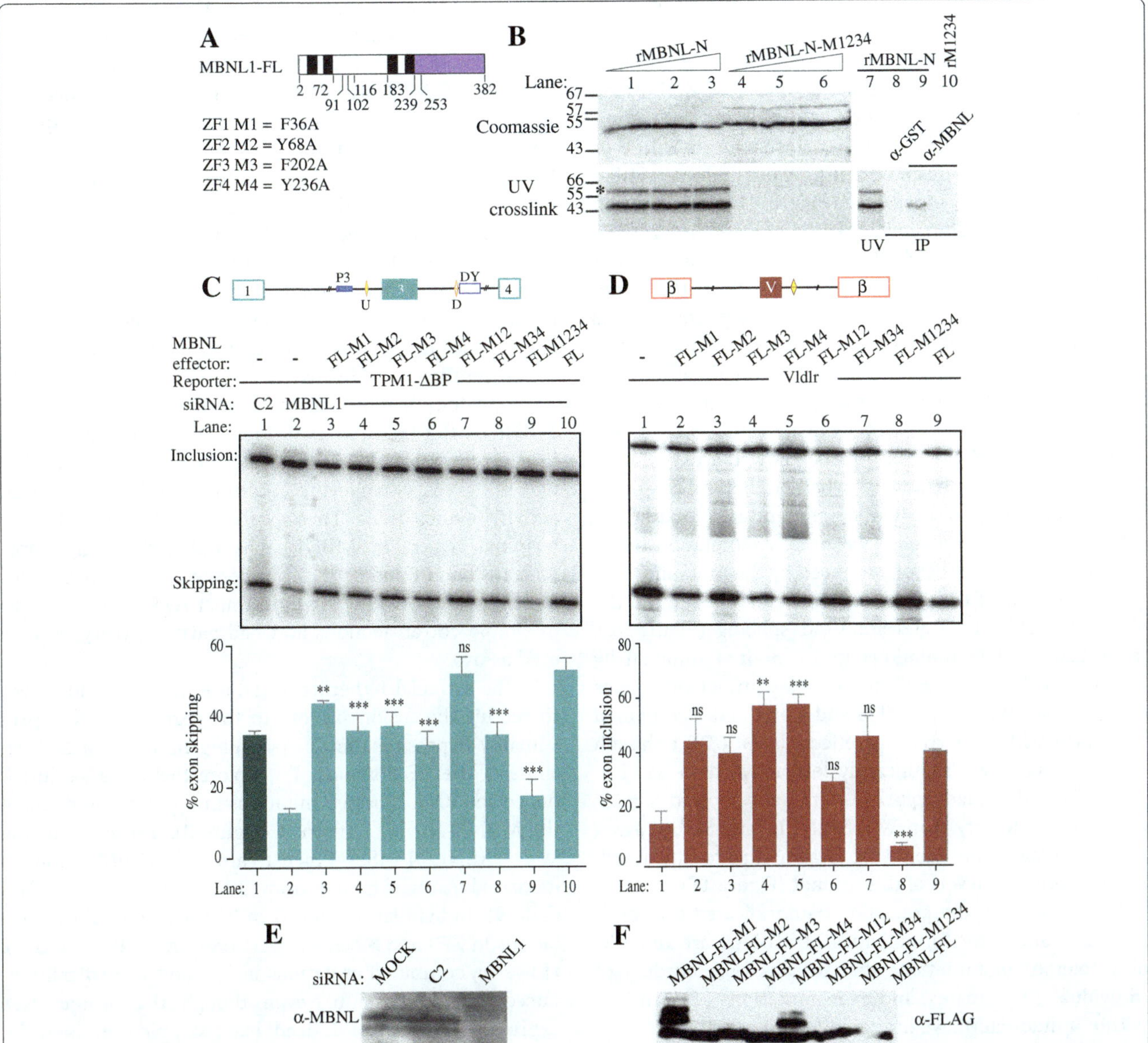

Figure 1 Effects of RNA binding mutations on MBNL1 splicing activity. A. Schematic representation of MBNL1. Zinc fingers are shown in black, the C-terminus in purple. Amino acid positions of deletion boundaries mutants and ZF domain inactivating mutations are indicated. The 382 aa MBNL1 isoform lacks sequences corresponding to exons 7 and 9. **B**. Comparison of wild type MBNL-N and MBNL-N-M1234 UV crosslinking to RNA. Upper panel, Coomassie blue stained gel; lower panel UV crosslinking. RNA used for crosslinking encompassed Tpm1 exon 3 and both upstream and downstream MBNL elements. The identity of crosslinked MBNL (lanes 1-3) was established by immunoprecipitation with anti-MBNL1 (lane 9) but not anti-GST antibodies (lane 8). The asterisked band is a contaminant that was not immunoprecipitated by anti-MBNL1; it does not correspond to the higher molecular weight contaminant in the Coomassie stained samples of mutant protein (lanes 4–6). **C**. Effects of MBNL1 knockdown and overexpression upon Tpm1 splicing in HeLa cells. The cartoon depicts the ΔBP minigene; the U and D MBNL binding elements and P3 and DY pyrimidine tracts are indicated. The minigene was co-transfected with control (C2, lane 1) or MBNL1 siRNAs (lanes 2–10), and with wild-type MBNL1 (lane 10) or MBNL1 mutants in the indicated ZF domains (lanes 3–9). Values significantly different from FL wild type MBNL1: **, $P < 0.01$; *** $P < 0.001$; ns, not significant. Values for the M1234 mutant are statistically significant, but lack of protein expression (panel F) prevents meaningful conclusions from being drawn. **D**. Effects of MBNL1 overexpression upon Vldlr splicing in HeLa cells. The cartoon depicts the Vldlr minigene; the MBNL1 binding site is indicated by the yellow diamond. The minigene was transfected alone (lane 1), with wild-type MBNL1 (lane 9) or mutants in ZF domains (lanes 2–8). **E**. Western blot of MBNL1 in mock transfected, control(C2) or MBNL1 siRNA treated cells. **F**. Anti-FLAG western blot showing expression of MBNL1 constructs from panels **C** and **D**.

exon inclusion from 14 to 39% (Figure 1D, lanes 1, 9). Note that in order to facilitate comparison of the repressor and activator activities of MBNL1 mutants, we refer throughout to percentage exon skipping of the repressed *Tpm1* exon but percentage exon inclusion of the activated *Vldlr* exon.

Compared to wild type MBNL1, all of the ZF domain single point-mutants had moderately reduced repressor activity, producing exon skipping levels of 32-44% (Figure 1C, lanes 3–6), as did the combined ZNF 3 and 4 mutant (lane 8). Surprisingly, the mutant with combined mutations in ZF 1 and 2 was as active as wild type MBNL1 (lane 7), despite being expressed at similar levels to the other constructs (Figure 1F). The mutant with all four ZF domains impaired showed no activity (Figure 1C, lanes 2 and 9). However, this mutant was consistently expressed at lower levels than the other constructs (Figure 1F), preventing strong conclusions about its activity. We noted that the MBNL proteins with mutations in ZF1 (MBNL-FL-M1 and MBNL-FL-M12) consistently showed the presence of additional slower migrating bands that were detected with FLAG antibodies (Figure 1F). We do not know the explanation for these additional bands, or whether they represent an active fraction of protein. It is therefore possible that higher total levels of active proteins with the M1 mutation might partially mask loss of activity induced by the mutation.

Mutations in ZF1 or 2 had no significant effect upon the ability of MBNL1 to activate Vldlr splicing (Figure 1D, lanes 2,3,9), while mutations in ZF 3 or 4 individually caused a small but significant increase in activity (lanes 4,5). Double mutations of ZF 3 and 4 or 1 and 2 combined were also without significant effect (lanes 6,7), although the 12 mutant was significantly less active than 34 ($P < 0.05$). Only the quadruple ZF1-4 mutant showed significantly lower activity than WT MBNL1 (lane 8), but again no firm conclusions could be drawn due to the much lower expression levels of this mutant (Figure 1F).

Taken together, the preceding data indicated that both repressor and activator activities of MBNL1 are remarkably tolerant of mutations that impair RNA binding of individual ZF domains, and even mutations of both ZFs within a didomain have limited effects.

MS2 tethering of MBNL1 activation and repression domains

We next compared the activities of deletion mutants of MBNL1 in simple cotransfection and tethered function assays (Figure 2). Consistent with previous data [22] in the knockdown/complementation assay with the *Tpm1* reporter, the N-terminal region of MBNL1 (aa 2–253) had similar repressor activity to the full length protein (Figure 2B, lanes 3,5). In contrast, a C-terminal fragment of MBNL1 (aa 239–382) had no activity (Figure 2B, lane 4). Similar effects were seen with the Vldlr reporter; the N-terminal fragment had indistinguishable activity to full length MBNL1 (Figure 2C, lanes 2,4), while the C-terminal fragment was devoid of activator activity (lane 3), despite being expressed to similar levels (Figure 2A).

As reported previously [22] replacement of the downstream MBNL1 binding element in *Tpm1* with a binding site for MS2 coat protein led to a ~3-fold decrease in exon skipping (Figure 2E, lanes 1,2). Addition of MS2 coat protein had little effect (lane 6), while fusion proteins of MS2 with full length MBNL1 or just the N-terminal region led to high levels of exon skipping (lanes 3,4). In contrast, the C-terminal region of MBNL1 fused to MS2 had a significant, but much more modest effect than full length MBNL1-MS2 (lane 5). Replacement of the reported MBNL1 binding site containing two GC motifs downstream of the Vldlr exon with a single MS2 site reduced exon inclusion from 14% to 2% (Figure 2F, lanes 1,2), consistent with the activity of this element as an MBNL-dependent splicing enhancer in mouse embryonic fibroblasts [10]. Co-transfection with MS2 protein had no effect (lane 6), while full length MBNL1-MS2 restored exon inclusion levels (lane 3). As in the repression assay, the N-terminal of MBNL1-MS2 had full activity, while the C-terminal region had partial activity (lanes 4,5). These data indicate that the N-terminal region of MBNL1 has full activity in simple co-transfection and artificial tethering repression and enhancing assays, while the C-terminal region was inactive in simple cotransfections and had partial activity in tethered assays.

In the artificial tethering assay, the MS2 domain serves to recruit the fusion protein to the regulated RNA, presumably bypassing the RNA-binding function of at least some of the ZF domains. To explore this issue we introduced the RNA binding mutations into the ZF domains of the MBNL-N-MS2 construct (Figure 3). Tethering of the WT MBNL-N-MS2 downstream of *Tpm1* exon 3 increased exon skipping from 20% (lanes 1,10) to 71% (lane 9). Individual mutations in ZF1-4 or combined mutations in ZF3 and 4 had no effect on activity (lanes 2–5, 7). However, combined mutations in ZF1 and 2 drastically reduced activity (lane 6), even though the protein was expressed (Figure 3B). Indeed, exon skipping levels in the presence of the ZF12 mutant were not significantly different from MS2 alone or no cotransfection (lane 6, compared to lanes 1 or 10). The quadruple mutant in ZF1-4 was also inactive, but again the protein was expressed at very low levels (lane 8 and Figure 3B). The complete loss of activity upon ZF12 mutation in the tethered repressor assay is in stark contrast to the more modest effects in the simple cotransfection assay (Figures 1C and 3A). In the tethered activation assay the single mutations in ZF1, 3 and 4, and the combined mutation of ZF3 and 4 led to a modest but significant increase in activity while the ZF2 mutation was without effect (Figure 3C, lanes 2–5,7 compared to 9). Only the dual ZF12 mutant showed decreased activity (lane 6) but the effect was modest compared to the loss of repressor activity.

MBNL1 is thought to dimerize through its C terminus [16,28]. However, the crystal structure of ZF34 revealed a

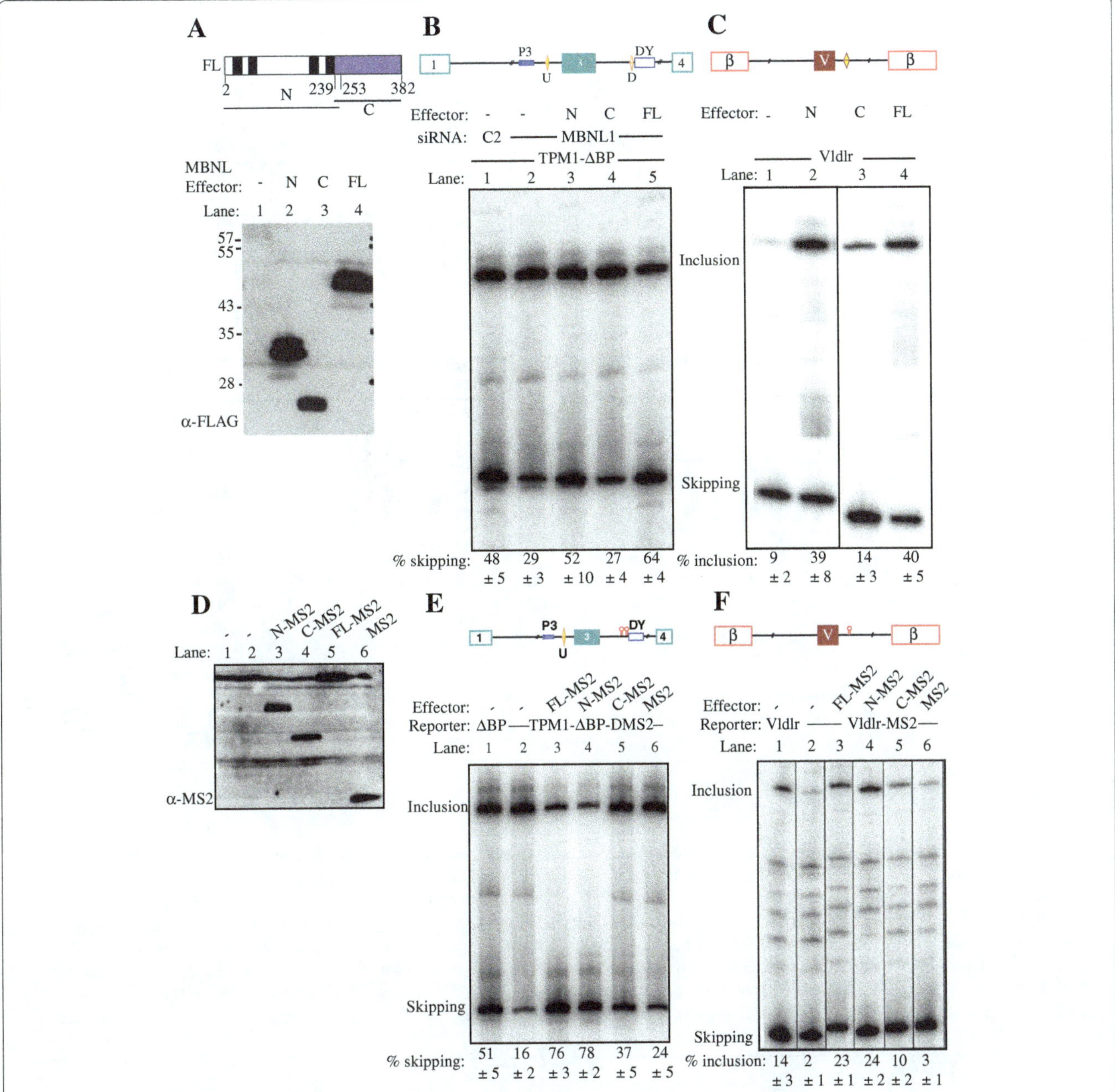

Figure 2 Activity of MBNL1 N and C-terminal domains in splicing repression and activation. A. Schematic of the regions of MBNL1 used in the experiments in this figure (top). Lower panel, anti-FLAG western blot showing expression of MBNL proteins. **B**. The Tpm1 ΔBP minigene was co-transfected with control (C2, lane 1) or MBNL1 targeting siRNAs (lanes 2–5). In addition, full-length MBNL1 (lane 5) or the MBNL1 N- or C-terminal domains (lanes 3–4) were cotransfected along with the MBNL1 siRNA. **C**. The Vldlr minigene was transfected alone (lane 1) or with full-length MBNL1 (lane 4) or the MBNL1 N- or C-terminal domains (lanes 2, 3). **D**. Anti-MS2 western blot of HeLa cells transfected with MS2 alone, lane 6, or MBNL1-MS2 fusion proteins: full length, lane 5; MBNL1 C-terminus, lane 4, MBNL1 N-terminus, lane 3; mock transfections lanes 1, 2. **E**. Schematic of the Tpm1 ΔBP DMS2 minigene in which the downstream MBNL site is replaced by a pair of MS2 hairpins (upper panel). Lower panel, RT-PCR of HeLa cells transfected with Tpm1 ΔBP (lane 1) or Tpm1 ΔBP DMS2 (lanes 2–6). Cotransfections with MS2 alone (lane 6) or the indicated MBNL1-MS2 fusion proteins (lanes 3–5). **F**. Schematic of the Vldlr-MS2 minigene in which the downstream MBNL site is replaced by an MS2 hairpin (upper panel). Lower panel, RT-PCR of HeLa cells transfected with Vldlr (lane 1) or Vldlr-MS2 (lanes 2–6). Cotransfections with MS2 alone (lane 6) or the indicated MBNL1-MS2 fusion proteins (lanes 3–5).

dimerization contact involving the RNA binding face of ZF4 in one subunit, with the reverse face of ZF4 in the other subunit [13]. We tested the effects of individual or combined mutations in Tyrosine 224 (Y224S) and Glutamine 244 (Q244N), which are predicted to impair the potential dimerization contact, but not RNA binding

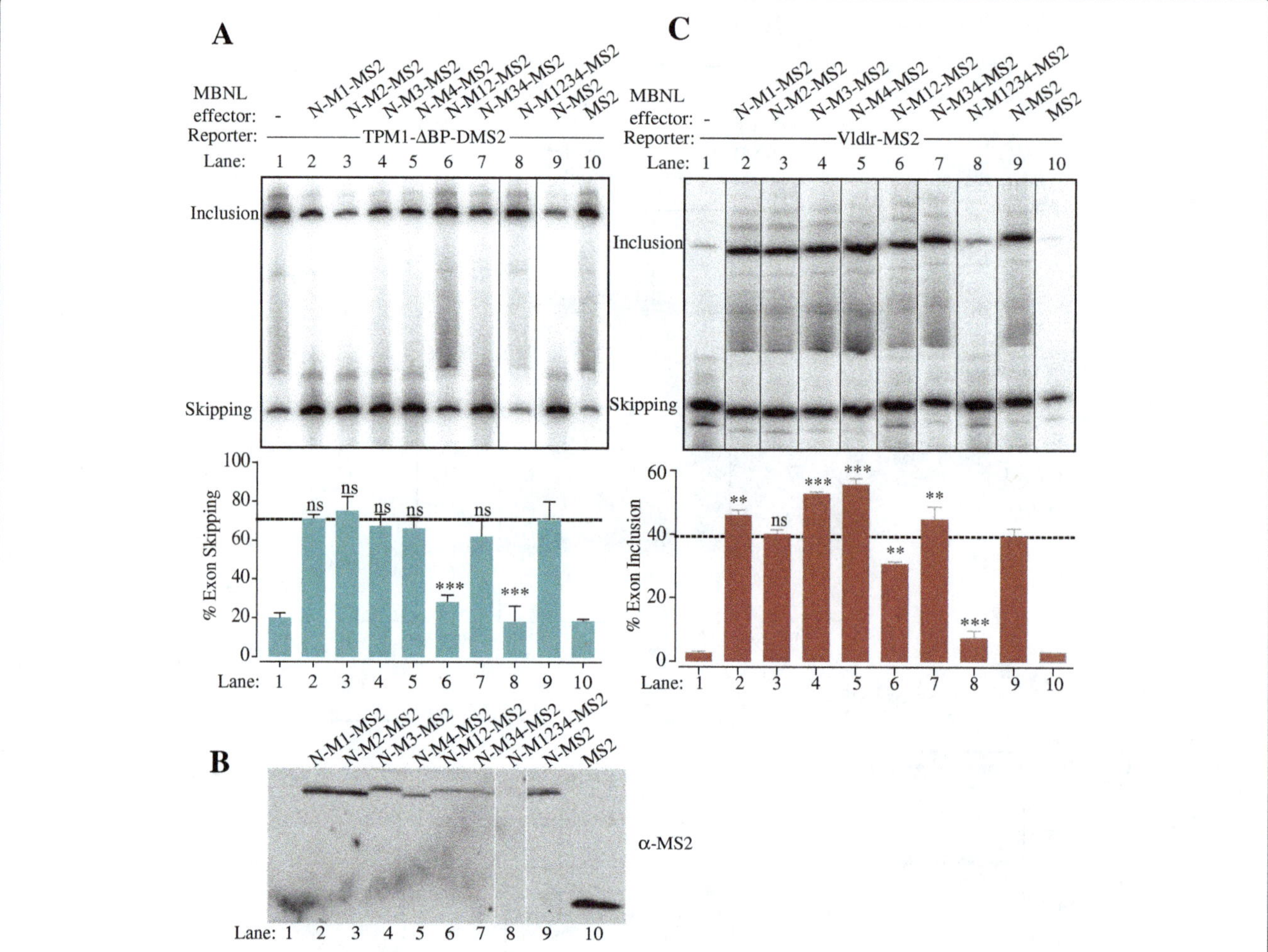

Figure 3 Effects of RNA binding mutations on tethered MBNL1 repressor and activator function. A. The Tpm1 ΔBP DMS2 minigene was transfected alone (lane 1), or co-transfected with MS2 (lane 10), or MS2 fused to the N-terminal (aa 2–253) of MBNL1 (lanes 2–9). The MBNL-MS2 fusion proteins were WT (lane 9) or had the indicated ZF mutations (lanes 2–8). The horizontal dashed lines indicate the activity of the wild type N-MS2 (lane 9). Values significantly different from wild type N-MS2 in panels **A** and **C**: **, $P < 0.01$; *** $P < 0.001$; ns, not significant. Note that although the values for the M1234 mutant in lane 8 of panels **A** and **C** are statistically significant, the lack of protein expression of the M1234 mutant (panel **B**) means that meaningful conclusions cannot be drawn. **B.** Anti-MS2 western blot of MBNL-MS2 fusion proteins used in panels **A** and **C**. **C.** The Vldlr MS2 minigene was transfected alone (lane 1), or co-transfected with MS2 (lane 10), or MS2 fused to the N-terminal (aa 2–253) of MBNL1 (lanes 2–9). The MBNL-MS2 fusion proteins were WT (lane 9) or had the indicated ZF mutations (lanes 2–8). The horizontal dashed lines indicate the activity of the wild type N-MS2 (lane 9).

(Figure 4A). These mutations had no effect upon the tethered repressor (4C) or activator (Figure 4D) activities of MBNL-N-MS2, or on the direct activation of Vldlr by full length MBNL1 (Figure 4B). These results suggest that the observed crystal contact between MBNL1 subunits is not important for MBNL1 function.

MBNL1 binding to RNA species from MBNL-regulated exons

Having investigated the role of the MBNL1 ZF domains in splicing repression and activation, we next tested the binding of MBNL1-N to RNAs containing the MBNL binding elements of Vldlr and Tpm1 by electrophoretic mobility shift assay (Figure 5). We compared binding of WT MBNL1-N with the mutants in ZF12 (M12) and ZF34 (M34). WT MBNL1 bound to the Vldlr and upstream Tpm1 elements, Tpm1 URE, with Kd in the 0.5 – 1 nM range, while binding to the downstream Tpm1 element Tpm1 Dugc, was approximately 10-fold lower affinity (Figure 5A, K_d 25–50 nM). With the Vldlr RNA a second binding event was also observed with a much lower affinity; we observed no additional binding events to either of the Tpm1 elements, even though their length is sufficient to accommodate multiple binding sites [15]. Mutation of ZF34 reduced the affinity of binding to all three RNAs by about ~20-fold (Figure 5B). In contrast, the effects of mutations in ZF12 were far more drastic; no

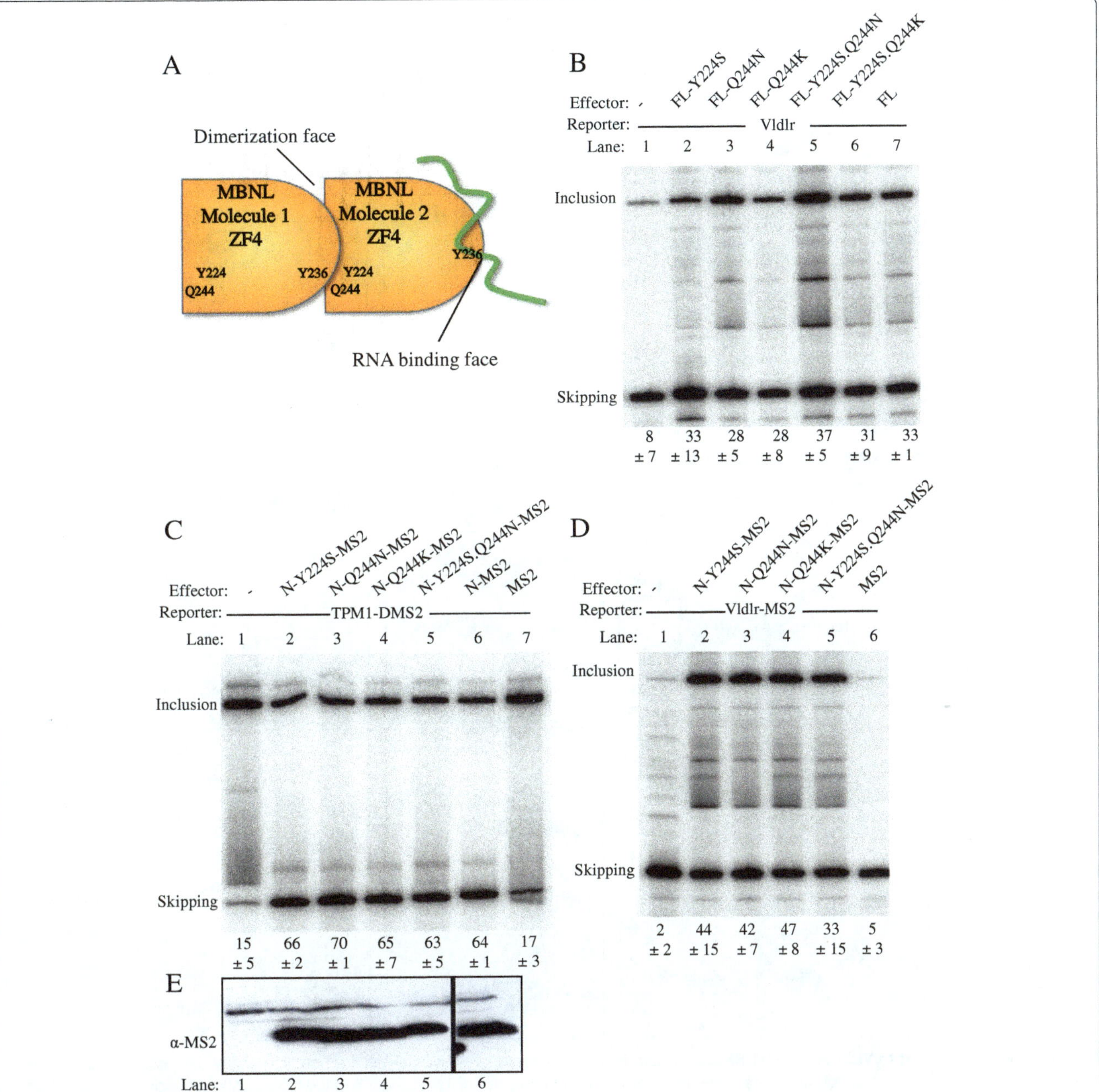

Figure 4 Mutations in a potential dimerization contact in ZF4 have no effect. A. Schematic of the potential ZF4-ZF4 dimerization contact identified in MBNL1 crystal structures. The RNA binding face of one ZF4 unit interacts with the opposite face of the second ZF4 unit. Mutation of Y224 and Q244 is predicted to interfere with the potential protein-protein interaction, but not with RNA binding by ZF4. **B**. The Vldlr minigene was transfected alone (lane 1), or with MBNL1 FL (lanes 2–7). The MBNL1 was WT (lane 7) or carried the indicated ZF4 mutations. **C**. The Tpm1 ΔBP DMS2 minigene was transfected alone (lane 1), or co-transfected with MS2 (lane 7), or MS2 fused to the N-terminal (aa 2–253) of MBNL1 (lanes 2–6). The MBNL-MS2 fusion proteins were WT (lane 6) or had the indicated ZF4 mutations (lanes 2–5). **D**. The Vldlr MS2 minigene was transfected alone (lane 1), or co-transfected with MS2 (lane 6), or MS2 fused to the N-terminal (aa 2–253) of MBNL1 (lanes 2–5). The MBNL-MS2 fusion proteins had the indicated ZF mutations (lanes 2–5). **E**. Anti-MS2 western blot of MBNL1-MS2 fusion proteins corresponding to lanes 1–6 of panel **C**.

stable complexes were observed on any of the RNAs, even when up to 2 μM MBNL protein was used (Figure 5C). Thus, the N-terminal ZF12 domains are more important for both binding to Tpm1 and Vldlr RNAs, as well as for tethered activity.

MS2 tethering of MBNL1 truncations

To analyse further the roles of the pairs of tandem zinc fingers we tested a series of deletion mutations based on MBNL-N-MS2. These included a C-terminal deletion series (previously tested on Tpm1 [22]), a natural deletion

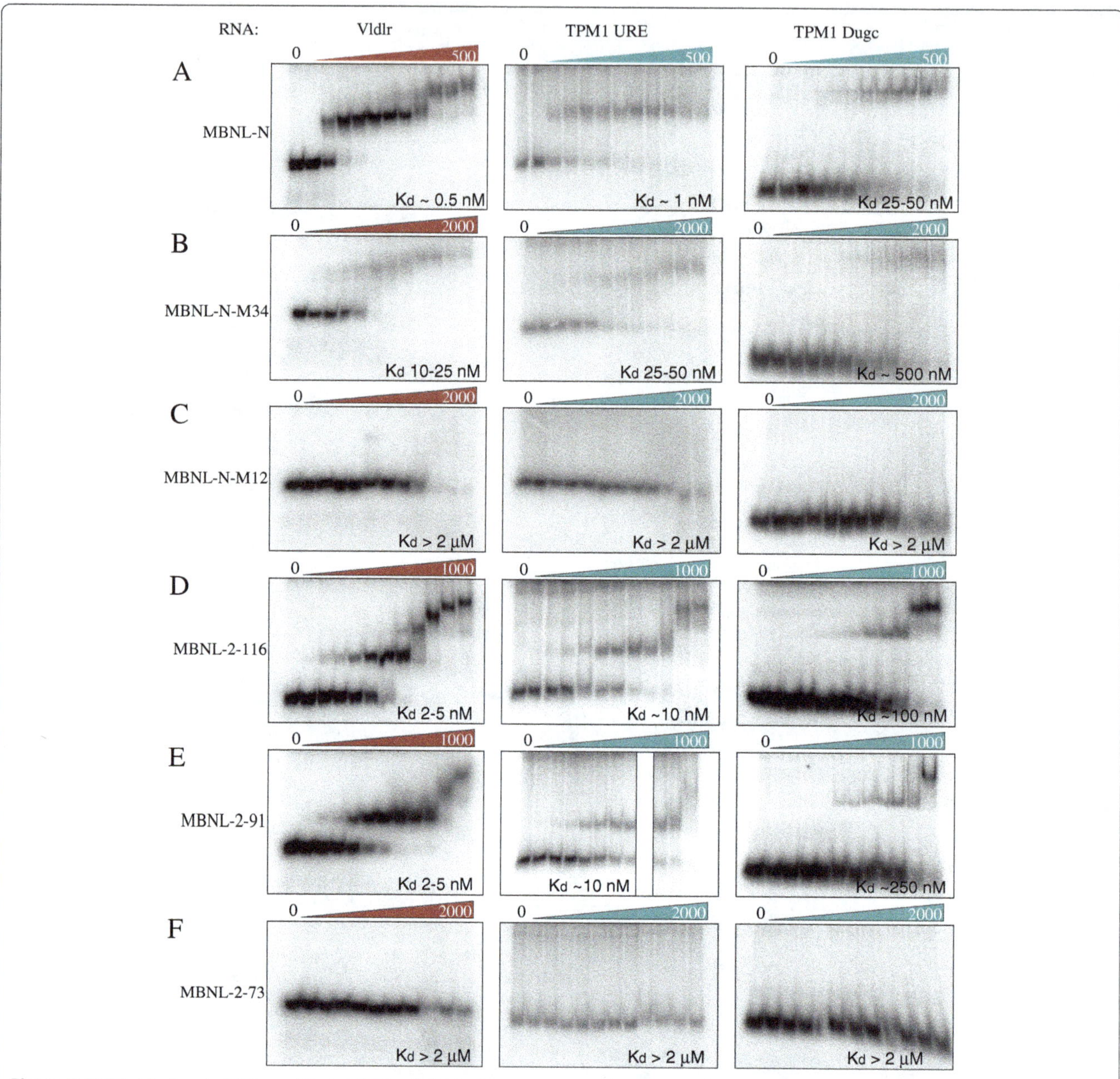

Figure 5 MBNL binding to Vldlr and Tpm1 RNAs. RNA binding was assessed by native gel electrophoretic mobility shift assay. RNAs are the MBNL-responsive element from Vldlr (left panels), the upstream MBNL binding site of Tpm1 exon 3 (middle panels), and the downstream MBNL binding site of Tpm1 exon 3 (right panels). Recombinant proteins used were: **A**, MBNL-N, the wild-type MBNL construct comprising amino acids 2–253 and containing all four zinc fingers, **B**, MBNL-N-M34, with ZF34 mutated, **C**, MBNL-N-M12 with ZF12 mutated, **D**, MBNL-2-116 **E**, MBNL-2-91 **F**, MBNL-2-72. Increasing protein concentrations are indicated by the wedges above each panel. Protein concentrations were 0, 0.1, 0.5, 1, 2, 5, 10, 25, 50, 100, 250, 500 nM for panel A, 0, 0.1, 0.5, 1, 2, 5, 10, 25, 50, 100, 250, 1000 nM for panels D and E, and 0, 1, 2, 5, 10, 25, 50, 100, 250, 500, 1000, 2000 nM for panels B, C and F. Estimated K_d's are indicated in the lower right corner of each panel.

variant lacking the C-terminal half of the linker (Δ116-183), an N-terminal deletion series, and the linker alone. The linker sequence is predicted to be unstructured, but parts of it are highly conserved and have been shown previously to have a role in MBNL activities [17,18,22]. We expressed these proteins as MS2-fusions (Figure 6B) and analysed their activity when recruited to either the downstream Tpm1 (repressed, Figure 6C) or Vldlr (activated, Figure 6D) sites.

When recruited downstream of the MBNL-repressed Tpm1 exon 3 (the Tpm1-ΔbpDMS2 minigene) zinc fingers 3 and 4 and the C-terminal part of the linker could be removed individually or in combination with no effect (Figure 6C, lanes 4–6). C-terminal truncations beyond

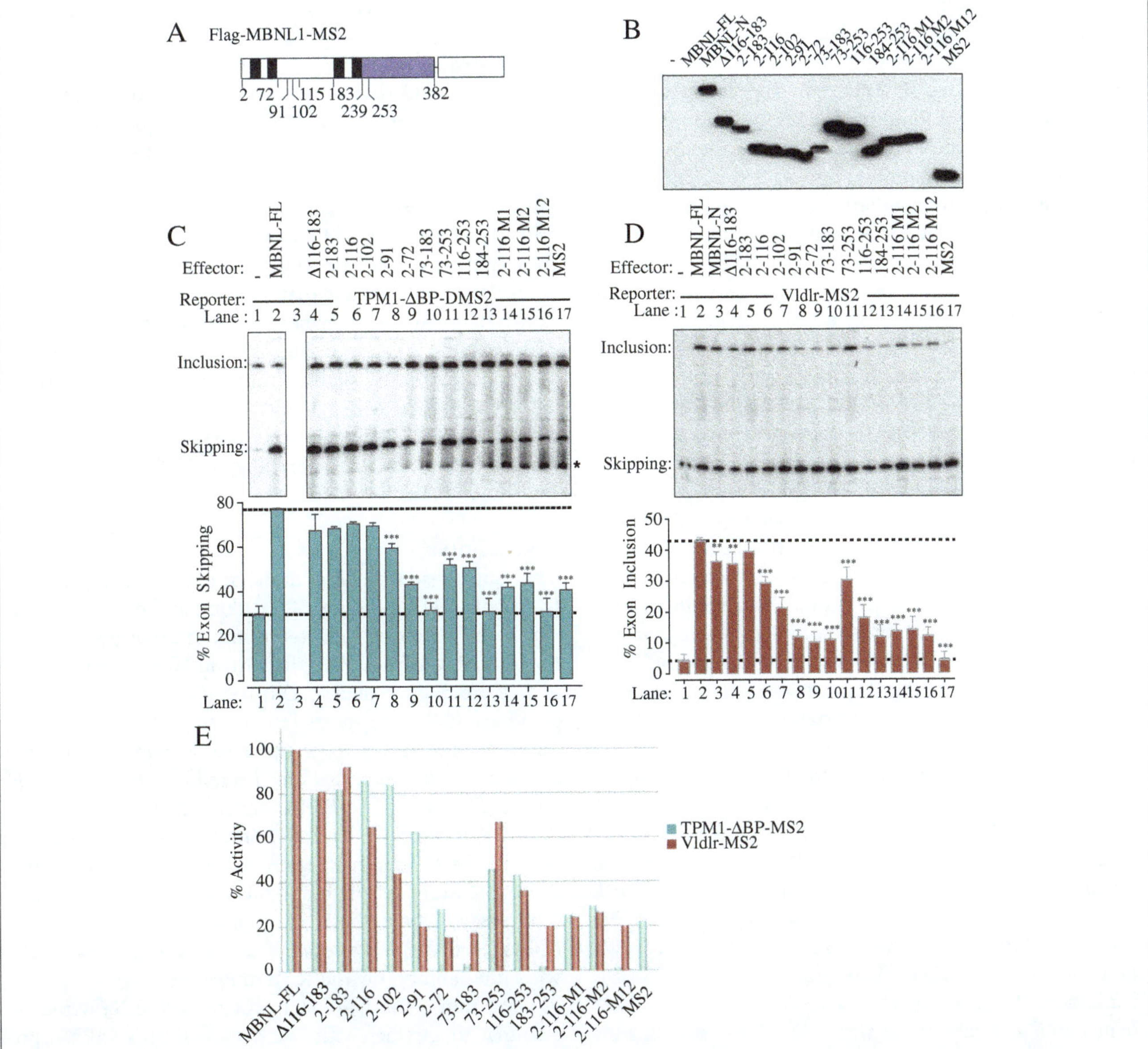

Figure 6 Comparison of MBNL1 deletions and point mutations upon tethered repression and activation. A. Schematic of the full-length MBNL1-MS2 construct with boundaries of various deletions indicated. **B**. Anti-MS2 western blot of proteins used in subsequent panels. **C**. The Tpm1 ΔBP DMS2 minigene was transfected alone (lane 1), or co-transfected with MS2 (lane 17), MS2 fused to full-length MBNL1 (lane 2), or the various deletion constructs indicated by the amino acid coordinates. Δ116-183 (lane 4) is a natural MBNL1 isoform resulting from exon skipping. Lanes 14–16 show the effects of the ZF1 and 2 mutations in the context of the 2-116-MS2 deletion mutant. In the experiment shown, the 2-253-MS2 was not expressed (lane 3); in other experiments (e.g. Figure 2E) its activity was similar to FL-MBNL1. The asterisked band is an artefactual band that does not appear in most experiments. Values significantly different from wild type MBNL1-FL-MS2 in panels C and D: **, $P < 0.01$; *** $P < 0.001$; ns, not significant. **D**. The Vldlr-MS2 minigene was co-transfected with the same effectors as panel C. **E**. Comparison of effects of mutations in panels C and D. The histogram indicates "% activity" for each of the MBNL1 constructs with the Tpm1 (green bars) and Vldlr (red bars) substrates. The values shown are relative to the activity of FL-MBNL1-MS2. 100% activity is defined by the difference in exon skipping/inclusion in the presence of FL-MBNL1-MS2 (upper horizontal line in histograms of panels C and D) and in the absence of co-transfection (lower horizontal lines). The comparison shows more rapid decline of activation than repression with C-terminal deletions into the ZF23 linker.

amino acid 116 led to diminished activity (lanes 6–9). Complete removal of the linker sequence and an alpha-helix of zinc finger 2 leaving only the first two zinc fingers results in an inactive protein (lane 9). Despite the importance of the linker region, when recruited alone it had no activity above MS2 alone (lane 10, 17). Zinc fingers 3 and 4 along with the complete preceding linker region, or with just the C-terminal part of the linker, were partially active (lanes 11,12). However, this activity required the linker sequence as ZF34 alone were inactive (lane 13). These data show that the N-terminal part of the protein comprising amino acids 2–102, encompassing the first two zinc fingers plus a third of the linker sequence, constitutes a minimal repressor domain. Introduction of RNA binding mutations into ZF1 or 2 individually drastically reduced activity of the 2–116 repressor domain, while combined mutation of ZF 1 and 2 abolished activity (compare lanes 14–16 with lane 6).

A similar, but not identical, response to the mutations was seen in the Vldlr context (Figure 6D). In this case, more of the linker sequence was required for full activity with progressively diminishing activity upon C-terminal deletion into the linker (Figure 6D, lanes 5–9, Figure 6E). Deletion of the second pair of zinc fingers (lane 5) or internal deletion of the C-terminal part of the linker (lane 4) had no effect on activity. As for repressor activity, the linker sequence alone was inactive (lane 10), but in combination with the second pair of zinc fingers the fusion protein retained substantial activity (lane 11), albeit less than the first pair (lane 5). This activity was reduced further by removal of the N-terminal part of the linker (lane 12) and abolished when only ZF34 remained (lane 13). Thus splicing activation is more dependent than repression upon the full linker sequence (Figure 6E). Although the activity of the 2-116 construct was already diminished, we tested the importance of RNA binding. Abrogating the RNA binding capacity of the 116 construct led to a severe, albeit not total reduction in activity (Figure 6D, lanes 14–16 compared to 6).

We analysed RNA binding of some of the C-terminal deletion fragments (Figure 5D-F). The miminal repressor domain MBNL1-2-116 bound to the Vldlr and Tpm1 RNAs with affinity reduced compared to the complete N-terminus but actually higher than the N-terminus with point mutations in ZF34 (Figure 5D compared to 5A,B). In addition, the 2–116 protein showed additional subsequent binding events on all three RNAs, consistent with the fact that it has only 2 ZFs that can contact the RNA. The 2–91 protein, which showed almost complete loss of tethered activation activity (Figure 5D lanes 6–8) bound to each RNA with affinity similar to 2–116 (Figure 5E), emphasizing that RNA binding is necessary but not sufficient for activity. Finally, the 2–72 protein, which lacks an experimentally observed C-terminal extension to the ZF2 domain [12,13] failed to bind RNA at any concentration, confirming the importance of the additional α-helix (Figure 5F).

PTB associates with Vldlr RNA but does not regulate its splicing

MBNL and PTB act as co-repressors of Tpm1 splicing [22]. Pull-downs with biotinylated Vldlr RNA indicated that PTB was one of the major binding proteins in HeLa nuclear extract (data not shown). We therefore asked whether PTB acted synergistically or antagonistically with MBNL1 in the regulation of Vldlr splicing. As shown earlier, overexpression of MBNL1 promoted skipping of Tpm1 exon 3 but inclusion of the Vldlr exon (Figure 7B, C lanes 1,2). Overexpression of PTB had little effect on Tpm1 splicing (Figure 7C lane 3), as PTB is not limiting in HeLa cells [29]. However, PTB/nPTB knockdown led to decreased exon skipping (lanes 1 and 4). In contrast, Vldlr splicing was unresponsive to either overexpression or knockdown of PTB (Figure 7C lanes 1,3,4), suggesting that binding of PTB to Vldlr is non-functional. Furthermore, the activating effect of MBNL was not reduced upon PTB knockdown, and actually appeared to be slightly increased (Figure 7C lane 5). Therefore, while MBNL1 and PTB cooperate to repress Tpm1 splicing, MBNL1 acts independently to activate Vldlr splicing, and PTB binding to Vldlr appears to be non-functional.

Discussion

The data presented here, drawing upon point mutations to impair RNA binding of ZF domains, deletion mutations, and artificial MS2 tethering, provide insights into the domains of MBNL1 that are involved in activation and repression of splicing. Our results are complementary to other published reports [16-18], and taken together the different studies converge upon some common themes. Among our key findings are the following. First, the N-terminal region of MBNL1 encompassing the four ZF domains is nearly fully active in most assays (Figure 2). Second the C-terminal region is inactive in conventional overexpression assays, but retains some activity in tethered assays where the MS2 domain recruits it to target RNAs (Figure 2). This residual activity might be associated with the ability of the C-terminal region to mediate dimerization [16,28], which might allow tethered C-terminal to interact with intact endogenous MBNL proteins, perhaps promoting their recruitment to the RNA. Third, MBNL1 is remarkably tolerant of RNA binding mutations to individual ZF domains (Figures 1, 3), consistent with previous results [18]. Fourth, pair-wise inactivation of ZF12 was in most cases more deleterious than inactivation of ZF34 (Figure 3). This is consistent with the effects of deletions that remove ZF12 or ZF34 (Figure 6, constructs 72–253 and 2–183 respectively), and with the effects of the didomain point mutations upon RNA binding (Figure 5A-C). Fifth, both repression of the Tpm1 exon and activation of the Vldlr exon required not just an intact ZF didomain, but also the linker sequence connecting ZF2 and 3 (Figure 6). Finally, there were some differences in the responses of the repressed Tpm1 and the activated Vldlr exons to different MBNL1 mutations. Activation of the Vldlr exon was more sensitive than repression of Tpm1 to deletions of the linker between ZF2 and

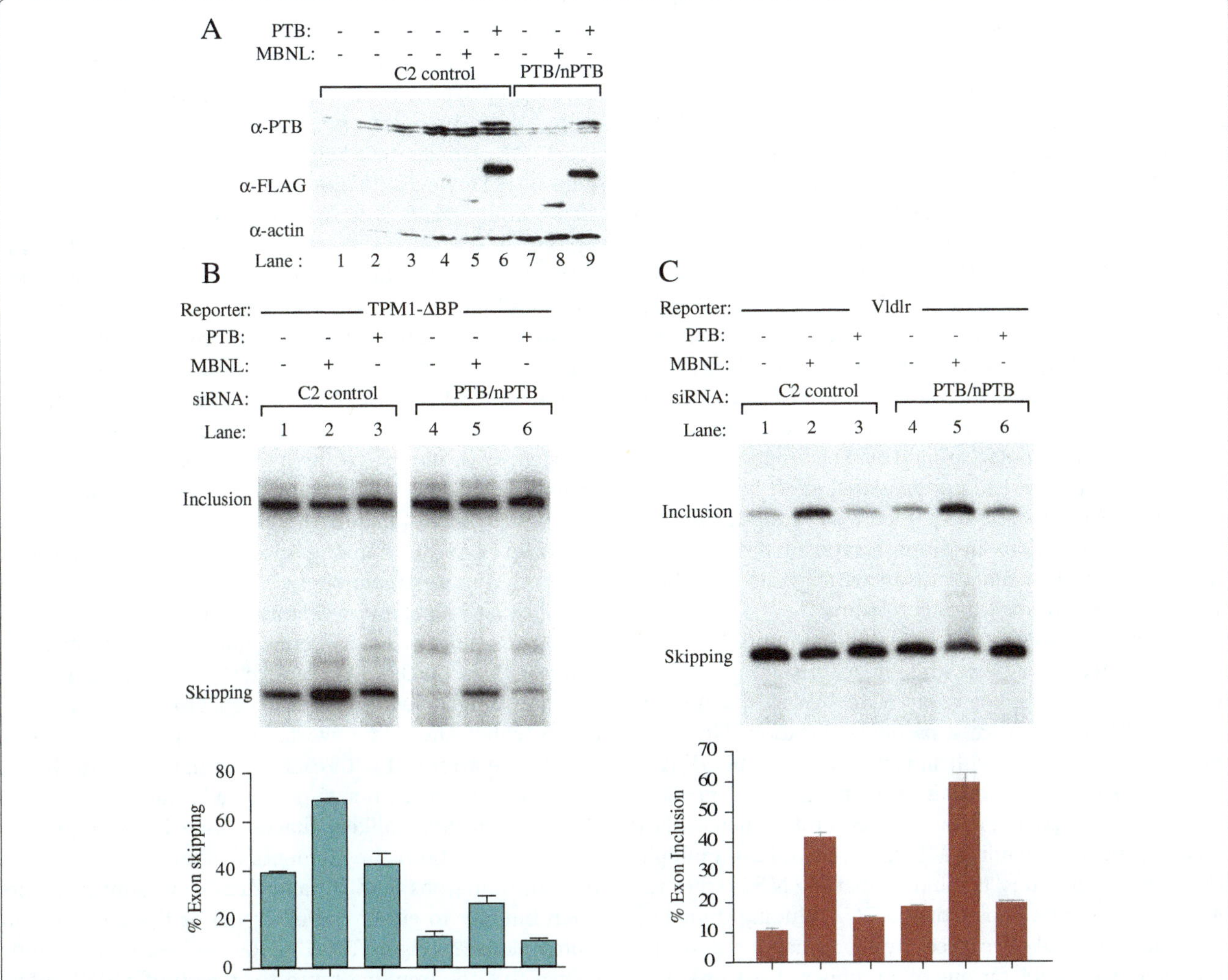

Figure 7 PTB co-regulates Tpm1 but not Vldlr splicing. A. Western blots with anti-PTB (upper), anti-FLAG (middle) and anti-actin (lower panel). Cells were treated with control C2 siRNA (lanes 1–6) or PTB/nPTB siRNAs (lanes 7–9). Lanes 4–1 show successive 2-fold dilutions of the C2 control to allow assessment of knockdown. In lanes 6 and 9, FLAG-PTB was overexpressed; in lanes 5 and 8, FLAG-MBNL1 was overexpressed. **B**. The Tpm1 ΔBP minigene was cotransfected with control C2 (lanes 1–3) or PTB/nPTB siRNAs (lanes 4–6). In addition, FLAG-PTB (lanes 3, 6) or FLAG-MBNL1 (lanes 2, 5) were also cotransfected. **C**. The Vldlr minigene was cotransfected with control C2 (lanes 1–3) or PTB/nPTB siRNAs (lanes 4–6). In addition, FLAG-PTB (lanes 3, 6) or FLAG-MBNL1 (lanes 2, 5) were also cotransfected.

3 (Figure 6), consistent with previous reports comparing activated and repressed exons [16,17]. The minimal repressor domain for MBNL1 when tethered encompasses zinc fingers 1 and 2, with a region of further linker sequence up to amino acid 102. The minimal tethered activation domain comprises ZF1 and 2 and the full linker sequence to amino acid 183. In contrast to the effects of truncation mutations, effects of the ZF12 RNA binding mutations were more pronounced upon Tpm1 than Vldlr. The ZF12 mutation led to complete loss of activity upon the Tpm1 exon, whereas with the Vldlr minigene this mutation led to only a slight reduction in activity (Figure 3). Likewise, within the context of the effector fragment 2-116-MS2, mutation of ZF12 abolished repressor activity, while retaining ~30% of activator activity (Figure 6). The differential effects of MBNL1 mutations upon the Tpm1 and Vldlr exons are interesting. However, they cannot be generalized for all activated or repressed targets of MBNL proteins. Indeed, a systematic analysis of combined MBNL1 ZF mutants with a panel of 6 MBNL regulated events, showed that the relationship between ZF mutations and effects upon activity upon different splicing substrates is quite complex, with at least two classes of target, each encompassing repressed and activated targets [18]. In agreement with our conclusions the same study showed that MBNL activity upon Vldlr did not correlate well with its RNA binding ability [18].

A surprising feature of our results is that the ZF RNA binding mutations had a greater effect in the MS2 tethering assays (Figure 3) than in the untethered assays

(Figure 1). One would anticipate that the tethered proteins would be less sensitive to mutations in their RNA binding domains, since the MS2 domain should bypass at least some of the RNA binding functions of the intact protein. However, the C-terminal domain, which mediates dimerization [16,28] and on its own has some activity in the tethering assays (Figure 2), was missing in most of the subsequent MS2-tethering assays (Figures 3, 6), which might account for the greater sensitivity to the ZF mutations. In addition to the RNA binding mutations of the ZF domains, we also tested a set of mutations in ZF4 designed to impair intra-molecular contacts formed between MBNL molecules in crystal structures. Although previous studies have implicated the variably spliced C-terminal region of MBNL in dimerization [16,28], it was possible that the ZF4 mediated contacts might also be functional. However, these mutations had no functional effect in a number of assays of MBNL1 activities (Figure 4). Although we did not test the mutations in direct assays for dimerization, the lack of effect in functional assays suggests that these contacts are not physiologically relevant.

The deleterious effects of RNA binding mutations in a tethered function assay is initially surprising, given that the starting point of the assay is to bypass the normal mode of RNA binding at a particular location. However, this has been observed with similar studies of RbFOX [23] and PTB [30], and provides insights into the possible mechanisms of splicing regulation. Given that many RNA binding proteins, including MBNL proteins, have multiple RNA binding domains, it could be that the MS2 tether replaces the role of a subset of the RNA binding domains, and that functional effects require the remaining domains to interact with RNA for one of a number of reasons. The protein might need to bind to more than one site in the target pre-mRNA to be functional, perhaps forming an RNA architecture conducive to regulated exon skipping or inclusion; the protein might have to interact with another RNA; or the protein might have to interact with RNA in order to interact with important partner proteins. Taken in order, we already know that there are at least two major MBNL1 binding sites flanking Tpm1 exon 3 [22,31]. We also identified additional YGCY motifs downstream of the Vldlr exon that mediate MBNL1 activity (data not shown). Alternatively, the additional RNA interactions could be with a distinct RNA species such as U1 snRNA; inhibition of the N1 exon of *CSRC* involves a PTB-U1 snRNA interaction [32]. Finally, RNA binding might be important to induce a conformation that facilitates interaction with co-regulatory proteins [22]. These different explanations are not all mutually exclusive; for example, a protein-RNA interaction might promote a necessary RNP architecture as well as inducing a conformation change promoting necessary protein-protein interactions.

Analysis of the relationship between position of binding on RNA substrates and the mode of action as a repressor or activator indicates that in general MBNL binding upstream on an exon is associated with repression while downstream binding leads to activation [10,33,34], similar to a number of other regulatory proteins [4]. However, in the case of repression there are also binding peaks downstream of the exon as well, suggesting a sub-set of events which are regulated negatively by MBNL sites flanking the exon. This suggests the possibility of two discrete types of MBNL-repressed event, which might operate in a mechanistically distinct manner or might share some common mechanistic elements. In the first case, MBNL binding sites immediately upstream of the regulated exon might be sufficient to interfere with binding or activity of constitutive splicing factors such as U2AF, as has been suggested in the cTNT transcript [35,36]. In the case of exons such as Tpm1 exon 3 the flanking sites might be necessary in order for the upstream sites to effectively interfere with 3′ splice site recognition factors, perhaps by cooperative binding of oligomers and consequent looping of intervening RNA [37,38]. Alternatively the flanking regulatory sites might need to act in a concerted way on the 3′ and 5′ splice sites. The two tandem zinc finger arrays of MBNL are arranged with each zinc finger 'back-to-back', which would cause an anti-parallel alignment of a bound RNA. It appears unlikely that a single MBNL1 protein could bind to both the elements flanking Tpm1 exon 3. Indeed, mutations in ZF12 and ZF34 have similar effects upon binding to either the upstream or the downstream Tpm1 element (Figure 5A-C). However, we have recently shown that the minimal repressor region of MBNL interacts with PTB protein, in an RNA-dependent manner [22], and there are PTB sites flanking exon 3, with two to three molecules binding either side [39]. Moreover MBNL has been shown to interact with itself [16,28]. Taken together this suggests a complex forming across the exon, with homotypic and heterotypic interactions between MBNL and PTB molecules acting to stabilise a looped structure which promotes exon skipping. Analysis of proteins binding to the Vldlr substrate indicated strong interaction of PTB, suggesting that it might act as a coactivator. However, overexpression and knockdown experiments clearly showed that PTB played no role in regulating Vldlr splicing (Figure 7). Indeed, we also found that PTB-MS2 had no activity when tethered downstream of the Vldlr exon, despite the fact that it can activate its own target exons from this location [40]. This clearly indicates that different activators have distinct molecular targets, even when binding at similar locations. An important future line of work will be to identify the molecular targets of the minimal MBNL1 activator domain.

Methods

Constructs

The *Tpm1* minigene reporters and Δbp mutation have been described previously [22,27,31]. The Vldlr wildtype minigene was a kind gift from Prof. Manny Ares (University of California Santa Cruz) [10]. This minigene was mutagenized to introduce a Pst1 site then an MS2 hairpin was inserted using the following oligo: 5'-gAGGATCACCctgca-3'. Expression plasmids for MBNL1 N and C terminal truncations have been described previously [22]. Mutagenesis was performed using standard protocols and the following primers: ZnF1: 5-CACGGAATGTAAAgcTGCACATCCT TCG-3, ZnF2: 5-GGAGAACTGCAAAgcTCTTCATCCA CC-3, ZnF3: 5-GAAAATGATTGTCGGgcTGCTCATCC TGC-3, ZnF4: 5-GGAAAAGTGCAAAgcCTTTCATCCC CC-3. MBNL1 truncations were cloned by inserting the coding sequence for the relevant portion of MBNL1 into the AvrII and MluI sites in the pCIMS2-NLS-FLAG vector [20,41]. Plasmids for bacterial expression were cloned by insertion of appropriate coding sequence into the pGEX-4-T3 vector. Sequences for expression of Tpm1 or Vldlr RNA were cloned into pGEM-4Z plasmids (Promega), and transcribed usingT7 polymerase. Sequences used were, Vldlr: GGGAGACAAGCTTTGCAAACTGT TAATCTCAACTAACTGCCGCTTAAATAATTAGTGC AGCTTTTAACTACTGGTTCTGTCCCAACTGGCTA CTTGTGCCTAAAGCCCAAAGAATT, Dugc: GGGA GACAAGCTTGAGCTGGATGCCGCCTCTGCTGCT GC, URE: gggagacaagcttaaGTCTACGCACCCTCAAcc CGCACCTTGCGGGATCACGCTGCCTGCTGCACC CCACCCCCTTCCCCCTTCCTTCCCCCCACCCCCG TACTCCACTGCCAACTCCCAG.

Cell culture

Cells were transfected a day after splitting to 10^5 – 2 × 10^5 cells per well in a 6 well plate. Transfections were performed using 400 ng effector construct unless otherwise stated, with 200 ng reporter, made up to 1 μg with empty pGEM4Z vector as necessary. Per well 1μg of DNA, 100 μl Optimem and 2 μl Lipofectamine (Invitrogen 18324–012) was used. Lipofectamine-DNA mix was incubated for 30min at room temperature, then diluted to 1ml in Opti-MEM-1 and applied to cell monolayer previously washed with PBS. Treated HeLa cells were incubated for 5 hr at 37°C, and then the transfection mix was replaced with 2 ml Dulbecco's Modified Eagles Medium (DMEM) supplemented with Glutamax & 10% fetal bovine serum (FBS). Cells were then incubated for a further 48 hours, then RNA and protein was harvested from the cultures using Trizol reagent (invitrogen) or boiled SDS loading buffer respectively.

For MBNL1 knockdown, the following target sequence was used: 5'-AACACGGAAUGUAAAUUUGCA-3' [42]. HeLa cells were split to a density of 2 × 10^5 in 1.7 ml DMEM +10% FBS medium in 6 well plates, and incubated at 37°C for 24 hours. Each well was treated with 10 nM siRNA (THH2 siRNA for MBNL1 knockdown or control C2 siRNA) and 15 μl Oligofectamine (Invitrogen 12252–011), diluted in 500 μl Optimem and incubated prior at room temperature for 20 minutes. Cells were then incubated for 24 hours at 37°C. After 24 hour incubation DNA transfections were performed as above using lipofectamine or lipofectamine 2000 reagent. Cells were incubated for 5 hours at 37°C, then the medium on them replaced with 1.5ml DMEM + 10% FBS. To each well 10 nM siRNA and 3 μl Lipofectamine 2000 reagent in 500 μl Optimem was added, which had been pre-incubated for 20 minutes at room temperature. Cells were incubated for a further 48 hours, then harvested.

RNA and protein analysis of transfections

For RNA analysis, cells in 6 well plates were washed using twice with 2 ml PBS, then to each well 1 ml trireagent (Sigma-Aldrich) was added, and purified according to the manufacturers' protocols. Samples were DNAse I treated using 2 units of rDNAse (Ambion) in 10 mM Tris pH 7.5, 2.5 mM $MgCl_2$, 0.5 mM $CaCl_2$ for 30 minutes – 1 hour at 37°C, phenol extracted and ethanol precipitated. PCR analysis used the following primers for Vldlr minigenes:

V4rt - 5'-GTGGCAAAGGTGCCCTTGAG-3' - (rt primer)
V1 - 5'-ACGTGGATGAAGTTGGTGGT-3' - (5' primer)
V3 - 5'-GGCACCGAGCACTTTCTTGC-3' - (3' primer)

and the following primers for Tpm1 minigenes:

SV3'RT: 5'-GCAAACTCAGCCACAGGT-3' - (rt primer)
SV5'2: 5'-GGAGGCCTAGGCTTTTGCAAAAAG-3' - (5' primer)
SV3'1: 5'-ACTCACTGCGTTCCAGGCAATGCT-3' - (3' primer)

Reverse transcriptions were performed using 2–3 μg of total RNA, and 100 ng of RT primer, in 50 mM Tris pH 6.3, 40 mM KCl, 8 mM MgCl2, 2 mM DTT. Samples were heated for 15 minutes at 55°C, then cooled to 42°C, and 2 μl 10 mM dNTP and 1 μl AMV-RT (Promega) enzyme added, and incubated at 42°C for 60 minutes. For the PCR reaction, the 3' primer was 5' end labeled with [^{32}P]-ATP,. Oligo primer (4 pmoles per PCR reaction) was incubated at 37°C for 60 minutes in 50 mM Tris 10 mM $MgCl_2$ T4 polynucleotide kinase enzyme (NEB) and 0.1 μl [α-^{32}P]-UTP per PCR reaction. After incubation the solution was phenol extracted, and purified on a G-50 spin column (GE Healthcare). The labelled oligo was made up to concentration of 1 pmole/μl. 2 μl of RT reaction was taken into fresh eppendorf in buffer (50 μM KCl, 10 μM

Tris pH 8.3, 1.5 mM MgCl2, 0.001% w/v gelatin) and 25 pmole of the reverse primer. Samples were heated to 92°C for 3 minutes, then cooled to 80°C, and 0.25 µl Taq polymerase (Roche) and 10 pmol ^{32}P-labeled probe added. The samples were then cycled for 30 cycles of 94°C for 30 seconds, 62°C for 30 seconds and 72°C for 60 seconds. RT-PCR products were analysed on denaturing 4% PAGE gels, using Sequagel (National Diagnostics EC-833) system. Samples were diluted in formamide loading buffer, heated to 90°C for 5 minutes, then loaded. Gels were run for 100 minutes at constant 38 W, the gel was dried and exposed on phosphorimager casette (Molecular Dynamics). The results were quantified using ImageQuant Software (GE Healthcare) and analysed using Excel (Microsoft) and Graphpad 5 (Prism Software). Statistical judgements were made using either students t-tests or, where multiple combinations tested, a post-ANOVA Tukey test, which is a variation of the students t-test which aims to eliminate type 2 errors stemming from multiple comparisons without Bonferroni corrections. Statistical significance is indicated by: ns, not significant; * $P < 0.05$; ** $P < 0.01$; ***, $P < 0.001$.

For protein analysis, the cell monolayer in 6 well plates was washed twice with 2 ml PBS, then directly to each well 150 µl of hot SDS buffer (pre-heated to 100°C for 5 minutes) was added. The cells were scrapped using upturned P-1000 tips, extracted into eppendorf tubes, and frozen on dry ice. Samples were heated again to 100°C for 5 minutes, separated using SDS-PAGE, analysed using standard western blotting techniques, and imaged using standard ECL techniques. For western blot analysis primary antibodies were in house anti-rabbit MS2 or anti-FLAG from Sigma (F1804). Protein loading was checked by Ponceau staining and, in some cases, by re-probing with anti-actin antibodies.

Recombinant protein expression

Recombinant MBNL1 protein was expressed and purified from *E.coli* BL21 cells. 400 ml cultures were induced at $OD_{600} = 0.5$ by the addition of 1 mM IPTG, and grown for 3 hours shaking at 225 rpm. The cultures were then pelleted, washed in MTPBS (150 mM NaCl, 16 mM Na_2HPO_4, 4 mM NaH_2PO_4, pH 7.3) and lysed using a French Press (Stansted Fluid Power) according to the manufacturer's instructions. The homogenised samples were centrifuged at 7741 rcf, 4°C, for 10 minutes. Samples were then purified using GST Sepharose 4B beads (GE Healthcare) according to manufacturers protocols. Briefly - the beads were pre-washed with 5–10 volumes of water, MTPBS and MTPBS + 1% Triton-X100. To the homogenised sample Triton-X100 added to concentration of 1%. This bacterial homogenate was then incubated with the GST beads at 4°C, for 1 hour. The beads were washed 4 times with 2.5 volumes of MPTBS + 1% Triton-X100, then loaded into a disposable biorad column at 4°C. The recombinant proteins were eluted from the column using 3×800 µl of 25 mM reduced glutathione in 100 mM HEPES (pH 8.9), followed by 3×800 µl of 50 mM reduced glutathione in 100 mM HEPES. All fractions of interest were pooled, and dialysed in 1.8 litres of Dignam Buffer E overnight at 4°C using Slide-A-Lyzer Dialysis Cassettes (Thermo Scientific), according to manufacturer's protocols. The concentration of the recombinant proteins was estimated with reference to BSA standards.

Electrophoretic-mobility shift and UV crosslinking assay

High specific activity [α-P^{32}] UTP labelled RNAs were made using standard protocols with either SP6 or T7 polymerase. Binding reactions were set up in microtitre plates (Corning) pre-lined with BSA. Mobility shift assays had a total 5 µl reaction volume, with 10 fmol RNA, 20 µg/ml rRNA, 10 mM HEPES pH 7.9, 10 µM ZnCl2, 3 mM MgCl2, 5% Glycerol, 1 mM DTT, and proteins at appropriate concentration. The reaction was incubated for 15 minutes at 30°C, then 0.5 µl of a 55 mg/ml Heparin (Sigma) added, and the samples incubated for a further 5 minutes. Before loading onto the gel 1 µl of 50% glycerol was added. 5% poly-acrylamide gels were used, with 30:1 bis:acrylamide ratio. Gels were run at 200 volts for ~ 2hr, 4°C after pre-running for 1–2 hours, at 200 volts. Dried gels were analysed by phosphorimager (Molecular Dynamics). After scanning on Typhoon scanner results were analysed using ImageQuant (GE Healthcare) and Photoshop (Adobe). Dissociation constants were estimated from the total protein concentration that produced 50% binding. For UV crosslinking, the total reaction volume was 10 µl. After addition of heparin, samples were subjected to 19200 $J.cm^{-2}$ UV light, followed by digestion with 50 µg RNase A and 140 U RNase T1. Samples were then separated by SDS gel electrophoresis and dried gels analyzed by phosphorimager.

Conclusions

Our results highlight the common and distinct domain requirements for activation or repression of splicing by MBNL1. Full length MBNL1 is relatively insensitive to inactivating mutations of individual ZF domains. However, when the protein is recruited to RNA by tethering with a heterologous RNA binding domain and deletion mutations are introduced, the dependency on functional ZF domains becomes more acute. Full tethered repressor and activator functions require ZF domains 1 and 2 that are able to bind RNA, suggesting that both types of activity require multivalent interactions with RNA. However, the ZF domains alone are insufficient for activity. Additional regions of the linker separating ZF domains 2 and 3 are required for splicing activity but not RNA binding. The additional regions differ for repression or activation, with more extensive regions of the linker required for full activation. This suggests the involvement of different sets of interacting cofactors for activation or repression of splicing by MBNL1.

Competing interest

The authors declare that they have no competing interests.

Authors' contributions

CWJS and CG conceived of the study and participated in its design and coordination. CE and CG carried out the experimental work and analyzed the data. All authors helped to draft the manuscript, and approved the final manuscript.

Acknowledgements

This work was supported by a Wellcome Trust Programme Grant to CWJS (092900). CE was supported by a studentship from the BBSRC. We thank Manny Ares for the Vldlr minigene, and Ben Luisi for help and advice in interpretation of MBNL structures and design of mutations.

References

1. Nilsen TW, Graveley BR: **Expansion of the eukaryotic proteome by alternative splicing.** *Nature* 2010, **463**(7280):457–463.
2. Barbosa-Morais NL, Irimia M, Pan Q, Xiong HY, Gueroussov S, Lee LJ, Slobodeniuc V, Kutter C, Watt S, Colak R, *et al*: **The evolutionary landscape of alternative splicing in vertebrate species.** *Science* 2012, **338**(6114):1587–1593.
3. Merkin J, Russell C, Chen P, Burge CB: **Evolutionary dynamics of gene and isoform regulation in Mammalian tissues.** *Science* 2012, **338**(6114):1593–1599.
4. Witten JT, Ule J: **Understanding splicing regulation through RNA splicing maps.** *Trends Genet* 2011, **27**(3):89–97.
5. Miller JW, Urbinati CR, Teng-Umnuay P, Stenberg MG, Byrne BJ, Thornton CA, Swanson MS: **Recruitment of human muscleblind proteins to (CUG)(n) expansions associated with myotonic dystrophy.** *EMBO J* 2000, **19**(17):4439–4448.
6. Kalsotra A, Xiao XS, Ward AJ, Castle JC, Johnson JM, Burge CB, Cooper TA: **A postnatal switch of CELF and MBNL proteins reprograms alternative splicing in the developing heart.** *Proceedings of the National Academy of Sciences of the United States of America* 2008, **105**(51):20333–20338.
7. Terenzi F, Ladd AN: **Conserved developmental alternative splicing of muscleblind-like (MBNL) transcripts regulates MBNL localization and activity.** *RNA Biol* 2009, **7**(1):43–55.
8. Botta A, Caldarola S, Vallo L, Bonifazi E, Fruci D, Gullotta F, Massa R, Novelli G, Loreni F: **Effect of the [CCTG]n repeat expansion on ZNF9 expression in myotonic dystrophy type II (DM2).** *Biochim Biophys Acta* 2006, **1762**(3):329–334.
9. Osborne RJ, Lin X, Welle S, Sobczak K, O'Rourke JR, Swanson MS, Thornton CA: **Transcriptional and post-transcriptional impact of toxic RNA in myotonic dystrophy.** *Human Molecular Genetics* 2009, **18**(8):1471–1481.
10. Du H, Cline MS, Osborne RJ, Tuttle DL, Clark TA, Donohue JP, Hall MP, Shiue L, Swanson MS, Thornton CA, *et al*: **Aberrant alternative splicing and extracellular matrix gene expression in mouse models of myotonic dystrophy.** *Nat Struct Mol Biol* 2010, **17**(2):187–193.
11. Kino Y, Washizu C, Oma Y, Onishi H, Nezu Y, Sasagawa N, Nukina N, Ishiura S: **MBNL and CELF proteins regulate alternative splicing of the skeletal muscle chloride channel CLCN1.** *Nucleic Acids Res* 2009, **37**(19):6477–6490.
12. He F, Dang W, Abe C, Tsuda K, Inoue M, Watanabe S, Kobayashi N, Kigawa T, Matsuda T, Yabuki T, *et al*: **Solution structure of the RNA binding domain in the human muscleblind-like protein 2.** *Protein Sci* 2009, **18**(1):80–91.
13. Teplova M, Patel DJ: **Structural insights into RNA recognition by the alternative-splicing regulator muscleblind-like MBNL1.** *Nat Struct Mol Biol* 2008, **15**(12):1343–1351.
14. Goers ES, Purcell J, Voelker RB, Gates DP, Berglund JA: **MBNL1 binds GC motifs embedded in pyrimidines to regulate alternative splicing.** *Nucleic Acids Res* 2010, **38**(7):2467–2484.
15. Cass D, Hotchko R, Barber P, Jones K, Gates DP, Berglund JA: **The four Zn fingers of MBNL1 provide a flexible platform for recognition of its RNA binding elements.** *BMC Mol Biol* 2011, **12**:20.
16. Tran H, Gourrier N, Lemercier-Neuillet C, Dhaenens CM, Vautrin A, Fernandez-Gomez FJ, Arandel L, Carpentier C, Obriot H, Eddarkaoui S, *et al*: **Analysis of exonic regions involved in nuclear localization, splicing activity, and dimerization of Muscleblind-like-1 Isoforms.** *Journal of Biological Chemistry* 2011, **286**(18):16435–16446.
17. Grammatikakis I, Goo YH, Echeverria GV, Cooper TA: **Identification of MBNL1 and MBNL3 domains required for splicing activation and repression.** *Nucleic Acids Res* 2011, **39**(7):2769–2780.
18. Purcell J, Oddo JC, Wang ET, Berglund JA: **Combinatorial Mutagenesis of MBNL1 Zinc Fingers Elucidates Distinct Classes of Splicing Regulatory Events.** *Mol Cell Biol* 2012, **32**(20):4155–4167.
19. Graveley BR, Maniatis T: **Arginine/serine-rich domains of SR proteins can function as activators of pre-mRNA splicing.** *Mol Cell* 1998, **1**(5):765–771.
20. Del Gatto-Konczak F, Olive M, Gesnel MC, Breathnach R: **hnRNP A1 recruited to an exon in vivo can function as an exon splicing silencer.** *Mol Cell Biol* 1999, **19**(1):251–260.
21. Robinson F, Smith CW: **A splicing repressor domain in polypyrimidine tract-binding protein.** *J Biol Chem* 2006, **281**(2):800–806.
22. Gooding C, Edge C, Lorenz M, Coelho MB, Winters M, Kaminski CF, Cherny D, Eperon IC, Smith CW: **MBNL1 and PTB cooperate to repress splicing of Tpm1 exon 3.** *Nucleic Acids Res* 2013, **41**(9):4765–4782.
23. Sun S, Zhang Z, Fregoso O, Krainer AR: **Mechanisms of activation and repression by the alternative splicing factors RBFOX1/2.** *Rna* 2011, **18**(2):274–283.
24. Shankarling G, Lynch KW: **Minimal functional domains of paralogues hnRNP L and hnRNP LL exhibit mechanistic differences in exonic splicing repression.** *Biochem J* 2013, **453**(2):271–279.
25. Hudson BP, Martinez-Yamout MA, Dyson HJ, Wright PE: **Recognition of the mRNA AU-rich element by the zinc finger domain of TIS11d.** *Nat Struct Mol Biol* 2004, **11**(3):257–264.
26. Fu Y, Ramisetty SR, Hussain N, Baranger AM: **MBNL1-RNA Recognition: Contributions of MBNL1 Sequence and RNA Conformation.** *Chembiochem* 2012, **13**(1):112–119.
27. Gooding C, Clark F, Wollerton MC, Grellscheid SN, Groom H, Smith CW: **A class of human exons with predicted distant branch points revealed by analysis of AG dinucleotide exclusion zones.** *Genome Biol* 2006, **7**(1):R1.
28. Yuan Y, Compton SA, Sobczak K, Stenberg MG, Thornton CA, Griffith JD, Swanson MS: **Muscleblind-like 1 interacts with RNA hairpins in splicing target and pathogenic RNAs.** *Nucleic Acids Res* 2007, **35**(16):5474–5486.
29. Wollerton MC, Gooding C, Robinson F, Brown EC, Jackson RJ, Smith CW: **Differential alternative splicing activity of isoforms of polypyrimidine tract binding protein (PTB).** *Rna* 2001, **7**(6):819–832.
30. Joshi A, Coelho MB, Kotik-Kogan O, Simpson PJ, Matthews SJ, Smith CW, Curry S: **Crystallographic analysis of polypyrimidine tract-binding protein-Raver1 interactions involved in regulation of alternative splicing.** *Structure* 2011, **19**(12):1816–1825.
31. Gromak N, Smith CW: **A splicing silencer that regulates smooth muscle specific alternative splicing is active in multiple cell types.** *Nucleic Acids Res* 2002, **30**(16):3548–3557.
32. Sharma S, Maris C, Allain FH, Black DL: **U1 snRNA directly interacts with polypyrimidine tract-binding protein during splicing repression.** *Mol Cell* 2011, **41**(5):579–588.
33. Wang ET, Cody NA, Jog S, Biancolella M, Wang TT, Treacy DJ, Luo S, Schroth GP, Housman DE, Reddy S, *et al*: **Transcriptome-wide regulation of pre-mRNA splicing and mRNA localization by muscleblind proteins.** *Cell* 2012, **150**(4):710–724.
34. Charizanis K, Lee KY, Batra R, Goodwin M, Zhang C, Yuan Y, Shiue L, Cline M, Scotti MM, Xia G, *et al*: **Muscleblind-like 2-mediated alternative splicing in the developing brain and dysregulation in myotonic dystrophy.** *Neuron* 2012, **75**(3):437–450.
35. Warf MB, Berglund JA: **MBNL binds similar RNA structures in the CUG repeats of myotonic dystrophy and its pre-mRNA substrate cardiac troponin T.** *Rna* 2007, **13**(12):2238–2251.
36. Warf MB, Diegel JV, von Hippel PH, Berglund JA: **The protein factors MBNL1 and U2AF65 bind alternative RNA structures to regulate splicing.** *Proc Natl Acad Sci U S A* 2009, **106**(23):9203–9208.
37. Nasim FU, Hutchison S, Cordeau M, Chabot B: **High-affinity hnRNP A1 binding sites and duplex-forming inverted repeats have similar effects**

on 5′ splice site selection in support of a common looping out and repression mechanism. *Rna* 2002, **8**(8):1078–1089.

38. Fisette JF, Toutant J, Dugre-Brisson S, Desgroseillers L, Chabot B: **hnRNP A1 and hnRNP H can collaborate to modulate 5′ splice site selection.** *Rna* 2010, **16**(1):228–238.
39. Cherny D, Gooding C, Eperon GE, Coelho MB, Bagshaw CR, Smith CW, Eperon IC: **Stoichiometry of a regulatory splicing complex revealed by single-molecule analyses.** *EMBO J* 2010, **29**(13):2161–2172.
40. Llorian M, Schwartz S, Clark TA, Hollander D, Tan LY, Spellman R, Gordon A, Schweitzer AC, de la Grange P, Ast G, *et al*: **Position-dependent alternative splicing activity revealed by global profiling of alternative splicing events regulated by PTB.** *Nat Struct Mol Biol* 2010, **17**(9):1114–1123.
41. Gromak N, Rideau A, Southby J, Scadden AD, Gooding C, Huttelmaier S, Singer RH, Smith CW: **The PTB interacting protein raver1 regulates alpha-tropomyosin alternative splicing.** *EMBO J* 2003, **22**(23):6356–6364.
42. Ho TH, Charlet BN, Poulos MG, Singh G, Swanson MS, Cooper TA: **Muscleblind proteins regulate alternative splicing.** *EMBO J* 2004, **23**(15):3103–3112.

5

Fidelity of end joining in mammalian episomes and the impact of Metnase on joint processing

Abhijit Rath[1], Robert Hromas[2] and Arrigo De Benedetti[1*]

Abstract

Background: Double Stranded Breaks (DSBs) are the most serious form of DNA damage and are repaired via homologous recombination repair (HRR) or non-homologous end joining (NHEJ). NHEJ predominates in mammalian cells at most stages of the cell cycle, and it is viewed as 'error-prone', although this notion has not been sufficiently challenged due to shortcomings of many current systems. Multi-copy episomes provide a large pool of genetic material where repair can be studied, as repaired plasmids can be back-cloned into bacteria and characterized for sequence alterations. Here, we used EBV-based episomes carrying 3 resistance marker genes in repair studies where a single DSB is generated with virally-encoded HO endonuclease cleaving rapidly at high efficiency for a brief time post-infection. We employed PCR and Southern blot to follow the kinetics of repair and formation of processing intermediates, and replica plating to screen for plasmids with altered joints resulting in loss of chloramphenicol resistance. Further, we employed this system to study the role of Metnase. Metnase is only found in humans and primates and is a key component of the NHEJ pathway, but its function is not fully characterized in intact cells.

Results: We found that repair of episomes by end-joining was highly accurate in 293 T cells that lack Metnase. Less than 10% of the rescued plasmids showed deletions. Instead, HEK293 cells (that do express Metnase) or 293 T transfected with Metnase revealed a large number of rescued plasmids with altered repaired joint, typically in the form of large deletions. Moreover, quantitative PCR and Southern blotting revealed less accurately repaired plasmids in Metnase expressing cells.

Conclusions: Our careful re-examination of fidelity of NHEJ repair in mammalian cells carrying a 3' cohesive overhang at the ends revealed that the repair is efficient and highly accurate, and predominant over HRR. However, the background of the cells is important in establishing accuracy; with human cells perhaps surprisingly much more prone to generate deletions at the repaired junctions, if/when Metnase is abundantly expressed.

Keywords: Accuracy of DSB repair in mammalian cells, Episomal model of NHEJ, End- processing and re-ligation, Metnase nuclease, Joint accuracy

Background

Double Stranded Breaks (DSBs), a serious type of chromosome damage are usually caused by exogenous agents such as ionizing radiation (IR), topoisomerase poisons, radiomimetic drugs like bleomycin and doxorubicin, or endogenous damages which arise as a result of cellular metabolism (free radicals and DNA replication errors or collapse of stalled forks). Programmed DSBs are also generated during processes such as V(D)J recombination of immunoglobulin genes and during allelic exchanges at meiosis or zygotic fusion. If misrepaired, DSBs can result in chromosomal translocations, oncogenesis, and genomic instability [1-3]. DSB repair can be broadly divided into two major pathways: homologous recombination (HR) and non-homologous end joining (NHEJ) [4]. HR requires the presence of an intact homologous template, often a sister chromatid, which allows accurate repair of DSBs via a replicative intermediate and is the preferred mode of repair in S and G2 phase of cell-cycle [5]. In contrast, NHEJ operates without the need for a template DNA; sometimes joining the ends with minimal processing, hence considered potentially erroneous/mutagenic. NHEJ is the predominant

* Correspondence: adeben@lsuhsc.edu
[1]Department of Biochemistry and Molecular Biology, Louisiana State University Health Sciences Center, 1501 Kings Highway, Shreveport, LA 71130, USA
Full list of author information is available at the end of the article

form of repair in mammalian system and had been shown to operate all throughout the cell cycle [6].

Currently used DSB repair assays have several shortcomings. Researchers employ several methods such as Ionizing radiation (IR), radiomimetic drugs, laser, and site-specific endonucleases for induction of DNA damage [7]. Of note, γ-irradiation and drug induced damage foci are randomly dispersed and non-homogenous in nature. Lasers create localized but exceedingly strong DNA damage. In addition, different wavelengths of the laser largely determine the damaged end structures subsequently resulting in widely different kinetics of repair with differential protein requirement (for a review see [7,8]). DSBs introduced with a nuclease are adequate for studies of the molecular/cellular mechanisms, but estimates of repair accuracy are problematic since these enzymes turn over slowly and thus selecting for erroneous processing of the ends that eliminate the target sequence. This is well-known for the Endonuclease I-SceI induced DSBs [9], a model that also suffers from the need of pre-integration of the target sequence, and often low cutting efficiency (either due to low transfection of the cell population or because of the chromosomal location of its target sequence). The Endonuclease I-PpoI also suffers from low cleavage efficiency because of its unique target location in the highly condensed and repetitive ribosomal DNA gene clusters. Furthermore, study of repair at defined genomic DSBs is problematic due to the difficulty in re-obtaining 'clean clonal' isolates for analysis in sufficient representation. Likewise, use of yeast as a model system to parlay mammalian NHEJ is not entirely acceptable. Yeast like *Saccharomyces cerevisiae* or *Saccharomyces pombe* lack many well established components of mammalian NHEJ, like DNA-PKcs and Artemis. To date, orthologs for newly discovered NHEJ components of human APLF, PNK, Metnase, and APTX have not been identified in yeast [10]. Furthermore, HR is the preferred repair pathway in yeast, whereas NHEJ is the predominant pathway in mammalian cells [11]. In yeast, IR-induced and endonuclease-induced DSBs are differentially processed in a cell cycle-dependent manner [12].

Hence, we set out to develop one model system to efficiently represent repair activities in mammalian cells in a physiological setting, with episomes. Repair of multicopy plasmids provides a large pool of mostly uniform genetic material where the repair process can be studied [13-15]. We utilized the yeast HO endonuclease (a key component of the mechanism for mating-type switch in yeast) for use in mammalian cells. In this system, the HO endonuclease is expressed by a recombinant adenovirus resulting in cleavage of its target site (on episomes) with high efficiency (depending on the MOI), and repair occurs via simple end-joining during a time course of infection. We previously utilized a mouse mammary cell line containing a single integrated HO target site at a defined genomic location [16]. Whereas, this has generated powerful and useful information, for example in terms of the effects of chromatin on generation of the DSB and in repair [16,17], for other purposes the system is limiting. For instance, damage induced by IR or genotoxins results in multiple simultaneous DSBs instead of a single break, which immediately raises questions in terms of similarity of activation and deactivation of the DNA damage response (DDR). It is of significant interest to the field of mammalian DSB repair to develop a model system, where there is synchronous induction of multiple 'homogeneous' site-specific DSBs in a population, and then be able to stop the nuclease activity rapidly [18].

Different assays, such as treatment of pre-cleaved plasmids with cellular extracts or transfection of linearized DNA templates into mammalian cells have been employed to study several different aspects of end-joining (efficiency of joining of different DNA ends, fidelity of repair [13,19-21]). However, technical limitations prevent a clear picture from emerging regarding the end-joining efficiency/fidelity and nature of sequence rearrangements. For instance, *Smith et al.*, using a transfection assay reported a similar efficiency of end-joining for different DNA ends. However, *Poplawski et al.*, using cellular extracts have reported widely varying degree of efficiency of end-joining with different end structures [13,22]. In fact, a recent report presented evidence for strikingly different repair efficiency using either lipid-based transfection vs. nucleofection [23]. This highlights the importance of conducting end-joining assays in a 'nearly unperturbed' cellular environment. Where looked at repair fidelity of a DSB at an integrated chromosomal locus [24] most studies were limited in terms of low cleavage efficiency, absence of multiple defined site-specific DSB sites, and also the persistent activity of the nuclease employed. We thus wanted to investigate the repair fidelity in an episomal population. It is well established that EBV-based episomes become chromatinized and behave as minichromosomes [15,25]. High density MNase maps for EBV-based episomes have been generated that have revealed highly positioned nucleosomes and putative origins of replication [26]. Hence, repair studies of minichromosomes can also bring into focus the contribution of chromatin, closely resembling genomic damage. Finally, we wanted to characterize the effect of a recently identified important component of human NHEJ, known as Metnase, as an endonuclease on episomal repair fidelity in intact cells. Metnase has been shown to be a general facilitator of NHEJ increasing both accurate and inaccurate repair [27]. Mutation of its nuclease domain has been shown to prevent this role, at least in a subset of end-joining events [28,29]. Using *in vitro* assays, Metnase has been shown to have a preference for single stranded DNA overhang of a partially duplex molecule with an effect more pronounced on a 3′ overhang [29]. Evidence from experiments done with plasmid-host cell transfection system, point to a significant role of Metnase in determining repaired junction fidelity for breaks with 3′ overhangs. Also, in a transfection assay coupled with integration, over-expression of Metnase did not significantly increase accurate NHEJ at the break site [27]. *Beck et al.*, reported a role of Metnase in promoting end-joining with non-compatible ends (both 5′ & 3′) in a cell free system [29]. However, a very recent study reported against any functional role of Metnase in promoting end-joining of modified DNA ends using a similar cell free system [30]. These contradictory findings hint towards a gap in knowledge regarding the role of Metnase as an endonuclease processing different ends [31], and even greater in intact cells. We wanted to determine the contribution of Metnase in altering repair fidelity (accurate vs. inaccurate repair) in our model system and to characterize the nature of nucleotide rearrangements to understand more in depth NHEJ fidelity in human cells.

Methods

Preparation of stable cell lines maintaining episomal HO-CAT plasmid

The 34 bp HO endonuclease target sequence (agatcttttagtttcagctttccgcaacagtata) was cloned in to the HindIII site of the pREP4/CAT shuttle vector just before the chloramphenicol acetyl transferase (CAT) coding sequence and the plasmid was renamed as HO-CAT plasmid. Further, HO-CAT plasmid was transfected in to 293 T cells which maintain the plasmid episomal due to presence of EBNA-1, under hygromycin selection (Hygromycin B/Calbiochem, Cat # 400052). The cell line was named 293 T-HO-CAT. Similarly HEK293 cells were transfected and selected to produce the 3-HO-CAT cell line. Metnase expressing pCAPP-Metnase-V5 plasmid which was prepared in Robert Hromas' lab was transfected in to 293 T cells and puromycin (Puromycin Dihydrochloride/MP biomedicals, Cat # 100552) selected to produce the Metnase over expressing 293 T stable cell line. Subsequently, HO-CAT plasmid was transfected into these cells to produce the Met-T-HO-CAT cell line and maintained under dual selection of hygromycin and puromycin. GenePORTER (Genlantis, Cat # T201007) was used as the transfection reagent for all the above mentioned transfections. All the three cell lines were cultured using DMEM with 10% FCS as growth medium at 5% CO2 and 37°C.

Assay for DNA cleavage and repair

In a typical assay, for each different time point, 150,000 cells were infected with the recombinant adeno virus encoding HO endonuclease (Gift from Dr. Hamish Young, Columbia University, NY, USA) at 10–30 MOI, referred to as pPF446::HO in a plasmid map in [32]. DNA from cells was collected using either Wizard SV genomic DNA purification system (Promega, Cat # A2360) or Wizard plus SV miniprep DNA purification (Promega,Cat # A1460) protocol. Equivalent amount of DNA was used in PCR and qPCR reactions for CAT and AMP regions using GoTaq DNA polymerase kit (Promega) and DynAmo Flash SYBR Green qPCR kit (Thermo Scientific, Cat # F-415S) respectively. 5′-CTACAACAAGGCAAGGCTTGACC-3′ and 5′-TCTAGTTGTGGTTTGTCCAAACTCATC-3′ were used as forward and reverse primers respectively for CAT amplicon. 5′-TTCCGTGTCGCCCTTATTCCC-3′ and 5′-GGCACCTATCTCAGCGATCTG-3′ were used as forward and reverse primers respectively for AMP region in PCR reactions. PCR conditions: 30 cycles of 94°C for 30 s, 55°C for 30 s, and 72°C for 39 s with a final extension of 10 minutes at 72°C. The CAT signal was normalized to AMP signal. For qPCR, 5′-GTACCAGCTGCTAGCAAGCT-3′ and 5′-TCAACGGTGGTATATCCAGTGAT-3′ were used as forward and reverse primers respectively for the CAT region (amplicon size 133 bp). 5′-CATCGAACTGGATCTCAACAGCG-3′ and 5′-GTCATGCCATCCGTAAGATGCT-3′ were used as forward and reverse primers respectively for AMP region.

Immunoblotting

Protein samples from different time points were collected by RIPA lysis buffer (50 mM Tris pH 8.0, 0.1% SDS, 1% Triton-X, 1 mM EDTA, 150 mM NaCl, and with 1X protease inhibitors – SIGMAFAST protease inhibitor cocktail tablets, Cat # S8820) and quantitated using BCA protein assay kit (Pierce, Cat # 23223). Equal amounts were run on a 12% polyacrylamide gel, blotted on to Immobilon-P (Millipore, Cat # IPVH08100), and subsequently probed with anti-E3-11.6 K antibody (fusion protein with HO). The antibody was a kind gift from Dr. William S.M. Wold, St. Louis school of medicine, Missouri, USA. The blot was re-probed with Phospho-(S/T)Q-ATM/ATR substrate antibody (Cell Signaling, Cat # 2909S). Anti-Rabbit-HRP conjugated antibody (Cell signaling, cat # 7074S) was used as the secondary antibody. Stripping was done using Re-Blot plus mild (Millipore, Cat # 2502). The blot was either developed using Pierce ECL reagent (Thermo Scientific, Cat # 32106) or Opti-4CN (Bio-Rad, Cat # 170–8235).

NheI or NotI screening and bacterial transformation

Plasmids recovered from each different time point were subjected to NheI (or NotI) digestion (Promega, Cat # R6501) which has its recognition sequence very close to HO induced cleavage site. Subsequently, the enzyme was heat inactivated and the miniprep DNA was used to transform XL-1 Blue supercompetent cells (Agilent, Cat # 200236) and plated on an Ampicillin plate (100 μg/mL).

Replica plating assay

Plasmid DNA isolated from different cell lines at different time points was used to transform was incubated for 16 h at 37°C and subsequently replica plated on a chloramphenicol plate (50 μg/mL). Colonies were counted with Bio-Rad ChemiDoc (Cat # 170–8265) machine using Quantity-One software.

Luciferase assay

The HO-Luc plasmid was transfected using GenePORTER (Genlantis, Cat # T201007) and concomitantly infected with adeno-HO virus in either 293 T-HO-CAT cells or 293 T cells. At each time point, cells were taken of the plate and one aliquot of it was used to extract the DNA to be used for PCR to assay cleavage and repair. The rest of the cells were assayed for luminescence using Luciferase Assay system (Promega, Cat # E1501). GFP fluorescence was used to normalize the luminescence values. Both luminescence and fluorescence were measured by Synergy 4 plate reader by BioTEK instruments.

Southern Blotting for plasmid DNA

One fully confluent T75 flask was used for each time point. Plasmids were recovered using Zyppy plasmid Kit (Zymo Research, Cat # D4019). 10ug of plasmid DNA from each time point was digested with either Csp45I and NotI or Csp45I alone. Heat inactivated samples were run in 1% agarose-gel. Blotting onto Immobilon-Ny + (Millipore, Cat # INYC00010) was achieved using Trans-Blot SD semi-dry transfer cell (Biorad, Cat # 170–3957). EKONO hybridization buffer (Research Products International, Cat # 248800) was used for both pre-hybridization and hybridization. Hybridization probe was synthesized by random priming using random hexamers (Invitrogen, Cat # 51709) using whole plasmid DNA as template. Finally, the blot was exposed to X-ray films and developed using standard developer.

Preparation of cell extracts

About 1 million cells were used to prepare the whole cell extract for each mentioned time point. Cells were collected off the plate and washed once with cold 1X PBS. Subsequently, the cells are washed once with hypotonic buffer (25 mM Tris pH 7.9, 1 mM MgCl2, 0.4 mM CaCl2, and 0.5 mM DTT), spun down, and resuspended again in 200 μL of hypotonic buffer and kept in ice for 20 minutes. The cells were homogenized in a tight fitting Dounce homogenizer (30 strokes) and the debris was spun down by spinning at 13000 rpm for 10 minutes at 4°C. The supernatant is collected and used in the DNA cleavage reaction.

In vitro DNA cleavage assay

The reaction was carried out in a 10 μL reaction mixture with 250 ng of pMat-Puro plasmid, 6 mM MgCl2, 50 mM NaCl, 5 units of XmnI enzyme (Promega, Cat# R727A) (0.5 μL), and 2 μL cell extract for each given time. The mixture was incubated at 30°C for 1 h. Then, the reactions were stopped by adding 5 μL of stop solution (100 mM EDTA, 1% SDS, 1 mg/mL Proteinase K) and 3 μL of 6× loading dye and incubated at 55°C for 30 min-1 h. Then, the mixture was run in a 1% agarose gel.

Results

Development of an episomal model system to study DSB repair

We initially established a cell line (human 293 T) transfected with a shuttle vector with the HO target site cloned into it (293 T-HO-CAT) (see methods for detailed description). The vector is maintained as a stable copy number episomal unit inside the mammalian cells because of the presence of EBNA-1 protein encoded by the vector (Figure 1A) [33]. This provides us with multiple copies of homogenous template which become chromatinized and have been shown to accurately reflect repair of genomic DNA [14]. Minichromosomes are physiologically relevant but simple substrates to study DNA repair. Being homogenous in nature and also the ability to isolate them from bulk chromatin are some other advantages of the episomal system. We are conducting studies by sucrose gradient sedimentation of isolated episomes that have revealed highly ordered, chromatinized plasmid species (to be published elsewhere).

In a typical cleavage and repair assay, a single site specific DSB was generated in the episomal population by infecting the cells with recombinant adeno-virus (Ad-HO) encoded HO endonuclease [16,32] that recognizes a 34 base pair (bp) specific sequence and produces a DSB with a 4 bp 3′ overhang. Repair was followed in a time course after isolating the population of plasmids and using endpoint PCR and q-PCR detecting an amplicon spanning the break site (Figure 1B, CAT amplicon, Figure 1C). Upon induction of the DSB, the PCR product is lost, but then recovers with time as the repair proceeds. The Ampicillin region on the vector was used as an internal normalization control for plasmid recovery. The cleavage efficiency of HO-site was consistently >90% as judged from multiple experiments at 10–30 MOI. We have previously shown robust DDR activation due to HO induced site-specific cleavage across the episomal population that matches the kinetics of repair in terms of appearance and disappearance of S1981

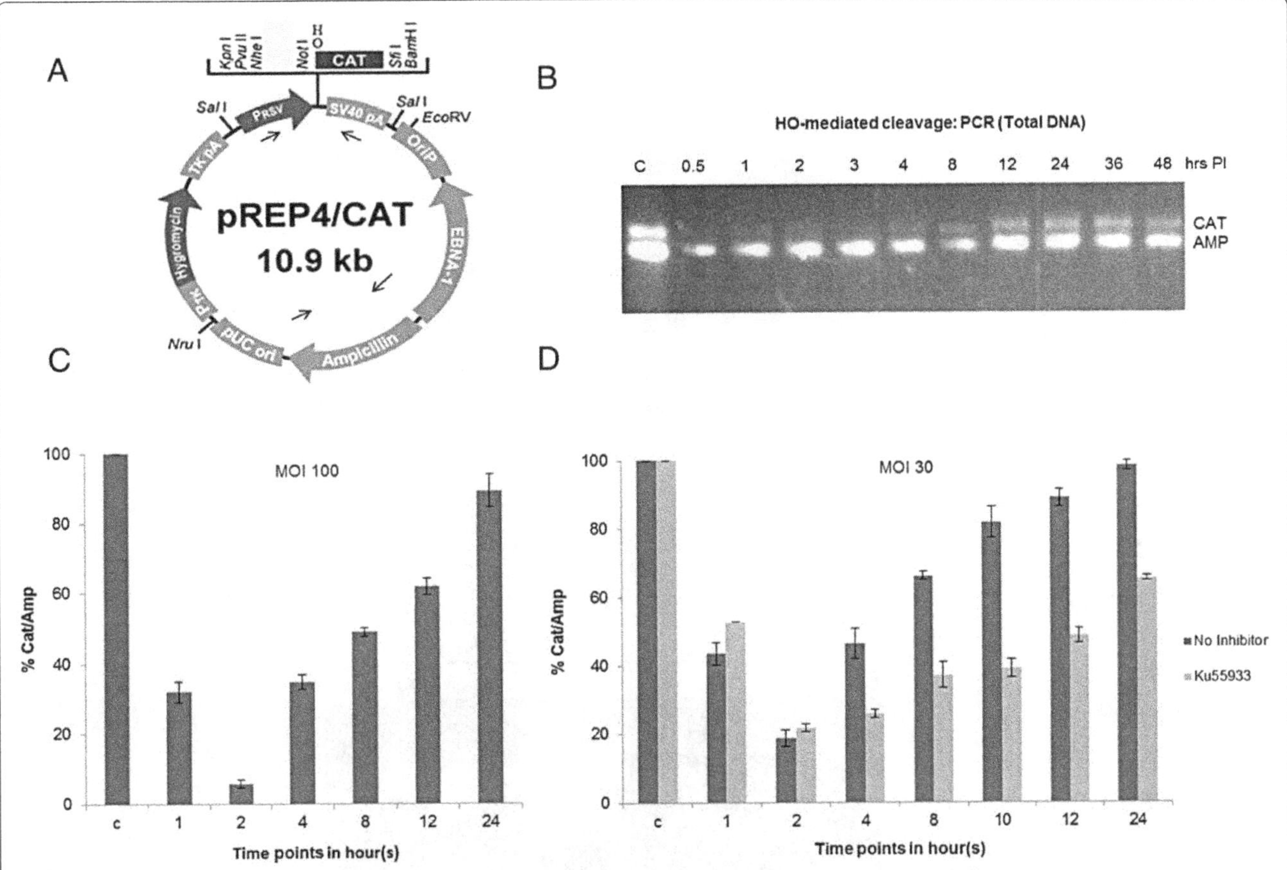

Figure 1 Typical episomal cleavage and repair assay. (A) Map of the pRep4-HO-CAT plasmid showing key features. HO recognition cassette is cloned in just before CAT gene. Arrows indicate location of the primers. **(B)** Episomal cleavage-and-repair assay at the HO site DSB during a time course of Ad-HO infection. T-HO-CAT cells were infected with adeno-HO (Ad-HO) virus, cells were collected at different time points, and episomes were recovered. PCR was performed by putting primers across the break site (CAT amplicon). AMP (Ampicillin) region on the episome acts as the positive internal control and used for normalization. In addition, a genomic product was also used to ensure equal total DNA recovery from the samples. **(C)** qPCR analysis for kinetics of cleavage and repair of epsiomes as explained in **(B)**. qPCR values are represented as (CAT/AMP)% over control. **(D)** Same as in **(C)** but in presence of the ATM inhibitior-KU55933 (10 μM).

phosphorylation of ATM (Figure 2 in [34]; reproduced with permission in Additional file 1: Figure S1). Such data confirm previous observations of DDR activation with as little as 10–20 DSBs or even with a single DSB [35-37]. To ensure that the episomal repair derives a significant contribution from the DDR, we treated the cells with the ATM inhibitor KU55933. The kinetics of repair in the drug treated cells was slower in comparison to untreated control (Figure 1D). These results show that in this system an active DDR pathway is elicited due to presence of multiple DSBs. Of interest, some previous studies have suggested the non-essential role of ATM during much of NHEJ events, based largely on evidence from mutant cell lines, except for instances of condensed heterochromatic sites [38]. In this episomal system, while we have abundant direct evidence that the plasmids are chromatinized and supercoiled, presence of heterochromatic regions are not believed to be likely. Since, for example, histone H1 is found only in small amounts in association with episomes (and their nucleosomes) after isolation on sucrose gradients [[25] and data not shown].

Kinetics of HO expression and inactivation

In Figure 2 we show a typical experiment with the kinetics of HO expression. In initial studies, we have observed a second cleavage cycle between 24-36 h Post-infection (PI) that could be explained by release of new virus and reactivation of the promoter driving HO expression in 293 T cells. For a study of repair fidelity, it is important to have the cleaving endonuclease transiently active. This reduces the bias toward inaccurate repair of the joint introduced in the system by the recurrent cutting and accurate restoration of the recognition sequence. In this context, *Kaplun et al.*, have shown that HO is a target of Mec1/ATR mediated phosphorylation leading to its subsequent ubiquitination and degradation in yeast [39]; an obvious mechanism for haploid yeast to avoid

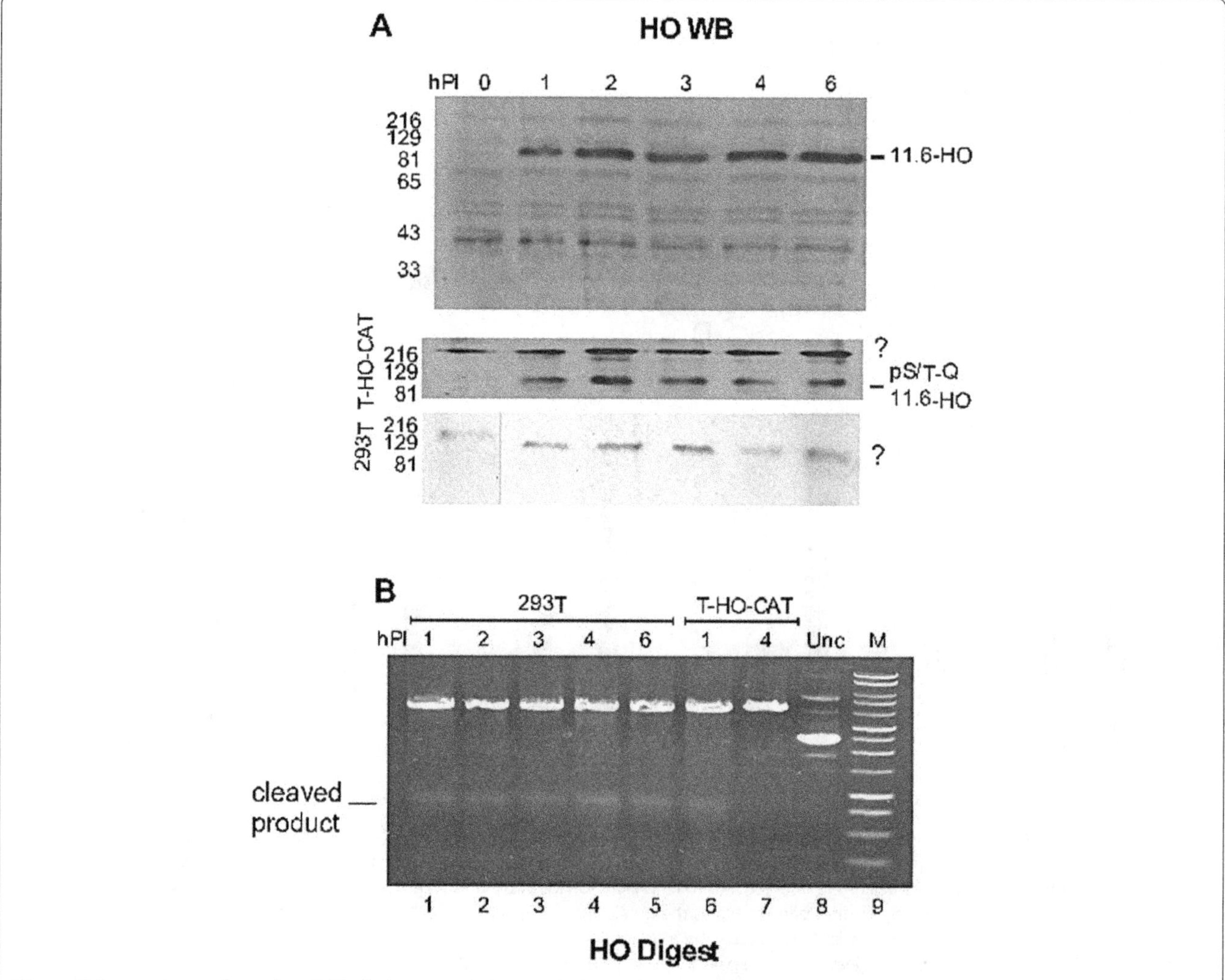

Figure 2 Expression and activity of HO-fusion protein in permissive cell system (293 T cells). (A) Depicts immunoblot of HO fusion protein upon infection with adeno-HO virus at different time points in 293 T-HO-CAT cells. A map of the virus showing the HO fusion is presented in [32], and generates a presumed chimeric protein with the 11.6 k adenovirus gene product. Cells were lysed using RIPA lysis buffer and equal amounts of proteins were loaded and run in a 10% SDS-PAGE gel and subsequently probed with anti-E3-11.6 K (fusion protein with HO) antibody. The blot in panel **(A)** was re-probed with phospho-(S/T) Q-ATM/ATR substrate antibody, and the region of the blot corresponding to the position of HO in panel (A) is shown. Also shown, the p(S/T)-Q banding pattern in 293 T cells which lack the episomes after Ad-HO infection, to serve as a control. ? denotes undetermined bands obtained after probing with p(S/T)-Q antibody. **(B)** HO activity was measured in an *in vitro* cleavage reaction. Cell extracts were prepared from 293 T cells (lanes 1–5) and cells carrying pRep4-HO-CAT episomes (T-HO-CAT cells) (lanes 6–7) after Ad-HO infection at given time points. The plasmid pMat-Puro, harboring the HO recognition site [16], was added exogenously at 250 ng per reaction. The plasmid was treated with cell extracts obtained from different time points from both 293 T cells and T-HO-CAT cells. HO activity was judged by presence/absence of a cleaved product. In reactions shown in lanes 1–7, we have added Xmn1 to produce the ~750 bp band and corroborate the cleavage at the HO site. Lane 8 shows the treatment of the pMat-Puro plasmid with extracts obtained from uninfected cells (Unc). "M" denotes the Promega 1 Kb DNA ladder as a molecular weight marker.

repeated mating-type switching until the next cell division. We have previously shown that in a non-permissive cell system, HO expression shuts down by 3 h of infection, ensuring (mostly) a single round of cleavage/repair implicating the existence of a similar mechanism in mouse cells [40]. In these cells (MM3MG mouse cells - non-permissive to human adeno virus) the HO protein, is expressed soon after infection leading to cleavage of its target site. But subsequently, ATM activation, and phosphorylation of HO results in its proteolytic loss with no new net synthesis [16], thus explaining absence of recurrent cutting of the restored target site and the enzyme is then degraded.

However, in a replication permissive system (human 293 T cells) HO nuclease persists throughout the infection (Figure 2A) due to a complex pattern of activation of early and late promoters usage [41] driving the E3/11.6-HO fusion [42]. We should caution that the identification of this band, clearly a viral product, as the E3/11.6-HO fusion remains uncertain. Unfortunately a published HO antiserum is not available, and the way to detect the HO was through its fusion with the 11.6 k protein [32,42], but previously presence of this product had been correlated with the nuclease activity [16]. Assuming we have correctly identified the HO protein, however, it was unclear whether the HO enzyme remained active after cutting its target(s) or if it became nonetheless inactivated via its ATM/ATR-mediated phosphorylation. Maintenance of HO activity would elicit multiple rounds of cleavage and repair of the episomal targets, and the fundamental question was whether the pattern of repair (recovery of the CAT PCR product) was the result of a single round of cutting and repair, or a more complex pattern of cycles of cleavage and religation. To answer this, first we determined that HO indeed became phosphorylated (by ATM/ATR) in this system, after we re-probed the HO-WB with a phospho-Ser/Thr (S/T) ATM/ATR substrate antibody (S/T-Q motif is the well characterized target of ATM/ATR substrates [43] and is also present in HO). Figure 2A shows rapid S/T-Q phosphorylation of the band co-migrating with HO that paralleled the cleavage of plasmid and the kinetics of ATM activation in this system [34] (Additional file 1: Figure S1). Treatment with λ phosphatase abolished the signal (not shown). In contrast, infection of plain 293 T cells showed a number of phospho-S/T-Q immunereactive bands, but none that corresponded to the HO position or that corresponded to phospho-proteins induced after doxorubicin treatment (Additional file 2: Figure S2). In short, a possible scenario is that HO becomes S/T-Q phosphorylated after (maximal) cleavage of the bulk of episomes; and no longer seems to cleave until viral replication is complete and the next cycle of infection begins (the cells do not lyse until ~5-7 days later).

To confirm that HO is indeed inhibited following a single round of cleavage and repair, we performed two types of experiments, one with extract of HO-infected cells supplemented with a reporter plasmid exogenously and one in intact cells. We already knew by WBs that the HO fusion protein varies little throughout the time course of infection, but its activity seems to be limited to the first hour, after which the enzyme is phosphorylated by ATM/ATR and appears to be inhibited. We postulated that the source of ATM/ATR activation is the HO induced cleavage of HO-CAT episomes as shown previously (Figure 2 in [34]; Additional file 1: Figure S1). Hence, preparing the cell extract from Ad-HO infected 293 T cells not carrying the episomes would effectively avoid ATM/ATR activation, thus maintaining an active HO endonuclease. We thus prepared cell extract from both 293 T cells and cells carrying the episomes (T-HO-CAT cells) at different time points post infection with Ad-HO virus. As shown in Figure 2B, addition of pMat-puro plasmid (carrying a HO recognition site, [16]) to uninfected cell extract (lane 8) results in no appreciable cleavage (after 1 h of incubation *in vitro*). However, incubation in Ad-HO infected 293 T cell extracts collected at different time points results in complete linearization of the plasmid even after the first hour PI, and the HO activity remains so for the next 6 h, as judged by the appearance of cleaved product (we also added Xmn1 to these reactions to enhance the distinction between the linearized plasmid and the mobility of the relaxed circular form). In contrast, in cells carrying the pRep4-HO-CAT plasmid (labeled T-HO-CAT), the enzyme is active only for the first hour PI, but not after 4 h (note presence/absence of the cleaved product) (Figure 2B), indicating that the HO is inactive following cleavage of the endogenous episomes. Treatment of this 'HO extract' with λ phosphatase gave some return of cleavage but with unclear results as there was more plasmid smear (not shown).

For the experiment *in vivo*, we transfected a Luciferase/GFP reporter [44] in Ad-HO-infected the 293 T-HO-CAT cells. In the Luciferase reporter, the HO target site separates the promoter from the ORF. The Luciferase expressed is short-lived, so that only sustained synthesis results in sufficient expression. New synthesis of Luciferase, measured as enzyme activity lags about an hour after DSB repair, when compared with PCR, which also showed a single prominent repaired band of correct size with no evidence of deleted products [44]. Luciferase was measured at 2 h intervals for 12 h, and the Ad-HO infected samples were compared to the uninfected samples for the same time points (Figure 3). In addition, DNA from each time point was recovered to determine the kinetics of cleavage and repair of the HO-CAT episomes. Figure 3, lower panel (CAT amplicon) shows that most of the target site is cleaved by 2 h with the repair product back to control level by 6-8 h post infection. The reporter plasmid also showed a similar pattern of Luc activity, an indication of cleavage/repair and then expression kinetics. If the HO was still active after 4 h of infection, then this would result in concomitant cutting of the Luc reporter, and be reflected as an initial delay in Luciferase expression (as evident in difference of slopes between uninfected and infected samples between 0-6 h) or slower accumulation kinetics (RLUs produced) in comparison to the uninfected control. Figure 3 shows that the pattern of Luciferase activity in the infected samples closely parallels the pattern as in the uninfected samples. Taken together, these results strongly suggest that even in a permissive system, HO gets inactivated by 4 h time point after infection, limiting its cleavage activity to just once in an experimental window of 0-12 h, or at least all the period in which ATM remains active. Of course, the situation changes once the viral

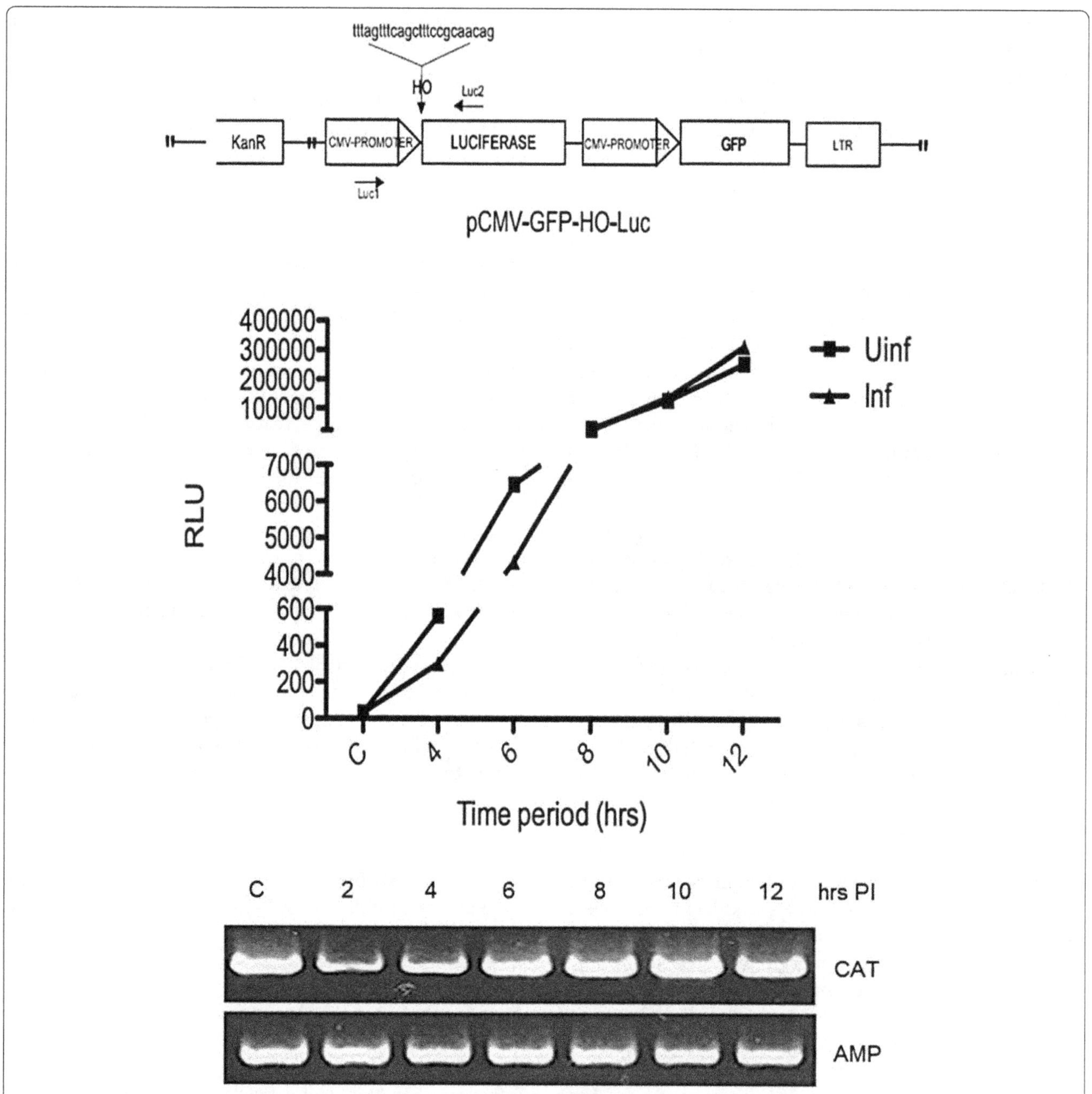

Figure 3 Luciferase expression in 293 T-HO-CAT cells at different time points. The cells were transfected with the vector construct with HO target site cloned in just before the Luciferase ORF (shown on top). Luciferase expression was measured in both uninfected and Ad-HO infected cells and normalized to GFP expression. Both infection and plasmid reporter transfection were carried out concomitantly. PCR shows the cleavage and repair kinetics across the time points in the HO-CAT episomes.

replication is complete and one must take into account multiple rounds of infection in the event of absence of cell lysis. Despite the appeal of this mechanism of ATM involvement in HO inactivation, we acknowledge that our results remain correlative. A definitive proof could be obtained with the generation of a new system in AT- cells, but then again the possibility exists that it is instead ATR that is involved in phosphorylation and inactivation of HO. However, it remains clear that only one round of cutting is obtained in the first few hours of Ad-HO infection.

High Fidelity repair of episomes via NHEJ

We wanted to study the repair fidelity of end-joining in this episomal system. The HO site is positioned just before the chloramphenicol resistance marker (CAT)

(35 bp away from the ATG start site) in the plasmid. So, any alteration in the form of deletions or insertions would profoundly affect expression of CAT and result in chloramphenicol sensitivity, detected by replica plating the LB-Ampicillin bacterial plate onto chloramphenicol Plates. 293 T-HO-CAT cells were infected with adeno-HO virus to induce the site-specific DSB. Plasmids were rescued from cells at each time point and were back transformed into supercompetent bacteria and plated on LB-Ampicillin. Somewhat to our surprise, almost all of the bacterial colonies could replica-plate on chloramphenicol and virtually all the rescued plasmids were 'unmutated'. Boil preps were analyzed from over 200 randomly chosen clones across different time points. As shown in Figure 4A, almost all of them revealed presence of right sized plasmids (10.9 Kb). *Smith et al.*, have previously reported an average deletion size of 250 bp for a 3' complementary overhang in a plasmid host-cell transfection assay [13]. To easily identify clones within a deletion range of 250–500 bp, we employed SalI digestion to screen random clones picked from two different time points. SalI restriction sites are present on both sides of the break site (refer plasmid map in Figure 1A) and upon digestion should give a 9 Kb fragment and a 1.9Kb fragment. We could not detect any obvious deletions among these random clones, none affecting SalI site(s), and all of them yielded correct sized fragments (Figure 4B). We also performed a PCR screen across the break using closely spaced primers (120 bp amplicon) in order to detect micro-deletions or insertions. Plasmids obtained from

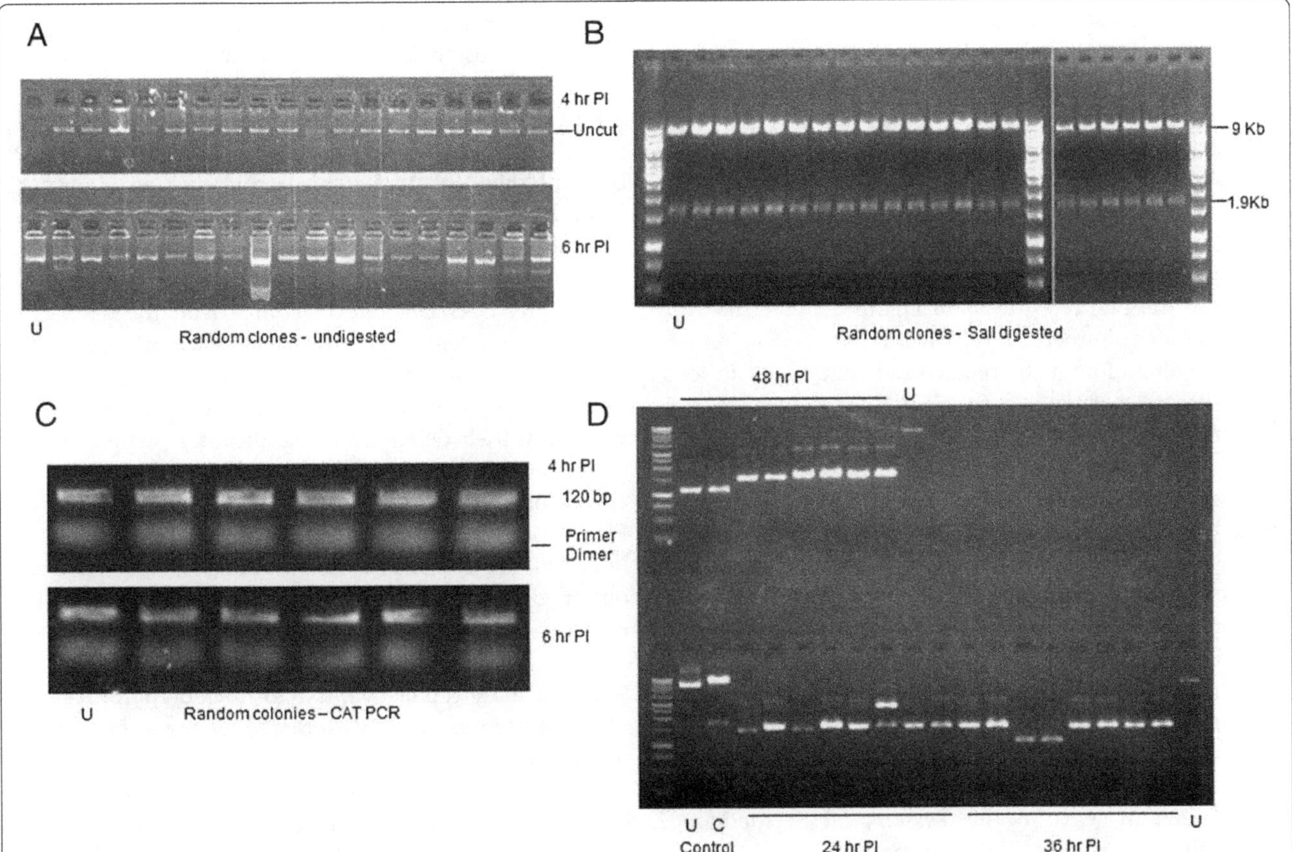

Figure 4 High fidelity repair in T-HO-CAT cells. (A) Episomes were recovered from cells after Ad-HO infection at 4 h and 6 h time point, subsequently used to transform bacteria, and plated on ampicillin plates. Boil prep of random clones from the 4 h and 6 h plates was performed to obtain the episomes and were run on 0.8% agarose gel to depict difference in size after repair (10.9 Kb in size). **(B)** SalI digestion of plasmids obtained from random clones of 4, 6, and 8 h plates PI was performed to reveal deletion in the range of 250–500 bp in size (Refer plasmid map and description in text). **(C)** PCR was performed across the DSB site using primers generating amplicon size of 120 bp with plasmids from random clones picked up from 4 h and 6 h time point plates and subsequently run in a 2.5% agarose gel to reveal micro deletions or insertions. **(D)** Accurately repaired clones were removed after isolating the episomes from cells at given time points PI by employing NheI digestion, thus enriching for inaccurately repaired plasmids (truncated ones). Episomes recovered from cells at indicated time points were digested with NheI overnight and subsequently transformed. Plasmids were subsequently obtained from random clones across different time points and were run on a 0.8% agarose gel. 'U' denotes uncut control plasmid. 'C' denotes SalI cleaved control plasmid.

dozens of randomly picked colonies were subjected to PCR using these primers. However, we were unable to detect any such clones with small deletions (Figure 4C) (we previously noted that resection of a 4 bp 5′-overhang with S1 nuclease or its fill-in with Klenow was visibly shifted on a 2.5% agarose gel). In fact, most instances of rearranged clones could only be identified after digesting the rescued plasmids with NheI which has its recognition sequence on both sides and very close to the HO site. Accurately repaired plasmids would maintain the NheI site and get linearized upon digestion and thus hardly transform bacteria [The background of colonies from NheI-cut plasmids not processed for repair in mammalian cells was ~4 logs less than uncut plasmids; while for plasmids with 3′ overhangs (as produced by digesting with *Pst*I or HO) was even less]. Forced selection of plasmids (deleted in mammalian cells) in this fashion; revealed presence of extensively truncated plasmids from different time points, upon transformation into bacteria (Figure 4D).

To summarize, deletions/insertions in the recovered plasmid population from T-HO-CAT cells at different time points during the course of repair was a very rare event and could only be detected in less than 10% of rescued plasmids.

As a means to quantitatively and independently estimate accurate/inaccurate repair, and also to be sure that cutting and repair was indeed taking place, we also employed the Luc reporter as in Figure 3. Since the HO site is cloned immediate upstream of the Luciferase ORF, any alteration at the repair site would result in loss of Luciferase normalized to GFP fluorescence at the undamaged site. In a transient transfection assay, 293 T cells were transfected with the reporter plasmid with or without concomitant infection with Ad-HO virus. Luciferase and GFP expression in these assays becomes clearly detectable by 4 h even with a transfection efficiency of 20-30% (Figure 5B). Maximal Luciferase accumulation (~10^6 photons/sec - set as 1) is achieved after ~16 h (Figure 5A). An initial reduction in Luciferase light units is indicative of DSB induction due to turn over of the enzyme about 3 h later, but subsequently accurate repair reveals full recovery of luminescence (Figure 5A). These results indicated that at least in this system, NHEJ mediated end-joining is generally highly accurate.

Metnase promotes end-processing and erroneous repair

In context of DNA repair proteins, a major difference between 293 T cells and HEK 293 cells is that Metnase is not expressed in 293 T cells but is expressed in HEK-293 cells [45]. To determine the effect of Metnase expression on episomal repair fidelity, we conducted an identical study using HEK 293 cells. Stable cell line (3-HO-CAT) was generated maintaining multiple copies of pREP4/HO-CAT plasmid. We also over-expressed Metnase in the 293 T-HO-CAT cell line that lacks endogenous Metnase (Figure 6A). DSB repair assays in all the 3 cell lines were carried out in parallel and replica plating was carried out to fish out clones with altered repair junction. As shown in Figure 6C and 6D, HEK 293 cells (express Metnase) revealed a population of plasmids (30-35%) harboring deletions. Similarly, over expression of Metnase in Met-T-HO-CAT cell line (293 T cells lack Metnase) resulted in high frequencies of deletions in 40-45% of total plasmid population (we analyzed typically at least 10 plates). In addition, there were fewer overall colonies obtained, suggesting more degradation in Metnase expressing cells. These results suggest a prominent role for Metnase in endonuclease processing at least in case of DSBs that leave a 4 bp 3′ overhang. Further, using qPCR we also found that the yield of plasmids with repaired joint, generating the CAT PCR product, was lower in cells overexpressing Metnase across different time points PI (Figure 6B).

Further, to assess the frequency and nature of rearrangements, we removed the accurately repaired clones from the rescued plasmid population (at 4 h PI) by NotI digestion (refer to plasmid map in Figure 7C) from both 293 T-HO-CAT cells and Met-T-HO-CAT cells. A large increase in number of colonies that did not replica-plate was obtained with episomes rescued from Met-T-HO-CAT cell line in comparison to T-HO-CAT cells, independently confirming results obtained in Figure 6C. As shown in Figure 7A (upper panel), presence of Metnase generated a population of plasmids widely varying in their deletions. A significantly smaller number of clones were recovered from T-HO-CAT cells for the corresponding time point (Figure 7A, lower panel and Figure 4D) and almost all of them are extensively deleted. Figure 7B shows distribution of random clones obtained from these two different cell lines with different extent of deletions.

Study of early repair events and nucleolytic processing of the episomes after HO induced cleavage by Southern blotting

To probe the cleavage and repair of episomes at a molecular level, we resorted to southern blotting at different time points post-Ad-HO infection. Such a method is expected to emphasize the most prominent cleavage products obtained in the HO-infected population, to generate prototypical episomal processing patterns. The control plasmid (pREP4-HO-CAT from non-infected cells) was digested with Csp45I and NotI to yield bands of size close to 9Kb and 2Kb (N.I. 1st lane, Figure 7D). Plasmids rescued at different time points from infected cells were digested with Csp45I only (refer plasmid map in Figure 7C). As the NotI site is adjacent the HO site,

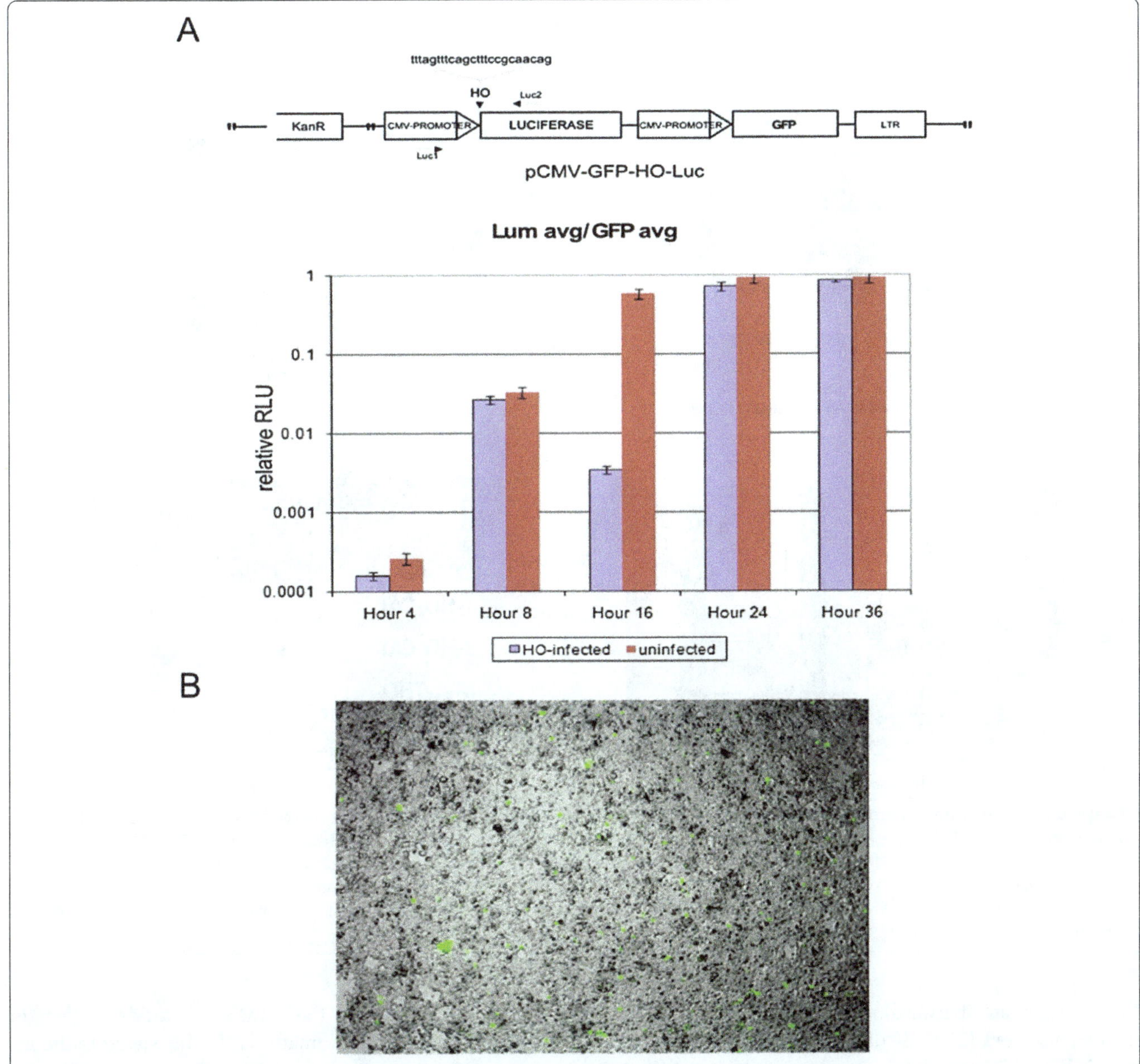

Figure 5 Assay for repair fidelity in T-HO-CAT cells using luciferase reporter. (A) Luciferase activity in 293 T cells upon transfection of HO-Luc reporter as in Figure 3 at different time points. **(B)** Demonstration of GFP-positive cells at 4 h post transfection as a measure of transfection efficiency.

we should expect to observe a similar banding pattern as for the control (1st lane) upon action by HO endonuclease. Figure 7D shows an expected cleavage pattern. Interestingly, we also observed a band migrating faster than the 9 Kb product (see 4 h time point, arrowhead) and a smaller species of approximately 1.5 kb that could represent further nucleolytic processing of DNA fragments following the initial HO cleavage.

In the presence of elevated Metnase expression (Met-T-HO-CAT cells), quantitatively the recovery of intact episomes after the given time window for repair (indicated as 11Kb band product) was less in comparison to episomes recovered from cells without Metnase (compare the 4 h and 6 h PI lanes; Figure 7D). Of note, we would like to remind that similar results were obtained with quantitative PCR showing that overexpression of Metnase delayed the kinetics of repair, resulting in diminished yield of repaired episomes for a given time point in comparison to control (Figure 6B), independently confirming the southern blot finding.

Sequencing of deleted clones obtained from Metnase over expressing cells revealed a recurrent pattern of DNA resection observed at the break site. Many clones lack a region of sequence, close to 200 bp in length, upstream of

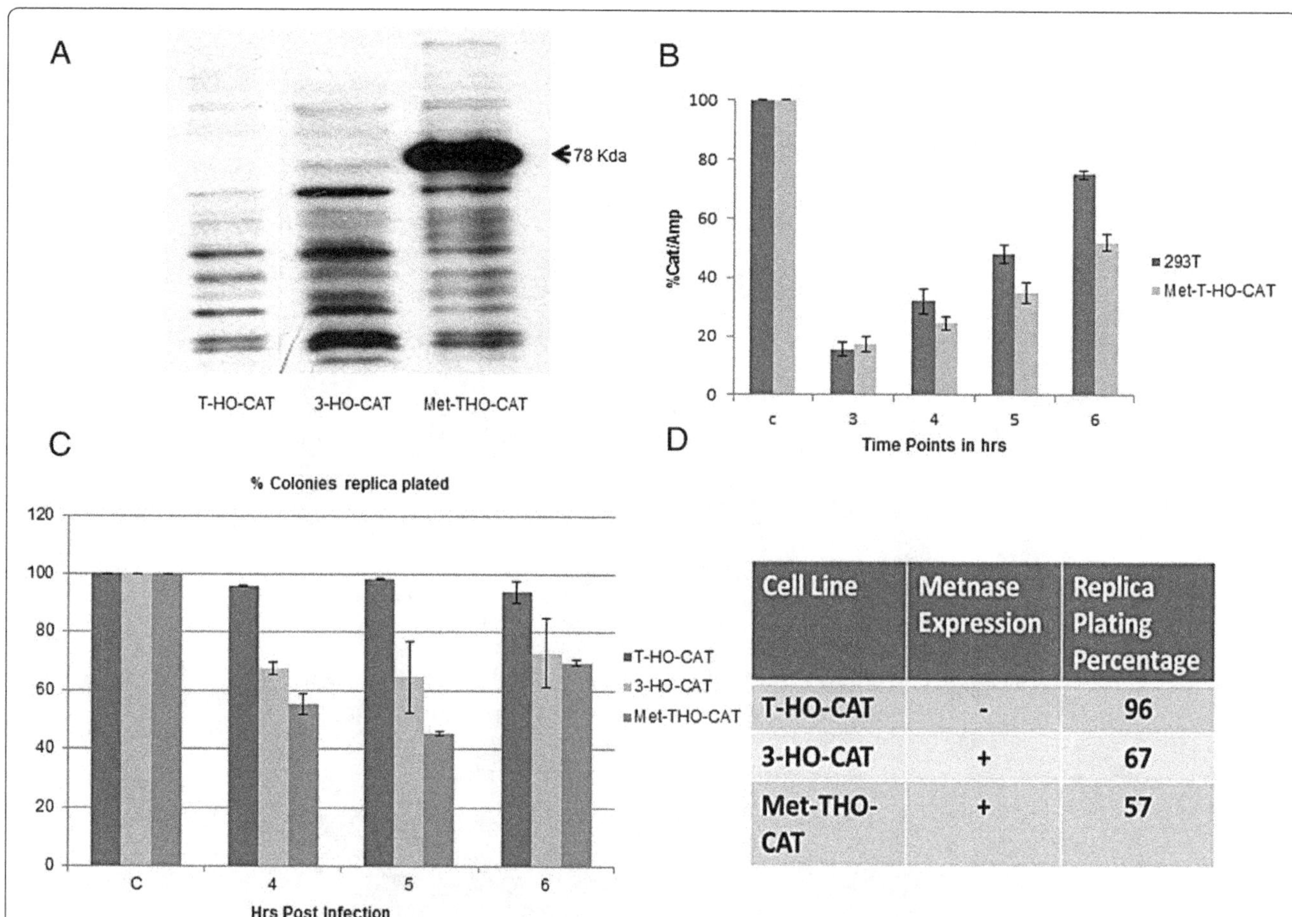

Cell Line	Metnase Expression	Replica Plating Percentage
T-HO-CAT	-	96
3-HO-CAT	+	67
Met-THO-CAT	+	57

Figure 6 Metnase promotes erroneous end-processing. (A) Immunoblot showing stable Metnase overexpression in Met -THO-CAT cells along with endogenous level of Metnase in 3-HO-CAT cells. Arrow marks the correct size band for Metnase in the indicated cells. **(B)** qPCR analysis of kinetics of episomal cleavage and repair in 293 T cells (without Metnase) and Met-T-HO-CAT cells (expressing Metnase) as described in Figure 1B & 1C. qPCR values are represented as (CAT/AMP)% over control. **(C)** Replica plating assay conducted for indicated time points in three different cell lines with varying Metnase expression. **(D)** Table showing average replica plating efficiency obtained from 3 different time points for each cell line and from three independent experiments.

the cleavage site (leftward) while getting more heterogeneously resected (5′ to 3′ direction) on the other side of the DSB (Additional file 3: Supplementary text files 1 & 2). This is consistent with a very similar observation previously reported in yeast for HO induced breaks [46]. There is remarkable similarity in terms of position and segment length of the missing region found on different clones, which are of different sizes. In this regard, it is worth mentioning a very recent report by Adkins *et al.*, involving studies done with nucleosomal substrates, presented evidence for requirement of a nucleosome-free gap region adjacent to the DSB for Sgs1-Dna2 dependent resection machinery [47]. Very recently, resection proteins Sgs1 and Exo1 have been also implicated in G1 checkpoint activation in budding yeast [48]. Of note, a nucleosome plus the linker region is approximately 200 bp of DNA sequence. We also had made similar observations for precise removal of nucleosomal-length fragments at the single genomic HO-DSB in the MM3MG model when in the context of heterochromatin [17]. This suggests the involvement of a common mechanism for providing a template more amenable for endonucleolytic processing, especially active in presence of Metnase. As a last observation we should note that the Southern blots also indicate that the repair mechanism appears to be predominantly accurate NHEJ (simple plasmid re-ligation) for these episomes, as we have rarely observed evidence of formation of concatemers, even when the population of plasmids was not completely cut, and thus offering intact template strands for HR between plasmid molecules, thus presumably resulting in Holliday junctions and concatemeric units that we were unable to detect on Southern blots. Even after presumably inhibiting NHEJ with a general inhibitor of PIKs (Wortmannin) or a more specific inhibitor of DNA-PKcs (KU55933) and hence shifting the balance more in favor of HRR, the majority of repair was again

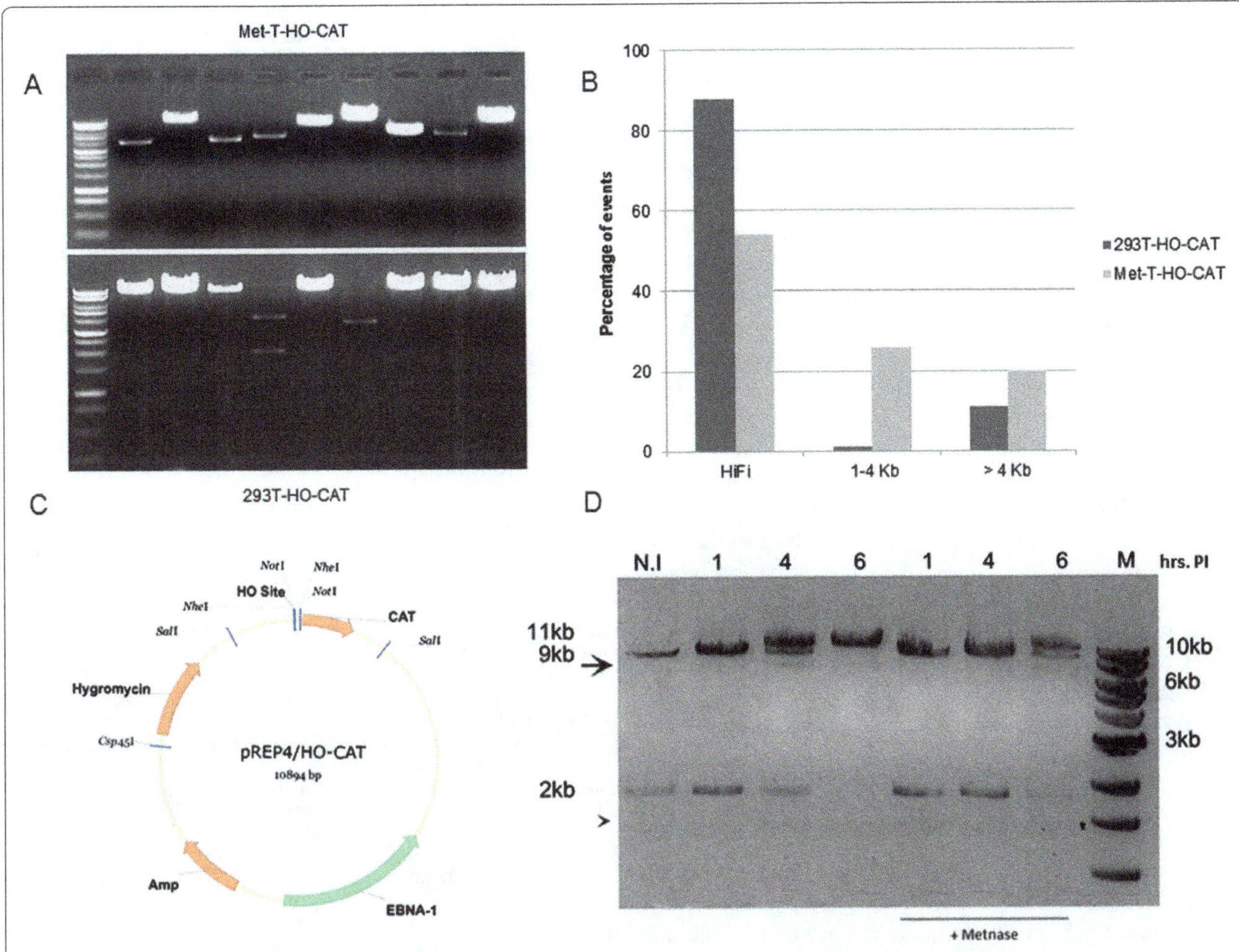

Figure 7 Characterization of effect of Metnase on end-processing. (A) Representation of deleted clones obtained from Met-T-HO-CAT cells and T-HO-CAT cells after NotI digestion of rescued plasmid population at 6 h time point post-infection. **(B)** Size distribution of deleted clones obtained from the two different cell lines. **(C)** Cartoon displaying position of different restriction enzymes on HO-CAT episome used for various digestions and southern blot. **(D)** Southern blot for pRep4-HO-CAT fragments recovered at given time points PI from cell lines in presence or absence of Metnase, after appropriate restriction digestion (see the description in text). N.I. denotes the episomes recovered from non-infected cells. Different arrow positions show processed episomal products. 11Kb size arrow denotes the singly-cut plasmid. M denotes the molecular weight marker, that is NEB 1 kb ladder labeled with PNK and γATP^{32}.

NHEJ and the proportion of concatemers (based on position of the gel) was hardly increased (Additional file 4: Figure S3). The Southern blot showed that NHEJ was inhibited by the KU55933 (reduction of the 11 kb religated product) but there was only a modest shift toward formation of concatemers (generated via HRR).

Discussion

In this report, we present results from an episomal model system to study end-joining process in intact mammalian cells upon induction of a DSB. This system is designed to overcome the limitations of existing DSB repair assays and gives the opportunity to evaluate end-joining fidelity of multiple DSBs in a single cell with 4 bp 3′ overhang *in vivo*. We also evaluated the role of Metnase and its contribution towards determining repair fidelity in this system. We found that > 90% of the recovered clones from the T-HO-CAT cells, which lack endogenous Metnase, are precisely rejoined indicating high fidelity of repair (Figure 4). This is in agreement with some previous reports indicating highly accurate NHEJ in mammalian cells [49-51]. In fact, a very recent report suggested the repair accuracy to be as high as 99% in a modified I-SceI mediated cleavage and repair assay and also showed precise ligation to be mediated by classical NHEJ components [52]. However, Metnase expression may play a unique role in determining the fidelity of repair specifically in humans. Metnase is only known to be expressed in anthropoid primates (humans and apes). It has both histone methylase and DNA endonuclease activity and has been shown to promote NHEJ-mediated erroneous repair of DSBs (for a review [53]). Previous reports using cell extract based assays have been inconclusive in determining the exact role of Metnase in processing of DSBs with overhangs [29,30]. Upon Metnase overexpression, we observed a 25-30% of inaccurately repaired plasmids by replica plating assay. Molecular analysis of recovered clones and direct southern blot results independently confirm this observation. This is suggestive of a prominent role of Metnase as an endonuclease for processing of DSBs. The southern blots specifically indicate a loss of full-length repaired products (compare 6 h time point of T-HO-CAT and Met-T-HO-CAT cells in Figure 7D). This is in accordance to a recent report using cellular extract where addition of Metnase did not increase yield of repaired products but resulted in increased resection and seemed to inhibit repair [30]. However, by Southern blotting we also did not find evidence of a role for Metnase in preventing larger deletions (Figure 7B) as previously reported [27]. These findings, along with evidence of clones with precise removal of a nucleosomal-length fragment at the left (5′ side) of the break may implicate a broader role of Metnase in determining the outcome of end-joining events in mammalian cells.

Conclusion

We have carried out a careful analysis of the fidelity of repair via NHEJ in mammalian cells using a high copy number episomal system that can be cleaved to generate homogeneous DSBs, and found that rather than error-prone, the repair was highly accurate in most cases. NHEJ was the predominant pathway of repair in this system, where the majority of plasmid was cleaved and leaving minimum number of intact copies for HR. The outcome of the accurate repair depended on presence of Metnase, a nuclease present in human cells, and which generated a pattern of discrete deletions near the DSB and resulting in much less accurate repair.

Additional files

Additional file 1: Figure S1. DNA damage response (DDR) activation in 293 T-HO-CAT cells upon Ad-HO infection. Infection with Ad-HO virus and generation of a single DSB in episomes (in T-HO-CAT cells) results in the phosphorylation of ATM at S1981 (activation), TLK1 (S695, inhibition), and H2AX (S139). Infection of 293 T cells that do not contain the HO-targeted episomes does not result in sufficient ATM activation (bottom panel). Cells were infected and whole cell lysates were collected at indicated time points and immunoblotted with appropriate antibody. "C" denotes the uninfected control. Drug combination (HU + TFP) was used as a positive control for ATM activation due to DNA damage (Reproduced with permission from [34]).

Additional file 2: Figure S2. pS/T-Q status in 293 T cells upon doxorubicin treatment. WB for pS/T-Q proteins of 293Tcells incubated or not with doxorubicin and then allowed to recover after removing the drug for different hours (R1, R3, R5).

Additional file 3: Supplementary text files 1 and 2. Sequencing data and alignment from selected deleted clones.

Additional file 4: Figure S3. Evidence for presence of rare concatemers after DN-PKcs inhbition. 293 T-HO-CAT cells were treated 2 h prior to Ad-HO infection either with Wortmannin (20 μM) or KU55933 (10 μM) (potent ATM inhibitor). Episomes recovered from cells at indicated time points and were processed as described previously in text and Figure 7. Concatamers can be observed as slow migrating high molecular weight forms (> 11Kb).

Competing interests

The authors declare that there are none with publication of this study.

Authors' contributions

ADB and AR and RH conceived the study and wrote the manuscript. AR carried out the experiments. All authors have approved the final manuscript.

Acknowledgements

We thank Ms. Sharon Ronald and Mr. Sanket Awate for technical help and Dr. Brent Reed for help with designing the vector map.

Funding

This work was supported by grant W81XWH-10-1-0120 IDEA Development Award from the Department of Defense Prostate Cancer Research Program.

Author details

[1]Department of Biochemistry and Molecular Biology, Louisiana State University Health Sciences Center, 1501 Kings Highway, Shreveport, LA 71130, USA. [2]Department of Medicine, College of Medicine, University of Florida & Shands, Gainesville, FL 32610-0277, USA.

References

1. Ciccia A, Elledge SJ: **The DNA damage response: making it safe to play with knives.** *Mol Cell* 2010, **40**(2):179–204.
2. Helleday T, Lo J, van Gent DC, Engelward BP: **DNA double-strand break repair: from mechanistic understanding to cancer treatment.** *DNA Repair (Amst)* 2007, **6**(7):923–935.
3. Jackson SP, Bartek J: **The DNA-damage response in human biology and disease.** *Nature* 2009, **461**(7267):1071–1078.
4. Pardo B, Gomez-Gonzalez B, Aguilera A: **DNA repair in mammalian cells: DNA double-strand break repair: how to fix a broken relationship.** *Cell Mol Life Sci* 2009, **66**(6):1039–1056.
5. Moynahan ME, Jasin M: **Mitotic homologous recombination maintains genomic stability and suppresses tumorigenesis.** *Nat Rev Mol Cell Biol* 2010, **11**(3):196–207.
6. Lieber M: **The mechanism of double-strand DNA break repair by the nonhomologous DNA end-joining pathway.** *Annu Rev Biochem* 2010, **79**:181–211.
7. Nagy Z, Soutoglou E: **DNA repair: easy to visualize, difficult to elucidate.** *Trends Cell Biol* 2009, **19**(11):617–629.
8. Reynolds P, Botchway SW, Parker AW, O'Neill P: **Spatiotemporal dynamics of DNA repair proteins following laser microbeam induced DNA damage - When is a DSB not a DSB?** *Mutat Res* 2013, **756**(1-2):14–20.
9. Perrin A, Buckle M, Dujon B: **Asymmetrical recognition and activity of the I-SceI endonuclease on its site and on intron-exon junctions.** *EMBO J* 1993, **12**(7):2939–2947.
10. Polo SE, Jackson SP: **Dynamics of DNA damage response proteins at DNA breaks: a focus on protein modifications.** *Genes Dev* 2011, **25**(5):409–433.
11. Sonoda E, Hochegger H, Saberi A, Taniguchi Y, Takeda S: **Differential usage of non-homologous end-joining and homologous recombination in double strand break repair.** *DNA Repair* 2006, **5**(9–10):1021–1029.
12. Barlow JH, Lisby M, Rothstein R: **Differential regulation of the cellular response to DNA double-strand breaks in G1.** *Mol Cell* 2008, **30**(1):73–85.

13. Smith J, Baldeyron C, De Oliveira I, Sala-Trepat M, Papadopoulo D: **The influence of DNA double-strand break structure on end-joining in human cells.** *Nucleic Acids Res* 2001, **29**(23):4783–4792.
14. Sen SP, De Benedetti A: **TLK1B promotes repair of UV-damaged DNA through chromatin remodeling by Asf1.** *BMC Mol Biol* 2006, **7**:37.
15. Kumala S, Fujarewicz K, Jayaraju D, Rzeszowska-Wolny J, Hancock R: **Repair of DNA strand breaks in a minichromosome in vivo: kinetics, modeling, and effects of inhibitors.** *PLoS One* 2013, **8**(1):e52966.
16. Sunavala-Dossabhoy G, De Benedetti A: **Tousled homolog, TLK1, binds and phosphorylates Rad9; TLK1 acts as a molecular chaperone in DNA repair.** *DNA Repair* 2009, **8**(1):87–102.
17. Kanikarla-Marie P, Ronald S, De Benedetti A: **Nucleosome resection at a double-strand break during Non-Homologous Ends Joining in mammalian cells - implications from repressive chromatin organization and the role of ARTEMIS.** *BMC Res Notes* 2011, **4**:13.
18. Deem AK, Li X, Tyler JK: **Epigenetic regulation of genomic integrity.** *Chromosoma* 2012, **121**(2):131–151.
19. North P, Ganesh A, Thacker J: **The rejoining of double-strand breaks in DNA by human cell extracts.** *Nucleic Acids Res* 1990, **18**(21):6205–6210.
20. Chen S, Inamdar KV, Pfeiffer P, Feldmann E, Hannah MF, Yu Y, Lee JW, Zhou T, Lees-Miller SP, Povirk LF: **Accurate in vitro end joining of a DNA double strand break with partially cohesive 3'-overhangs and 3'-phosphoglycolate termini: effect of Ku on repair fidelity.** *J Biol Chem* 2001, **276**(26):24323–24330.
21. Verkaik NS, Esveldt-van Lange RE, van Heemst D, Bruggenwirth HT, Hoeijmakers JH, Zdzienicka MZ, van Gent DC: **Different types of V(D)J recombination and end-joining defects in DNA double-strand break repair mutant mammalian cells.** *Eur J Immunol* 2002, **32**(3):701–709.
22. Poplawski T, Pastwa E, Blasiak J: **Non-homologous DNA end joining in normal and cancer cells and its dependence on break structures.** *Genet Mol Biol* 2010, **33**(2):368–373.
23. Magin S, Saha J, Wang M, Mladenova V, Coym N, Iliakis G: **Lipofection and nucleofection of substrate plasmid can generate widely different readings of DNA end-joining efficiency in different cell lines.** *DNA Repair* 2013, **12**:148–160.
24. Guirouilh-Barbat J, Huck S, Bertrand P, Pirzio L, Desmaze C, Sabatier L, Lopez B: **Impact of the KU80 pathway on NHEJ-induced genome rearrangements in mammalian cells.** *Mol Cell* 2004, **14**(5):611–623.
25. Reeves R, Gorman CM, Howard B: **Minichromosome assembly of non-integrated plasmid DNA transfected into mammalian cells.** *Nucleic Acids Res* 1985, **13**(10):3599–3615.
26. Papior P, Arteaga-Salas JM, Günther T, Grundhoff A, Schepers A: **Open chromatin structures regulate the efficiencies of pre-RC formation and replication initiation in Epstein-Barr virus.** *J Cell Biol* 2012, **198**(4):509–528.
27. Hromas R, Wray J, Lee SH, Martinez L, Farrington J, Corwin LK, Ramsey H, Nickoloff JA, Williamson EA: **The human set and transposase domain protein Metnase interacts with DNA Ligase IV and enhances the efficiency and accuracy of non-homologous end-joining.** *DNA Repair (Amst)* 2008, **7**(12):1927–1937.
28. Lee SH, Oshige M, Durant ST, Rasila KK, Williamson EA, Ramsey H, Kwan L, Nickoloff JA, Hromas R: **The SET domain protein Metnase mediates foreign DNA integration and links integration to nonhomologous end-joining repair.** *Proc Natl Acad Sci U S A* 2005, **102**(50):18075–18080.
29. Beck BD, Lee SS, Williamson E, Hromas RA, Lee SH: **Biochemical characterization of metnase's endonuclease activity and its role in NHEJ repair.** *Biochemistry* 2011, **50**(20):4360–4370.
30. Mohapatra S, Yannone SM, Lee SH, Hromas RA, Akopiants K, Menon V, Ramsden DA, Povirk LF: **Trimming of damaged 3′ overhangs of DNA double-strand breaks by the Metnase and Artemis endonucleases.** *DNA Repair (Amst)* 2013, **12**(6):422–432.
31. Povirk LF: **Processing of Damaged DNA Ends for Double-Strand Break Repair in Mammalian Cells.** *ISRN Molecular Biology* 2012, **2012**:16.
32. Nicolas AL, Munz PL, Falck-Pedersen E, Young CS: **Creation and repair of specific DNA double-strand breaks in vivo following infection with adenovirus vectors expressing Saccharomyces cerevisiae HO endonuclease.** *Virology* 2000, **266**(1):211–224.
33. Middleton T, Sugden B: **Retention of plasmid DNA in mammalian cells is enhanced by binding of the Epstein-Barr virus replication protein EBNA1.** *J Virol* 1994, **68**(6):4067–4071.
34. Ronald S, Awate S, Rath A, Carroll J, Galiano F, Dwyer D, Kleiner-Hancock H, Mathis JM, Vigod S, De Benedetti A: **Phenothiazine Inhibitors of TLKs Affect Double-Strand Break Repair and DNA Damage Response Recovery and Potentiate Tumor Killing with Radiomimetic Therapy.** *Genes Cancer* 2013, **4**:39–53.
35. Deckbar D, Birraux J, Krempler A, Tchouandong L, Beucher A, Walker S, Stiff T, Jeggo P, Lobrich M: **Chromosome breakage after G2 checkpoint release.** *J Cell Biol* 2007, **176**(6):749–755.
36. Huang LC, Clarkin KC, Wahl GM: **Sensitivity and selectivity of the DNA damage sensor responsible for activating p53-dependent G1 arrest.** *Proc Natl Acad Sci U S A* 1996, **93**(10):4827–4832.
37. Ben-Yehoyada M, Wang LC, Kozekov ID, Rizzo CJ, Gottesman ME, Gautier J: **Checkpoint signaling from a single DNA interstrand crosslink.** *Mol Cell* 2009, **35**(5):704–715.
38. Goodarzi A, Jeggo P, Lobrich M: **The influence of heterochromatin on DNA double strand break repair: Getting the strong, silent type to relax.** *DNA Repair* 2010, **9**:1273–1282.
39. Kaplun L, Ivantsiv Y, Kornitzer D, Raveh D: **Functions of the DNA damage response pathway target Ho endonuclease of yeast for degradation via the ubiquitin-26S proteasome system.** *Proc Natl Acad Sci* 2000, **97**(18):10077–10082.
40. Canfield C, Rains J, De Benedetti A: **TLK1B promotes repair of DSBs via its interaction with Rad9 and Asf1.** *BMC Mol Biol* 2009, **10**:110.
41. Pahl HL, Sester M, Burgert HG, Baeuerle PA: **Activation of transcription factor NF-kappaB by the adenovirus E3/19 K protein requires its ER retention.** *J Cell Biol* 1996, **132**(4):511–522.
42. Tollefson AE, Ryerse JS, Scaria A, Hermiston TW, Wold WS: **The E3-11.6-kDa adenovirus death protein (ADP) is required for efficient cell death: characterization of cells infected with adp mutants.** *Virology* 1996, **220**(1):152–162.
43. Kim ST, Lim DS, Canman CE, Kastan MB: **Substrate specificities and identification of putative substrates of ATM kinase family members.** *J Biol Chem* 1999, **274**(53):37538–37543.
44. Ronald S, Sunavala-Dossabhoy G, Adams L, Williams B, De Benedetti A: **The expression of tousled kinases in CaP cell lines and its relation to radiation response and DSB repair.** *Prostate* 2011, **71**:1367–1373.
45. Wray J, Williamson EA, Sheema S, Lee SH, Libby E, Willman CL, Nickoloff JA, Hromas R: **Metnase mediates chromosome decatenation in acute leukemia cells.** *Blood* 2009, **114**(9):1852–1858.
46. Lee SE, Moore JK, Holmes A, Umezu K, Kolodner RD, Haber JE: **Saccharomyces Ku70, mre11/rad50 and RPA proteins regulate adaptation to G2/M arrest after DNA damage.** *Cell* 1998, **94**(3):399–409.
47. Adkins NL, Niu H, Sung P, Peterson CL: **Nucleosome dynamics regulates DNA processing.** *Nat Struct Mol Biol* 2013, **20**:836–842.
48. Balogun FO, Truman AW, Kron SJ: **DNA resection proteins Sgs1 and Exo1 are required for G1 checkpoint activation in budding yeast.** *DNA Repair* 2013, **12**(9):751–760.
49. Adams BR, Hawkins AJ, Povirk LF, Valerie K: **ATM-independent, high-fidelity nonhomologous end joining predominates in human embryonic stem cells.** *Aging (Albany NY)* 2010, **2**(9):582–596.
50. Jiang G, Plo I, Wang T, Rahman M, Cho JH, Yang E, Lopez BS, Xia F: **BRCA1-Ku80 protein interaction enhances end-joining fidelity of chromosomal double-strand breaks in the G1 phase of the cell cycle.** *J Biol Chem* 2013, **288**(13):8966–8976.
51. Honma M, Sakuraba M, Koizumi T, Takashima Y, Sakamoto H, Hayashi M: **Non-homologous end-joining for repairing I-SceI-induced DNA double strand breaks in human cells.** *DNA Repair (Amst)* 2007, **6**(6):781–788.
52. Lin WY, Wilson JH, Lin Y: **Repair of chromosomal double-strand breaks by precise ligation in human cells.** *DNA Repair (Amst)* 2013, **12**(7):480–487.
53. Shaheen M, Williamson E, Nickoloff J, Lee SH, Hromas R: **Metnase/SETMAR: a domesticated primate transposase that enhances DNA repair, replication, and decatenation.** *Genetica* 2010, **138**(5):559–566.

6

Three-stage biochemical selection: cloning of prototype class IIS/IIC/IIG restriction endonuclease-methyltransferase TsoI from the thermophile *Thermus scotoductus*

Piotr M Skowron[1], Jolanta Vitkute[2], Danute Ramanauskaite[3], Goda Mitkaite[2], Joanna Jezewska-Frackowiak[1], Joanna Zebrowska[1], Agnieszka Zylicz-Stachula[1] and Arvydas Lubys[2,3*]

Abstract

Background: In continuing our research into the new family of bifunctional restriction endonucleases (REases), we describe the cloning of the *tsoIRM* gene. Currently, the family includes six thermostable enzymes: TaqII, Tth111II, TthHB27I, TspGWI, TspDTI, TsoI, isolated from various *Thermus* sp. and two thermolabile enzymes: RpaI and CchII, isolated from mesophilic bacteria *Rhodopseudomonas palustris* and *Chlorobium chlorochromatii*, respectively. The enzymes have several properties in common. They are large proteins (molecular size app. 120 kDa), coded by fused genes, with the REase and methyltransferase (MTase) in a single polypeptide, where both activities are affected by S-adenosylmethionine (SAM). They recognize similar asymmetric cognate sites and cleave at a distance of 11/9 nt from the recognition site. Thus far, we have cloned and characterised TaqII, Tth111II, TthHB27I, TspGWI and TspDTI.

Results: TsoI REase, which originate from thermophilic *Thermus scotoductus* RFL4 (*T. scotoductus*), was cloned in *Escherichia coli* (*E. coli*) using two rounds of biochemical selection of the *T. scotoductus* genomic library for the TsoI methylation phenotype. DNA sequencing of restriction-resistant clones revealed the common open reading frame (ORF) of 3348 bp, coding for a large polypeptide of 1116 aminoacid (aa) residues, which exhibited a high level of similarity to Tth111II (50% identity, 60% similarity). The ORF was PCR-amplified, subcloned into a pET21 derivative under the control of a T7 promoter and was subjected to the third round of biochemical selection in order to isolate error-free clones. Induction experiments resulted in synthesis of an app. 125 kDa protein, exhibiting TsoI-specific DNA cleavage. Also, the wild-type (wt) protein was purified and reaction optima were determined.

Conclusions: Previously we identified and cloned the *Thermus* family RM genes using a specially developed method based on partial proteolysis of thermostable REases. In the case of TsoI the classic biochemical selection method was successful, probably because of the substantially lower optimal reaction temperature of TsoI (app. 10-15°C). That allowed for sufficient MTase activity *in vivo* in recombinant *E. coli*. Interestingly, TsoI originates from bacteria with a high optimum growth temperature of 67°C, which indicates that not all bacterial enzymes match an organism's thermophilic nature, and yet remain functional cell components. Besides basic research advances, the cloning and characterisation of the new prototype REase from the *Thermus* sp. family enzymes is also of practical importance in gene manipulation technology, as it extends the range of available DNA cleavage specificities.

* Correspondence: arvydas.lubys@thermofisher.com
[2]Thermo Fisher Scientific, V.A. Graiciuno 8, LT-02241, Vilnius, Lithuania
[3]Department of Botany and Genetics, Vilnius University, M.K. Ciurlionio 21/27, LT-03101, Vilnius, Lithuania
Full list of author information is available at the end of the article

Background

Subtype IIS REases, unlike classic Type II REases which recognise palindromic DNA sequences and cleave within those sites, bind an asymmetric DNA sequence and cleave outside it at a defined distance, up to 21 nt, regardless of the sequence within the cleavage site [1,2]. Further structural and functional complications include atypical IIS enzymes, which are a fusion of REase and MTase in a single polypeptide (Subtype IIC) and/or require/are stimulated by SAM (Subtype IIG). Further diversity within Subtypes IIS/IIC/IIG [3-8] is a family of enzymes grouping REases originally found in *Thermus* sp., which includes TsoI. Since we are aiming at studying the *Thermus* sp. IIS/IIC/IIG enzymes family, we have undertaken cloning, expression and characterisation of TsoI. The *Thermus* sp. family enzymes recognise 5–6 bp DNA sequences which show certain similarities: TspGWI [5′-ACGGA-3′ (11/9) [3], TaqII [(5′-GACCGA-3′ (11/9) [8] or 5′-GACCGA-3′ and 5′-CACCCA-3′ (11/9) [7,9], TspDTI [(5′-ATGAA-3′ (11/9) [4]], Tth111II/TthHB27I isoschizomers [(5′-CAARCA-3′ (11/9) [2,8,10]] and TsoI [5′-TARCCA-3′ (11/9) [2,8], (this work)]]. As detected by bioinformatics analysis, enzymes from mesophilic bacteria – RpaI, recognising the 7-bp degenerate sequence 5′-GTYGGAG-3′ (11/9) and CchII, recognizing 5′-GGARGA-3′ (11/9) apparently belong to the *Thermus* sp. family as well [2,11]. The family shares common biochemical features (summarized in Table one in reference [2]), such as a large molecular size of approximately 120 kDa, REase activity affected by SAM or its analogues, similarities in amino acid (aa) sequences despite distinct specificities, an identical cleavage distance of 11/9 nt downstream from the recognition site and the domain architecture related to simplified Type I REases. All characterised family members originate from the genus *Thermus*, suggesting that they have evolved from a common ancestor [4]. Further comparison has revealed that the group is further internally diversified [6,8]. Bioinformatics analysis and site-directed mutagenesis have led to the differentiation between two subfamilies of closely related enzymes: the TspDTI-subfamily, containing TspDTI, Tth111II/TthHB27I, TsoI, CchII and the TspGWI-subfamily, which includes TspGWI, TaqII and RpaI [2,3,6,9]. Besides aa sequence homologies, the subfamilies are also differentiated by the types of their catalytic motifs: the TspDTI-subfamily has atypical REase catalytic motif D-EXE (also detected previously in typical Type II BamHI REase [12]) and cysteine+serine containing SAM binding motif (D/P) PACGSG, while the TspGWI-subfamily has typical REase catalytic motif PD-(D/E)XK and SAM binding motif (DPA(V/M)GTG [6,8]. Moreover, the TspGWI-subfamily exhibits an interesting feature: a novel phenomenon of REase specificity change, induced by a cofactor analogue. Both TspGWI and TaqII can be converted to very frequent app. 3 bp cutters from canonical 5–6 bp sites by replacing SAM with its analogue sinefungin (SIN), with reversed charge distribution ([5,13]; in press). These chemically-induced changes in the recognition sequence differ from the well-known "star activity" phenomenon. They are apparently a result of SIN interaction with an allosteric pocket on the protein surface which binds the stimulatory SAM molecule. Hence, two more novel prototype specificities have been generated by chemical means [5]. Because of the very high frequency of DNA cleavage, they are uniquely suited for use as molecular tools for generating quasi-random genomic libraries [14].

Results and discussion

Cloning, sequencing and analysis of the *tsoIRM* gene

Following our studies of the *Thermus* sp. family of enzymes [4], we cloned the genes coding for TspGWI [6], TspDTI [8], TaqII [GenBank: AY057443, AAL23675; manuscript submitted], TthHB27I/Tth111II [manuscript in preparation] and TsoI [this work; GenBank: KC503938]. Previously, as a result of the enzymes' feature of incomplete DNA digestion, complicating the application of known methods of biochemical selection both for the methylation phenotype [15] or the related 'white-blue' screen for DNA damage/modification [16-18], we developed an *in vitro* approach for thermophilic REase cloning [6,8]. TsoI has a substantially lower reaction optimum than other *Thermus* sp. REases - app. 10-15°C. Thus, according to the chemical rule whereby reaction speeds decrease 2–3 fold per 10°C drop in reaction temperature, it was assumed that TsoI would retain a substantially higher activity *in vivo* in recombinant *E. coli* cells at 37°C than other *Thermus* sp. REases. Accordingly, we expected specific methylation *in vivo* of plasmids carrying the cloned TsoI MTase gene at a high enough level to allow the classic biochemical selection method to be used [15]. Due to the partial cleavage feature of TsoI, however, selection difficulties were anticipated. Thus, a new variant of the classic method of selecting for the methylation phenotype of recombinant "positive" clones was developed. An integral part of the procedure was the use of the proprietary positive selection vector pSEKm′-MCS (Thermo Fisher Scientific/Fermentas). The vector features resistance for both ampicillin and kanamycin, possesses five TsoI targets and selects for any cloned insert, thus decreasing the cloning background which may have resulted from self-ligated vector molecules. Non-specific selection of recombinant plasmids was combined with two rounds of the classic biochemical selection method for the TsoI methylation phenotype. Since the vector included five target recognition sites for TsoI, it was expected to be an efficient substrate for TsoI REase. Notwithstanding, transformation yielded 8 × 10^5 colonies, thus providing an over 1000-fold genome coverage. Such a high coverage seems preferable in order to increase the chances of obtaining intact REase genes, as when cloning highly toxic genes, such as those coding for REases. Owing to the negative selection pressure in the recombinant *E. coli* host against detrimental plasmids, libraries tend to be non-representative. Considering the partial digestion feature of TsoI and the lower enzyme activity *in vivo* at 37°C, a departure from the standard biochemical selection procedure was made: following plasmid DNA isolation from colonies collected as a pool and digestion with excess TsoI, re-transformation and repeated plating, no analysis of the surviving colonies was performed. Instead, the next round of pooled *in toto* colonies from the first round of selection was subjected to repeated plasmid DNA isolation, excess TsoI cleavage, re-transformation and re-plating. After two biochemical selection rounds, 50 clones were analysed by colony PCR to screen for plasmids carrying fragments larger than 3 kb (the expected size of the *tsoIRM* gene). Figure 1 shows that 14 HindIII-digested large recombinant plasmids isolated during the screening procedure which share common cloned fragments of 1.2 and 1.5 kb in size, suggesting that all contain the same genetic locus from *T. scotoductus*.

The initial verification of clones, based on the detection of common restriction fragments (Figure 1), was followed by functional analysis. The putative *tsoIRM* gene was expressed at a detectable level in *E. coli*, since its presence was high enough to ensure protection against TsoI digestion in the biochemical selection procedure. DNA of five individual recombinant plasmids were digested in excess TsoI and analysed using agarose gel electrophoresis (Figure 2). As shown in Figure 2, only traces of completely protected plasmid DNA were observed. Inclusion of λ DNA into parallel reactions served as an internal control which revealed that (*i*) DNA of isolated plasmids do not interfere with the activity of TsoI, and (*ii*) the same minute amount of completely protected plasmid DNA was observed in both reactions with/without λ DNA, which suggest that the TsoI-specific methylation was probable cause of plasmid DNA resistance to cleavage. These *in vivo* results were subjected to further validation by *in vitro* assays for the native TsoI methylation activity using homogeneous enzyme. Under a variety of conditions tested, the MTase specific activity was substantially lower than that of the REase (not shown). Thus, the MTase activity *in vivo* was probably enhanced by cytoplasmic environment and/or extended incubation of cells prior to plasmid isolations. This has provided enough reaction time to significantly protect DNA from further TsoI REase cleavage. Taken together, selection difficulties were possibly related to both the partial cleavage feature of TsoI and incomplete DNA protection by TsoI MTase activity. What's worthy of note, a relatively low level of protection explains why two rounds of biochemical selection were required in order to enrich the library by plasmids carrying the cloned *tsoIRM* gene. The clones shown in Figure 2 were subjected to sequencing of inserted fragments, first by using vector-specific primers and then by insert-specific primers. As a result, the combined and cross-checked 4365-bp long genomic contig from *T. scotoductus* RFL4 was determined [GenBank: KC503938]. The fragment, as analysed by DNASIS MAX software, contained 3348 bp ORF, which encoded a putative long polypeptide, exhibiting a high level of similarity with the Tth111II bifunctional REase-MTase (50% identity, 60% similarity). Since the ORF sequence-based predicted molecular weight of putative 1116 aa TsoI is 127.6 kDa (DNASIS MAX and Vector NTI calculations) which matches very well our previously published SDS/PAGE results of native TsoI (app. 120 kDa) [8], we concluded that the detected cloned ORF indeed codes for the *tsoIRM* gene.

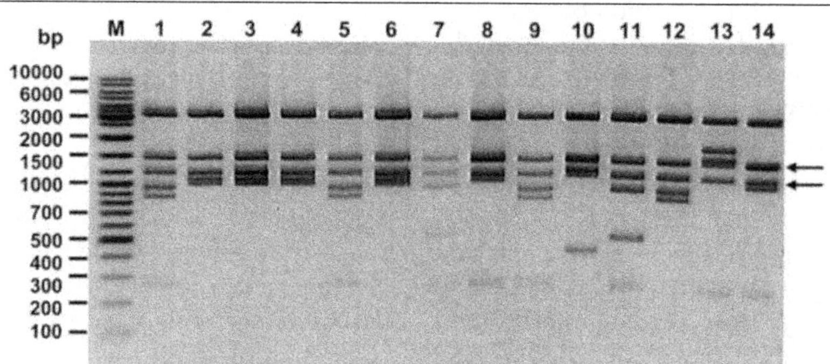

Figure 1 Restriction analysis of large recombinant plasmids isolated during the biochemical selection of *T. scotoductus* library. Patterns of DNA fragments resulting from the HindIII cleavage of recombinant plasmids, isolated after two rounds of biochemical selection using TsoI, visualized on electrophoretic agarose/Tris-borate-EDTA (TBE) gel (1%), stained with ethidium bromide (EtBr). Arrows show cloned fragments of 1.75 and 2 kb in size, which originate from *T. scotoductus* and are common for all TsoI-selected plasmids examined. Lane M - DNA standards (Gene Ruler™ Ladder Mix (Thermo Fisher Scientific (Fermentas); selected bands marked).

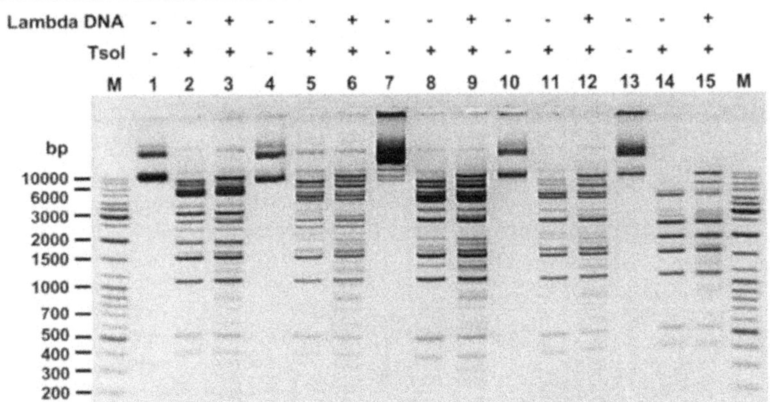

Figure 2 Evaluation of resistance to TsoI digestion of plasmids, obtained after two rounds of biochemical selection. Lanes M – DNA standards (Gene Ruler™ Ladder Mix (Thermo Fisher Scientific (Fermentas); selected bands marked); lanes 1, 4, 7, 10 and 13 - DNAs of individual plasmids, untreated; lanes 2, 5, 8, 11 and 14 - DNAs of individual plasmids treated with TsoI; lanes 3, 6, 9, 12 and 15 - DNAs of individual plasmids supplemented with λ DNA and treated with TsoI. Samples were resolved on 1.2% agarose gel in TBE buffer, stained with EtBr.

Further bioinformatics analysis has revealed that TsoI is moderately basic, with a calculated pI of 8.06-8.11 (Vector NTI and DNASIS MAX calculations, respectively) (the only basic member of the *Thermus* sp. enzyme family). No sequence similarity of TsoI to any MTase or DNA-binding protein was found in the flanking regions of the ORF. There is only a single TsoI recognition site present within the ORF. The ORF begins with the ATG START codon and contains 3 putative upstream RBSs: -7 AG, -10 AGAA and -16 GGGA. Note therefore, that within the ORF there is a second potential ATG START codon, located at residue 48, with a ribosome-binding site upstream GAGGAG, located at a sub-optimal distance of −12 (Figure 3). Translation from the second start codon would result in slightly smaller protein of 1069 aa and 122.1 kDa. The ORF of 1116 aa is GC rich (56.54%); nevertheless, it is markedly lower than other *Thermus* sp. family coding genes, except *tspDTIRM* [8]. Thus, like *tspDTIRM*, *tsoIRM* may have been acquired/evolved differently than other *Thermus* sp. genes, which may have included horizontal gene transfer from a lower GC content bacteria. According to the previously published bioinformatics analysis [8], TsoI exhibited similarity to several known and putative Type IIC/IIG enzymes, including the previously characterised nucleases TthHB27I/Tth111II isoschizomer pair and TspDTI with alignment covering essentially the whole length of the polypeptide. Despite such a high sequence similarity between TthHB27I/Tth111II and TspDTI enzymes, TsoI has a different sequence specificity – 5'-TARCCA-3' [2,8] to TspDTI (5'-ATGAA-3') and TthHB27I/Tth111II (5'-CAARCA-3') – while cleaving at the same distance of 11/9 nt from the recognition site. Nevertheless, all of these asymmetric cognate sequences share two common adenine residues, located at the same positions. In contrast, two other Type IIC/IIG enzymes from *Thermus*, i.e. TspGWI [GenBank: EF095488, ABO26710] and TaqII [GenBank: AY057443, AAL23675], showed very low sequence similarity between TspDTI, TthHB27I/Tth111II and TsoI in pairwise comparisons, dividing the *Thermus* sp. enzymes family into two sub-families. Nevertheless, both sub-families share a common organisation scheme, sharing a modular structure with the same linearly located functional/physical domains of very similar sizes [8]. This scheme is followed by TsoI, which has consecutive fused segments, starting from the N-terminus: (*i*) DNA cleavage/Mg^{2+} -binding (including the atypical D-EXE motif, approximate aa 1–160), (*ii*) a helical region/interaction between domains (fusion subunits link and potentially regulatory region, app. aa 160–360) - recently, the crystal structure of a Type IIC/IIG bifunctional BpuSI REase was established and an alpha-helical domain connecting the REase and MTase domains was suggested to link and regulate structure as well as domain communication - furthermore, it may determine the cleavage distance from the recognition site [19], (*iii*) DNA m6A methylation (including SAM binding motif PPACGSG and methylation catalytic motif NPPW, app. aa 360–790) and (*iv*) DNA sequence recognition region (possibly including two Target Recognition Domains (TRDs), app. aa 790-C-terminus) [8] (Figure 3). Overall, this organisation resembles the simplified (fused in the same polypeptide) HsdR, HsdM and HsdS subunit domain architecture of Type I REases [8].

Expression analysis of the cloned *tsoIRM* gene

Since the primary clones expressed the TsoI REase-MTase at a very low level, as judged by recombinant plasmid partial protection *in vivo*, further constructs were made by subcloning of PCR-amplified *tsoIRM* gene into a pET-derivative vector pET21NS. The minor Multiple Cloning Site (MCS) modification (not shown) was

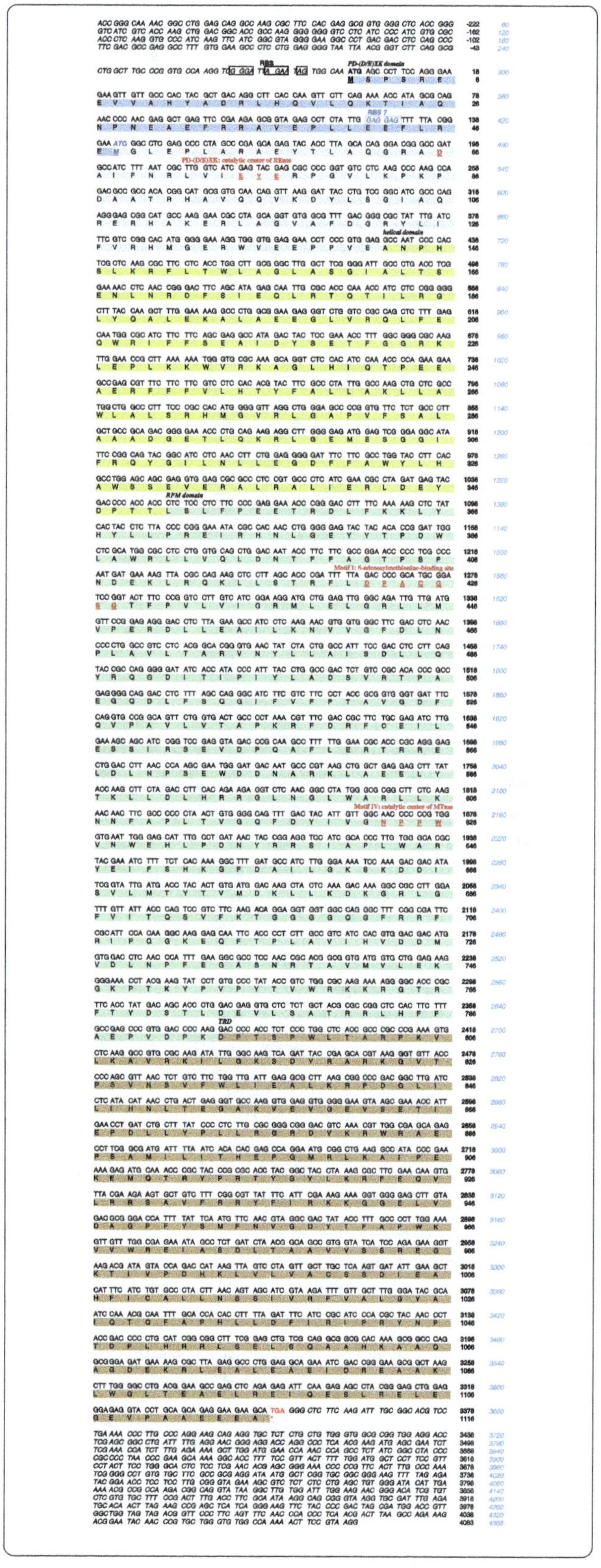

Figure 3 DNA, amino acid sequence and functional motifs of the *tsoIRM* gene and its flanking regions and schematic organization of the enzyme domains. The predicted amino acid sequence of the 127.6 kDa TsoI protein is indicated in capital letters. The DNA sequence is numbered in two styles: (*i*) black numbering starts with negative values, with +1 nt marking the beginning of the *tsoIRM* ORF, (*ii*) the blue numbering corresponds to the entire DNA sequence deposited in the GenBank under acc. number KC503938] (includes *tsoIRM* gene and flanking regions). The ATG START codon is in black bold. The TGA stop codon is shown in red. The potential *tsoIRM* ORF Ribosome Binding Sites (RBS) are boxed. The potential internal ATG START is in blue. The potential *tsoIRM* internal ORF RBS is boxed in blue. The crucial amino acids of the catalytic centres are dark red, bold and underlined. The functional protein domains are marked as follows: REase domain in blue, helical domain in light green, MTase domain in dark green, TRD in brown.

made to allow for directional cloning of the NotI-SmiI cleaved PCR fragment encompassing the full-length TsoI-coding gene. The TsoI ORF was placed under the control of the T7 promoter and strong RBS. IPTG-induction was used to evaluate whether the cloned *tsoIRM* gene was indeed expressed. Crude cell extracts from small scale induced cultures of individual clones revealed a trace amounts of TsoI REase activity, while SDS-PAGE demonstrated abundant amounts of a large enzyme (app. 120 kDa) that appeared after induction (Figures 4,5). Testing of the TsoI protection level of 15 individual plasmids, chosen from PCR subcloning, revealed that they are all unprotected before induction and only partially protected after induction, suggesting that either the chosen PCR conditions may have been favourable for the appearance of errors, or there may have been selective pressure for growth of only those colonies which contain plasmids coding for mutants of TsoI with reduced either the REase, MTase or both activities. Sequencing results of a few of these clones revealed the presence of multiple mutations, which most probably appeared during PCR and were further spontaneously selected *in vivo*, due to lower toxicity to *E. coli* host. In order to isolate clones, which encode highly active TsoI, 5000 ampicillin-resistant colonies obtained after the transformation of expression host *E. coli* ER5266 with a ligation mixture of expression vector and PCR amplified *tsoIRM* gene were subjected to a third round of biochemical selection. Transformants were pooled without separate cultivation of single clones and used directly to inoculate 100 ml LB media (supplemented with ampicillin). Cells were grown at 37°C until the mid-log phase, the T7 promoter was induced by the addition of IPTG, the culture was grown further at 37°C for 4 hours and then used for isolating total plasmid DNA. The latter was cleaved with TsoI and the reaction mixture was introduced back to *E. coli* ER2566. The 20 resulting colonies were again tested for TsoI activity in crude cell extracts and for protection level against TsoI cleavage of

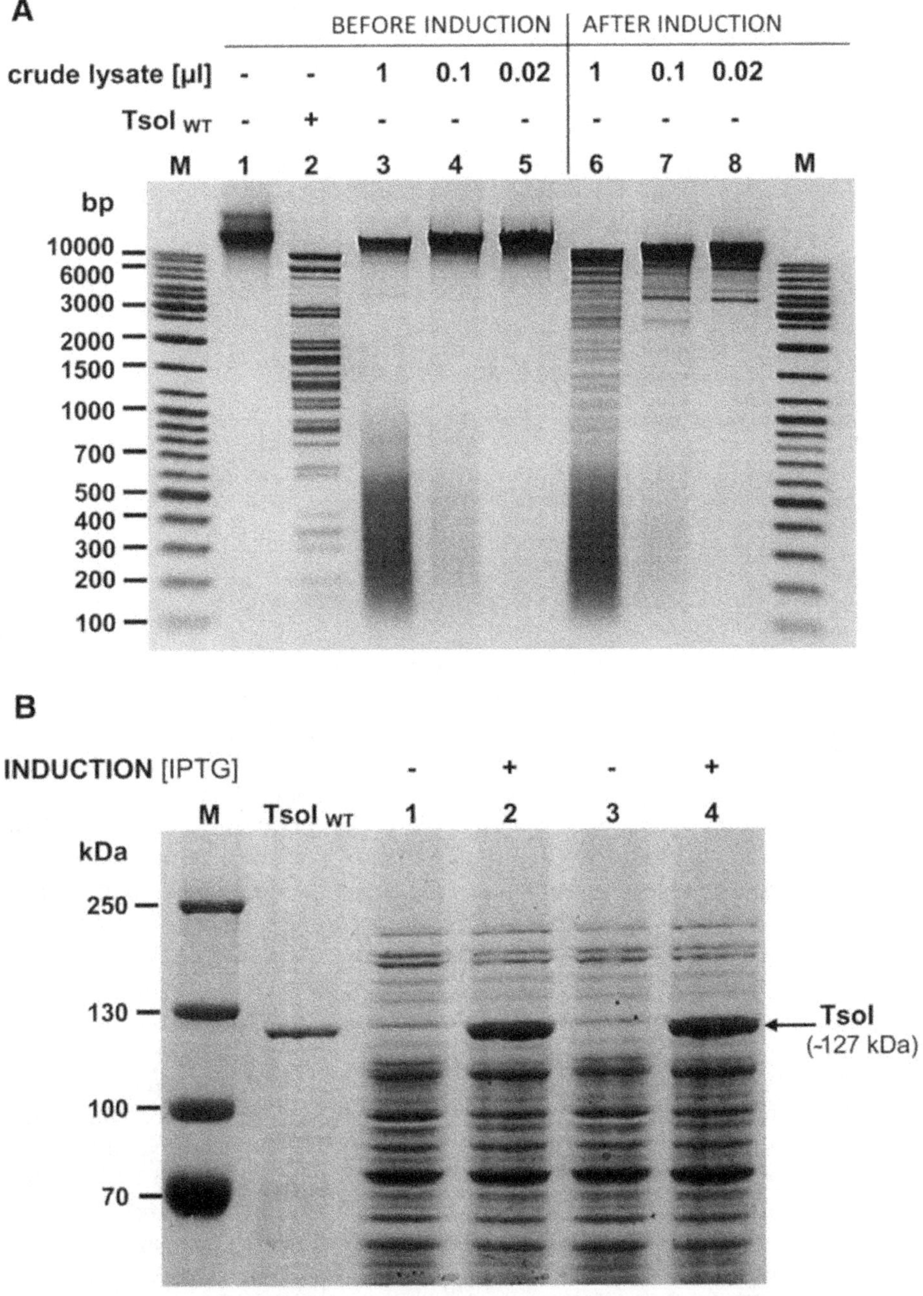

Figure 4 Analysis of the induction pattern of recombinant TsoI REase. (A) Enzymatic activity in crude *E. coli* ER2566/pET21NS-*TsoIRM* extracts. Two selected clones of the confirmed *tsoIRM* gene nucleotide sequence after the 3rd round of biochemical selection were subjected to expression experiments, giving similar TsoI induction patterns. For clarity, the expression experiment of only one selected clone was shown. Lanes M, DNA size standards (Gene Ruler™ Ladder Mix (Thermo Fisher Scientific (Fermentas); selected bands marked); lane 1, control (untreated) λ DNA; lane 2, λ DNA cleaved with native (wt) TsoI; lanes 3, 4, 5, various amounts (1, 0.1, 0.02 μl, respectively) of cell extracts prior to induction incubated with λ DNA in TsoI digestion buffer; lanes 6, 7, 8, various amounts (1, 0.1, 0.02 μl, respectively) of induced cell extracts incubated with λ DNA in TsoI digestion buffer. Samples were resolved on 1.2% agarose gel in TBE buffer, stained with EtBr. **(B)** SDS/PAGE evaluation of TsoI induction in *E. coli* ER2566/pET21NS-*TsoIRM*. Two selected clones as in A were analysed for the presence of TsoI protein bands on 8% SDS/PAGE gels. Electrophoresis was subjected to an extended run in order to clearly visualize protein bands in the high molecular weight range. Lanes M, protein standards (Gene Ruler™ Prestained Protein Ladder Plus (Thermo Fisher Scientific (Fermentas); lane $TsoI_{WT}$, native (wt) TsoI (1 unit); lanes 1 and 3, samples prepared from two colonies of *E. coli* ER2566/pET21NS-*TsoIRM* clones as in A, before induction; lanes 2 and 4, as in lanes 1 and 3, after IPTG induction.

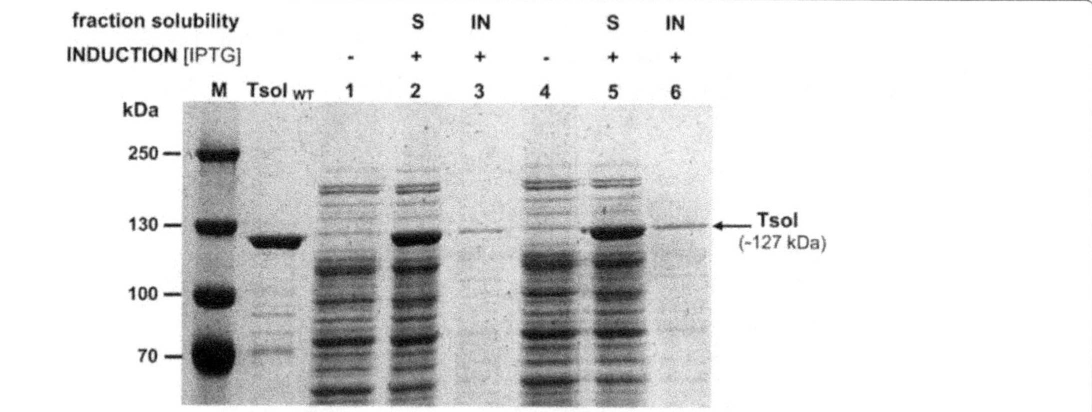

Figure 5 Analysis of the solubility of expressed recombinant TsoI REase SDS/PAGE (8%) analysis of induced versus uninduced and soluble versus insoluble fractions. Lane M, protein standard (PageRuler™ Prestained Protein Ladder Plus (Thermo Fisher Scientific (Fermentas); lane TsoIWT, wt TsoI (2 units); lanes 1 and 4, samples prepared directly from two *E. coli* ER2566 colonies resulting from transformation with the same pET21NS-*TsoIRM* plasmids, before induction; lanes 2 and 5, the same as in lanes 1 and 3, but after IPTG induction (soluble fraction); lanes 3 and 6, the same as in lanes 2 and 5 (insoluble fraction).

recombinant plasmids isolated from induced cultures. In contrast to previous experiments, the crude cell extracts in this case exhibited a much higher TsoI REase activity (Figure 4A), while plasmids, containing the cloned *tsoIRM* gene and isolated from induced strains, were almost completely protected from TsoI cleavage, indicating an adequate MTase activity at the same time (Figure 6, lanes 4, 5 and 6). Four of the selected plasmids were subjected to insert sequencing, and two of them were found to have no mutations in the PCR-amplified TsoI-coding gene, while expressing large amounts of TsoI protein (Figure 4B). One of these plasmids (pET21NS-TsoIRM) was used for further protein isolation and characterisation experiments.

Characterisation of TsoI protein

As shown in the previous section, the expression experiments resulted in the appearance of specific DNA cleavage activity as well as large amounts of a high molecular weight recombinant polypeptide band of app. 120 kDa, which was in fact the dominant protein band in the recombinant *E. coli* lysate. The observed protein location on the SDS/PAGE gel is in very good agreement with the predicted protein size of 127.6 kDa coded by the *tsoIRM* gene. The band on SDS/PAGE, corresponding to the recombinant protein, also perfectly matches the position on the gel of purified wt TsoI from *T. scotoductus* (Figure 4B, lane TsoIWT). On the other hand, preliminary estimation of the TsoI activity in crude cell extract allowed the conclusion to be drawn that it is much lower than it could be expected based on the amount of enzyme synthesised, and suggested that either the recombinant enzyme is insoluble or has lower specific activity. In order to identify the cause for the discrepancy between TsoI activity and its intracellular amount, solubility studies of recombinant TsoI were conducted. The results in Figure 5 clearly indicate that TsoI, when expressed in *E. coli*, apparently retains a soluble conformation. Therefore, disproportionally low endonucleolytic activity of the induced TsoI may be either due to the slower turnover of recombinant enzyme compared with the wt isolate, or higher MTase activity, which (theoretically) could dominate and modify substrate DNA, preventing it from undergoing TsoI endonucleolytic cleavage. If so, dominant MTase activity could explain the apparently lower REase activity of recombinant TsoI. In order to test this idea, bacteriophage λ DNA was incubated in the presence of SAM and Mg^{2+} with cleared lysates prepared from induced and uninduced cultures (Figure 7A). The use of cleared lysates was not problematic, owing to the relatively high concentration of expressed recombinant TsoI. Subsequently, reaction products were purified by chloroform extraction and isopropanol precipitation, following dissolution in TsoI reaction buffer (supplemented with SAM), which were incubated with wt TsoI. The experiment clearly demonstrated that the induced culture not only has quite a weak endonucleolytic activity (Figure 7A, lanes 9 and 12), but also a substantial TsoI MTase activity, which is manifested by the conversion of bacteriophage λ DNA into a partially cleaved yet completely modified form that is resistant to cleavage by the subsequent addition of wt (*T. scotoductus* – isolated) TsoI (Figure 7A, lanes 10 and 13). The additional amount of unmodified λ DNA was subsequently cleaved by adding wt TsoI either completely (Figure 7A, lanes 8 and 11) or partially (lanes 5 and 14), thus suggesting that chloroform extracted and isopropanol precipitated substrates had inhibitory effect on wt TsoI in some cases. Taken together, conclusion could be made that the substrate was completely methylated during incubation with

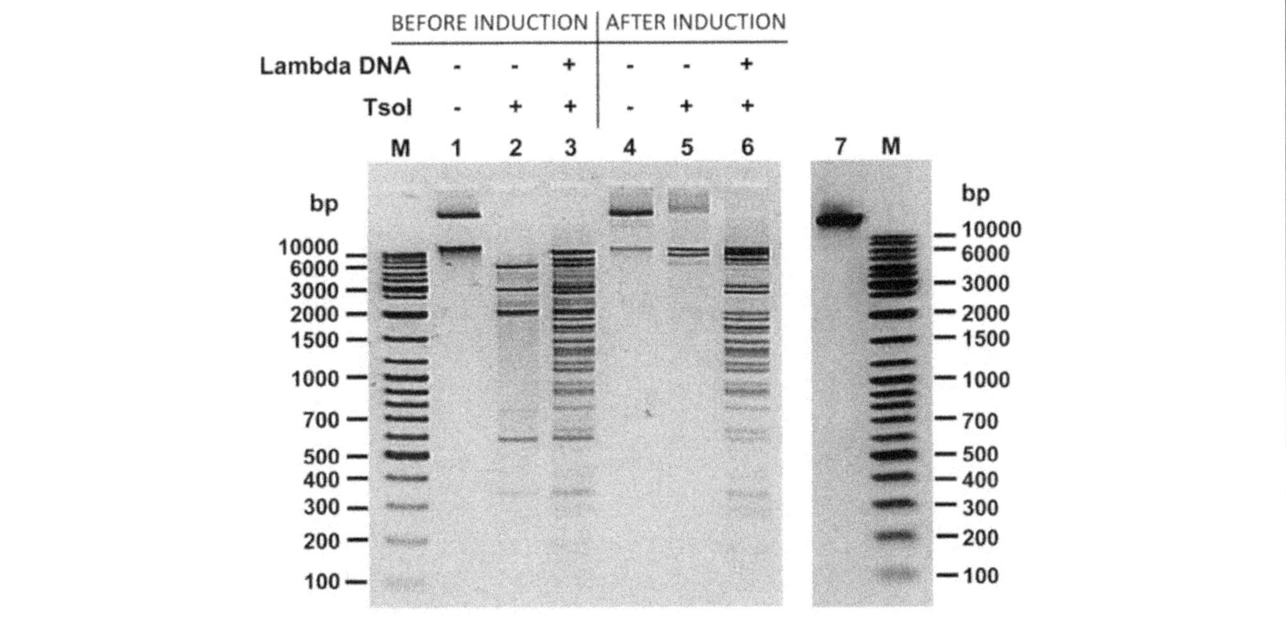

Figure 6 Evaluation of the TsoI-resistance of plasmids isolated from the same culture before and after TsoI synthesis induction. Lane M, DNA standard (Gene Ruler™ Ladder Mix (Thermo Fisher Scientific (Fermentas); selected bands marked); lanes 1 and 4, uncleaved pET21NS-*TsoIRM* plasmid DNA; lanes 2 and 5, pET21NS-*TsoIRM* after incubation with TsoI; lanes 3 and 6, pET21NS-*TsoIRM* supplemented with λ DNA and then incubated with TsoI.

cleared lysate of induced culture in the presence of SAM (Figure 7A). Furthermore, methylation appears to be complete even when a very small amount of cleared lysate is used (Figure 7A, lane 13). These results were compared with further assays using purified, homogeneous wt TsoI, which in turn failed to show that the specific activity of TsoI MTase was higher than that of TsoI REase, regardless of the variety of conditions tested (not shown). However, the standard MTase assay, based on *in vitro* protection against a subsequently added cognate REase, while yielding quantitative results in the case of classic Type II REases, is not perfect for analysis of Subtype IIG/IIC REases. In the case of TsoI, the assay was complicated due to at least three factors: (*i*) fusion of both activities in the same polypeptide, thus allowing for concurrent actions, when divalent cations were present, (*ii*) the REase and MTase protein domains interactions, including SAM binding/allosteric effect and (*iii*) incomplete cleavage by TsoI REase, which prevents from precise distinguishing protected DNA from uncleaved DNA. Other reasons which may explain the difference between higher TsoI MTase activity *in vivo* and in crude lysates as compared to purified wt TsoI, may include: (*i*) protective effect of the high concentration of cellular proteins present *in vivo* and in crude lysates stabilising TsoI MTase, (*ii*) an important cytoplasmic component for methylation is missing or (*iii*) selective inactivation of the MTase domain during purification, while the TsoI REase domain remains functionally intact. This interwound REase-MTase activities' relationship is further complicated by the fact that TsoI is a very slow enzyme. To test whether TsoI REase is a multi- or single turnover enzyme, the serial dilutions under controlled enzyme:recognition sites molar ratios were performed, both in the presence and absence of SAM (Figure 8). Reactions were performed for a prolonged time (overnight), to allow consecutive cleavage reaction cycles. Results shown in Figure 8ABD clearly show that at a molar ratio 1:1 and lower, a single TsoI REase molecule on average performs less than 1 cleavage per single cognate site. Considering that a competitive TsoI MTase reaction may not proceed in an experiment without SAM (Figure 8A), unless some tightly bound SAM is carried over during purification, the conclusion can be drawn that either TsoI REase is a single turnover enzyme or a majority of TsoI molecules were inactivated during purification, diluting functional enzyme molecules with non-active ones. Thus the turnover issue was not conclusively resolved, nevertheless it was confirmed that TsoI is a very "slow" enzyme, as even a very long reaction time did not result in substrate digestion exceeding a 1:1 molar ratio. TsoI approaching the characteristics of a single-turnover may represent an intermediate evolutionary stage.

Based on the fact that the MTase activity of TsoI is entirely dependent on SAM, whereas the activation of the REase function by SAM is weak, although noticeable both during prolonged digestions (Figure 8AB lanes 0.5:1 and Figure 8D) and time-limiting conditions (Figure 8C), the

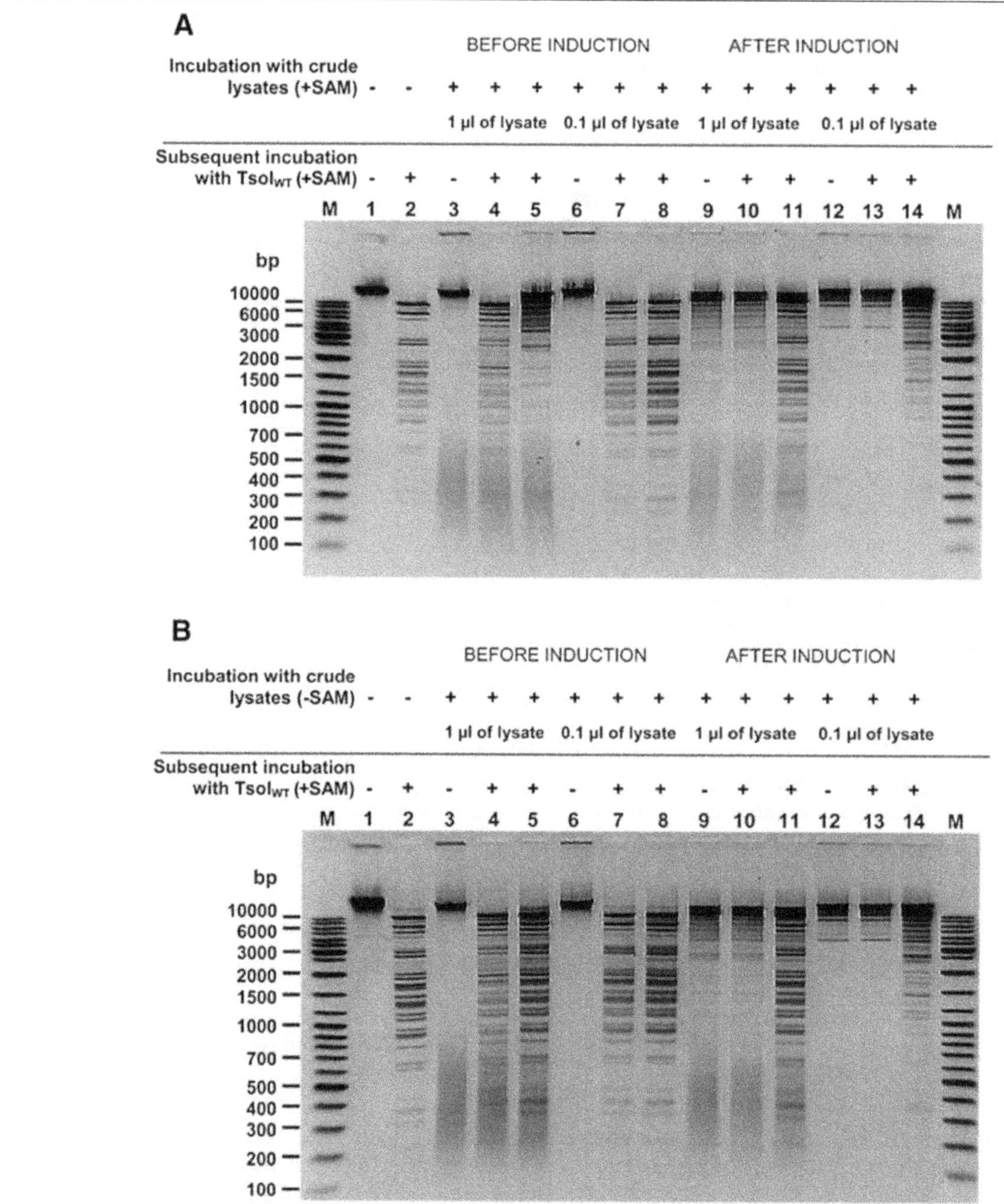

Figure 7 Evaluation of TsoI REase/MTase activities in cleared lysates. (A) Evaluation in the presence of SAM. Lanes M, DNA standard (Gene Ruler™ Ladder Mix (Thermo Fisher Scientific (Fermentas); selected bands marked); lane 1, λ DNA; lane 2, λ DNA, cleaved with wt TsoI; lanes 3, 6, 9, 12, λ DNA after incubation with indicated amounts of cleared lysates prepared from cultures: (*i*) before TsoI induction (3, 6) or (*ii*) after induction (9, 12) in the standard reaction buffer supplemented with SAM; lanes 4, 7, 10, 13, the same as in lanes 3, 6, 9, 12, but after chloroform extraction, isopropanol precipitation and subsequent incubation of dissolved samples with wt TsoI; lanes 5, 8, 11, 14, the same as in lanes 4, 7, 10, 13, except that λ DNA was added before incubation with wt TsoI in order to assess the putative inhibitory effect of the chloroform-extracted DNA on wt TsoI. **(B)** Evaluation in the absence of SAM. Description of lanes as in **A**, with the exception that SAM was not included into the reaction mixtures used to evaluate activities of crude lysates.

idea was proposed that the MTase activity of recombinant TsoI might use TsoI-bound (carried over) SAM which potentially could make a difference between activities of wt and recombinant TsoI variants. In order to test this idea, the same experiment as in Figure 7A was repeated, without the addition of SAM (Figure 7B). Bearing in mind the absolute prerequisite of the presence of SAM for the MTase DNA modification reaction, we expected to obtain

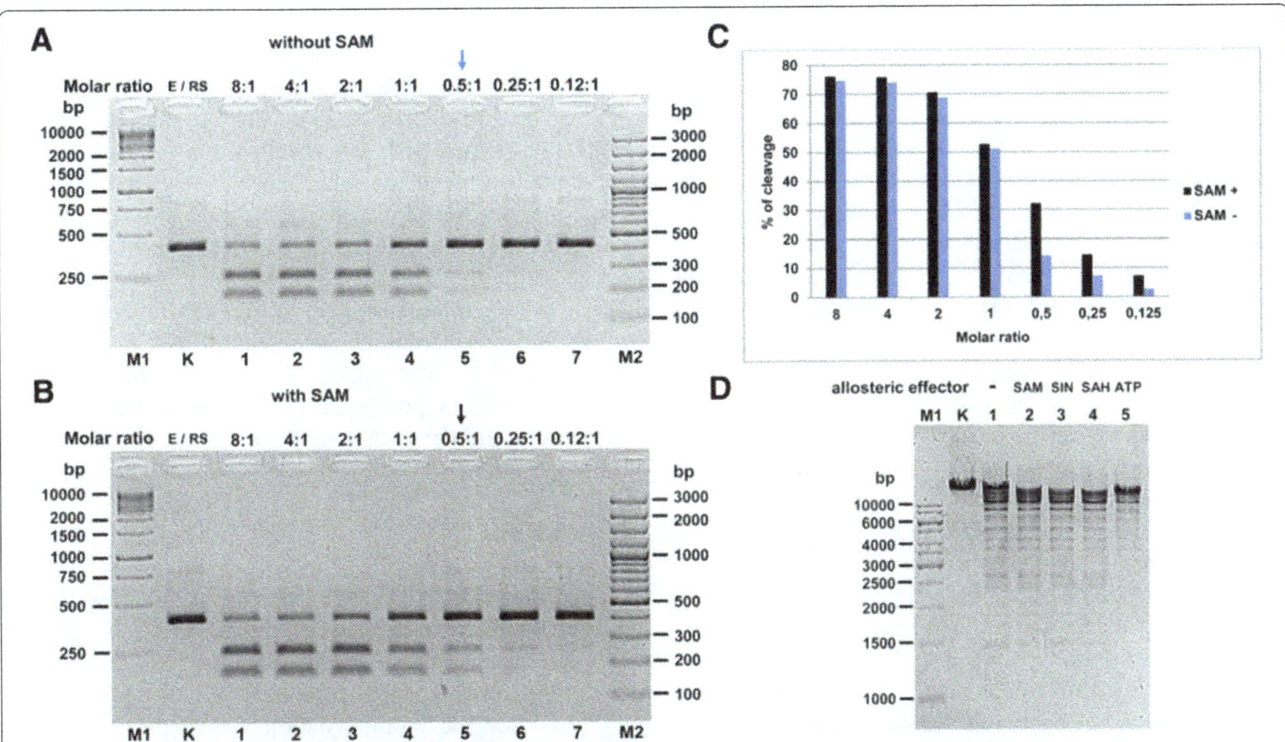

Figure 8 Evaluation of the effect of cofactor SAM and its analogues on TsoI activity. 0.3 μg (= 0.6 pmol recognition sites) single site PCR substrate (390 bp) was digested with decreasing amounts of wt TsoI for 16 h at 55°C in the standard reaction buffer: 10 mM Tris–HCl, pH 7.5, 10 mM NaCl, 10 mM $MgCl_2$, 0.01 mg/ml BSA, 0.5 mM DTTn the presence or absence of SAM. **(A)** DNA cleavage in the absence of SAM. Lane M1, GeneRuler™ 1 kb DNA Ladder (Thermo Fisher Scientific (Fermentas); selected bands marked); lane K, undigested PCR fragment; lanes 1–7, digested PCR fragment: lane 1, with 4.8 pmol; lane 2, with 2.4 pmol; lane 3, with 1.2 pmol; lane 4, with 0.6 pmol; lane 5, with 0.3 pmol; lane 6, with 0.15 pmol; lane 7, with 0.075 pmol; lane M2, GeneRuler™ 100 bp Plus DNA Ladder (Thermo Fisher Scientific (Fermentas); selected bands marked). **(B)** DNA cleavage in the presence of 50 μM SAM. Lane M1, GeneRuler™ 1 kb DNA Ladder (Thermo Fisher Scientific (Fermentas); selected bands marked); lane K, undigested PCR fragment; Lanes 1–7, samples were digested with decreasing amounts of TsoI as described in **(A)**; lane M2, GeneRuler™ 100 bp Plus DNA Ladder (Thermo Fisher Scientific (Fermentas); selected bands marked). **(C)** The influence of enzyme to recognition site ratio on TsoI DNA cleavage in the presence or absence of SAM. **(D)** Effect of allosteric cofactors on wt TsoI REase activity. Three putative cofactors or analogues (SAM, SAH, SIN) as well as ATP were compared for their influence on TsoI DNA digestion activity. 0.5 μg of λ DNA (= 0.016 pmol recognition sites) was digested with 0.7 pmol (0.048 u) of TsoI in standard TsoI buffer supplemented with 50 μM of the appropriate effector and 0.5 mM DTT for 30 min at 55°C. Lane M, GeneRuler™ 1 kb DNA Ladder (Thermo Fisher Scientific (Fermentas); selected bands marked); lane K, untreated λ DNA; lane 1, λ DNA cleaved with wt TsoI (no cofactors, except Mg^{2+}); lane 2, (+TsoI wt, +SAM); lane 3, (+TsoI wt, +SIN); lane 4, (+TsoI wt, +SAH); lane 5, (+TsoI wt, +ATP). DNA was treated with low amount of TsoI to pinpoint differences in the stimulatory effect. The reaction products were resolved on 1.2% agarose gel in TBE buffer and stained with EtBr.

the same result if the enzyme used bound SAM, and less methylation if this was not the case. The results shown in Figure 7B are nearly identical to those shown in Figure 7A, suggesting that either the cleared lysate of induced culture has a sufficiently high concentration of SAM which is enough to promote efficient DNA methylation even when a very small amount of the cleared lysate was used (Figure 7B, lanes 12 and 13), or that the recombinant enzyme has SAM already bound. Previously, we suggested that the *Thermus* sp. family enzymes might have two physically separate binding sites for SAM: one for allosteric stimulation of REase activity and another for typical SAM binding/methylation [8]. However, the important conclusion from these two experiments is that, regardless of the source of the necessary cofactor for the MTase reaction, the MTase activity of recombinant TsoI is predominant over REase activity under the same reaction conditions, when tested on crude lysates, which mimics reaction conditions *in vivo*. Such an REase-unfavourable equilibrium between the two activities of TsoI raises the question of how effective is DNA cleavage by TsoI *in vivo* in its natural host *T. scotoductus* and whether indeed the primary function of this bifunctional REase-MTase is defence against invading foreign DNA. It is possible that other functions, such as participation in recombination, by rare DNA cleavage, area primary goal of this system. Further characterization studies were conducted on a wt *T. scotoductus*-isolated homogeneous TsoI preparation. Initial tests on the newly found wt TsoI prototype enzyme have indicated that SAM slightly stimulates TsoI REase activity [2]; (Lubys Arvydas, personal communication).

Hence, three potential effectors, adjudged from our previous work [3-8], were compared for their influence on TsoI activity: SAM, a natural and obligatory co-substrate for MTase activity and an allosteric stimulator for *Thermus* sp. REase activities; SIN, a SAM analogue, which apparently causes subtle changes in tertiary *Thermus* sp. REases structures, either stimulating DNA cleavage [6-8] and/or causing substrate specificity changes towards much more frequent cleavage [5]; (BMC Genomics, in press) and S-adenosylhomocysteines (SAH), the methylation reaction by-products. These results as well as TsoI digestions performed for a prolonged time under various enzyme: substrate molar ratios show that the activation of REase function by SAM, SIN and SAH is weak, but detectable (Figure 8). Nevertheless, this may lead to indirect conclusion that the enzyme retained some capability for allosteric interaction of the TsoI protein with SAM and its analogues, even though this interaction is not fully functional. A more precise answer to the interesting problem of the pleiotropic effect of SAM as well as the enzyme's inability to conduct multiple cleavage reactions signalled in Figure 7 and Figure 8 would come from detailed *in vitro* studies using DNA-band-shift-assay, radiolabelled SAM and DNA as well as TsoI mutants with an inactivated REase catalytic motif and/or MTase catalytic motifs (work in progress). ATP was also tested as potential effector, even though it is chemically more distant molecule as compared to SAM. However, since the *Thermus* sp. enzymes resemble "streamlined halves" of Type I REases [6,8], the possibility of ATP effect, was evaluated. In addition, unusual Type II REase of eukaryotic origin – CviJI is stimulated by ATP while its specificity changes [14]. Nevertheless, ATP had no effect on TsoI (Figure 8, lane 5).

Further evaluation of wt TsoI properties included molecular sieving in the reaction buffer 'G' (without SAM and BSA) (not shown) [4]. The experiment showed that the native REase elutes as a monomer, just like other *Thermus* sp. family enzymes, confirming their common organisation scheme [3-8]. Activity temperature profiling has shown somewhat surprising results, indicating that TsoI is the least thermostable enzyme in the *Thermus* sp. family, with an optimum at 55°C and retaining only 18.4% activity at 65°C (Figure 9A). We previously showed that the typical optimum temperature for *Thermus* sp. enzymes is 65-75°C [3-8]. The reaction optimum of TsoI is considerably lower, by app. 10-15°C, even though its natural host *T. scotoductus* grows optimally at 67°C [2] (Lubys Arvydas, personal communication). This indicates that certain cellular components may be much more sensitive than an organism as a whole entity and are still able to fulfil their function. Also, such a property may be reminiscent of a past acquisition of TsoI coding genes by *T. scotoductus* from less thermophilic bacteria. The pH influence on the TsoI REase activity was determined. Trace REase activity was detected in the 4.0-5.5 range; while activity increased from 95.5 to 100% in the 7.0-7.5 range, and decreased from 20.8 to 7.8% in the 8.0-8.5 range (Figure 9B). The results obtained were similar to those described for other *Thermus* sp. family members. At optimised pH (=7.5), a buffer with variable concentrations of NaCl was used to determine optimal ionic strength. The maximum activity was close to 100% in the relatively wide 10–30 mM concentration range, and decreased gradually to 60 mM NaCl (Figure 9C). For practical applications of TsoI in DNA manipulations, 50 mM NaCl is preferred, as a compromise between maximum enzymatic activity versus enzyme stability during the reaction and lowered "star" activity.

Conclusions

i. The prototype TsoI REase gene was cloned in *E. coli* and sequenced using a new modification of a classic biochemical selection method, where the positive selection vector was combined with two rounds of selection for the methylation phenotype.

ii. Expression of a cloned *tsoIRM* gene under a T7 promoter, yielding enzymatically active bifunctional TsoI REase-MTase, required an additional round of the biochemical selection of expression subclones to eliminate abundant spontaneous mutants.

iii. TsoI is a member of the *Thermus* sp. modular enzyme family and the TspDTI-subfamily, exhibiting a rare phenomenon among REases – relatively high homologies to TspDTI, Tth111II and TthHB27I, even though they recognise distinct cognate sites.

iv. Within the recombinant TsoI bifunctional enzyme, MTase dominates REase activity both *in vivo* and *in vitro* in crude lysates assays. This may suggest the existence of an additional biological role different than the restriction of invading DNA.

v. Reaction parameters and cofactor requirements were determined, including a surprisingly low temperature optimum of 55°C and lower than expected *tsoIRM* ORF GC content, which suggests the occurrence of horizontal gene transfer in the past.

Methods

Bacterial strains, plasmids, media and reagents

The TsoI-producing strain *Thermus scotoductus* RFL4 was obtained from the Thermo Fisher Scientific Fermentas (Vilnius, Lithuania) collection and cultivated at 67°C in modified Luria broth either in flasks or in a fermenter (details on fermentation and detailed composition of broth are available under request). *E. coli* ER2566 {*fhuA2 lacZ:: T7 gene1 [lon] ompT gal sulA11 R(mcr-73::miniTn10–*

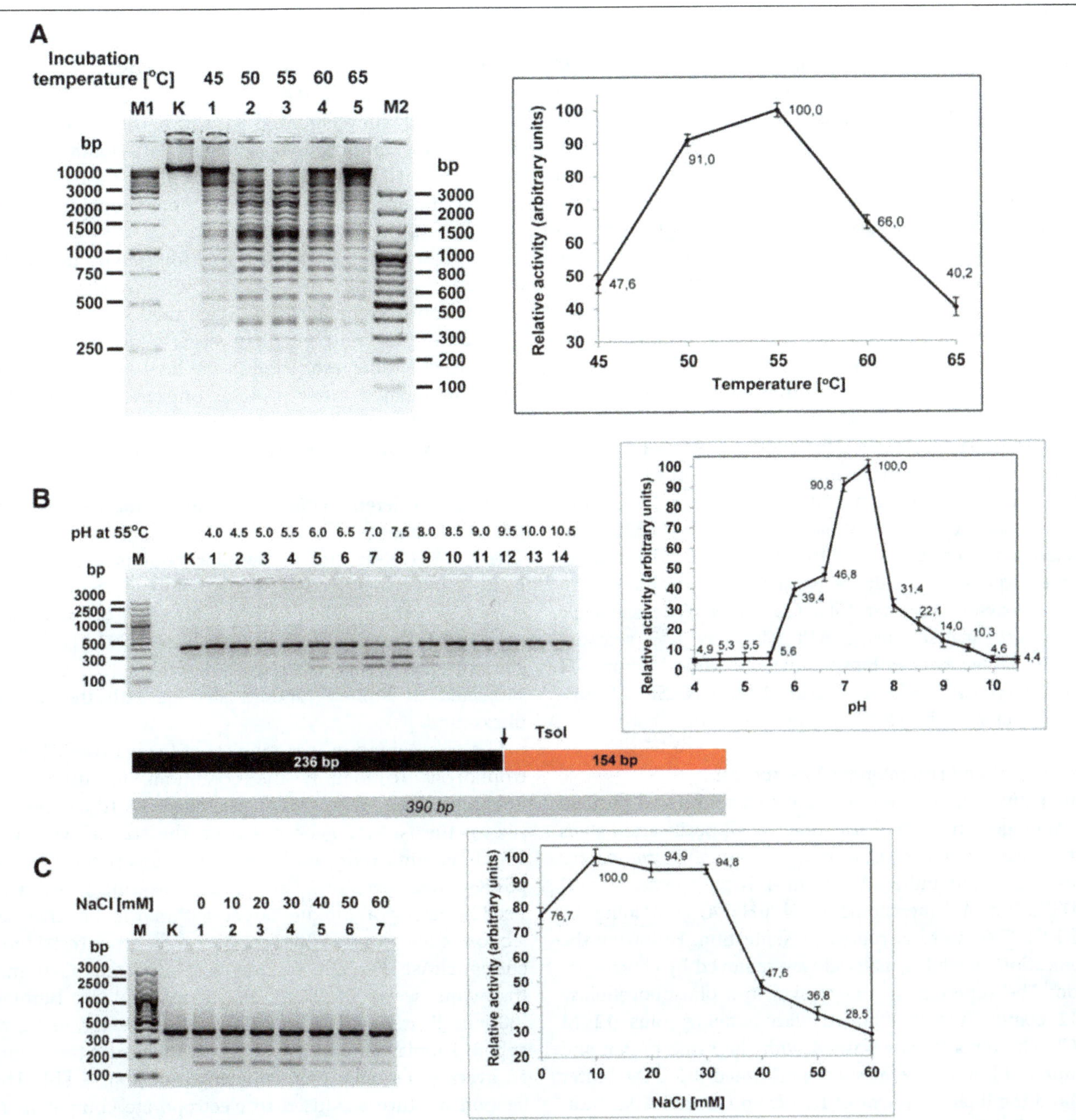

Figure 9 Evaluation of temperature, pH and salt concentration effect on TsoI REase activity. (A) The optimum temperature range of TsoI REase. 0.5 µg of T7 DNA (= 0.02 pmol recognition sites) was digested with 1 pmol (0.05 u) of TsoI in standard buffer supplemented with 50 µM SAM for 30 min in the temperature range from 45 to 65°C. Lane M1, GeneRuler™ 1 kb DNA Ladder (Thermo Fisher Scientific (Fermentas); selected bands marked); lane K, undigested T7 DNA; lane 1, 45°C; lane 2, 50°C; lane 3, 55°C; lane 4, 60°C; lane 5, 65°C; lane M2, GeneRuler™ 100 bp Plus DNA Ladder (Fermentas), selected bands marked. **(B)** The pH activity range of TsoI REase. 0.3 µg (= 0.6 pmol recognition sites) single site PCR substrate (390 bp) was digested with 0.33 pmol TsoI (0.016 u) in the pH range from 4.0 to 10.5 for 30 min at 55°C as described in the Methods section. GeneRuler™ 100 bp Plus DNA Ladder (Thermo Fisher Scientific (Fermentas); selected bands marked); lane K, undigested PCR fragment; lanes 1–14, PCR fragments after incubation with wt TsoI at indicated pH values: lane 1, at pH 4.0; lane 2, 4.5; lane 3, 5.0; lane 4, 5.5; lane 5, 6.0; lane 6, 6.5; lane 7, pH 7.0; lane 8, 7.5; lane 9, 8.0; lane 10, 8.5; lane 11, 9.0; lane 12, 9.5; lane 13, 10.0; lane 14, 10.5. **(C)** The influence of ionic strength on TsoI REase activity. 0.3 µg (= 0.6 pmol recognition sites) single site PCR substrate (390 bp) was digested with 0.33 pmol TsoI (0.016 u) in the NaCl concentration range from 0 to 60 mM for 30 min at 55°C in the standard reaction buffer, devoid of initial NaCl content: 10 mM Tris–HCl, pH 7.5, 10 mM $MgCl_2$, 0.01 mg/ml BSA, 0.5 mM DTT, 50 µM SAM. Lane M, GeneRuler™ 100 bp Plus DNA Ladder (Thermo Fisher Scientific (Fermentas); selected bands marked); lane K, undigested PCR fragment; lanes 1–7, digested PCR fragment: lane 1, without NaCl; lane 2, with 10 mM; lane 3, 20 mM; lane 4, 30 mM; lane 5, 40 mM; lane 6, 50 mM; lane 7, 60 mM.

TetS)2 [dcm] R(zgb-210::Tn10–TetS) endA1 Δ(mcrC-mrr) 114::IS10} (New England Biolabs, MA, USA) were used for all cloning and expression procedures. *E. coli* bacteria were grown in LB medium [20]. Media were supplemented with 50 mg/mL kanamycin and 50 mg/mL ampicillin for pSEKm'-MCS vector, and 50 mg/mL ampicillin for pET 21NS vector. Difco media were from Becton- Dickinson (Franklin Lakes, NJ, USA), DNA ladders and protein size standard, DNA purification kits, restriction enzymes, λ DNA, T4 DNA polymerase, T4 DNA ligase, Taq DNA polymerase, alkaline phosphatase and PCR primers were from Thermo Fisher Scientific/Fermentas (Vilnius, Lithuania). T7 DNA was from Vivantis Technologies Sdn. Bhd. (Malaysia). The expression vector pET21NS (Fermentas) was a modification of the pET21 vector (Novagen, WI, USA) (AmpR, MCS, *col* E1 *ori*, f1 *ori*, and T7-lac promoter), containing NotI and SmiI restriction sites introduced into MCS. All other reagents were purchased from Sigma-Aldrich (St. Louis, MO, USA). Sequencing was carried out using the ABI Prism 310 automated sequencer with the ABI Prism BigDye Terminator Cycle Sequencing Ready Reaction Kit (Perkin Elmer Applied Biosystems, Foster City, CA, USA). The sequence data were analysed using ABI Chromas 1.45 software (Perkin Elmer Applied Biosystems) and either Vector NTI (Invitrogen, CA, USA) or DNAIS MAX /DNASIS 2.5 software (Hitachi Software, San Bruno, CA, USA).

Native (wt) and recombinant TsoI sources

The native (wt) TsoI enzyme was first found and purified to homogeneity from *T. scotoductus*. All purification steps were carried out at +4°C. Frozen biomass of *T. scotoductus* was thawed in buffer A (10 mM K-phosphate, 1 mM EDTA, 7 mM 2-mercaptoethanol, pH 7.4) containing 0.1 M KCl. Cells were disrupted by sonication. Following the sonication, insoluble material was removed by centrifugation. The supernatant was applied to a phosphocellulose P11 column pre-equilibrated with buffer A plus 0.1 M KCl. The column was washed with the same buffer, and elution of bound enzymes was performed using the buffer A and the linear gradient of KCl from 0.1 to 0.8 M. Individual fractions were tested for the REase activity. The TsoI enzyme was eluted at between 0.4 and 0.6 M KCl. The peak fractions were pooled, dialyzed against 20 volumes of buffer A supplemented with 0.2 M NaCl, and applied to a Blue Sepharose column. After the washing of the column with the A buffer (+ 0.2 M NaCl) the elution of bound enzymes was accomplished using 0.2–1.0 M gradient of NaCl in buffer A. The REase activity was eluted from the column between 0.6 and 0.8 M NaCl. The peak fractions were pooled, dialysed against 20 volumes of buffer A supplemented with 0.2 M KCl, and the TsoI pool from Blue Sepharose was then fractionated on a Heparin Sepharose column pre-equilibrated with buffer A plus 0.2 M KCl. The column was washed with the same buffer and the enzyme was eluted using a linear gradient between 0.2 to 0.9 M KCl in buffer A. REase activity was found in fractions eluted at 0.40-0.60 M KCl. These fractions were pooled, supplemented with BSA (final concentration 0.05 mg/ml) and dialysed against 10 volumes of 10 mM Tris–HCl (pH 7.4), 100 mM KCl, 1 mM EDTA, 1 mM DTT and 50% glycerol. The final preparation was stored at −20°C.

Recombinant TsoI was prepared for activity testing by sonication-mediated disruption of *E. coli* cells followed by cell debris centrifugation. Since the TsoI protein expression in plasmid constructs was at a high level, where the corresponding TsoI band in the final expression construct dominated other proteins present in the cleared lysates, such a partial purification procedure was sufficient to conduct the experiments described.

Cloning and determination of the nucleotide sequence of the *tsoIRM* gene

The *tsoIRM* gene and its flanking regions were cloned using a positive selection vector combined with a two-stage biochemical selection procedure [15] of the library prepared from *T. scotoductus* genomic DNA. [15]. The genomic DNA was isolated as described [20] and was subjected to limited random sharing with the use of ultrasound.

The fragmentation was monitored by agarose gel electrophoresis to identify the conditions favourable for obtaining DNA fragments larger than 3 kb (the expected size of the *tsoIRM* gene based on the size of wt TsoI). The fragments were used for library construction in the pSEKm'-MCS vector. The vector was linearized with Eco32I having a unique target within the positive selection gene *eco47IR* which codes for the restriction endonuclease Eco32I, and dephosphorylated. Genomic fragments were T4 DNA Polymerase/dNTPs blunted [20] and ligated with a Eco32I-linearised vector, using an app. 3: 1 molar ratio of insert: vector molecules (assuming an average genomic fragment length of app. 5 kb). The ligation mixture was used to electroporate competent *E. coli* ER2566 cells and transformants were plated onto LB/kanamycin+ampicillin plates. From the obtained library of 8×10^5 clones recombinant plasmids were isolated *in toto* from the pooled library and were subjected to two consecutive rounds of biochemical selection. The plasmid pool (1 μg) was digested overnight with an excess (5 units) of native, *T. scotoductus* -isolated TsoI REase; the reaction mixture was purified using phenol/chloroform extraction [20], DNA products precipitated with the ethanol, dissolved in 5 mM Tris–HCl pH 8.0 and transformed by electroporation back to the same ER2566 strain. The resulting 2800 colonies were pooled again, the plasmids were isolated *in toto* and were subjected again to TsoI biochemical selection. Fifty colonies out of 120 obtained after the second round of biochemical selection/transformation were subjected to individual analysis with colony PCR, used for screening (using the vector's flanking standard primers) for plasmids carrying fragments larger than 3 kb, and 14 plasmids among 50 analysed were found to fulfil this criterion. Selected plasmids were subjected to HindIII digestion to locate the same insert-derived DNA fragments, which would indicate that they originated from the same genomic region. Restriction mapping revealed seven types of plasmids all possessing the same cloned HindIII fragments of 1.2 and 1.5 kb in size. Of these, plasmids representing five smallest isolates were used for both methylation analysis and sequencing of cloned regions. The combination of both strand sequences resulted in a 4365 bp genomic DNA segment, where TsoI ORF was detected.

Overexpression of the *tsoIRM* gene under T7-lac promoter in *E. coli*

A specially designed pET21 derivative pET21NS was used for the directional cloning of the NotI-SmiI cleaved PCR fragment encompassing the full-length *tsoIRM* gene. The 15 resulting recombinant plasmids were introduced into *E. coli* ER2566 for expression trials. Recombinant strains were grown overnight at 37°C in 5 ml LB, supplemented with ampicillin. 100 μl of the overnight culture were used to inoculate 5 ml of ampicillin-supplemented fresh LB, grown at 37°C until OD600 = 0.5-0.8, induced with 1 mM IPTG and allowed to grow further at 37°C for 3 hours. Crude cell extracts exhibited only trace the amounts of TsoI REase activity, while SDS-PAGE demonstrated abundant amounts of a large enzyme that appeared after induction. Testing of the TsoI protection level of all 15 plasmids before and after induction showed that all were unprotected before induction and only partially protected after induction. Thus, considering the possibility of the appearance and selection for PCR errors in the large, toxic *tsoIRM* gene, the few expression constructs were subjected to DNA sequencing, which revealed the presence of multiple mutations. To select for mutation-free plasmids, 5000 ampicillin-resistant colonies obtained after the transformation of ER5266 by (pET21NS + *tsoIRM*-PCR fragment) ligation mixture were pooled and used to inoculate 100 ml LB supplemented with ampicillin. Cells were grown at 37°C until OD_{600} = 0.7, TsoI expression was induced by the addition of 1 mM IPTG. The culture was further grown at 37°C for 4 hours and then used for the isolation of total plasmid DNA. The latter has 7 TsoI targets and thus can be enriched for more active MTase variants using the same biochemical selection approach. Total plasmid DNA was cleaved with TsoI and the reaction mixture transformed back to *E. coli* ER2566 (3rd round of biochemical selection). The 20 individual colonies resulting from the transformation were again tested for TsoI activity in crude cell extracts and for the TsoI protection level of plasmids carrying the *tsoIRM* gene, isolated from induced cultures. Crude cell extracts exhibited much higher TsoI activity, while expression plasmids of producing strains were almost completely protected from TsoI action when isolated after induction. Four of these plasmids were sequenced, and two were found to have no mutations in the PCR-amplified TsoI-coding gene. One of these two plasmids was called pET21NS-TsoIRM and was used for further experiments.

REase and MTase assays

REase assays

For REase assays various modifications of standard TsoI reaction conditions were used, which provide a compromise between enzyme stability, lowest "star" activity and DNA cleavage efficiency. Typically, reactions were performed in 50 μl of the reaction buffer, containing 10 mM Tris–HCl pH 7.5 at 37°C, 10 mM $MgCl_2$, 50 mM NaCl, 0.5 mM DTT, 0.1 mg/ml BSA and supplemented with 50 μM SAM. The following DNAs were used as test substrates: λDNA, T7 bacteriophage DNA, PCR fragment with a single TsoI site and DNA of recombinant plasmids (for biochemical selection and protection assays). One unit of TsoI REase is defined as the amount of enzyme required to hydrolyse 1 μg of λ DNA in an hour at 55°C in 50 μl of standard TsoI buffer, enriched with 50 μM SAM, resulting in a stable partial DNA cleavage pattern.

Quantitative evaluations of temperature, pH, salt concentrations were determined using DNA cleavage reactions under enzyme-limiting conditions. Comparative densitometry was performed on selected reference DNA bands from photographs of ethidium bromide and/or Sybr Green stained gels, taken under various exposure times. The temperature reaction optimum was determined in standard TsoI buffer. The NaCl concentration optimum was determined in standard TsoI buffer, devoid of initial salt addition. The pH reaction optimum was evaluated in three buffer systems, each dedicated to the maximum buffering capacity range: sodium acetate-acetic acid of pH from 4.0 to 5.5, HEPES-KOH from 6.0 to 7.0, and Tris–HCl buffer from 7.5-10.5. The pH of the reaction buffers was adjusted at 55°C after all the buffer components had been dissolved. The cleavage reactions were performed under enzyme-limiting conditions, (0.55: 1 molar ratio of the enzyme to cognate sites for 30 min).

Methyltransferase assays

The *in vitro* methylation activity of the TsoI enzyme was tested by the DNA protection assay, in 50 μl of TsoI standard buffer (without Mg^{2+}) supplemented with 50 μM SAM. After the addition of recombinant TsoI protein present in cleared lysates, the reaction mixtures were incubated at 55°C. The cleavage products visible after the incubation with crude cell extract, resulted from the resident TsoI REase activity in the presence of Mg^{2+} ions (Figure 7, lanes 9,12). Samples were purified to remove TsoI protein by chloroform extraction and then DNA precipitated with isopropanol. Modified DNA was challenged with an excess of wt TsoI (2 units, app. 2: 1 M ratio of enzyme to recognition sites) for an hour in 50 μl of standard TsoI buffer supplemented with 50 μM SAM at 55°C. The reaction products were then resolved by agarose gel electrophoresis.

Competing interests

Arvydas Lubys, Jolanta Vitkute, Goda Mitkaite are affiliated with Thermo Fisher Scientific Inc. (USA), Fermentas branch (Vilnius, Lithuania) and provided scientific information concerning *tsoIRM* gene cloning, TsoI amino acid sequence and selected enzyme features. The authors declare that they have no competing interest.

Authors' contributions

AL conceived and coordinated the TsoI cloning project. PS coordinated the native TsoI characterisation experiments, came up with the concept of the new *Thermus* sp. enzyme family and drafted the manuscript. DR and GM performed the TsoI gene cloning and preliminary expression experiments. JZ and JF participated in the enzyme characterisation experiments. AZS performed some enzyme characterisation experiments, participated in the design and interpretation of the experimental analyses, prepared all figures and drafted the manuscript. All authors read and approved the final manuscript.

Acknowledgements

The authors thank Katarzyna Maczyszyn for the digital picture imaging and her valuable technical assistance. Also, the authors thank Audra Ruksenaite for preparing and loading of sequencing reactions. This work was supported by DS/530-8170-D201-12 (PMS, AZS), Gdansk University, Chemistry Department fund.

Author details

[1]Division of Molecular Biotechnology, Department of Chemistry, Institute for Environmental and Human Health Protection, University of Gdansk, Wita Stwosza 63, 80-952, Gdansk, Poland. [2]Thermo Fisher Scientific, V.A. Graiciuno 8, LT-02241, Vilnius, Lithuania. [3]Department of Botany and Genetics, Vilnius University, M.K. Ciurlionio 21/27, LT-03101, Vilnius, Lithuania.

References

1. Szybalski W, Kim SC, Hasan N, Podhajska AJ: **Class-IIS restriction enzymes – a review.** *Gene* 1991, **100:**13–26.
2. *The Restriction Enzyme Database.* http://rebase.neb.com.
3. Zylicz-Stachula A, Harasimowicz-Slowinska RI, Sobolewski I, Skowron PM: **TspGWI, a thermophilic class-IIS restriction endonuclease from *Thermus* s., recognizes novel asymmetric sequence 5'-ACGGA(N11/9)-3'.** *Nucleic Acids Res* 2002, **e33:**30.
4. Skowron PM, Majewski J, Zylicz-Stachula A, Rutkowska SM, Jaworowska I, Harasimowicz-Slowinska RI: **A new *Thermus* sp. class-IIS enzyme sub-family: isolation of a 'twin' endonuclease TspDTI with a novel specificity 5'-ATGAA(N(11/9))-3', related to TspGWI, TaqII and Tth111II.** *Nucleic Acids Res* 2003, **31:**e74.
5. Żylicz-Stachula A, Żołnierkiewicz O, Jeżewska-Frąckowiak J, Skowron PM: **Chemically-induced affinity star restriction specificity: a novel TspGWI/sinefungin endonuclease with theoretical 3-bp cleavage frequency.** *Biotechniques* 2011, **50:**397–406.
6. Zylicz-Stachula A, Bujnicki JM, Skowron PM: **Cloning and analysis of bifunctional DNA methyltransferase/nuclease TspGWI, the prototype of a *Thermus* sp. family.** *BMC Mol Biol* 2009, **10:**52.
7. Żylicz-Stachula A, Żołnierkiewicz O, Śliwińska K, Jeżewska-Frąckowiak J, Skowron PM: **Bifunctional TaqII restriction endonuclease: redefining the prototype DNA recognition site and establishing the Fidelity Index for partial cleaving.** *BMC Bioch* 2011, **12:**62.
8. Żylicz-Stachula A, Żołnierkiewicz O, Lubys A, Ramanauskaite D, Mitkaite G, Bujnicki JM, Skowron PM: **Related bifunctional restriction endonuclease-methyltransferase triplets: TspDTI, Tth111II/TthHB27I and TsoI with distinct specificities.** *BMC Mol Biol* 2012, **13:**13.
9. Barker D, Hoff M, Oliphant A, White R: **A second type II restriction endonuclease from *Thermus aquaticus* with unusual sequence specificity.** *Nucleic Acids Res* 1984, **12**(14):5567–5581.
10. Shinomiya T, Kobayashi M, Sato S: **A second site specific endonuclease from *Thermus thermophilus* 111, Tth111II.** *Nucleic Acids Res* 1980, **8:**3275–3285.
11. Furuta Y, Abe K, Kobayashi I: **Genome comparison and context analysis reveals putative mobile forms of restriction-modification systems and related rearrangements.** *Nucleic Acids Res* 2010, **38:**2428–2443.
12. Newman M, Strzelecka T, Dorner LF, Schildkraut I, Aggarwal AK: **Structure of restriction endonuclease BamHI phased at 1.95 A resolution by MAD analysis.** *Structure* 1994, **2**(5):439–452.
13. Madhusoodanan UK, Rao DN: **Diversity of DNA methyltransferases that recognize asymmetric target sequences.** *Crit Rev Biochem Mol Biol* 2010, **45**(2):125–145.
14. Fitzgerald MC, Skowron PM, Van Etten JL, Smith LM, Mead DA: **Rapid shotgun cloning utilizing the two base recognition endonuclease CviJI.** *Nucleic Acid Res* 1992, **20:**3753–3762.
15. Szomolanyi E, Kiss A, Venetianer P: **Cloning the modification methylase gene of *Bacillus sphaericus* R in *Escherichia coli*.** *Gene* 1980, **10:**219–225.
16. Piekarowicz A, Yuan R, Stein DC: **A new method for the rapid identification of genes encoding restriction and modification enzymes.** *Nucleic Acids Res* 1991, **19**(8):1831–1835.
17. Piekarowicz A, Weglenska A: **Improvement of the strain for the rapid identification of genes encoding restriction and modification enzymes.** *Acta Microbiol Pol* 1994, **43**(2):229–231.
18. Kiss A, Posfai G, Keller CC, Venetianer P, Roberts RJ: **Nucleotide sequence of the BsuRI restriction-modification system.** *Nucleic Acids Res* 1985, **13**(18):6403–6421.
19. Shen BW, Xu D, Chan SH, Zheng Y, Zhu Z, Xu SY, Stoddard BL: **Characterization and crystal structure of the type IIG restriction endonuclease RM.BpuSI.** *Nucleic Acids Res* 2011, **39**(18):8223–8236.
20. Sambrook J: Fitsch EF, Molecular Cloning MT: *A Laboratory Manual.* 2nd edition. Cold Spring Harbour NY: CSH Press; 1989.

7

Insight into the cellular involvement of the two reverse gyrases from the hyperthermophilic archaeon *Sulfolobus solfataricus*

Mohea Couturier[1,2,3,4], Anna H Bizard[1,2,5], Florence Garnier[1,2*†] and Marc Nadal[1,2,6*†]

Abstract

Background: Reverse gyrases are DNA topoisomerases characterized by their unique DNA positive-supercoiling activity. *Sulfolobus solfataricus*, like most Crenarchaeota, contains two genes each encoding a reverse gyrase. We showed previously that the two genes are differently regulated according to temperature and that the corresponding purified recombinant reverse gyrases have different enzymatic characteristics. These observations suggest a specialization of functions of the two reverse gyrases. As no mutants of the TopR genes could be obtained in Sulfolobales, we used immunodetection techniques to study the function(s) of these proteins in *S. solfataricus in vivo*. In particular, we investigated whether one or both reverse gyrases are required for the hyperthermophilic lifestyle.

Results: For the first time the two reverse gyrases of *S. solfataricus* have been discriminated at the protein level and their respective amounts have been determined *in vivo*. Actively dividing *S. solfataricus* cells contain only small amounts of both reverse gyrases, approximately 50 TopR1 and 125 TopR2 molecules per cell at 80°C. *S. solfataricus* cells are resistant at 45°C for several weeks, but there is neither cell division nor replication initiation; these processes are fully restored upon a return to 80°C. TopR1 is not found after three weeks at 45°C whereas the amount of TopR2 remains constant. Enzymatic assays *in vitro* indicate that TopR1 is not active at 45°C but that TopR2 exhibits highly positive DNA supercoiling activity at 45°C.

Conclusions: The two reverse gyrases of *S. solfataricus* are differently regulated, in terms of protein abundance, *in vivo* at 80°C and 45°C. TopR2 is present both at high and low temperatures and is therefore presumably required whether cells are dividing or not. By contrast, TopR1 is present only at high temperature where the cell division occurs, suggesting that TopR1 is required for controlling DNA topology associated with cell division activity and/or life at high temperature. Our findings *in vitro* that TopR1 is able to positively supercoil DNA only at high temperature, and TopR2 is active at both temperatures are consistent with them having different functions within the cells.

Keywords: Archaea, Hyperthermophile, Topoisomerase, Supercoiling, Topology, Low temperature, Cytometry, TopR, Quantification

* Correspondence: florence.garnier@igmors.u-psud.fr; marc.nadal@igmors.u-psud.fr
†Equal contributors
[1]Université Versailles St-Quentin, 45 avenue des Etats-Unis, Versailles 78035, France
[2]Institut de Génétique et Microbiologie, UMR 8621 CNRS, Université Paris-Sud, Bât. 409, Orsay Cedex 91405, France
Full list of author information is available at the end of the article

Background

DNA topoisomerases are enzymes responsible for changing the DNA topological state. They are necessarily involved in all DNA processes (transcription, replication, recombination, repair and chromosome segregation) and the resulting appropriate topoisomerase activity modifies the DNA linking number locally and thereby eliminate excess of negative and positive supercoils generated upstream and downstream from the corresponding machinery [1]. Topoisomerases are classified as type I or type II according to whether the transient break in the DNA during their activity is a single-strand (type I) or double-strand (type II) [2,3]. They are further classified as type IA or IB and type IIA or IIB according to the presence of particular motifs in the amino-acid sequence [3].

Reverse gyrase is a particular type IA topoisomerase and it is the only known DNA topoisomerase to introduce positive supercoils into DNA [4-9]. Reverse gyrase was initially discovered in hyperthermophilic and thermophilic *Archaea* and *Bacteria* [10,11]. It was further reported that reverse gyrase was the marker of hyperthermophily and that the corresponding gene(s) may be essential for life at high temperature [12,13]. The positive supercoiling activity of reverse gyrase was proposed to stabilize the DNA duplex against denaturation in extreme temperature environments [9,14], and thereby avoid DNA melting. Positive supercoiling of DNA prevents formation of open complexes [15] as demonstrated by *in vitro* assays of *Sulfolobus* DNA transcription [16]. Homeostatic control of DNA supercoiling involving reverse gyrase has been suggested in hyperthermophilic archaea [17], as it was previously reported for mesophilic bacteria [18,19]. Reverse gyrase acts *in vitro* as a heat-protective DNA chaperone, independently of its supercoiling activity [20]. The helicase-topoisomerase IA chimeric structure of the reverse gyrase [9,21,22] is reminiscent of the physical and functional interaction between the RecQ-like protein and topoisomerase III. This protein pair is found in *Bacteria* and *Eukarya*, and is involved in the DNA repair and recombination needed for genome stability [23-25]. It has therefore been suggested that the reverse gyrase in hyperthermophilic archaea has a role in the maintenance of genome stability [9,26]. Reverse gyrase efficiently anneals complementary single-stranded circles and introduces positive supercoils into DNA containing a bubble and may thus act as a renaturase, contributing to the genome stability by eliminating impaired regions [27,28]. There is indeed diverse evidence that reverse gyrase is involved in recombination and repair. It is specifically recruited to DNA after UV irradiation [29]. The positive supercoiling reaction of reverse gyrase *in vitro* is stimulated by the single-strand DNA binding protein (SSB) [30], a protein that binds to single-strand DNA to prevent its premature annealing during various DNA metabolism processes including replication, recombination and repair [31]; a functional interaction between these two proteins has been demonstrated *in vivo* in the presence of DNA [30]. SSB also enhances the binding and cleavage of UV-irradiated substrates by reverse gyrase, further implicating reverse gyrase in DNA repair [30]. Reverse gyrase inhibits the activity of the translesion DNA polymerase PolY/Dpo4 *in vitro*, possibly thereby preventing the potential high mutational effect of PolY/Dpo4 [32]. Finally, reverse gyrase shows unwinding activity of substrates containing helical junctions, consistent with its involvement in recombination and repair [33].

A gene encoding a reverse gyrase has been also discovered in some moderately thermophilic bacteria: for example in *Nautilia profundicola*, which grows optimally at 45°C [34], the expression of this gene increases substantially at higher temperature. This may confer a selective advantage for such organisms which live close to hydrothermal vents and are therefore subject to frequent and rapid temperature fluctuations [34]. Possibly, reverse gyrase may be important for thermoadaptation rather than the hyperthermophilic lifestyle as such. The situation seems to differ between hyperthermophilic organisms with one and those with two reverse gyrase genes. *Thermococcus kodakaraensis* is a hyperthermophilic organism belonging to the Euryarchaeota *phylum*; it has a single reverse gyrase gene that was shown to be not essential for hyperthermophilic life, except at temperatures above 90°C [35]. By contrast, in the crenarchaeon *Sulfolobus islandicus*, both *topR1* and *topR2* genes were recently demonstrated to be essential [36]. Thus, the two reverse gyrase genes in Sulfolobales, and possibly in all Crenarchaeota containing two, seem to be linked either to the hyperthermophilic lifestyle and/or to other essential functions.

Reverse gyrases clearly have several functions in the cell, probably involving interactions with different partners according to the cellular process. The redundancy of reverse gyrase genes in most members of the Crenarchaeota *phylum* strongly suggests specialization of the two reverse gyrases with TopR1 and TopR2 having different functions. The *topR1* and *topR2* genes in *S. solfataricus* P2 are differently regulated, with different expression patterns according to the growth phase and temperature, and TopR1 is probably involved in the control of the topological state of DNA [17]. Experiments *in vitro* with the two purified recombinant reverse gyrases from *S. solfataricus* showed that they exhibit different enzymatic characteristics and in particular different behaviors with respect to temperature [37].

As both genes are essential in Sulfolobales, we looked for culture conditions revealing differential regulation of the two enzymes to study further their respective roles. We determined the lowest temperature at which only one of the two reverse gyrases exhibits significant enzymatic activity *in vitro*. Then, we tested whether *S. solfataricus* P2 cells at this temperature contain one or both reverse gyrases, to assess whether one or both reverse gyrases are tightly linked to the hyperthermophilic lifestyle. We report that TopR1 is not active at 45°C whereas TopR2 exhibits significant positive supercoiling activity at this temperature. We also report for the first time the number of reverse gyrase molecules per cell: *S. solfataricus* contains approximately 50 molecules of TopR1 and 125 molecules of TopR2 per cell when cells are actively dividing at 80°C. After three weeks at 45°C, there was no cell division nor replication initiation; the abundance of TopR1 was very much lower than at 80°C whereas that of TopR2 is largely unaffected. The cultures were returned to 80°C, and growth ability and replication activity are fully restored and the amounts of TopR1 characteristic of actively dividing cells were recovered. These quantitative findings contribute to elucidating the different roles of reverse gyrase.

Results and discussion

TopR2, but not TopR1, remains active at 45°C

The two reverse gyrases of *S. solfataricus* exhibit different enzymatic properties [37]. Although both enzymes are able to introduce positive supercoils into DNA, TopR2 introduces a higher density of positive supercoils and at a higher rate than does TopR1. This difference is mainly due to the very high processivity of TopR2, whereas TopR1 is distributive. The activity of TopR1 is strictly dependent on the temperature: from relaxation at 60°C the linking number increases progressively with increasing temperature with maximum positive supercoiling being reached at 90°C. By contrast, TopR2 is not active at high temperature *in vitro* (its optimal temperature is around 70°C), but exhibits a significant positive supercoiling activity at 60°C [37]. Thus, the two reverse gyrases can be distinguished according to their activities at different temperatures. Several studies report that various protein machineries in hyperthermophilic organisms of the *Sulfolobus* genus are functional at low temperature: the proton pump [38], the transcription machinery [16], replication and repair DNA polymerases [39,40] and DNA topoisomerase(s) [4]. Consequently, we investigated whether both reverse gyrases of *S. solfataricus* are active at temperatures below 60°C, the lowest temperature previously tested [37]: we performed topoisomerase assays at temperatures from 45°C to 80°C (Figure 1). Enzymes are generally less active at low temperature, so we used a topoisomerase:DNA molecular ratio of 4 to allow weak activities to be detected. The main characteristics of the two enzymes were observed at control temperatures (60°C to 80°C; Figure 1): distributivity for TopR1 and processivity for TopR2 [37]. A significant

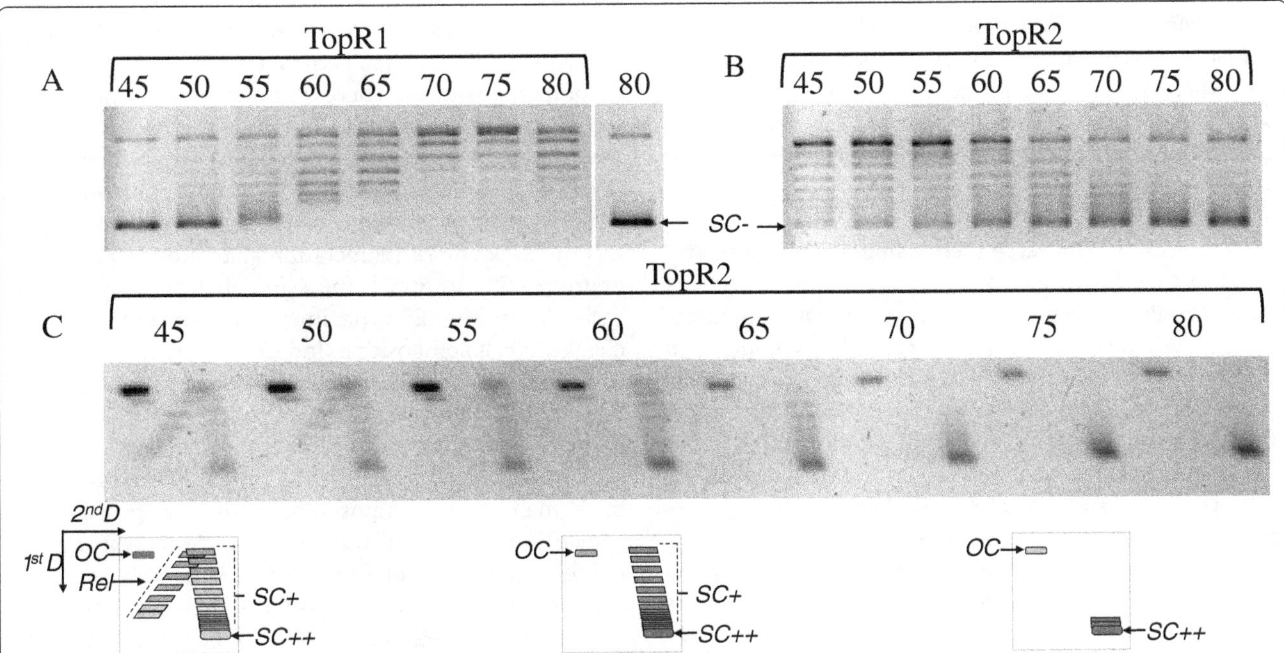

Figure 1 Effect of low temperature on the activities of TopR1 and TopR2. Standard positive supercoiling reactions were performed in the presence of a limiting amount of TopR1 **(A)** or TopR2 **(B and C)**. The reaction products were analyzed by one-dimensional gel electrophoresis **(A and B)** and for TopR2 also by two-dimensional gel electrophoresis **(C)**. In panels **A**, **B** and **C**, the incubation temperature is indicated above the gels. Pictures in panel C below the 2D gels schematize the position of the various forms of DNA in 2D gel electrophoresis. SC + and SC ++ indicate the positions of positively supercoiled and highly positively supercoiled DNA, respectively, Rel the position of the relaxed DNA, oc the position of nicked DNA, and SC- the position of negatively supercoiled DNA substrate.

positive supercoiling activity of TopR1 was observed only at temperatures above 70°C (Figure 1A). TopR1 displayed a weak relaxation activity at 55°C and 50°C but was not active at 45°C (Figure 1A). By contrast, TopR2 was active at 45°C as revealed by the topoisomer profile obtained in one-dimensional gel electrophoresis (Figure 1B). The two-dimensional gel electrophoresis experiments revealed that TopR2 was able to introduce large numbers of positive supercoils into DNA at 45°C (Figure 1C). Although the positive supercoil density at 45°C seemed to be similar to that obtained at 70°C, the population of highly positively supercoiled plasmids was smaller at 45°C (Figure 1C). There was also a wider range of intermediate topoisomers from more or less highly positively supercoiled to negatively supercoiled at lower (particularly at 45°C and 50°C) than higher temperatures (Figure 1C). These observations suggest that TopR2 is less processive at lower temperature. Two DNA polymerases from *S. solfataricus*, PolB1/Dpo1 and PolY/Dpo4 involved in DNA replication and repair, respectively, show also temperature-dependent processivity [39]. In addition, between 45°C and 60°C, we observed that TopR2 produces more nicked DNA than at higher temperatures (65°C-80°C) (Figure 1C). This clearly indicates that low temperature does not inhibit binding or DNA cleavage but does affect the DNA strand passage or religation. Interestingly, overproduction of nicked DNA is not observed for TopR1 (Figure 1A), further evidence of the difference between these two enzymes. In conclusion, in our conditions, TopR2 preserves a significant highly positive supercoiling activity at temperatures down to 45°C whereas TopR1 is not active at such low temperatures. We then tested whether *S. solfataricus* is still alive at this low temperature and studied the reverse gyrase content *in vivo*.

S. solfataricus cells preserve their membrane integrity at 45°C, but cells do not divide

We first evaluated cell density, cell size, membrane integrity, and cell distribution according to DNA content in cultures at 45°C. We transferred cultures of exponentially growing cells at 80°C to 45°C and maintained the cultures at 45°C for three weeks. Cell density was determined both by measuring optical density at 600 nm (OD_{600nm}) and by flow cytometry, and we confirmed that the OD_{600nm} accurately reflects the number of cells as counted by flow cytometry. Despite a small increase of OD_{600nm} during the first two days, the cell density remained roughly constant during the three weeks at 45°C (Figure 2A). The same result was obtained when we changed the culture medium every week (data not shown), showing that this was not due to any modification of the growth medium composition over the long period at 45°C. Thus, the constant OD_{600nm} over a period of three weeks at 45°C was due to the temperature.

Control cells cultured at 80°C and cells transferred to 45°C were examined by phase contrast microscopy: that both presented the same round shape (Figure 2B). Average cell sizes were calculated from measurements of at least 200 cells in each condition. In both conditions, the cells were similarly small: 1.60 μm ± 0.30 for control cells at 80°C and 1.53 μm ± 0.27 for cells transferred to 45°C, independent of the time at which the cells are collected (Figure 2B, panels b-d compared with panel a). This is consistent with a previous report that *Sulfolobus* cell size is not affected by transfer to room temperature or even ice-water [41].

When cells are exposed to an environmental condition different from their optimal growth conditions, cell membrane integrity may be affected. We used the LIVE/DEAD® *Bac*Light™ Bacterial Viability Kit [42], which has been validated for archaea [43] to test membrane integrity. As expected, the exponentially growing cultures at 80°C contained few damaged cells (Figure 2C, panel a); after transfer from 80°C to 45°C and incubation for 21 days, the numbers of cells with preserved membrane integrity were similar to control values (Figure 2C, panels b-d). This indicates that the cells do not lyse at this low temperature, and are resistant to it for a long period.

We used flow cytometry to investigate cell size and cell composition (forward light scattering; FSC) and cell distribution according to DNA content. Cells transferred to 45°C became slightly more heterogeneous, exhibiting a wider range of FSC than control cells (Figure 2D, b, c and d compared with a). There was a symmetric decrease of the FSC range for cells transferred back to 80°C after three weeks at 45°C (Figure 2D, e compared with b, c and d). As microscopy indicated that there was no significant change in the cell size at any time during the experiment, regardless the direction of temperature shift (from 80°C to 45°C or from 45°C to 80°C), the symmetric and reproducible variation of the FSC parameter presumably reflects a change in cell composition. Indeed, the protein concentration declined when cells were transferred to 45°C to half that in 80°C control cells. This probably reflects a change in the balance between degradation, stability and basal synthesis for a particular set of proteins. Changes in the lipid membrane composition is also a plausible explanation because the number of cyclopentane rings in the tetraether lipids of Sulfolobale membranes varies with temperature [44]. The composition and/or the properties of the flexible cell wall may also be modified by the temperature change and affect the FSC.

The 80°C control cells could be separated into three sub-populations according to the DNA content (Figure 2D, panel a) as previously described [45]: a minor population containing one genome equivalent; a significant population with a DNA content between one and two genome

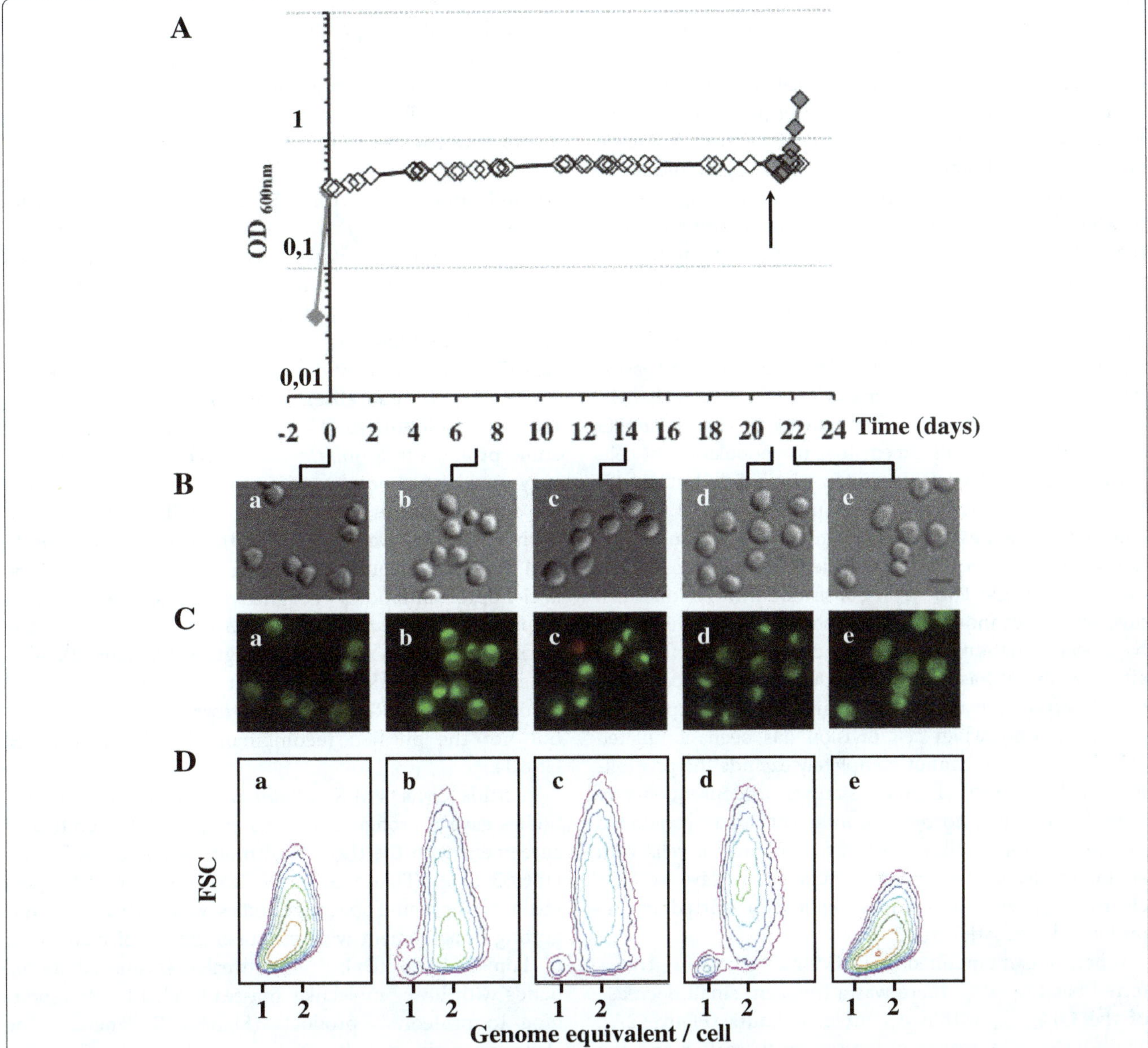

Figure 2 Cell size, cell integrity and DNA content of *S. solfataricus* transferred from 80°C to 45°C. Samples of cells growing exponentially at 80°C (0 h), or after being transferred to 45°C (time points: 7d, 14d and 21d), or transferred back to 80°C (24 h) were collected. The cell density was monitored by measuring the OD $_{600nm}$ **(A)** and by flow cytometry **(D)**. The time point (0 h) indicates when 80°C exponentially growing cells (closed gray diamond) were transferred from 80°C to 45°C (open black diamond). The arrow (↑) indicates when cells were transferred back to 80°C (closed gray diamond) after 21 days at 45°C. **(B)** Cell size was measured by phase-contrast microscopy **(a-e)**. Bar, 2 μm. **(C)** Cell membrane integrity was analyzed by fluorescence microscopy. Merged images of Syto 9 (green) and propidium iodide (red) are shown **(a-e)**. The same cells were studied by phase-contrast **(B)** and by fluorescence microscopy **(C)**. DNA content distribution and cell size of *S. solfataricus* cells transferred from 80°C to 45°C for three weeks and back to 80°C **(D)**. DNA content distribution was analyzed as the fluoresence of propidium iodide and the cell size determined by forward light scattering in flow cytometry **(a-e)**.

equivalents; and the largest cell population containing two genome equivalents. The proportions of the different populations in *S. solfataricus* cultures growing exponentially at 80°C were consistent with the relative lengths of cell cycle phases as previously reported for actively dividing *Sulfolobus* cells [45]. This distribution was drastically modified after a prolonged incubation at 45°C: the cytograms after 7, 14 and 21 days at 45°C evidence the disappearance of cells with a DNA content between one and two genome equivalents; cells containing one genome continue to constitute a small part of the cell population; and the proportion of cells with two genome equivalents increases even further (Figure 2D, panels b-d). The absence of cells with intermediate amounts of DNA reflects the absence of active replication. As the cell density remains constant, it is reasonable to correlate the absence of replication with the absence of cell division at 45°C. The disappearance of cells with intermediate amounts of DNA cannot be explained by degradation of partially replicated DNA because this process would imply either a large increase of the cell population harboring one genome or the lysis of these cells. These two possibilities are clearly not in accordance with our results. Thus, we conclude that the cells having initiated replication, finish their replication, leading cells with two genome equivalents. This completion of replication may be slow, that is consistent with the reduced processivity of the DNA replicases PolB1/Dpo1 and PolY/Dpo4, shown *in vitro* with decreasing temperatures [39,40]. Further evidence for this is that the cell distribution 48 hours after transfer to 45°C is similar to that of control cells except that the population of cells with an intermediate DNA content is slightly but significantly shifted towards a DNA content between 1.5 and 2 equivalent genomes and away from 1 to 1.5 genomes (data not shown). Moreover, the cell density increases slightly during these first two days, consistent with residual cell division. We cannot exclude the possibility that there may have been initiation of replication during the first few days after transfer. It has been shown that during the cell cycle of *Sulfolobus* cultivated at 80°C, a new round of replication can occur only when cell division has been completed [41]. However, we cannot completely exclude the possibility that the absence of cell division was as consequence of there having been no replication activity upon prolonged incubation at 45°C. In any case, the absence of cells with an intermediate DNA content after seven days at 45°C clearly indicates that no new replication started during prolonged incubation at 45°C.

When cells maintained at 45°C for 21 days were transferred back to 80°C, there was a transient small decrease of the OD_{600nm} within a couple of hours (Figure 2A). The OD_{600nm} increased progressively thereafter and the growth curve of the cultures transferred from 45°C to 80°C was similar to that for control cultures at 80°C (Figure 2A). Cell size did not change following the return to 80°C (1.6 μm ± 0.27) (Figure 2B, panel e) and the proportion of damaged cells remained very low, and similar to control values (Figure 2C, panel e). In flow cytometry, the cytogram of cells transferred back to 80°C and collected 24 hours after the up-shift was similar to that of the 80°C control cells (Figure 2D, panel e and panel a, respectively). Indeed, the presence of cells with an intermediate DNA content is fully restored, indicating that normal replication activity was recovered. Hence, the *Sulfolobus* cells maintained at 45°C retained their ability to grow again actively at 80°C even after three weeks at the low temperature. A similar phenomenon has been reported for *Sulfolobus* cells kept for a short time at room temperature and transferred back to their optimal growth temperature [41].

In conclusion, *S. solfataricus* cells are resistant to long periods at 45°C and are able to recover a normal cell activity, *i.e.* cell division and replication, when they are transferred back to 80°C.

Quantification of TopR1 and TopR2 per *S. solfataricus* cell

We tested whether the resistance of *S. solfataricus* to relatively low temperatures was associated with changes to the reverse gyrase content. To distinguish between TopR1 and TopR2 in crude extracts, we obtained two specific antibodies (see Materials and methods), each raised against two peptides only found in one of the two reverse gyrases and absent from all other putative proteins of *S. solfataricus*. The pre-immune *sera* recognized some proteins in *S. solfataricus* extracts, but did so only very weakly, and none were similar in molecular mass to the reverse gyrases (data not shown). The specificity of both antibodies was checked with recombinant TopR1 and TopR2, previously purified in our lab [37]. Thus, the anti-TopR1 antibodies recognized purified recombinant TopR1 of *S. solfataricus* (Figure 3A, lane 1) but not the purified recombinant TopR2 (Figure 3A, lane 2); likewise, the anti-TopR2 antibodies recognized the purified recombinant TopR2 of *S. solfataricus* (Figure 3B, lane 2) but not the purified recombinant TopR1 (Figure 3B, lane 1).

In crude extracts of *S. solfataricus*, the anti-TopR1 antibodies gave a strong signal for a single band migrating in agreement with the theoretical molecular mass of TopR1 (146.65 kDa) (Figure 3A, lane 3). The strongest signal given by the anti-TopR2 antibodies was with a band migrating in agreement with the theoretical molecular mass of TopR2 (132.6 kDa); it also revealed several additional bands with lower molecular masses, probably corresponding to proteolysis products (Figure 3B, lane 3). The electrophoretic mobility of the protein detected in the *S. solfataricus* crude extract was slightly slower, indicating a higher apparent molecular mass, than that of recombinant TopR2 (Figure 3B, lane 3 versus lane 2). This difference was also observed with antibodies raised against the same specific oligopeptides but produced in another rabbit (data not shown). We conclude that the protein in crude extracts recognized by the antibodies raised against the TopR2 peptides is indeed TopR2 which, like reverse gyrase [46] and numerous *Sulfolobales* proteins [47,48], has post-translational modifications which remain to be elucidated. We used the TopR1- and TopR2-specific *sera* to determine the numbers of the two reverse gyrase molecules present in the different crude *S. solfataricus* cell extracts. We established a calibration curve with known amounts of purified recombinant proteins to allow determination of the absolute number of molecules of each

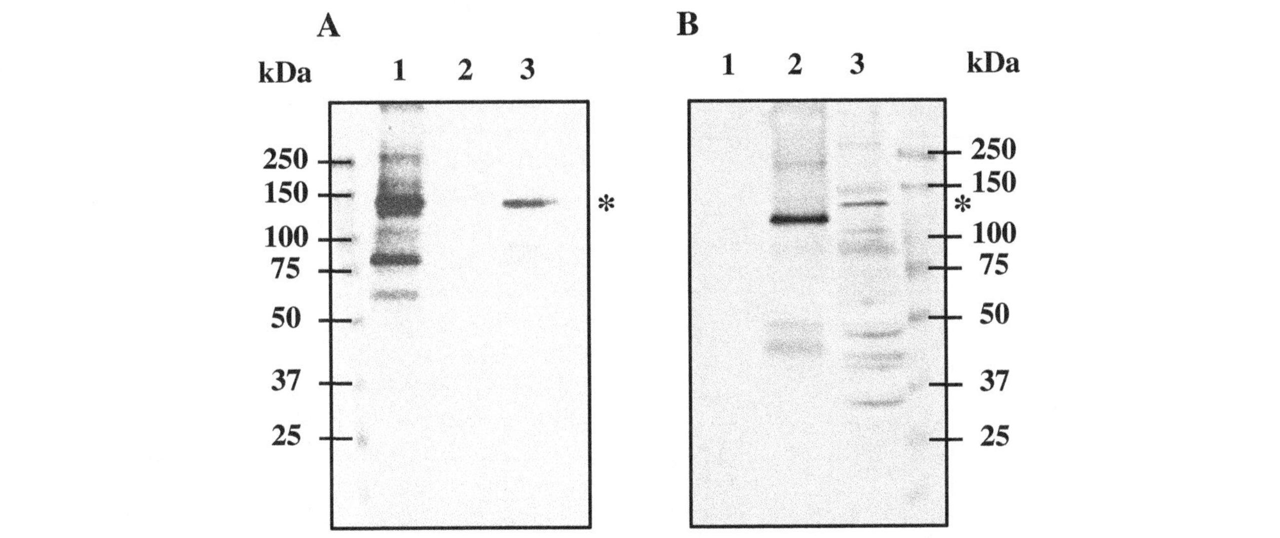

Figure 3 Specificity of the antibodies raised against the reverse gyrases of *S. solfataricus*. Western blot analysis of recombinant proteins TopR1 and TopR2 of *S. solfataricus* in lanes 1 and 2, respectively, and of *S. solfataricus* crude extracts in lane 3. Lanes 1-3 were probed with either anti-TopR1 *serum* diluted 1/6000 **(A)** or anti-TopR2 *serum* diluted 1/4000 **(B)**. Molecular weight standards in kDa (Kaleidoscope, BioRad) are indicated near panels **A** and **B**. Lanes 1 in panels A and B contain 0.3 ng and 50 ng of the recombinant TopR1, respectively. Lanes 2 in panels **A** and **B** correspond to 50 ng of the recombinant TopR2. Lane 3 in panel A and lane 3 in panel B contain 10 μg and 50 μg of *S. solfataricus* crude extracts, respectively. Asterisks indicate the signals for TopR1 and TopR2 in crude extracts.

reverse gyrase per cell (Additional file 1: Figure S1). The signal intensity in the various crude extracts was in the same range as that of the calibration, we were able to determine the precise number of TopR1 and TopR2 molecules per *S. solfataricus* cell (Figure 4).

Number of TopR1 and TopR2 per *S. solfataricus* cell at 80°C

This is the first study reporting the specific detection of two reverse gyrases within the same organism. The immunodetection signal obtained for cells growing exponentially at 80°C (Figure 4A and B, lane 1) corresponded to 54 ± 6 molecules of TopR1 and 127 ± 18 molecules of TopR2 per cell (Figure 4C). These are very low values, but nevertheless in agreement with the very small amounts of the corresponding transcripts reported previously [17].

It has been shown that reverse gyrase has a heat-protective DNA chaperone activity *in vitro*. This activity requires a high protein/DNA mass ratio, of at least 10 [20]. The results we report here indicate a total of approximately 180 molecules of reverse gyrase per cell; if reverse gyrases are regularly distributed along the chromosome, this corresponds to a protein/DNA mass ratio approximately 38-fold lower than required for the *in vitro* chaperone activity. These observations can be reconciled by putative recruitment of reverse gyrase to, for example, sites of damaged DNA as previously discussed [9,29,49]. Similarly, our evidence that there is little reverse gyrase (either TopR1 or TopR2) in each cell is discordant with the requirement for a saturating concentration of the protein for the unwinding activity of TopR1 of four-way DNA junctions [33].

By contrast, our estimations of the amounts of reverse gyrase *in vivo* are consistent with the renaturase activity evidenced in *in vitro* [28]. Indeed the introduction of positive supercoils into DNA containing a bubble of denaturation or the annealing of complementary single strand DNA circles have been evidenced *in vitro* with low protein/DNA mass ratios (0.5 and 0.8, respectively) [28]. These activities are in agreement with thermoprotection of DNA by reverse gyrase: its positive supercoiling activity limits the formation of single-strand regions at high temperature [16]. All *in vitro* experiments have been performed with DNA substrate devoid of any other bound protein; however, *in vivo*, DNA is obviously stabilized by the binding of chromatin proteins, such as Sul7d. The activities of reverse gyrase evidenced *in vitro* may be modulated by DNA binding proteins. Indeed, Sul7d inhibits reverse gyrase activity [50], and the single-strand binding protein, SSB, stimulates both DNA binding and the positive supercoiling of reverse gyrase [30].

Loss of TopR1 from *S. solfataricus* cells maintained at 45°C

When cells were transferred from 80°C to 45°C for three weeks, the amount of TopR2 per cell remained approximately constant throughout the period at 45°C (Figure 4B and C, conditions 2-4) suggesting either particularly high stability and/or a basal synthesis. The concentration of TopR1 decreased (Figure 4A and C, conditions 2-4): the

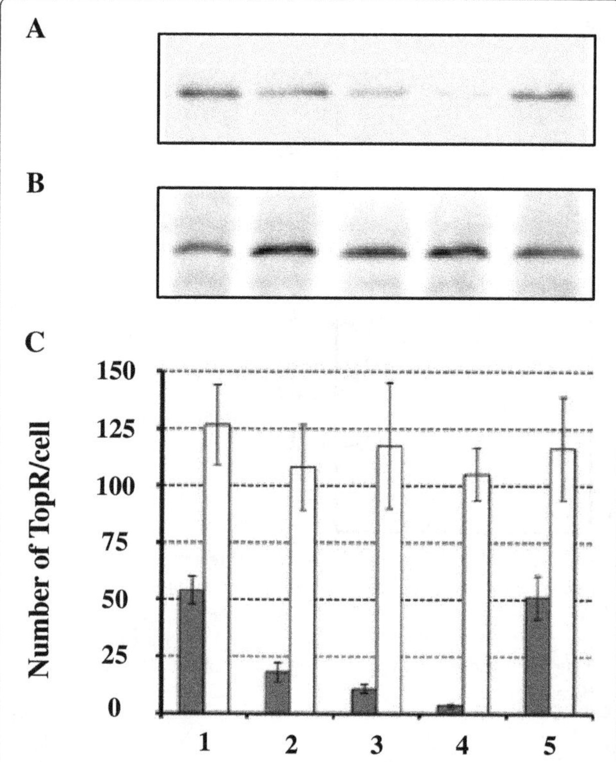

Figure 4 Reverse gyrase content in *S. solfataricus* cells. Crude protein extracts were prepared from cells growing exponentially at 80°C (lane 1), transferred to 45°C for 7 days (lane 2), 14 days (lane 3) and 21 days (lane 4) or back to 80°C for 24 hours (lane 5). Aliquots of 10 µg and 50 µg of crude extract were used for detection of TopR1 and TopR2, respectively, by western blotting with anti-TopR1 **(A)** or anti-TopR2 **(B)** antibodies. The specific signals were converted to the number of TopR molecules per cell, shown in panel **C**. The quantification data shown in panel C represent the average of at least three independent experiments.

number of TopR1 per cell declined by two thirds after 7 days, four fifths after 14 days, and to close to the detection threshold after 21 days at 45°C (Figure 4C, conditions 2-4). There was no cell division at 45°C such that the cell number remained unchanged over the three weeks at 45°C, so the loss of TopR1 was necessarily due to a specific but slow degradation. The protein concentration in the crude extracts from cells transferred to 45°C was systematically half that in extracts from 80°C control cells. Consequently, the number of TopR1 molecules per unit protein declined even more than the number per cell, whereas the number of TopR2 molecules per unit protein doubled. Thus, the amounts of the two reverse gyrases were regulated differently in this long-term down-shift experiment.

Cells maintained at 45°C for 21 days were able to resume active cell division when shifted back to 80°C. We assayed the two reverse gyrases 24 hours after the shift back to 80°C. The concentration of TopR1 (molecules per cell) returned to baseline control values and that of TopR2 remained unchanged (Figure 4C, condition 5 compared with condition 1). Thus, the ability of the cells to divide correlated with there being appropriate amounts of both TopR1 and TopR2.

The amounts of TopR1 and TopR2 did not change detectably within the first 6 hours following the transfer from 80°C to 45°C (data not shown). The loss of TopR1 observed was slow, and may therefore have been related to the reduced cell activity rather than directly to the temperature change itself. We found that TopR1 is not active at low temperature *in vitro*, consistent with its loss from cultures at low temperature. When the cell activity is reduced, DNA transaction processes are less active, and therefore rigorous control of the topological state of DNA is less critical. At 45°C there was no cell division and no replication, and therefore few or no topological modification of the DNA, so presumably TopR1 is not needed. This implies that TopR1 is the topoisomerase mostly responsible for the DNA topological state in *S. solfataricus*. However, it is also possible that, at low temperature, TopR2 complements the TopR1 deficiency. TopR2 exhibits DNA positive supercoiling activity, which may be sufficient for the residual DNA topological regulation at low temperature. In cells growing at high temperature, TopR1 is presumably required to resolve the frequent modifications of DNA topological state triggered by the cell division activity and enhanced by the high temperature itself.

TopR2 is present both at high temperature (actively dividing cells) and low temperature (non-dividing cells) and it is active at 45°C *in vitro* (Figure 1B and C). The resumption of normal cell activity following the shift back to 80°C after three weeks at 45°C reveals that genome stability was preserved. Although there is little rigorous evidence, reverse gyrase has long been proposed to be involved in maintenance of genome stability [9,26]. As only TopR2 is the only reverse gyrase maintained in cells at 45°C, it may have this function either by stabilizing particular regions of the DNA or by participating in DNA metabolism pathways. After a long period at 45°C, cultures contain few or no cells with between 1 and 2 genome equivalent, although such cells with 1-2 genome equivalent are observed during the 48 h following the transfer to 45°C. TopR2 may contribute to the residual replication activity in cells with 1-2 genome equivalent, such that they progress to containing a whole number (two) of genomes. Transcriptional analysis has shown that the abundance of the *topR2* transcript increases during the G1/S transition phase, suggesting an implication of TopR2 in replication [51]. At 45°C, TopR2 may have short-term functions in both genome stability and replication; subsequently, only the function contributing to genome stability appears to be relevant in the long term as there is no new replication.

TopR1 from *S. solfataricus* inhibits the *in vitro* activity of the translesion DNA PolY/Dpo4, a DNA polymerase involved in response to DNA damage [32]. TopR1 and PolY/Dpo4 interact both *in vivo* and *in vitro*, and presumably this physical interaction mediates the inhibition. However, the interaction was demonstrated with anti-TopR1 antibodies that recognize both TopR1 and TopR2, so it is not clear whether one or both reverse gyrases co-immunoprecipitated with PolY/Dpo4. The topoisomerase domain of TopR1 was reported to be responsible for its affinity for PolY/Dpo4. The topoisomerase domains of *Sulfolobus* TopR1 and TopR2 exhibit a high identity [37], so it is plausible that TopR2 may exhibit significant affinity for PolY/Dpo4. If this were true, it would imply that TopR2, the only reverse gyrase present and active at 45°C, inhibits PolY/Dpo4. At 80°C, there would be competition between the two reverse gyrases, the results of which would depend on their abundance and their respective affinities for PolY/Dpo4.

Sulfolobus cells at 45°C have no TopR1 although it is required at 80°C. Presumably, TopR1 is inessential at low temperatures and indispensable at high temperature because the mechanical response of DNA is temperature dependent. *N. profundicola* grows optimally at 45°C and has only one reverse gyrase gene; following exposure to 65°C for two hours, the abundance of the reverse gyrase transcript increases substantially, suggesting that more reverse gyrase is required [34]. This is consistent with the increased frequency of DNA melting at high temperature leading to a greater requirement for reverse gyrase. Possibly, the *Sulfolobus* TopR1 is the functional homologue of the reverse gyrase of *Nautilia*, both proteins exhibiting quantitative variations with temperature. *T. kodakaraensis* also has a single reverse gyrase: a *T. kodakaraensis* mutant deleted for the corresponding gene is viable at temperatures of 60-90°C but not at temperatures higher than 90°C [35]. Possibly, there are compensatory mutations, or other topoisomerases, present in *T. kodakaraensis*, compensating for the absence of reverse gyrase but only at temperatures lower than 90°C. In *S. islandicus*, a strain very closely related to *S. solfataricus*, no mutant viable at 75°C could be obtained for either TopR1 or TopR2 encoding genes, suggesting that both enzymes are essential [36]. In this work, we found that in *S. solfataricus* TopR1 is the major reverse gyrase for the control of the topological state DNA. However, the absence of TopR1 can be compensated at least partially by TopR2, its activity being sufficient to control the minor low topological changes at low temperature. At 75°C, TopR2 cannot complement for the absence of TopR1 because DNA melting is much more extensive, and problematic, at this temperature. As TopR2 may have functions not displayed by TopR1, TopR2 mutants may be lethal. These various observations are concordant in implying that TopR1 is required at high temperature and/or for thermoadaptation.

In this report, we demonstrate that *S. solfataricus* is able to survive at a low temperature (45°C) for a long period, without dividing but with most cells in the culture containing two fully replicated genomes. They are ready to resume a normal cell activity with active cell division as soon as favorable conditions are restored. This property of "cold" resistance may facilitate the spread of *S. solfataricus* to new niches. We also show that the two reverse gyrases are not regulated in the same way indicating that they do have different and possibly overlapping functions in the cells. We provide here evidence, in addition to our previous results [17,37], that TopR1 is important for the regulation of the supercoiling density of the genome, which is affected by replication, transcription and recombination, particularly active in dividing cells at high temperature.

Conclusions

We report the first quantification of the numbers of reverse gyrase molecules per cell in *S. solfataricus*: in actively growing cultures at 80°C, there are approximately 50 molecules of TopR1 and 125 molecules of TopR2 per cell. At 45°C, *S. solfataricus* does not grow and the reverse gyrase content changes: the amount of TopR1 decreases substantially, although that of TopR2 remains unchanged. These findings *in vivo* are in agreement with the activities of the two enzymes *in vitro*. TopR2 exhibits significant positive supercoiling activity at 45°C, a temperature at which TopR1 is not active. TopR1 is inessential at low temperature but required at high temperature and therefore probably involved in thermoadaptation and/or in DNA transaction processes during active division of *S. solfataricus* cells at 80°C. By contrast, TopR2 is at a constant concentration at both 80°C and 45°C suggesting that TopR2 may be involved in the maintenance of genome stability, particularly in the long term at 45°C when there is no cell division and no replication.

Methods

Materials

Tris, glycine, SDS, dimethylsulfoxide (DMSO), acrylamide and bis-acrylamide were purchased from Euromedex. Ponceau S, p-coumaric acid, bromophenol blue, brilliant blue R250, $MgSO_4$, $Ca(NO_3)_2$, $COSO_4$, $CuCl_2$, $ZnSO_4$, $Na_2M_0O_4$, $Na_2B_4O_7$, $FeSO_4$, $VOSO_4$, vitamins, HEPES, hydrogen peroxide, Tween 20, luminol, Triton X-100, sucrose, acetic acid, dithiothreitol (DTT), RNase A, and propidium iodide (PI) were obtained from Sigma. KCl, H_2SO_4 and KH_2PO_4 were from Merck. NaCl was from Fischer scientific. Tryptone peptone and yeast extract were obtained from Difco (Becton Dickinson). Ammonium persulfate (APS), N,N,N',N'-tetra methyl ethylenediamine

(TEMED), bovine *serum* albumin and Bradford reagents were purchased from BioRad. Propan-2-ol, ethanol, sorbitol, HCl, $MgCl_2$, $MnCl_2$ and $(NH_4)_2SO_4$ were from Prolabo. Glycerol was obtained from Acros organique.

Strain

Sulfolobus solfataricus strain P2 (DSMZ 1617) was purchased from the Deutsche Sammlung von Mikroorganismen und Zelkulturen in Braunschweig, Germany.

Reverse gyrase assays

Enzymatic assays

The standard reaction mixture contained 50 mM Tris-HCl pH 8.0, 0.5 mM DTT, 0.5 mM EDTA, 20 mM $MgCl_2$, 100 mM NaCl, 1.25 mM ATP and 0.15 µg of negatively supercoiled pTZ18R DNA. The purified enzymes [37] were added to the mixture at a molar ratio topoisomerase/DNA of 4, and the mixture was incubated at the indicated temperature for 20 min. The reaction was then stopped by cooling on ice; 0.1% SDS, 25 mg/mL bromophenol blue and sucrose to 15% were added before loading onto the agarose gel.

One-dimensional gel electrophoreses

Electrophoresis was performed in 1.2% agarose gels at room temperature in TEP buffer (36 mM Tris, 30 mM NaH_2PO_4, 1 mM EDTA, pH 7.8) and run at 3 V/cm for 6 h. Gels were washed in TEP buffer for 15 min, stained with ethidium bromide (2 µg/mL for 30 min) and digitalized under UV light.

Two-dimensional gel electrophoreses

TopR2 activity was analyzed after a two-dimensional gel electrophoresis. The first dimension was run at room temperature in a 1.2% agarose gel in TEP buffer at 3 V/cm for 150 min. The gel was then soaked for 30 min in TEP buffer containing 10 µg/mL chloroquin. The second dimension was run in the same buffer, perpendicularly to the first, at 0.9 V/cm for 14 h. The gel was washed in TEP buffer for 30 min and then stained.

Culture of *Sulfolobus solfataricus* P2 in liquid medium

Sulfolobus solfataricus P2 was cultured as previously described [17] at 80°C or 45°C. The cell density was monitored by measuring the optical density at 600 nm (OD_{600nm}) and by flow cytometry. Cells cultivated at 80°C and having reached an exponential growth phase ($0.3 < OD_{600nm} < 0.6$) were transferred to 45°C and maintained at this low temperature for three weeks either in the same medium, or resuspended every six days with fresh medium pre-heated at 45°C. After three weeks at 45°C, the cultures were transferred back to 80°C.

Phase-contrast and fluorescence microscopy

S. solfataricus cells (1.2×10^9 cells) were collected by centrifugation at room temperature at 5000 × *g* for 5 min from cultures at 80°C before the shift to 45°C, at 45°C at various times for three weeks, and again when cells were transferred back to 80°C. The cells were washed with medium then centrifuged at 10000 × *g* for 6 min. Syto 9 and propidium iodide of the LIVE/DEAD *Bac*Light bacterial viability kit (Molecular Probes) were diluted in the culture medium to a final concentration of 7 µM and 10 µM respectively and used to resuspend cell samples to a final concentration of 10^7 cells/µL. These samples were incubated for 15 minutes at room temperature in the darkness. The cells were observed in a three-dimensional deconvolution microscope (DMIRE2; Leica) equipped with an HCxPL APO 100 × oil CS objective, NA = 1.40 (Leica). The images were captured on a 10-MHz Cool $SNAP_{HQ\ 2}$ CCD camera (Roper Instruments), with a Z-optical spacing of 0.2 µm. METAMORPH software (Universal Imaging Corp.) was used to acquire Z-series, deconvoluted or not, and treat the images. Only one image with sharp fluorescence from the numerous acquired undeconvoluted Z-series is shown for each condition. In all samples, all the cells incorporated one or both stains and at least 200 cells/sample were examined to determine cell size and cell viability. A minimum of three independent measurements were performed for each condition.

Flow cytometry

Aliquots of *S. solfataricus* cells were fixed by adding five volumes of 70% ethanol, then diluted with 70% ethanol to obtain a final concentration of 10^8 cells/mL. The cells were then washed twice with TE buffer (10 mM Tris-HCl pH 7.4 and 10 mM EDTA, pH 8) by centrifugation at 4°C at 10000 x *g* for 6 min. Samples were kept at 4°C during all steps. The cells were then resuspended in TE buffer containing RNase A (10 µg/mL) to a final concentration of 10^8 cells/mL. The cell preparations were incubated at 37°C for 120 min and centrifuged at 4°C at 10000 × *g* for 6 min. The cells were resuspended in the same volume of TE buffer containing propidium iodide (10 µg/mL) and kept at 4°C overnight in the darkness and analyzed with a PAS III Partec Flow cytometer as previously described [52]. Results are shown as representations of combined forward light scatter (FSC) and DNA content distributions.

Crude extract preparation

Cell samples (8.25×10^9 cells) were collected as described for microscopy but were cooled immediately on ice, and then centrifuged at 4°C at 5000 × *g* for 5 min. The cells were washed twice with buffer containing 20 mM HEPES and 1 M sorbitol then centrifuged at 4°C at 10000 × *g* for 6 min. The cells were resuspended to a final cell concentration of 4 × 10^{10} cells/mL in extraction buffer (50 mM Tris-HCl pH 7, 15 mM $MgCl_2$, 50 mM NaCl, 1 mM DTT, 400 mM sorbitol) and were gently disrupted by the addition of 0.5% Triton X-100 and moderate agitation for 15 min at 4°C. Protein concentrations in the resulting crude extracts were determined by the Bradford method with bovine *serum* albumin as the standard. Aliquots of these protein extracts were denatured with Laemmli buffer and heated at 95°C for 5 min, and resolved by SDS-PAGE.

Detection of reverse gyrases

Primary antibodies were produced by Eurogentec against two epitopes specific for each of the two reverse gyrases of *S. solfataricus*: PRILYNKQSPTQTEN and EDIQTTMKLL-RENIG for anti-TopR1 and GRSKLNIKKYVEDL and YFSEKRKVEEYINNL for anti-TopR2. The choice of these epitopes was based on both the amino acid sequences deduced from the sequences of the genes in *S. solfataricus* [53] and the published structure of reverse gyrase [54]. These peptides are absent from all other CDS of *S. solfataricus*. The peptides were synthesized, linked to hemocyanine and used to immunize rabbits. Proteins were separated by SDS-PAGE (10% acrylamide/0.13% bis-acrylamide), then electroblotted onto nitrocellulose membranes (Whatman Protran BA79) over 1 hour at 4°C in transfer buffer (25 mM Tris-HCl, 192 mM glycine, 15% propan-2-ol). Membranes were stained with Ponceau S, washed with TBS -Tween buffer (20 mM Tris-HCl pH 7.6, 13.7 mM NaCl, 0.1% Tween 20) and blocked with 3% milk in TBS-Tween. Membranes were then probed overnight at 4°C with either 1/6000 anti-TopR1 antibodies or 1/4000 anti-TopR2 antibodies. Membranes were then washed with TBS-Tween and probed for 1 hour at 4°C with 1/20000 horseradish-peroxidase-conjugated anti-rabbit IgG from donkey (GE Healthcare). All dilutions of antibodies were in TBS-Tween. Bound antibodies were visualized with ECL mix (100 mM Tris-HCl pH 8.45, 0.009% H_2O_2, 0.225 mM p-coumaric acid, 1.25 mM luminol). The chemiluminescence and Ponceau S signals were captured on a CCD camera (Image Quant LAS 4000, GE Healthcare) and analyzed with the ImageQuant TL (v 7.0) package.

Estimation of TopR1 and TopR2 number per *S. solfataricus* cell

To determine the number of copies of the two reverse gyrases in *S. solfataricus* cells, we established a relation between ECL signal intensity in the crude extracts and the amount of reverse gyrase. Purified recombinant TopR1 and TopR2 [37] were used as standards. Series of quantities from 0.3 to 50 ng were loaded onto gels with the various crude extracts such that the ECL signal intensities can be compared. The ECL signal intensities obtained for various crude extracts for each condition were compared with the TopR1 and TopR2 calibration curves (Additional file 1: Figure S1) and converted into amounts of reverse gyrase. The amount of protein (in ng) was converted into a number of molecules on the basis of the theoretical molecular weight of the corresponding reverse gyrase. Only membranes for which the Ponceau S staining of the various crude extracts loaded was homogeneous were used for these analyses. Each experiment was performed at least three times independently, and the results reported are the mean values.

Additional file

Additional file 1: Figure S1. Calibration curves of recombinant proteins TopR1 and TopR2. The purified recombinant TopR1 and TopR2 proteins of *S. solfataricus* were used as standards and the western blots obtained with anti-TopR1 **(A)** or anti-TopR2 **(B)** antibodies are shown. Calibration curves, **C** and **D**, were generated by plotting the ECL signal intensity against the amount of the purified recombinant loaded: from 0.3 to 1.2 ng for TopR1 and from 12.5 to 50 ng for TopR2. The ECL signal intensity obtained for each crude extract was within the range of the ECL signal intensities for three amounts of the purified recombinant TopR, so only these three points are shown. The protein concentrations of the purified fractions of TopR1 and TopR2 were measured by using the Bradford method (BioRad). The purity of the band corresponding either to TopR1 or TopR2 was estimated after SDS-PAGE and gel Coomassie-blue staining. With the specific molecular mass of TopR1 and TopR2 and the % of purity of the corresponding fraction, the precise quantity of the full length proteins, TopR1 or TopR2, was determined as previously reported [37].

Competing interests

The authors declare that they have no competing interests.

Authors' contributions

MC participated in the design of the study, carried out cell culture, microscopy, flow cytometry and western blotting, and helped draft the manuscript. AB carried out reverse gyrase assays. FG participated in the design of the antibodies and establishing the western blot methods, participated in the design of the study and its coordination, and drafted the manuscript. MN participated in the design of the antibodies and establishing the western blot methods, participated in the design of the study and its coordination, and drafted the manuscript. All the authors have read and approved the final manuscript.

Acknowledgments

This work was supported by CNRS, EDF (RB-2003-16) and IFR 115.
We thank Magali Prigent for her help with fluorescent microscopy and Jean-Luc Ferat for his help with flow cytometer experiments. We thank Hélène Débat, Florence Constantinesco, Christiane Elie, Adrienne Kish and Martine Mathieu for helpful discussions. We are indebted to Patrick Forterre for support during this work.

Author details

[1]Université Versailles St-Quentin, 45 avenue des Etats-Unis, Versailles 78035, France. [2]Institut de Génétique et Microbiologie, UMR 8621 CNRS, Université Paris-Sud, Bât. 409, Orsay Cedex 91405, France. [3]Université d'Evry-Val d'Essonne, Boulevard François Mitterrand, Evry 91025, France. [4]Present address: Department of Molecular Biosciences, The Wenner-Gren Institute, Stockholm University, Stockholm, Sweden. [5]Present address: Institute of Cellular and Molecular Medicine (ICMM), Center for Healthy Ageing (CEHA),

University of Copenhagen, Blegdamsvej 3B, København N DK-2200, Denmark.
[6]Université Paris Diderot, 5 rue Thomas Mann, Paris 75013, France.

References

1. Liu LFL, Wang JCJ: **Supercoiling of the DNA template during transcription.** *Proc Natl Acad Sci U S A* 1987, **84:**7024–7027.
2. Champoux JJ: **DNA topoisomerases: structure, function, and mechanism.** *Annu Rev Biochem* 2001, **70:**369–413.
3. Chen SH, Chan N-L, Hsieh T-S: **New Mechanistic and Functional Insights into DNA Topoisomerases.** *Annu Rev Biochem* 2013, **82:**139–170.
4. Mirambeau G, Duguet M, Forterre P: **Atp-Dependent Dna Topoisomerase From the Archaebacterium Sulfolobus-Acidocaldarius - Relaxation of Supercoiled Dna at High-Temperature.** *J Mol Biol* 1984, **179:**559–563.
5. Kikuchi A, Asai K: **Reverse gyrase–a topoisomerase which introduces positive superhelical turns into DNA.** *Nature* 1984, **309:**677–681.
6. Forterre P, Mirambeau G, Jaxel C, Nadal M, Duguet M: **High positive supercoiling in vitro catalyzed by an ATP and polyethylene glycol-stimulated topoisomerase from Sulfolobus acidocaldarius.** *EMBO J* 1985, **4:**2123–2128.
7. Nakasu S, Kikuchi A: **Reverse gyrase; ATP-dependent type I topoisomerase from Sulfolobus.** *EMBO J* 1985, **4:**2705–2710.
8. Nadal M, Jaxel C, Portemer C, Forterre P, Mirambeau G, Duguet M: **Reverse gyrase of Sulfolobus: purification to homogeneity and characterization.** *Biochemistry* 1988, **27:**9102–9108.
9. Nadal M: **Reverse gyrase: an insight into the role of DNA-topoisomerases.** *Biochimie* 2007, **89:**447–455.
10. Bouthier de la Tour C, Portemer C, Nadal M, Stetter KO, Forterre P, Duguet M: **Reverse gyrase, a hallmark of the hyperthermophilic archaebacteria.** *J Bacteriol* 1990, **172:**6803–6808.
11. Bouthier de la Tour C, Portemer C, Huber R, Forterre P, Duguet M: **Reverse gyrase in thermophilic eubacteria.** *J Bacteriol* 1991, **173:**3921–3923.
12. Forterre P: **A hot story from comparative genomics: reverse gyrase is the only hyperthermophile-specific protein.** *Trends Genet* 2002, **18:**236–237.
13. Brochier-Armanet C, Forterre P: **Widespread distribution of archaeal reverse gyrase in thermophilic bacteria suggests a complex history of vertical inheritance and lateral gene transfers.** *Archaea* 2007, **2:**83–93.
14. Nadal M, Mirambeau G, Forterre P, Reiter WD, Duguet M: **Positively Supercoiled Dna in a Virus-Like Particle of an Archaebacterium.** *Nature* 1986, **321:**256–258.
15. Lim HM, Lee HJ, Jaxel C, Nadal M: **Hin-mediated inversion on positively supercoiled DNA.** *J Biol Chem* 1997, **272:**18434–18439.
16. Bell SD, Jaxel C, Nadal M, Kosa PF, Jackson SP: **Temperature, template topology, and factor requirements of archaeal transcription.** *Proc Natl Acad Sci U S A* 1998, **95:**15218–15222.
17. Garnier F, Nadal M: **Transcriptional analysis of the two reverse gyrase encoding genes of Sulfolobus solfataricus P2 in relation to the growth phases and temperature conditions.** *Extremophiles* 2008, **12:**799–809.
18. Menzel R, Gellert M: **Regulation of the genes for E. coli DNA gyrase: homeostatic control of DNA supercoiling.** *Cell* 1983, **34:**105–113.
19. Peter BJ, Arsuaga J, Breier AM, Khodursky AB, Brown PO, Cozzarelli NR: **Genomic transcriptional response to loss of chromosomal supercoiling in Escherichia coli.** *Genome Biol* 2004, **5:**R87.
20. Kampmann M, Stock D: **Reverse gyrase has heat-protective DNA chaperone activity independent of supercoiling.** *Nucleic Acids Res* 2004, **32:**3537–3545.
21. Confalonieri F, Elie C, Nadal M, Bouthier de La Tour C, Forterre P, Duguet M: **Reverse Gyrase: A Helicase-Like Domain and a Type 1 Topoisomerase in the Same Polypeptide.** *Volume* 1993, **90:**4753–4757.
22. Jaxel C, Bouthier de la Tour C, Duguet M, Nadal M: **Reverse gyrase gene from Sulfolobus shibatae B12: gene structure, transcription unit and comparative sequence analysis of the two domains.** *Nucleic Acids Res* 1996, **24:**4668–4675.
23. Gangloff S, McDonald JP, Bendixen C, Arthur L, Rothstein R: **The yeast type I topoisomerase Top3 interacts with Sgs1, a DNA helicase homolog: a potential eukaryotic reverse gyrase.** *Mol Cell Biol* 1994, **14:**8391–8398.
24. Harmon FG, DiGate RJ, Kowalczykowski SC: **RecQ helicase and topoisomerase III comprise a novel DNA strand passage function: a conserved mechanism for control of DNA recombination.** *Mol Cell* 1999, **3:**611–620.
25. Mankouri HW, Hickson ID: **The RecQ helicase-topoisomerase III-Rmi1 complex: a DNA structure-specific 'dissolvasome'?** *Trends Biochem Sci* 2007, **32:**538–546.
26. Perugino G, Valenti A, D'amaro A, Rossi M, Ciaramella M: **Reverse gyrase and genome stability in hyperthermophilic organisms.** *Biochem Soc Trans* 2009, **37:**69–73.
27. Slesarev AI, Kozyavkin SA: **DNA substrate specificity of reverse gyrase from extremely thermophilic archaebacteria.** *J Biomol Struct Dyn* 1990, **7:**935–942.
28. Hsieh T-S, Plank JL: **Reverse gyrase functions as a DNA renaturase: annealing of complementary single-stranded circles and positive supercoiling of a bubble substrate.** *J Biol Chem* 2006, **281:**5640–5647.
29. Napoli A, Valenti A, Salerno V, Nadal M, Garnier F, Rossi M, Ciaramella M: **Reverse gyrase recruitment to DNA after UV light irradiation in Sulfolobus solfataricus.** *J Biol Chem* 2004, **279:**33192–33198.
30. Napoli AA, Valenti AA, Salerno VV, Nadal MM, Garnier FF, Rossi MM, Ciaramella MM: **Functional interaction of reverse gyrase with single-strand binding protein of the archaeon Sulfolobus.** *Nucleic Acids Res* 2004, **33:**564–576.
31. Wadsworth RIR, White MFM: **Identification and properties of the crenarchaeal single-stranded DNA binding protein from Sulfolobus solfataricus.** *Nucleic Acids Res* 2001, **29:**914–920.
32. Valenti A, Perugino G, Nohmi T, Rossi M, Ciaramella M: **Inhibition of translesion DNA polymerase by archaeal reverse gyrase.** *Nucleic Acids Res* 2009, **37:**4287–4295.
33. Valenti A, Perugino G, Varriale A, D'Auria S, Rossi M, Ciaramella M: **The archaeal topoisomerase reverse gyrase is a helix-destabilizing protein that unwinds four-way DNA junctions.** *J Biol Chem* 2010, **285:**36532–36541.
34. Campbell BJ, Smith JL, Hanson TE, Klotz MG, Stein LY, Lee CK, Wu D, Robinson JM, Khouri HM, Eisen JA, Cary SC: **Adaptations to submarine hydrothermal environments exemplified by the genome of Nautilia profundicola.** *PLoS Genet* 2009, **5:**e1000362.
35. Atomi H, Matsumi R, Imanaka T: **Reverse gyrase is not a prerequisite for hyperthermophilic life.** *J Bacteriol* 2004, **186:**4829–4833.
36. Zhang C, Tian B, Li S, Ao X, Dalgaard K, Gökce S, Liang Y, She Q: **Genetic manipulation in Sulfolobus islandicus and functional analysis of DNA repair genes.** *Biochem Soc Trans* 2013, **41:**405–410.
37. Bizard A, Garnier F, Nadal M: **TopR2, the Second Reverse Gyrase of Sulfolobus solfataricus, Exhibits Unusual Properties.** *J Mol Biol* 2011, **408:**839–849.
38. Lübben M, Schäfer G: **Chemiosmotic energy conversion of the archaebacterial thermoacidophile Sulfolobus acidocaldarius: oxidative phosphorylation and the presence of an F0-related N, N'-dicyclohexylcarbodiimide-binding proteolipid.** *J Bacteriol* 1989, **171:**6106–6116.
39. Lin H-KH, Chase SFS, Laue TMT, Jen-Jacobson LL, Trakselis MAM: **Differential temperature-dependent multimeric assemblies of replication and repair polymerases on DNA increase processivity.** *Biochemistry* 2012, **51:**7367–7382.
40. Choi J-Y, Eoff RL, Pence MG, Wang J, Martin MV, Kim E-J, Folkmann LM, Guengerich FP: **Roles of the four DNA polymerases of the crenarchaeon Sulfolobus solfataricus and accessory proteins in DNA replication.** *J Biol Chem* 2011, **286:**31180–31193.
41. Hjort K, Bernander R: **Changes in Cell Size and DNA Content in Sulfolobus Cultures during Dilution and Temperature Shift Experiments.** *J Bacteriol* 1999, **181:**5669–5675.
42. Boulos L, Prévost M, Barbeau B, Coallier J, Desjardins R: **LIVE/DEAD® BacLight™: application of a new rapid staining method for direct enumeration of viable and total bacteria in drinking water.** *J Microbiol Methods* 1999, **37:**77–86.
43. Leuko S, Legat A, Fendrihan S, Stan-Lotter H: **Evaluation of the LIVE/DEAD BacLight kit for detection of extremophilic archaea and visualization of microorganisms in environmental hypersaline samples.** *Appl Environ Microbiol* 2004, **70:**6884–6886.
44. De Rosa M, Gambacorta A, Nicolaus B, Sodano S, Bu'lock JD: **Structural regularities in tetraether lipids of Caldariella and their biosynthetic and phyletic implications.** *Phytochemistry* 1980, **19:**833–836.
45. Bernander RR, Poplawski AA: **Cell cycle characteristics of thermophilic archaea.** *J Bacteriol* 1997, **179:**4963–4969.

46. Esser D, Pham TK, Reimann J, Albers S-V, Siebers B, Wright PC: **Change of Carbon Source Causes Dramatic Effects in the Phospho-Proteome of the Archaeon Sulfolobus solfataricus.** *J Proteome Res* 2012, **11**:4823–4833.
47. Barry RC, Young MJ, Stedman KM, Dratz EA: **Proteomic mapping of the hyperthermophilic and acidophilic archaeon Sulfolobus solfataricus P2.** *Electrophoresis* 2006, **27**:2970–2983.
48. Reimann J, Esser D, Orell A, Amman F, Pham TK, Noirel J, Lindås A-C, Bernander R, Wright PC, Siebers B, Albers S-V: **Archaeal Signal Transduction: Impact of Protein Phosphatase Deletions on Cell Size, Motility, and Energy Metabolism in Sulfolobus acidocaldarius.** *Mol Cell Proteomics* 2013, **12**:3908–3923.
49. Perugino G, Vettone A, Illiano G, Valenti A, Ferrara MC, Rossi M, Ciaramella M: **Activity and regulation of an archaeal DNA-alkyltransferase: conserved protein involved in repair of DNA alkylation damage.** *J Biol Chem* 2012, **287**:4222–4231.
50. Napoli A, Zivanovic Y, Bocs C, Buhler C, Rossi M, Forterre P, Ciaramella M: **DNA bending, compaction and negative supercoiling by the architectural protein Sso7d of Sulfolobus solfataricus.** *Nucleic Acids Res* 2002, **30**:2656–2662.
51. Lundgren M, Bernander R: **Genome-wide transcription map of an archaeal cell cycle.** *Proc Natl Acad Sci U S A* 2007, **104**:2939–2944.
52. Saifi B, Ferat J-L, Marinus MG: **Replication Fork Reactivation in a dnaC2 Mutant at Non-Permissive Temperature in Escherichia coli.** *PLoS One* 2012, **7**:3613.
53. She QQ, Singh RKR, Confalonieri FF, Zivanovic YY, Allard GG, Awayez MJM, Chan-Weiher CCC, Clausen IGI, Curtis BAB, De Moors AA, Erauso GG, Fletcher CC, Gordon PMP, Jong IIH-D, Jeffries ACA, Kozera CJC, Medina NN, Peng XX, Thi-Ngoc HPH, Redder PP, Schenk MEM, Theriault CC, Tolstrup NN, Charlebois RLR, Doolittle WFW, Duguet MM, Gaasterland TT, Garrett RAR, Ragan MAM, Sensen CWC, *et al*: **The complete genome of the crenarchaeon Sulfolobus solfataricus P2.** *Proc Natl Acad Sci U S A* 2001, **98**:7835–7840.
54. Rodríguez AC, Stock D: **Crystal structure of reverse gyrase: insights into the positive supercoiling of DNA.** *EMBO J* 2002, **21**:418–426.

8

Cooperative activation of *Xenopus* rhodopsin transcription by paired-like transcription factors

Sarah E Reks, Vera McIlvain, Xinming Zhuo and Barry E Knox*

Abstract

Background: In vertebrates, rod photoreceptor-specific gene expression is regulated by the large Maf and Pax-like transcription factors, Nrl/LNrl and Crx/Otx5. The ubiquitous occurrence of their target DNA binding sites throughout rod-specific gene promoters suggests that multiple transcription factor interactions within the promoter are functionally important. Cooperative action by these transcription factors activates rod-specific genes such as rhodopsin. However, a quantitative mechanistic explanation of transcriptional rate determinants is lacking.

Results: We investigated the contributions of various paired-like transcription factors and their cognate cis-elements to rhodopsin gene activation using cultured cells to quantify activity. The *Xenopus* rhodopsin promoter (*XOP*) has a bipartite structure, with ~200 bp proximal to the start site (RPP) coordinating cooperative activation by Nrl/LNrl-Crx/Otx5 and the adjacent 5300 bp upstream sequence increasing the overall expression level. The synergistic activation by Nrl/LNrl-Crx/Otx5 also occurred when *XOP* was stably integrated into the genome. We determined that Crx/Otx5 synergistically activated transcription independently and additively through the two Pax-like cis-elements, BAT1 and Ret4, but not through Ret1. Other Pax-like family members, Rax1 and Rax2, do not synergistically activate XOP transcription with Nrl/LNrl and/or Crx/Otx5; rather they act as co-activators via the Ret1 cis-element.

Conclusions: We have provided a quantitative model of cooperative transcriptional activation of the rhodopsin promoter through interaction of Crx/Otx5 with Nrl/LNrl at two paired-like cis-elements proximal to the NRE and TATA binding site. Further, we have shown that Rax genes act in cooperation with Crx/Otx5 with Nrl/LNrl as co-activators of rhodopsin transcription.

Background

Rhodopsin is an abundantly expressed gene specifically found in rods, and its expression is primarily regulated at the level of transcription (for a recent review see [1]). The principal transcriptional control sequences, called the rhodopsin proximal promoter (RPP), reside within ~200 bp proximal to the transcription start site [2,3]. Additional regulatory sequences in mammalian genes are upstream of the RPP [4,5]. Alignment of 33 tetrapod RPP sequences reveals an extraordinarily high degree of conservation not observed with fish or invertebrates (Figure 1), with four conserved cis-elements (Ret1, BAT1, Ret4 and NRE) immediately upstream of the TATA box. NRE, a Maf family recognition element (MARE) site forty nucleotides upstream of the TATA box, is a target for Nrl (neural retina-specific leucine zipper protein), a transcription factor expressed exclusively in rods [6]. Flanking the NRE are three highly conserved paired-like homeodomain binding sites, Ret4, BAT1 and Ret1. These are target sites for several paired-like homeodomain transcription factors, most notably Crx but also Otx2 and Rax [7-11] Crx (cone rod homeobox protein) is expressed in both rods and cones. Nrl and Crx are highly conserved among vertebrates and are essential for rhodopsin expression. Nrl knockout mice have no rhodopsin transcripts and in fact, are missing most rod-specific genes [12,13]. Crx knockout mice have greatly reduced rhodopsin expression [14]. In both knockout mice, rod differentiation is also severely affected, indicating that these genes play multiple roles in development and photoreceptor maintenance. The importance of Nrl for rhodopsin transcription is further demonstrated in transgenic *Xenopus* rods harboring NRE

* Correspondence: knoxb@upstate.edu
Departments of Neuroscience & Physiology, Ophthalmology and Biochemistry & Molecular Biology, State University of New York Upstate Medical University, Syracuse, NY 13210, USA

mutant promoters, which have drastically reduced expression of a GFP reporter gene [15]. By contrast, individual deletion of the conserved paired-like homeobox sites does not have a major effect on reporter gene expression. Nevertheless, a mechanistic basis for the combined transcriptional activity of Nrl and Crx (and their modulators) is yet to be established.

In vitro studies have largely supported those in transgenic animals, namely that Nrl and Crx are the key regulators of rhodopsin transcription [1]. Nrl stimulates transcription of the rhodopsin promoter via NRE in transfected non-retinal cells [15,16]. Furthermore, mutagenesis of Nrl has revealed multiple domains required for rhodopsin promoter transactivation [17,18]. Similarly, Crx alone stimulates transcription of the rhodopsin promoter in transfected non-retinal cells [7]. Although it has a more promiscuous target, Crx binds to a core sequence, TAAT [7,19]. Rax, paired-like homeobox transcription factors, bind to the Ret1 and BAT1 sites [8,20,21] and alone weakly activates the rhodopsin promoter [8,22]. Crx and Nrl interact physically and functionally *in vitro*. Crx binds to Nrl through its homeodomain [23]. The binding site on Nrl is not as well defined [17,18]. Both Crx and Nrl interact with other transcription factors, such as TBP [17], Fiz1 [24] and Qrx [21], as well as chromatin modifiers [25]. However, the most intriguing feature of rhodopsin transcriptional activation *in vitro* is its cooperative (or synergistic) activation by Crx and Nrl [15,23,26,27]. Together they transactivate the rhodopsin promoter an order of magnitude or more than the additive effect of the individual transcription factors. This has led to the hypothesis that in rods rhodopsin transcription is activated by the cooperative activity of Crx and Nrl.

Current models for transcriptional activation describe the complex interplay between distal and proximal promoter regulatory elements, transcription factors and the basal transcription machinery (for example [28-30]). The usual hypothesis postulates that expression levels are determined by accessibility of the cis-elements (regulated by chromatin environment [31]) and the availability of functionally active transcription factors [28]. This leads to the localization of transcription factors on the promoter for a significant length of time and thus promotes interaction with other general transcription factors assembling on the promoter [32]. Thus, cis-elements are thought to increase the local concentration of transcription factors and to promote protein-protein interactions. Therefore, to account for rhodopsin transcription levels, it is important to quantitatively examine the contribution of DNA binding site architecture to activation of the rhodopsin promoter by Crx and Nrl. Our aim was to characterize the arrangement of DNA binding motifs in the RPP, i.e.to determine the requirements for cooperative (synergistic) activation by Nrl-Crx. Employing an *in vitro* system coupled with mutational analysis of the RPP, we characterized the specific role of the paired-like cis-elements in activation of the *Xenopus* rhodopsin promoter by Crx and its *Xenopus* homolog, Otx5, Rax and a novel *Xenopus* Rax family member, Rax2b [20,33].

Results

Rhodopsin proximal promoter activity in transfected HEK293 cells

HEK cells (human embryonic kidney cells transformed by adenovirus 5) are a versatile and rapid system in which to characterize rhodopsin promoter activation, since these cells do not express Nrl, Crx or the endogenous (human) rhodopsin gene. We compared two common lines, HEK293 (ATCC CRL-1573) and HEK293T [34,35], the latter of which expresses the SV40 large T-antigen to drive episomal replication of plasmids containing the SV40 origin of replication [36]. To quantify rhodopsin promoter activation, we transiently transfected cells with a luciferase reporter under control of various promoters and plasmids encoding Crx and/or Nrl. The activity in the absence of added transcription factors for either the promoter-less (pGL2) or a *Xenopus* rhodopsin promoter (XOP, -503/+41) containing plasmid was 1.5-fold higher in HEK293 cells compared to HEK293T cells (Figure 2A). The concentration of promoter-luciferase plasmid for all experiments was in the linear range (*data not shown*); therefore this difference in activity between the two cell lines reflects higher basal transcriptional activity independent of the specific promoter upstream of the luciferase gene.

Both cell lines exhibited robust transactivation of the XOP promoter by the combination of human Crx (hCrx) and Nrl (hNrl) or *Xenopus* homologs of Crx (Otx5) and Nrl (LNrl) (Figure 2B, [15]). However, the *Xenopus* homologs stimulated luciferase activity more strongly than their human counterparts. The higher levels of stimulation are in general agreement with our previous results, although the relative stimulation by transcription factors is somewhat higher here, mostly likely due to differences in experimental conditions.

Surprising, the activity in the presence of hNrl-hCrx was 2-fold higher in HEK293T cells compared to HEK293 cells, and contrary to the corresponding basal promoter activities (Figure 2A). It is typical to express activation by transcription factors as the relative activity, i.e. the ratio (or fold) of the activity in the presence and absence of transcription factors. For example, the activation of XOP by hNrl-hCrx is 100-fold in HEK293T compared to only 20-fold activation in HEK293 cells (Figure 2A). Thus, there is a 5-fold discrepancy in the estimated synergistic activation by hNrl-hCrx in HEK293T cells compared to HEK293 cells. This difference is a result of the higher non-specific activity of XOP as well as the lower maximal activation by hNrl-hCrx in HEK293 cells. For clarity and to facilitate comparisons between different experiments, we present all transcriptional activity data in relative light units (RLU).

Taken together, the low basal transcription and the higher Nrl-Crx stimulation in HEK293T cells are better for quantifying transcriptional stimulation of rhodopsin promoters. If we assume that the same phenomenon underlies the transcriptional differences between the two cell lines in basal activity and maximal stimulation. Therefore, we can eliminate differences in transfection efficiencies, plasmid copy number or rate of plasmid degradation. More likely are mechanisms which alter the luciferase template for transcription. One possibility is a difference in extrachromosomal plasmid packaging leading to alterations in plasmid conformation or episomal structure. Alternatively, there could be differences in the plasmid subcellular localization that lead to differences in transcription efficiency. Finally, the cells could have differences in transcription factors that lead to recognition of cryptic promoters in the plasmid backbone. It is possible that plasmids are differentially occupied by transcription complexes that are not readily displaced by hNrl-hCrx. These could account for the higher basal and lower stimulated activity in the HEK293 cells compared to HEK293T.

Xenopus rhodopsin promoter 5'upstream sequence

The only region of extended conservation in tetrapod rhodopsin promoters is the RPP (Figure 1). Comparison of the 5′ upstream sequence (5′US) of *Xenopus* with other vertebrate rhodopsin promoters did not reveal any significant conserved regions (> ~ 10 bp). However, there are many core homeobox sites (ATTA, Additional file 1: Table S3), including several Crx binding sites, present in the 5′US that could contribute to transcriptional activation. Crx binding sites have also been identified in the 5′ US of mouse rhodopsin by chromatin immunoprecipitation and chromosome confirmation capture [11,19]. In *Xenopus* transgenic frogs both −5361 and −503 genomic fragments drive rod-specific reporter gene expression [3]. To examine the potential modulation of transcriptional activity by regions upstream of the *Xenopus* RPP (xRPP, -145/-10, [3]), we compared two different 5′US sequences: (−5361/-146 and −503/-146) to the activity of the xRPP. We controlled for plasmid size by placing the rhodopsin structural gene upstream or downstream of the xRPP ((Figure 2D). Both 5′US sequences increased the synergistic stimulation of the xRPP by LNrl-Otx5 (Figure 2D). However, XOP(−5361/+41) stimulation was 10-fold while XOP(−503/+41) was 5-fold higher than that of the xRPP. Plasmid size does not account for this difference, since the activity of the -503UP and -503DWN plasmids, which are similar in size to the −5361 plasmid, was closer to the −503 plasmid than the −5361 plasmid (Figure 2D). The basal activity decreased 2.4-fold with increased 5′US length. We also found that while human Nrl-Crx can strongly stimulate synergistic activation, there is no difference in the level of activity between the two promoter lengths containing upstream sequence and the xRPP as compared to the *Xenopus* homologs (Figure 2B). Taken together, these results show that the RPP plays the principal quantitative role in determining transcriptional synergy by Nrl-Crx. Furthermore, the 5′US enhancement of synergistic activation appears to be specific to *Xenopus* transcription factors. Qualitatively, the relative activity of the promoters containing these 5′US sequences (−503, -5361) is similar in transgenic *Xenopus* ([3,37] and M. Haeri, *personal communication*). However, position effects and copy number variation prevent a direct comparison between transgenics and *in vitro* experiments.

Stable integration of *Xenopus* rhodopsin promoter

We examined whether synergistic activation of rhodopsin promoters by Nrl-Crx would occur if the promoters were integrated into the host genome. We utilized the Flp-In system, which allows for the expression of a single copy of the XOP controlled luciferase gene per cell inserted into a specific location in the genome. We generated individual clones which were hygromycin resistant, zeocin sensitive and contained XOP(−503/+41)-luciferase or XOP(−5361/+41)-luciferase. The activity in the absence of transcription factors between the four clones tested differed widely and by as much as 13 fold, with higher basal activities observed in the shorter promoter (Figure 3). LNrl individually stimulated activity with folds ranging from 5.2 to 6.2 as compared to 4.4 fold for XOP(−503/+41) and 11 fold for XOP(−5361/+41) in transiently transfected cells. Otx5 individually stimulated activity with folds ranging from 5.7 to 15.3 as compared to 6.8 for XOP(−503/+41) and 52 fold for XOP (−5361/+41) in transiently transfected cells. The human homologs, hNrl and hCrx, stimulated individually but not as robustly (data not shown). All four stable cell lines were synergistically activated by LNrl-Otx5 with increased activity ranging from 2.4 to 6.6 fold over basal + LNrl + Otx5 activity and for hNrl-hCrx, 1.7-5.1 fold. While the synergistic activation by Nrl-Crx is higher in transiently expressing HEK293T cells, this is largely due to higher basal activity of the integrated promoters. These differences could be due to the chromatin organization of the promoter DNA in an episomal environment versus an integrated state. Nevertheless, the synergistic activation of these stably integrated promoters provides support for similar behavior in rod cells.

Role of XOP cis-elements in Otx5 activation

Our goal is to determine the quantitative contribution of the conserved cis-elements to synergistic activation of the rhodopsin promoter. Previous results suggested that all four conserved elements in the RPP serve a functional role in transcriptional activation. The NRE cis-element is essential for transcriptional activation of the rhodopsin promoter by Crx-Nrl [15]. Deletion analysis in HEK293T cells demonstrated that all three paired-like cis-elements, Ret1, BAT1 and Ret4, contribute to the transcriptional activation by LNrl-Otx5 (Additional file 2: Figure S1, [15]). To determine more precisely the specific site(s) of Otx5 interaction, we generated single and multiple cis-element mutants in the XOP promoter and measured Otx5-LNrl stimulated reporter activity in HEK293T cells. The conserved core sequences of Ret1 and BAT1 are ATTA and of Ret4 , CTTA (Figure 1). The core sites were changed to AGGA and CGGA, respectively, to disrupt binding of paired-like transcription factors (Figure 4A). Mutation of Ret1 (m1) in XOP (-503) increased activity 17% compared to wild type (WT) (Figure 4B). The BAT1 site has two conserved ATTA sites. Mutation of the first site (m2) decreased activity by 58%, while mutation of the second BAT1 site (m3) increased activity by 26% compared to WT (Figure 4B). Mutation of Ret4 (m4) significantly decreased activity by 68% (Figure 4B). Mutation of both the Ret1 site and the first BAT1 core site (m5) increased activity in the short promoter fragment by 35%. Mutation of both BAT1 sites (m6) decreased the activity by 50%. Mutation of both Ret1 and Ret4 (m7) decreased the activity by 45%. Mutation of the two BAT1 sites in combination with Ret4 (m8) dramatically decreased the activity by 94%. Mutation of the BAT1 sites in combination with the Ret1 site (m9) decreased the activity by 61%. Mutations in all four conserved core sites (m10) strongly inhibited Otx5-LNrl stimulated activity 92%. None of the mutants were able to inhibit the activity to basal levels. Similar results were obtained for the longer XOP promoter (-5361) containing identical combinations of mutations (Additional file 3: Figure S2). These mutational studies show that Otx5 utilizes both the BAT1 and Ret4 sites equally and additively to activate the promoter in concert with LNrl. The Ret1 site and second BAT1 site do not appear to be necessary for activation but contribute to the overall level of transcriptional activation. In addition, Otx5 does not utilize Ret1 for transactivation. Furthermore, while the cis-elements in the proximal promoter are required for maximal Otx5-LNrl activation, the 5′US contributes to XOP activation. We can conclude that Ret1 has a functional role distinct from BAT1 and Ret4 in this assay. Only one of the BAT1 AATA sites is important for transactivation despite the strong conservation of both.

Identification of a novel Rax family member

Using a yeast one hybrid screen with the Ret1 cis-element (GCCAATTAA) as bait and a *Xenopus* laevis adult retina cDNA library, we identified a novel gene, which we name Rax2b (Accession number: JN392465). Rax2b is a member of the Rax paired-like homeodomain transcription factor family and in sequence comparisons is the paralog of Rax2a [20] (89.9% amino acid). Rax2b was detected by RT-PCR in embryos at stage 35/36 (Additional file 4: Figure S3A). All Rax family genes were expressed in the adult retina with Rax1a the most highly expressed followed by Rax2b, Rax2a and Rax1b (Additional file 4: Figure S3B).

Co-activation of rhodopsin transcription by the Rax family

We measured the activity of two rhodopsin promoter fragments (−5361/+41, -503/+41) co-expressed with individual members of the Rax family alone or in combination with LNrl and/or Otx5 in transfected HEK293T cells. Individually, each Rax transcription factor decreased basal activity of XOP(−503/+41). In combination with Otx5 or LNrl, Rax did not significantly increase activity (Figure 5A-C). All four Rax transcription factors in combination with Otx5-LNrl increased activity (Figure 5D) by 17% (Rax1a), 26% (Rax1b), 20% (Rax2a) and 31% (Rax2b). Similar results were observed for the longer promoter (−5361/+41) [22% (Rax1a), 23% (Rax1b), 45% (Rax2a) and 29% (Rax2b)] (Figure 6A-C). These data demonstrate that Rax family TFs are co-activators with Nrl and Crx of rhodopsin transcription.

Role of xRPP cis-elements in Rax co-activation

To determine the xRPP cis-element(s) utilized by Rax, we measured Otx5-LNrl-Rax2b stimulated promoter activity in the deletion and the XOP promoter mutants in transfected HEK293T cells. Rax2b increased Otx5-LNrl stimulated activity by 30% in the BAT1Δ mutant but had no effect on the level of activity of the Ret1Δ or Ret1/BAT1Δ mutants (Additional file 5: Figure S4). Rax2b co-activation was abolished in the XOP promoters with a mutated Ret1 site (net decrease: m1(32%), m5 (35%), m7 (22%), m9 (13%), m10 (killed) but preserved in the promoters containing mutations in the BAT1 or the Ret4 sites (net increase: m2 (22%), m3 (12%), m6 (23%), m8 (6%)) (Figure 7, Additional file 6: Figure S5). These mutant data suggest that Rax2b co-activates transcription specifically through the Ret1 site.

Rax co-activates preferentially through the Ret1 site

To examine further the apparent preference of Rax for the Ret1 cis-element, we compared Otx5-LNrl stimulated activity with titrated concentrations of Rax1b or Rax2b using XOP (−145/+41) and a XOP fragment truncated just before the Ret1 site (−128/+41) (Figure 8A).

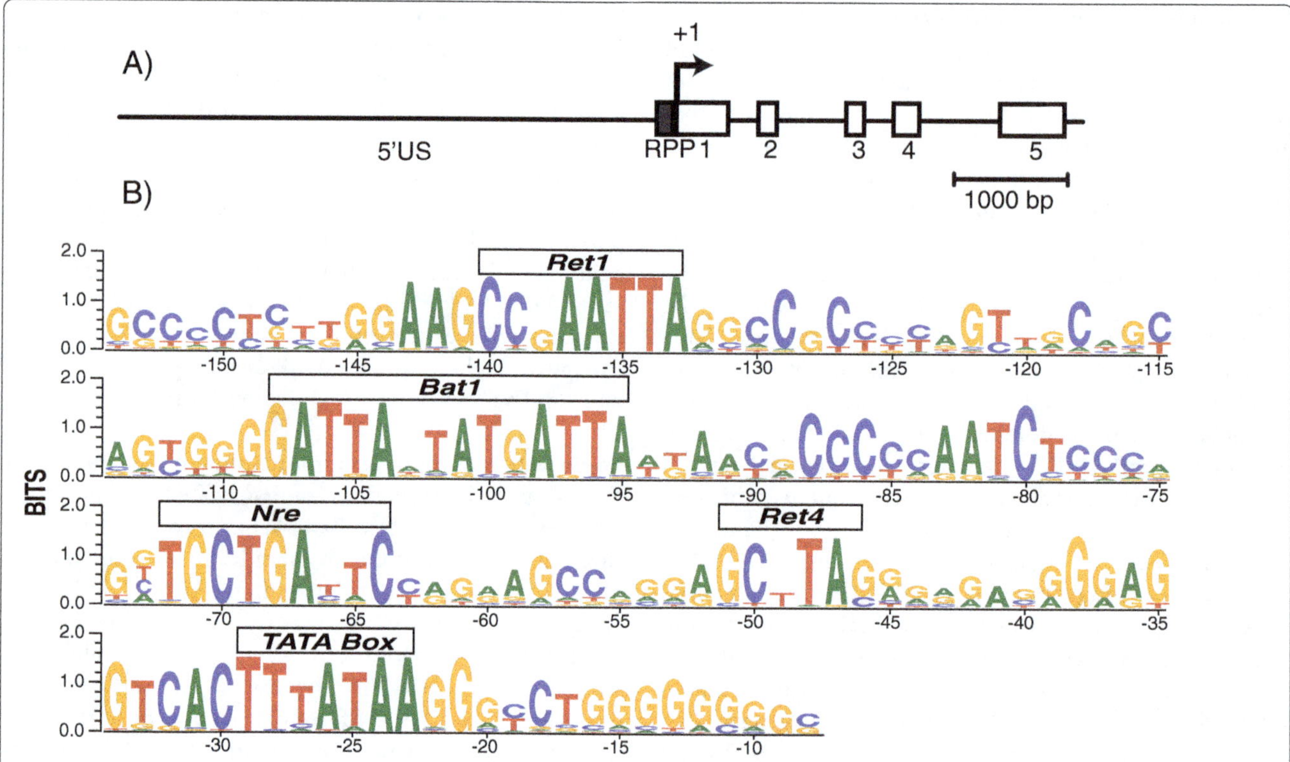

Figure 1 Highly conserved vertebrate RPP. (A) A schematic representation of the rhodopsin gene, which consists of 5′ upstream sequence (5′ US), the proximal promoter (RPP, black box), the transcription initiation site (+1) and 5 exons (white boxes). **(B)** Sequence logo representation of an alignment of 33 vertebrate proximal promoters. Approximately 200 bp around the NRE were aligned using CLUSTALW and logos created using WEBLogo3.0. The species used included seven primates, five rodents, one erinaceid, two lagomorphs, three ungulates, one camelid, one loxodonta, two carnivores, one cetacean, one tenrec, one megabat, two marsupials, one monotreme, one reptile, two birds and two *Xenopus*. Rhodopsin promoters from fish did not fit this alignment.

Rax1b increased the Otx5-LNrl stimulated activity of XOP(−145/+41) maximally by 3-fold at a 1:1 ratio, plateauing at higher concentrations (Figure 8B). The Rax2b titration results were comparable (Figure 8C). These data suggest that Rax does not compete with Otx5 for the same cis-element but preferentially selects the Ret1 site for co-activation. Interestingly, a similar result was obtained titrating Otx5 in the presence of LNrl and measuring XOP(−503/+41) activity (data not shown), suggesting that Otx5 does not utilize the Ret1 site for activation. Addition of Rax1b decreased the Otx5-LNrl stimulated activity of the truncated XOP fragment (−128/+41) by 2-fold at a 0.25:1 ratio and maximally at a ratio of 4:1. Comparable results were obtained for Rax2b with an initial 1.2-fold decrease steadily dropping to below stimulatory levels of Otx5-LNrl alone. These results suggest that when the Ret1 site is abolished, Rax can compete with Otx5 for the BAT1 and/or Ret4 sites as indicated by the decrease in Otx5-LNrl stimulation. Importantly despite the ability of Rax to compete with Otx5, it cannot replace Otx5 to partner with Nrl for synergistic activation of the promoter. Thus, while Rax is able to interact with the BAT1 sight, it still preferentially selects Ret1 in the complete xRPP.

Discussion

Nrl-Crx synergy

We used HEK293T cells for quantitatively measuring transcriptional activation of the rhodopsin promoter by Nrl, Crx and Rax. The transcriptional activity of the *Xenopus* rhodopsin gene is regulated by a combination of these transcription factors interacting with multiple conserved sequence motifs. The transiently transfected cultured cells are a powerful tool for the quantitative measurement of promoter activation by individual or combinations of TFs, although they do not mimic the photoreceptor cell/native chromatin environment. We demonstrated that the *Xenopus* rhodopsin proximal promoter (RPP) is solely responsible for regulating the synergistic activation by Nrl-Crx/Otx5. These results build on previous published studies (summarized in [1]) which initially identified the essential regulatory sites in the *Xenopus* RPP: NRE, Ret1, BAT1 and Ret4 [3,15]. It has been clearly established that the NRE cis-element, which

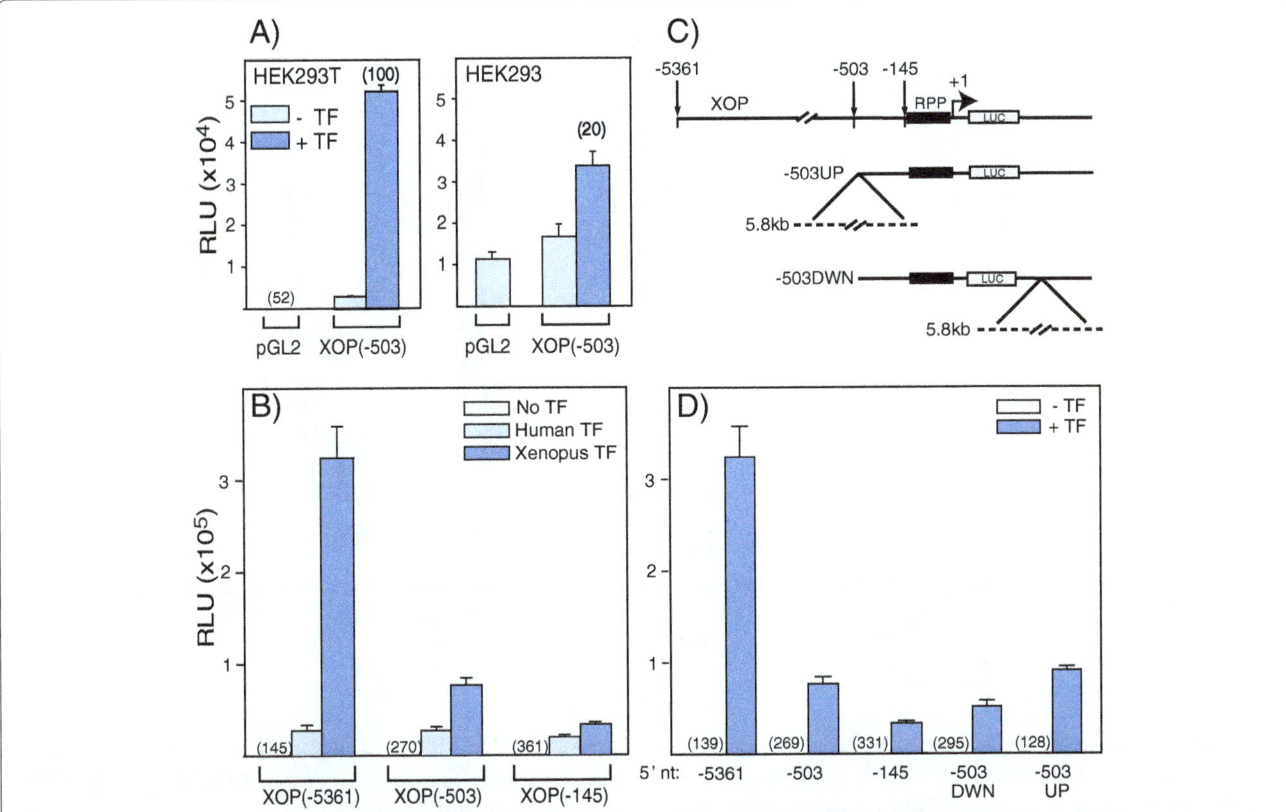

Figure 2 Effect of 5′ flanking sequences on *Xenopus* rhodopsin promoter activity in transfected 293 cells. **(A)** Comparison of luciferase activity (relative light units, RLU) from lysates of HEK293 or HEK293T cells transfected with a promoter-less pGL2 or XOP(−503/+41) containing luciferase plasmid in the absence (−TF) or presence (+TF) of pCS2-hCrx and pCS2-hNrl (+TF). The fold stimulation of XOP by hCrx-hNrl is indicated in parentheses. Data are mean ± S.E.M. (n = 2-6) Promoter basal activity (RLU) is shown in parentheses. **(B)** Comparison of luciferase activity in 293 T cell lysates transfected with plasmids containing different lengths of XOP 5′ US sequence: -5361, -503 and −145 all having the same 3′ sequence at +41. Cells were co-transfected with either human hCrx-hNrl or *Xenopus* LNrl-Otx5. Data are presented as mean RLU ± S.E.M. (n = 2-6). Promoter basal activity (RLU) is shown in parentheses. **(C)** Schematic of reporter constructs with additional *Xenopus* genomic DNA to control for the effect of plasmid size. All plasmids included the rhodopsin proximal promoter (RPP) upstream of the luciferase gene (LUC). XOP (−5361) contained 5.3 kb 5′ US sequence, -503UP and -503DWN contained a 5.8 kb fragment of the rhodopsin structural gene 5′ or 3′ to the RPP, respectively. The dotted line represents the inserted 5.8 kb rhodopsin gene. **(D)** Comparison of luciferase activity directed by different XOP constructs (−5361/+41, -503/+41, -145/+41, -503UP, -503DWN) in the absence (−TF) or presence (+TF) of Otx5 and L-Nrl. Activities are presented in RLU. The solid line represents the activity of samples transfected with empty pGL2 vector alone. The dotted line is a reference line for XOP (−503) to aid comparison. Data are presented as mean ± S.E.M. (n = 2-12). Promoter basal activity (RLU) is shown in parentheses.

binds Nrl, can modestly activate transcription, alone, and is required for synergistic activation of rhodopsin transcription in concert with Crx/Otx5 [3,6,7,15,23]. To date, little is known about the contribution of the highly conserved paired-like cis-elements, Ret1, BAT1, and Ret4, to synergistic activation by Nrl-Crx/Otx5. Otx5 activated through the BAT1 and Ret4 sites, and both were utilized equally for synergistic activation in concert with Nrl. The Ret1 site was not utilized by Otx5 and thus, did not contribute to the synergistic activation. Based on these results and those of previous studies [21,23,38], these data support a model of rhodopsin transcriptional regulation by Nrl and Crx/Otx5 (Figure 9). An Nrl site and a Crx site within 40 bp of each other are found in many photoreceptor genes and constructs containing these cis-regulatory elements were capable of driving high levels of expression in mouse retina explants [10]. Interestingly, synthetic promoters, which contained different arrangements of Nrl/Crx motifs but lacked overall sequence homology, where able to drive robust expression in mouse retinal explants. Many photoreceptor genes are regulated by Nrl and Crx and synergistic activation has been observed often [3,7,11,13,15,25-27,39]. Thus, the arrangement and spacing of these promoter cis-elements appears to be important for generating high levels of gene expression.

A recent study quantifying bovine rhodopsin promoter activity in mouse retinal explants reported that the BAT1 site was not only essential (required) for promoter activation but was the most efficient site in driving

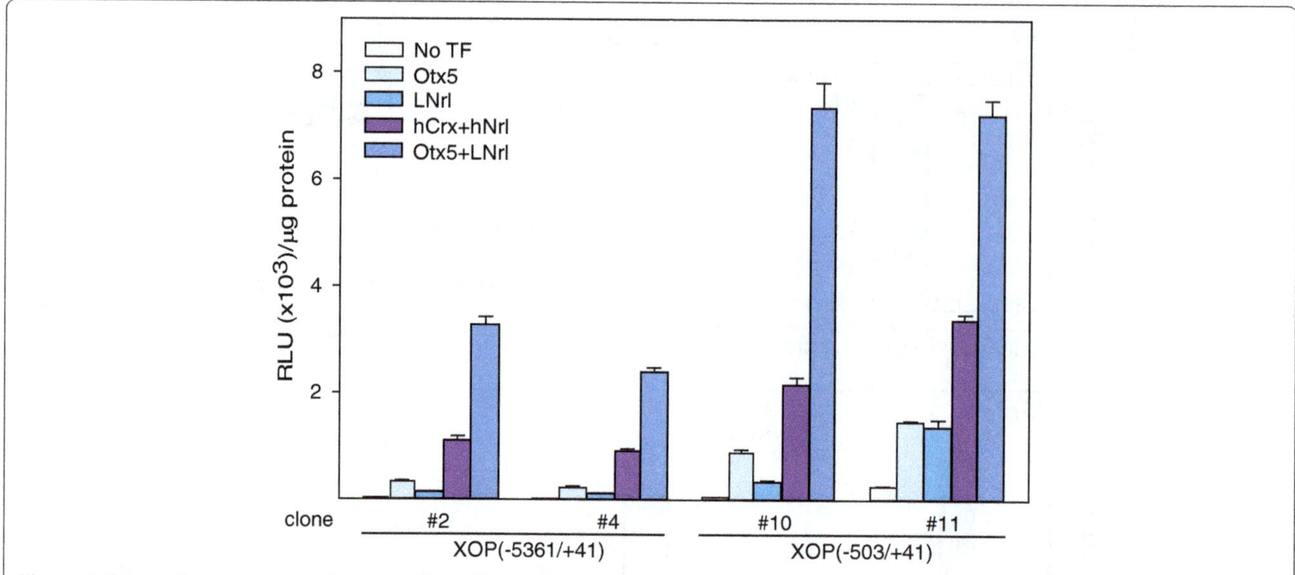

Figure 3 Comparison of transcriptional activity of *Xenopus* rhodopsin promoters stably integrated into the genome of 293 cells. Stable cell lines with integrated XOP (−5361/+41)-luciferase (#2 and #4) or XOP (−503/+41)-luciferase (#10 and #11) DNA were transfected in the absence or presence of Otx5, LNrl, Otx5-LNrl or hCrx-hNrl, and assayed for luciferase activity. Activities are presented in RLU normalized to total protein (µg) and represent mean ± S.E.M. (n = 4).

activity [40]. The Ret4 site appeared to be less important in transcriptional activation in this live mouse cell system. Substitution of the mouse Ret1 and Ret4 sites with a BAT1 consensus sequence motif increased activity by 10-fold. Replacing the Ret4 sequence with BAT1 consensus sequence increased activity approximately 4-fold. These differences suggest that there may be species dependent differences in opsin gene regulation, either from subtle sequence variation or transcription factor complexes functional across different cells.

Rax is a co-activator of rhodopsin

We identified a novel retinal homeobox transcription factor, Rax2b, which is a paralog of Rax1a [20]. Rax2 belongs to the aristaless-related paired-like homeobox gene family, which includes Rax1a and Rax1b. In promoter activity assays employing mutant promoter constructs, we demonstrated that while the Ret1 cis-element is not utilized by Otx5 for cooperative activation by Nrl-Otx5, it is utilized by Rax to enhance the level of synergistic activation. Furthermore, Rax activates, selectively and preferentially, through Ret1 and cannot substitute for Otx5 as a co-activator with Nrl of synergistic activation. These results suggest that the Ret1 site is used by Rax to fine tune synergistic activation by Nrl-Otx5. Based upon our results, we suggest that Rax provides the cell with an additional means of increasing rhodopsin expression levels. An unidentified repressor might also modulate transcription through the Ret1 site to inhibit transcription. Rax has been shown to bind to Ret1, BAT1 and Ret4 motifs in *in vitro* binding assays and to Ret1 by ChIP [8,20,21,41]. Previously published studies have also reported that Rax, alone, can weakly activate rhodopsin transcription and in combination with Nrl-Crx, can enhance synergistic activation [8,20,41]. The Rax family (Rax1a and b, Rax2a and b), which is found in most vertebrate species, has been implicated in eye formation, playing a critical role in cell-fate determination and development [33,41,42]. Rax1 and Rax2 continue to be expressed in the adult retina. However, the functional role of Rax in the mature retina is not clear. Qrx, a paired-like homeodomain TF, appears to be the mammalian ortholog to Rax2 [21]. It is expressed in both the outer and inner nuclear layers of the retina. In transient tranfection reporter activity assays, it activated rhodopsin expression through the Ret1 site, was a weak activator alone but enhanced Nrl-Crx synergy. Interestingly, mice do not appear to carry the Rax2 gene. However, the regulatory machinery for Rax2 seems to be intact, since transgenic mice strongly expressed an EGFP-tagged Qrx promoter fragment in photoreceptors.

Role of rhodopsin 5′ US

The transcriptional activity assays performed on different lengths of the *Xenopus* rhodopsin promoter provide additional insights into promoter activation. While the RPP was responsible for the cooperative synergistic activation by Nrl, Crx and Rax, the 5′ upstream promoter sequence (5′ US) distal to the RPP contributed to the activation as well. The level of activity was dependent on the length of the promoter as well as the species of TFs used to activate. This observation was reinforced by the

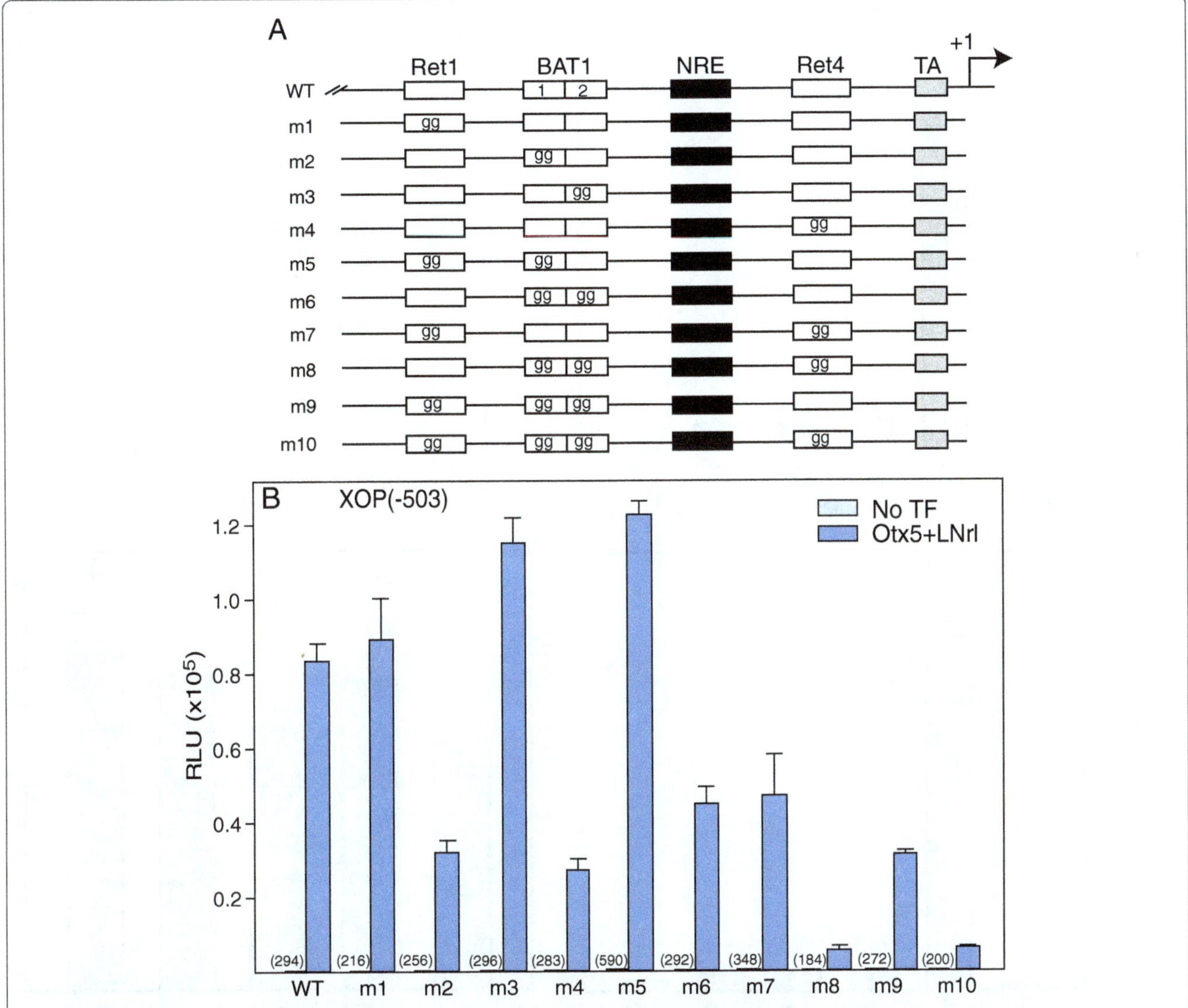

Figure 4 Mutational analysis of *Pax*-like *cis*-elements in the RPP: Effect on Otx5-LNrl activation. (A) Diagram indicating the location of the Ret1, BAT1, NRE, Ret4 and TATA box (TA) elements and the mutations (TT to GG) constructed in these sequences. The transcription start site is marked by +1. **(B)** Cells were transiently transfected with a plasmid containing a wild type promoter (XOP(−503/+41)) or one of the mutants (m1-m10) in the absence (No TF) or presence of Otx5-LNrl. Activities in cell lysates are presented as mean ± S.E.M (n = 6-8). Promoter basal activity (RLU) is shown in parentheses.

promoter activity assay data in which all three paired-like cis-elements were mutated (m6), resulting in a dramatic but not complete loss of activation. These results suggest that there are regulatory elements upstream of the *Xenopus* RPP that enhance transcription and are species-specific. This seems plausible since a distal regulatory element, the rhodopsin enhancer element (RER), which activates transcription, has been described for the mammalian promoter and is required for maximal activation of expression *in vivo* [4,5]. Although we could not identify any conserved sequence element similar to the mammalian RER, there are multiple core ATTA/CTTA sites present in the 5′US, including core Ret1 (CAATTA), BAT1 (GGATTA), Ret4 (GCTTA) and NRE (TGCTGA) motifs which could potentially bind Crx, Rax, other paired-like TFs or bZIP TFs, such as Nrl, which might contribute to the increased activity.

Conclusions

In the present study, we characterized regulatory regions of the *Xenopus* rhodopsin promoter and their contribution to transcriptional activation. The RPP is solely responsible for synergistic activation, however, the 5′US contributes to overall transcriptional activity. We defined quantitatively the individual contributions of *Xenopus* rhodopsin promoter cis-elements to synergistic activation of transcription

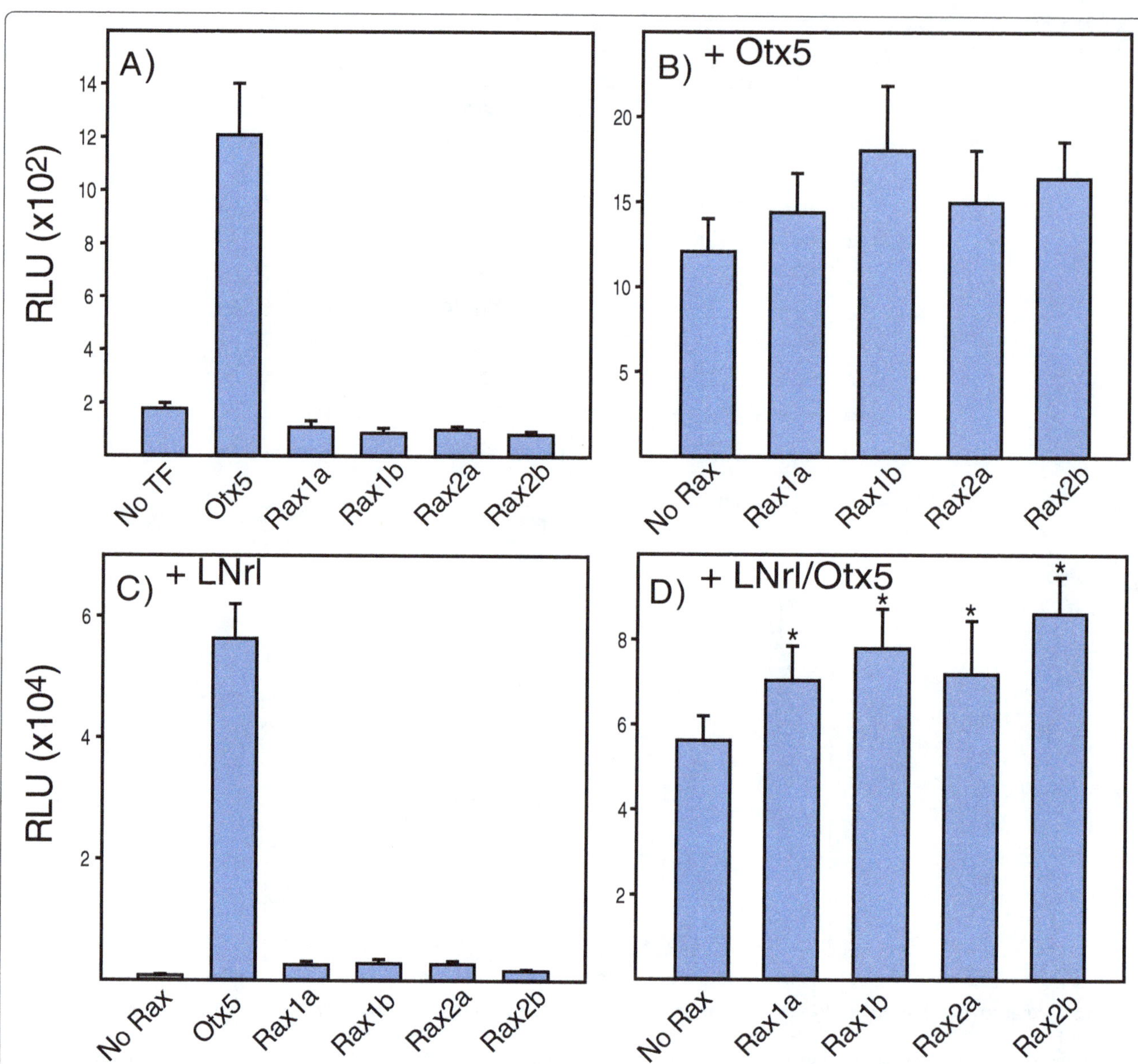

Figure 5 *Xenopus* Rax family transcription factors co-activate the rhodopsin promoter with Otx5-LNrl. Comparison of luciferase activity in lysates from cells transfected with XOP(−503/+41) alone or individually with **(A)** Otx5, Rax1a, Rax1b, Rax2a, Rax2b; **(B)** Otx5 in combination with Rax1a, Rax1b, Rax2a or Rax2b; **(C)** LNrl in combination with Rax1a, Rax1b, Rax2a or Rax2b; **(D)** LNrl/Otx5 in combination with Rax1a, Rax1b, Rax2a or Rax2b. All activities are presented as mean RLU ± S.E.M. (n = 6).

by the paired like transcription factors, Rax and Crx/Otx5. We showed specificity of the interaction between the transcription factors and cis-elements. Rax preferentially binds to the Ret1 cis-element and augments synergy, while Otx5/Crx binds to both the BAT1 and Ret4 cis-elements, contributing equally to synergistic activation. These studies characterizing the rhodopsin promoter architecture will advance our understanding of the role of transcriptional network machinery in rod photoreceptor development and homeostasis.

Methods

Vertebrate rhodopsin proximal promoter consensus sequence

Approximately 200 bp around the NRE were aligned using CLUSTALW and then manually adjusted to optimize Pax-like cis-elements. Logos were created using WEBLogo3.0. The accession numbers for the sequences used in the alignment are described in Additional file 7: Table S1 and the multiple sequence alignment is in Additional file 8.

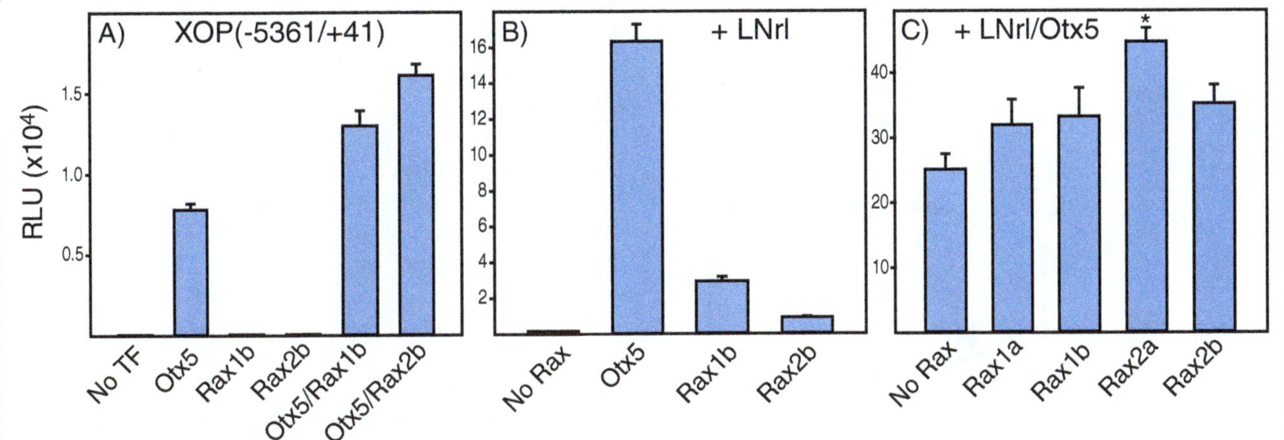

Figure 6 Sequences upstream of the rhodopsin RPP enhance co-activation by *Xenopus* Rax family transcription factors. Comparison of luciferase activity in lysates from cells transfected with XOP(−5361/+41) alone or individually with **(A)** Otx5, Rax1b, Rax2b or in combination; **(B)** LNrl in combination with Otx5, Rax1b or Rax2b; **(C)** LNrl-Otx5 in combination with Rax1a, Rax1b, Rax2a or Rax2b. All activities are presented as mean RLU ± S.E.M (n = 6).

DNA expression constructs

pCS2-Otx5 was obtained from A. Viczian. pCS2-LNrl was subcloned from pMT-LNrl [15]. The pCS2-Rax2b expression construct was generated by releasing the full length cDNA from the yeast one hybrid vector pGADT7-Rec2 with EcoRI and XhoI and ligating into the same sites of pCS2. pCS2-Rax2a (Rx-L; NM_001095716) was generated by PCR from a cDNA clone (XL073a16) obtained from NIBB and subcloned into pCS2. pCS2-Rax1a (Rx1; NM_001088218) and pCS2-Rax1b (Rx2a; AF001049) were generated by RT-PCR from *Xenopus* laevis adult retina total RNA and subcloned into pCS2. pCS2-hNrl and pCS2-hCrx were generated by PCR from cDNA clones obtained from A. Swaroop and S. Chen, respectively. hNrl was subcloned into the BamHI and Xba I sites, and hCrx into the BamHI and Not I sites in the pCS2 vector. All constructs were verified by sequencing. See Additional file 9: Table S2 for a list of primers.

Promoter reporter constructs

Four WT *Xenopus* rhodopsin promoter luciferase reporter constructs, pGL2-XOP (−5361/+41, -503/+41, -145/+41, -128/+41), were described previously [3]. The deletion constructs containing disrupted regions within the XOP (−503/+41) proximal promoter (Ret1Δ, BAT1Δ) were described previously [3]. The Ret1/BAT1Δ (−136/-91)

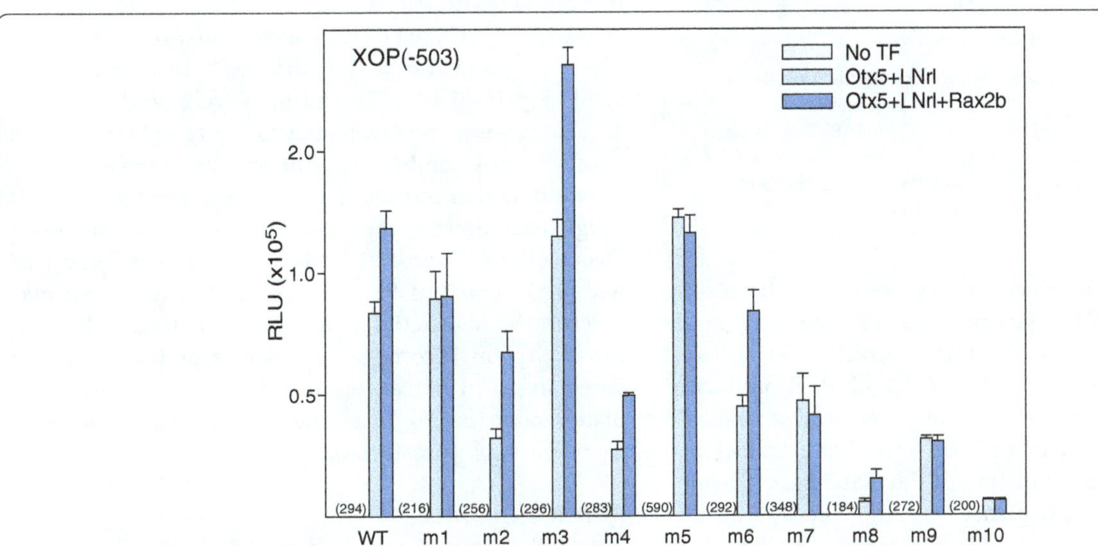

Figure 7 Mutational analysis of *Pax*-like *cis*-elements in the RPP: Effect on Rax co-activation. Comparison of luciferase activity in lysates from cells transfected with XOP(−503/+41) or RPP mutants *m1-m10*alone, with Otx5-LNrl or with Otx5-LNrl-Rax2b. Activities are presented as mean RLU ± S.E.M. (n = 6-8). Promoter basal activity (RLU) is shown in parentheses.

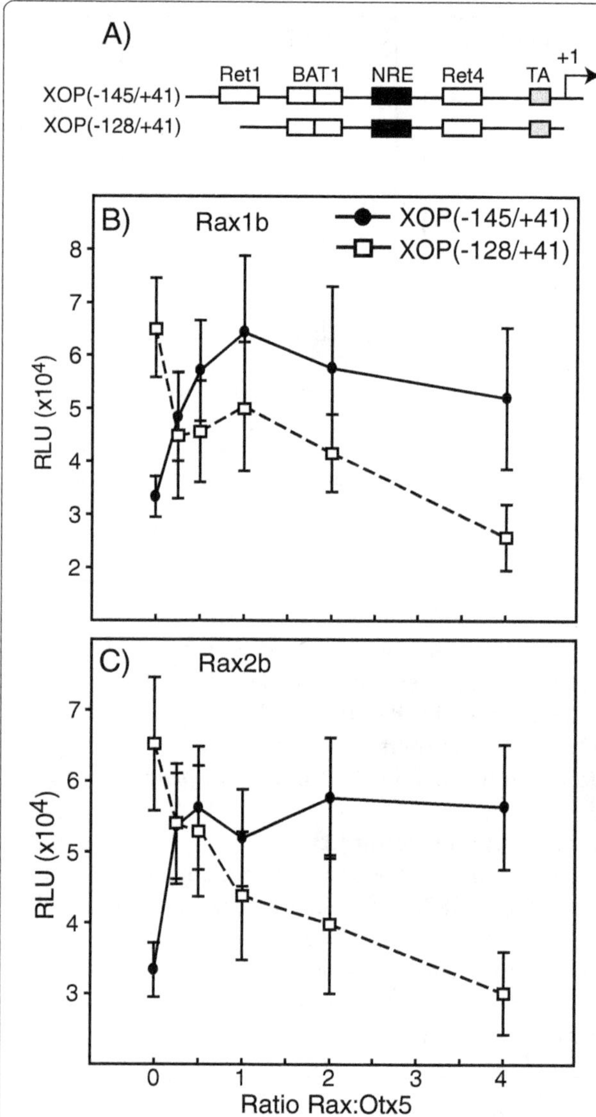

Figure 8 Effect of Rax concentration on co-activation of the RPP. (A) Schematic diagram of the plasmids containing RPP (XOP (−145/+41) or RPP missing Ret1 (XOP(−128/+41). **(B, C)** Cells were transiently transfected with either XOP(−145/+41) or XOP(−128/+41), LNrl-Otx5 and various concentrations of Rax1b **(B)** or Rax2b **(C)**. Activities are presented as mean RLU ± S.E.M. (n = 6).

deletion mutant was generated as described previously [3]. The pGL2 -503DWN plasmid was described previously [pXOP(−503/+41)luc(5800)] [3]. The pGL2 -503UP construct was generated as follows. A 5.8 kb PCR product of the *Xenopus* rhodopsin exons and 3′ region was amplified using the -503DWN plasmid as a template and cloned into the Xma I and Mlu I sites of the pGL2 basic plasmid. The XOP(−503/+41) sequence was amplified by PCR and inserted into the Nhe I and Bgl II sites just downstream of the 5.8 kb rhodopsin sequence. Mutations in the cis-elements (ATTA to AGGA) were introduced in both XOP (−5361/+41) and XOP(−503/+41) following the QuikChange XL Site Directed Mutagenesis protocol (Stratagene). The promoter constructs were verified by sequencing. The primers used in cloning are found in Additional file 9: Table S2.

Flp-In stable cell lines

A DNA fragment that included the FRT sites and hygromycin gene was PCR amplified using the pcDNA5 FRT TOPO vector as template (Invitrogen). The purified PCR fragment was cloned into the Sal I and Bam HI sites of the pGL2-basic vector (Promega), which we named pGL2-FRT. The XOP(−503/+41) and XOP(−5361/+41) promoter fragments were then excised from their pGL2 reporter constructs with Xma I and Bam HI and cloned into the same restriction sites in the modified pGL2/FRT vector. Flp-In 293 T-REx cells (Invitrogen), containing the stably integrated FRT target sites upstream of the Zeocin gene, were co-transfected with the pOG44 plasmid, containing the Flp recombinase gene (Invitrogen), and the pGL2-FRT-XOP(−503/+41) or pGL2-FRT-XOP (−5361/+41) plasmids using Fugene 6 following the manufacturer's protocol (Roche Diagnostics). Cells were passaged 48 h after transfection. Hygromycin B (150 μg/ml) was added to the culture medium the next day. Twelve individual stable clones expressing XOP(−503/+41)-Luciferase or XOP(−5361/+41)-Luciferase were picked 14 days after selection and expanded. Cell lines were also tested for loss of Zeocin (100 μg/ml) resistence. Only the dually selected clones were used for reporter assays. The promoter constructs were verified by sequencing. The primers used in cloning are found in Additional file 9: Table S2.

HEK293(T) transfections

HEK293 or HEK293T cells were seeded into 24 wells at 2.4×10^5 and 1.6×10^5 per well, respectively. The stable Flp-In 293 T-REx cell lines were seeded at 2.4×10^5. Cells were co-transfected the next day with Fugene 6 and various combinations of reporter gene and/or expression constructs at a total DNA concentration of 1 μg/well following the manufacturer's protocol (Roche Diagnostics). After 48 h cells were harvested and lysed with 100 μl/well of Passive Lysis Buffer following manufacturer's instructions for the Luciferase Reporter Assay System (Promega). Luciferase activity was measured in 10 μl of cell lysate using a Biotek Synergy II plate reader (Biotek Instruments, Inc.). Luciferase activity from mock transfected.

RT-PCR

Total RNA was isolated from *Xenopus* embryo heads (stages 35/36, 39/40) or adult retinas using an RNAeasy kit (Qiagen) and analyzed using a 2100 Bioanalyzer (Agilent Technologies). Complementary DNA was synthesized from

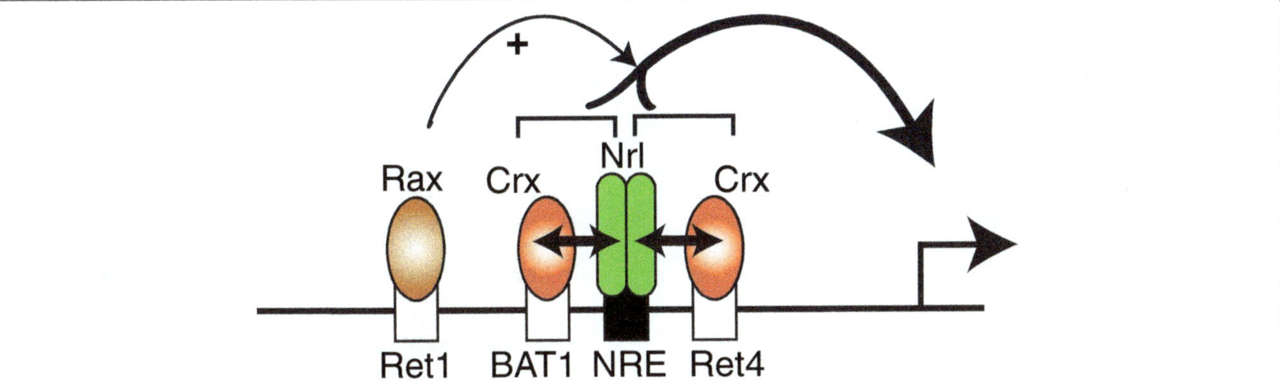

Figure 9 Model of *Xenopus* rhodopsin transcriptional activation. Nrl binds to the rhodopsin promoter via a highly conserved NRE site (TGCTGAnnC) that is immediately upstream of the TATA binding site. Crx binds independently to two adjacent sites (G/CTTA). This combination synergistically activates transcription with each Crx contributing equally to the overall activity. The Ret1 site (CCAATTA) mediates Rax protein binding to stimulate transcription.

1 μg total RNA with Superscript II RT polymerase and random hexamers (Invitrogen). cDNA was amplified over 33 cycles using Advantage 2 DNA polymerase (Clontech) and the appropriate primers for histone H4 and Rax family members (Additional file 9: Table S2).

Real time RT-PCR

Total RNA was isolated from harvested adult *Xenopus* retinas using an RNAeasy kit (Qiagen) and analyzed using a 2100 Bioanalyzer (Agilent Technologies). Complementary DNA was synthesized from 1 μg total RNA using the Quantitect Reverse Transcription kit (Qiagen). Real time PCR reactions were comprised of 1 μl of cDNA (diluted 1:2), 5 μl of SYBR green mix (Lightcycler 480 SYBR Green I Master kit, Roche Diagnostics), 0.5 μl each of sense and antisense primers (10 pmol) to target (see Additional file 9: Table S2 for list of primers). Real time PCR was performed in a Lightcycler 480 instrument (Roche Diagnostics). The cycling parameters were 95°C for 10 min followed by 46 cycles at 95°C, 60°C, 72°C and 80°C for 15 s, 15 s, 15 s and 5 s, respectively.

Yeast one-hybrid assay

The MATCHMAKER one-hybrid system (Clontech) was used to isolate *Xenopus* transcription factors which might bind to a highly conserved sequence found in tetrapod L-opsin promoters, the ROP2 cis-element [43]. The bait vector was generated by ligation of an oligonucleotide containing four tandem repeats of ROP-2 (5′-GCCAATTAAGAGAT-3′) into the Xma I and Sac II sites of the pHIS2 vector. The construct was confirmed by sequencing. A cDNA library of the adult *Xenopus* retina was prepared using poly A RNA and oligo(dT) primers according to MATCHMAKER protocol except that the denaturation time during PCR was increased to 25 s. A fusion library in pGADT7-Rec(2) vector was constructed by homologous recombination in yeast. Yeast were sequentially transformed first with the bait plasmid and then the library. Transformants (6×10^6) were screened on SD/-His/-Ura/-Leu supplemented with 10 mM 3-amino-1,2,4-triazole (3-AT). The screen yielded 30 primary transformants. The plasmid DNA containing the activation domain and cDNA inserts was isolated using standard techniques. The Y190 strain was transformed with these rescued library plasmids and used for mating the secondary screen. Of the 30 transformants, seven survived after mating with the original ROP2-containing Y187 strain on selective plates containing 40 mM 3-AT, but not in plates containing Y187 strain containing scrambled ROP2 sequence bait. The library cDNA inserts were rescued following standard protocols and sequenced.

Additional files

Additional file 1: Table S3. Characterization of *Xenopus* promoter sequences. Nucleotide composition and number of ATTA sites of the different *Xenopus* rhodopsin promoter lengths used in this study: XOP (−5361/+41), XOP(−503/+41) and XOP(−145/+41).

Additional file 2: Figure S1. Effect of *cis*-element deletions on *Xenopus* rhodopsin promoter activity. Comparison of luciferase activities in lysates from cells transfected with WT XOP(−503/+41), Ret1Δ(−503/+41), BAT1Δ (−503/+41) or Ret1/BAT1Δ (−503/+41) alone or with LNrl-Otx5. Activities are presented as mean RLU ± S.E.M. (n = 6).

Additional file 3: Figure S2. Mutational analysis of *Pax*-like *cis*-elements in the RPP with 5′ upstream sequences: Effect on Otx5-LNrl activation. Cells were transiently transfected with a plasmid containing a wild type promoter (XOP(−5361/+41)) or a mutation in cis-element (s) (m1-m10, *see Figure 4A*) in the absence (No TF) or presence of Otx5-LNrl. Activities in cell lysates are presented as mean ± S.E.M (n = 6-8). Promoter basal activity (RLU) is shown in parentheses.

Additional file 4: Figure S3. Expression of Rax transcription factors in *Xenopus laevis*. (A) RT-PCR was performed using total RNA (1 μg) from embryo heads (stages 35/36 and 39/40) or adult retinas and primers to amplify histone H4 and the three Rx paralogs (xRax1b, 2a and 2b). (B) Quantitative analysis of Rax expression levels in adult retinas by qRT-PCR.

Additional file 5: Figure S4. Deletion analysis of *Pax*-like *cis*-elements in the RPP: Effect on Rax2b co-activation. Comparison of luciferase activities in lysates from cells transfected with WT XOP(−503/+41), Ret1Δ (−503/+41), BAT1Δ(−503/+41) or Ret1/BAT1Δ (−503/+41) alone or with LNrl-Otx5-Rax2b. Activities are presented as mean RLU ± S.E.M. (n = 6).

Additional file 6: Figure S5. Mutational analysis of *Pax*-like *cis*-elements in the RPP with 5′ upstream sequences: Effect on Rax2b co-activation. Comparison of luciferase activity in lysates from cells transfected with XOP(−5361/+41) or XOP(−5361/+41) containing RPP mutants (*see Figure 4A*) alone with Otx5-LNrl or with Otx5-LNrl-Rax2b. Activities are presented as mean RLU ± S.E.M. (n = 6-8). Promoter basal activity (RLU) is shown in parentheses.

Additional file 7: Table S1. Accession numbers of tetrapod rhodopsin proximal promoters.

Additional file 8: Alignment of tetrapod rhodopsin proximal promoters.

Additional file 9: Table S2. List of primers used in this study.

Abbreviations

TF: Transcription factor; RPP: Rhodopsin proximal promoter; TA: TATA box; XOP: *Xenopus* opsin promoter; 5′ US: 5′ upstream sequence; LNrl: *Xenopus* laevis neural retinal leucine zipper; Crx: Cone rod homeobox; Otx5: Orthodenticle-related homeobox; hNrl: Human neural retinal leucine zipper; hCrx: Human cone rod homeobox; Rax: Retina and anterior neural fold homeobox; RCOP: Red cone opsin promoter; RLU: Relative light unit; BAT1: Ret1, NRE, cis-elements in the XOP promoter.

Competing interests

The authors declare that they have no competing interests.

Authors' contributions

SER carried out the molecular biology and cell transfection studies, VM performed the one-hybrid analysis and cloned Rax2b cDNA, XZ prepared the stable cell lines and contributed to their characterization, BEK carried out the sequence analysis. BEK conceived of the study. All authors participated in various aspects of the study design. SER and BEK performed the statistical analysis. The manuscript was drafted by SER and BEK, and edited by all authors. All authors read and approved the final manuscript.

Acknowledgements

This work was supported by the National Institutes of Health Grants R01-EY11256 and R01-EY12975 (BEK.), Research to Prevent Blindness (Unrestricted Grant to SUNY UMU Department of Ophthalmology), Fight For Sight (FFS), Lions of CNY, F32-EY016644 (VM). We thank Drs. D. Amberg and B. Haerer for valuable assistance in the design and performance of the yeast one-hybrid screen, Y. Milgrom for critical evaluation of data analysis and members our lab for helpful discussions.

References

1. Swaroop A, Kim D, Forrest D: **Transcriptional regulation of photoreceptor development and homeostasis in the mammalian retina.** *Nat Rev Neurosci* 2010, **11**:563–576.
2. Chen S, Zack DJ: **Ret 4, a positive acting rhodopsin regulatory element identified using a bovine retina in vitro transcription system.** *J Biol Chem* 1996, **271**:28549–28557.
3. Mani SS, Batni S, Whitaker L, Chen S, Engbretson G, Knox BE: **Xenopus rhodopsin promoter. Identification of immediate upstream sequences necessary for high level, rod-specific transcription.** *J Biol Chem* 2001, **276**:36557–36565.
4. Nie Z, Chen S, Kumar R, Zack DJ: **RER, an evolutionarily conserved sequence upstream of the rhodopsin gene, has enhancer activity.** *J Biol Chem* 1996, **271**:2667–2675.
5. Tummala P, Mali RS, Guzman E, Zhang X, Mitton KP: **Temporal ChIP-on-Chip of RNA-Polymerase-II to detect novel gene activation events during photoreceptor maturation.** *Mol Vis* 2010, **16**:252–271.
6. Rehemtulla A, Warwar R, Kumar R, Ji X, Zack DJ, Swaroop A: **The basic motif-leucine zipper transcription factor Nrl can positively regulate rhodopsin gene expression.** *Proc Natl Acad Sci USA* 1996, **93**:191–195.
7. Chen S, Wang QL, Nie Z, Sun H, Lennon G, Copeland NG, Gilbert DJ, Jenkins NA, Zack DJ: **Crx, a novel Otx-like paired-homeodomain protein, binds to and transactivates photoreceptor cell-specific genes.** *Neuron* 1997, **19**:1017–1030.
8. Kimura A, Singh D, Wawrousek EF, Kikuchi M, Nakamura M, Shinohara T: **Both PCE-1/RX and OTX/CRX interactions are necessary for photoreceptor-specific gene expression.** *J Biol Chem* 2000, **275**:1152–1160.
9. Onorati M, Cremisi F, Liu Y, He R-Q, Barsacchi G, Vignali R: **A specific box switches the cell fate determining activity of XOTX2 and XOTX5b in the Xenopus retina.** *Neural Dev* 2007, **2**:12.
10. Hsiau TH-C, Diaconu C, Myers CA, Lee J, Cepko CL, Corbo JC: **The cis-regulatory logic of the mammalian photoreceptor transcriptional network.** *PLoS One* 2007, **2**:e643.
11. Hennig AK, Peng G-H, Chen S: **Regulation of photoreceptor gene expression by Crx-associated transcription factor network.** *Brain Res* 2008, **1192**:114–133.
12. Mears AJ, Kondo M, Swain PK, Takada Y, Bush RA, Saunders TL, Sieving PA, Swaroop A: **Nrl is required for rod photoreceptor development.** *Nat Genet* 2001, **29**:447–452.
13. Yoshida S, Mears AJ, Friedman JS, Carter T, He S, Oh E, Jing Y, Farjo R, Fleury G, Barlow C, Hero AO, Swaroop A: **Expression profiling of the developing and mature Nrl−/− mouse retina: identification of retinal disease candidates and transcriptional regulatory targets of Nrl.** *Hum Mol Genet* 2004, **13**:1487–1503.
14. Furukawa T, Morrow EM, Li T, Davis FC, Cepko CL: **Retinopathy and attenuated circadian entrainment in Crx-deficient mice.** *Nat Genet* 1999, **23**:466–470.
15. Whitaker SL, Knox BE: **Conserved transcriptional activators of the Xenopus rhodopsin gene.** *J Biol Chem* 2004, **279**:49010–49018.
16. Kumar R, Chen S, Scheurer D, Wang QL, Duh E, Sung CH, Rehemtulla A, Swaroop A, Adler R, Zack DJ: **The bZIP transcription factor Nrl stimulates rhodopsin promoter activity in primary retinal cell cultures.** *J Biol Chem* 1996, **271**:29612–29618.
17. Friedman JS, Khanna H, Swain PK, Denicola R, Cheng H, Mitton KP, Weber CH, Hicks D, Swaroop A: **The minimal transactivation domain of the basic motif-leucine zipper transcription factor NRL interacts with TATA-binding protein.** *J Biol Chem* 2004, **279**:47233–47241.
18. Kanda A, Friedman JS, Nishiguchi KM, Swaroop A: **Retinopathy mutations in the bZIP protein NRL alter phosphorylation and transcriptional activity.** *Hum Mutat* 2007, **28**:589–598.
19. Corbo JC, Lawrence KA, Karlstetter M, Myers CA, Abdelaziz M, Dirkes W, Weigelt K, Seifert M, Benes V, Fritsche LG, Weber BHF, Langmann T: **CRX ChIP-seq reveals the cis-regulatory architecture of mouse photoreceptors.** *Genome Res* 2010, **20**:1512-1515.
20. Pan Y, Nekkalapudi S, Kelly LE, El-Hodiri HM: **The Rx-like homeobox gene (Rx-L) is necessary for normal photoreceptor development.** *Invest Ophthalmol Vis Sci* 2006, **47**:4245–4253.
21. Wang Q-L, Chen S, Esumi N, Swain PK, Haines HS, Peng G, Melia BM, McIntosh I, Heckenlively JR, Jacobson SG, Stone EM, Swaroop A, Zack DJ: **QRX, a novel homeobox gene, modulates photoreceptor gene expression.** *Hum Mol Genet* 2004, **13**:1025–1040.
22. Chen S, Wang Q-L, Xu S, Liu I, Li LY, Wang Y, Zack DJ: **Functional analysis of cone-rod homeobox (CRX) mutations associated with retinal dystrophy.** *Hum Mol Genet* 2002, **11**:873–884.
23. Mitton KP, Swain PK, Chen S, Xu S, Zack DJ, Swaroop A: **The leucine zipper of NRL interacts with the CRX homeodomain. A possible mechanism of transcriptional synergy in rhodopsin regulation.** *J Biol Chem* 2000, **275**:29794–29799.
24. Mali RS, Peng G-H, Zhang X, Dang L, Chen S, Mitton KP: **FIZ1 is part of the regulatory protein complex on active photoreceptor-specific gene promoters in vivo.** *BMC Mol Biol* 2008, **9**:87.
25. Peng G-H, Chen S: **Crx activates opsin transcription by recruiting HAT-containing co-activators and promoting histone acetylation.** *Hum Mol Genet* 2007, **16**:2433–2452.
26. Lerner LE, Gribanova YE, Whitaker L, Knox BE, Farber DB: **The rod cGMP-phosphodiesterase beta-subunit promoter is a specific target for Sp4 and is not activated by other Sp proteins or CRX.** *J Biol Chem* 2002, **277**:25877–25883.

27. Lerner LE, Peng G-H, Gribanova YE, Chen S, Farber DB: **Sp4 is expressed in retinal neurons, activates transcription of photoreceptor-specific genes, and synergizes with Crx.** *J Biol Chem* 2005, **280**:20642–20650.
28. Pan Y, Tsai C-J, Ma B, Nussinov R: **Mechanisms of transcription factor selectivity.** *Trends Genet* 2010, **26**:75–83.
29. Bulger M, Groudine M: **Functional and mechanistic diversity of distal transcription enhancers.** *Cell* 2011, **144**:327–339.
30. Levine M: **Transcriptional enhancers in animal development and evolution.** *Curr Biol* 2010, **20**:R754–R763.
31. Hager G, McNally J, Misteli T: **Transcription dynamics.** *Mol Cell* 2009, **35**:741–753.
32. Bai L, Charvin G, Siggia ED, Cross FR: **Nucleosome-depleted regions in cell-cycle-regulated promoters ensure reliable gene expression in every cell cycle.** *Dev Cell* 2010, **18**:544–555.
33. Wu HY, Perron M, Hollemann T: **The role of Xenopus Rx-L in photoreceptor cell determination.** *Dev Biol* 2009, **327**:352–365.
34. Graham FL, Smiley J, Russell WC, Nairn R: **Characteristics of a human cell line transformed by DNA from human adenovirus type 5.** *J Gen Virol* 1977, **36**:59–74.
35. DuBridge RB, Tang P, Hsia HC, Leong PM, Miller JH, Calos MP: **Analysis of mutation in human cells by using an Epstein-Barr virus shuttle system.** *Mol Cell Biol* 1987, **7**:379–387.
36. Cooper MJ, Lippa M, Payne JM, Hatzivassiliou G, Reifenberg E, Fayazi B, Perales JC, Morrison LJ, Templeton D, Piekarz RL, Tan J: **Safety-modified episomal vectors for human gene therapy.** *Proc Natl Acad Sci USA* 1997, **94**:6450–6455.
37. Knox BE, Schlueter C, Sanger BM, Green CB, Besharse JC: **Transgene expression in Xenopus rods.** *FEBS Lett* 1998, **423**:117–121.
38. Cheng H, Khanna H, Oh ECT, Hicks D, Mitton KP, Swaroop A: **Photoreceptor-specific nuclear receptor NR2E3 functions as a transcriptional activator in rod photoreceptors.** *Hum Mol Genet* 2004, **13**:1563–1575.
39. Pittler SJ, Zhang Y, Chen S, Mears AJ, Zack DJ, Ren Z, Swain PK, Yao S, Swaroop A, White JB: **Functional analysis of the rod photoreceptor cGMP phosphodiesterase alpha-subunit gene promoter: Nrl and Crx are required for full transcriptional activity.** *J Biol Chem* 2004, **279**:19800–19807.
40. Lee J, Myers CA, Williams N, Abdelaziz M, Corbo JC: **Quantitative fine-tuning of photoreceptor cis-regulatory elements through affinity modulation of transcription factor binding sites.** *Gene Ther* 2010, **17**:1390–9.
41. Pan Y, Martinez-De Luna RI, Lou C-H, Nekkalapudi S, Kelly LE, Sater AK, El-Hodiri HM: **Regulation of photoreceptor gene expression by the retinal homeobox (Rx) gene product.** *Dev Biol* 2010, **339**:494–506.
42. Bailey TJ, El-Hodiri H, Zhang L, Shah R, Mathers PH, Jamrich M: **Regulation of vertebrate eye development by Rx genes.** *Int J Dev Biol* 2004, **48**:761–770.
43. Babu S, McIlvain V, Whitaker SL, Knox BE: **Conserved *cis*-elements in the Xenopus red opsin promoter necessary for cone-specific expression.** *FEBS Lett* 2006, **580**(5):1479–1484.

9

Molecular characterization of the *piggyBac*-like element, a candidate marker for phylogenetic research of *Chilo suppressalis* (Walker) in China

Guang-Hua Luo[1], Xiao-Huan Li[1], Zhao-Jun Han[2], Hui-Fang Guo[1], Qiong Yang[1], Min Wu[2], Zhi-Chun Zhang[1], Bao-Sheng Liu[1], Lu Qian[3] and Ji-Chao Fang[1*]

Abstract

Background: Transposable elements (TEs, transposons) are mobile genetic DNA sequences. TEs can insert copies of themselves into new genomic locations and they have the capacity to multiply. Therefore, TEs have been crucial in the shaping of hosts' current genomes. TEs can be utilized as genetic markers to study population genetic diversity. The rice stem borer *Chilo suppressalis* Walker is one of the most important insect pests of many subtropical and tropical paddy fields. This insect occurs in all the rice-growing areas in China. This research was carried out in order to find diversity between *C. suppressalis* field populations and detect the original settlement of *C. suppressalis* populations based on the *piggyBac*-like element (PLE). We also aim to provide insights into the evolution of PLEs in *C. suppressalis* and the phylogeography of *C. suppressalis*.

Results: Here we identify a new *piggyBac*-like element (PLE) in the rice stem borer *Chilo suppressalis* Walker, which is called *CsuPLE1.1* (GenBank accession no. JX294476). *CsuPLE1.1* is transcriptionally active. Additionally, the *CsuPLE1.1* sequence varied slightly between field populations, with polymorphic indels (insertion/deletion) and hyper-variable regions including the identification of the 3′ region outside the open reading frame (ORF). *CsuPLE1.1* insertion frequency varied between field populations. Sequences variation was found between *CsuPLE1* copies and varied within and among field populations. Twenty-one different insertion sites for *CsuPLE1* copies were identified with at least two insertion loci found in all populations.

Conclusions: Our results indicate that the initial invasion of *CsuPLE1* into *C. suppressalis* occurred before *C. suppressalis* populations spread throughout China, and suggest that *C. suppressalis* populations have a common ancestor in China. Additionally, the lower reaches of the Yangtze River are probably the original settlement of *C. suppressalis* in China. Finally, the *CsuPLE1* insertion site appears to be a candidate marker for phylogenetic research of *C. suppressalis*.

Keywords: Transposon, piggyBac, Molecular characterization, Evolution, *Chilo suppressalis*

Background

Transposable elements (TEs, transposons) are mobile genetic DNA sequences, and they are found in the genomes of nearly all eukaryotes [1,2]. TEs can insert copies of themselves into new genomic locations and they have the capacity to multiply. Therefore, TEs make up a significant portion of the eukaryotic genome and have driven genome evolution in many ways, including gene expression alterations, gene deletions and insertions, chromosome rearrangements and others [3-6]. TEs are divided into two major classes based on their transposition intermediate and distinct structural features [7]. Class I TEs, which are also called retrotransposons, use a "copy-and-paste" mechanism that involves an RNA intermediate. This intermediate is reverse transcribed before its reintegration into a new position. Class II TEs, which are also called DNA transposons, use a DNA-mediated mode of "cut-and-paste" transposition.

* Correspondence: fangjc@jaas.ac.cn
[1]Institute of Plant Protection, Jiangsu Academy of Agricultural Sciences, Nanjing 210014, China
Full list of author information is available at the end of the article

The *piggyBac* element, which is a class II transposon, was originally discovered in the TN-368 cell line of the cabbage looper moth *Trichoplusia ni* [8,9]. It transposes via a "cut-and-paste" mechanism, inserting exclusively at 5′-TTAA-3′ tetranucleotide target sites and excising with precision, leaving no footprint [10]. Transposons similar to the original functional *piggyBac* IFP2 called *piggyBac*-like elements (PLEs) have been found in diverse organisms, including fungi, plants, insects, crustaceans, urochordates, amphibians, fishes and mammals [1,11-15]. PLEs are highly divergent and can be classified into three main classes, namely by high sequence similarity to *IFP2*, moderate sequence similarity to *IFP*2 and very distantly related ancient elements [14].

The rice stem borer *Chilo suppressalis* Walker is one of the most important insect pests of many subtropical and tropical paddy fields in Asia, North Africa and southern Europe. This insect occurs in all the rice-growing areas in China, and it colonizes a wide range of hosts such as rice (*Oryza sativa*), water-oat (*Zizania aquatica*) and chufa (*Eleocharis tuberosa*) [16]. It is assumed that all *C. suppressalis* field populations in China have a common ancestor. However, there is no clear evidence of this. We want to know if *C. suppressalis* field populations have a common ancestor, and if so, where this common ancestor originated.

For this paper, we isolated a group of endogenous PLEs from the *C. suppressalis* genome, which were designated as *CsuPLE1*s. The *CsuPLE1* copy with an intact open reading frame (ORF) was named *CsuPLE1.1*. The frequency of *CsuPLE1.1* insertion at a specific locus in the *C. suppressalis* genome varied among populations. This study will contribute to our understanding of the distribution and characteristics of the *piggyBac* family. In addition, the analysis *CsuPLE1*s sequence variants identified in *C. suppressalis* from different field populations provides insights into the evolution of *CsuPLE1*s. Based on the insertion sites and sequence variations of *CsuPLE1*s, the phylogeography of *C. suppressalis* is discussed.

Results

Characterization of *piggyBac*-like element (PLE) in *C. suppressalis*

A full-length PLE from *C. suppressalis* was obtained and named *CsuPLE1.1* (GenBank accession no. JX294476, Figure 1). This PLE is 2406 bp in length and contains all the characteristic structures of a PLE, including 13 bp inverted terminal repeats (ITRs), asymmetrically located 25 bp sub-terminal inverted repeats and a single open reading frame (ORF) encoding a transposase of 505 amino acids. The putative transposase contains all the aspartate residues of the "DDD" motif, which correspond to D268, D346, D447 and D450 in *T. ni IFP2* transposase. As in other PLEs, the *CsuPLE1.1* was inserted into typical tetranucleotide target-site TTAA duplications and flanked by a sequence (912 bp at 5′-end and 576 bp at 3′) that was not significantly homologous to any gene sequences in the GenBank. Notably, *CsuPLE1.1* also has a putative CAAT site, and a TATA site exists at nt 392–395 and nt 478–481. There is also a polyadenylation signal site at nt 2188–2193, which is characteristic of an actively translated protein. Alignments showed that, among the known PLEs, the putative transposase of *CsuPLE1.1* shared the highest similarity (53%) with *HsaPGBD3* transposase (Figure 2), and belongs to a class that is moderately similar to IFP2.

RACE amplification using a cDNA template revealed that *CsuPLE1.1* was expressed as a 1748 bp transcript with a 111 bp 5′ untranslated region (UTR) and a 119 bp 3′ UTR containing a 24 bp poly (A) tail (Figure 1).

Insertion site, TSDs and ITR variations

The 5′ TE display showed that the insert sites of *CsuPLE1*s varied between populations (Additional file 1: Figure S1). Further sequencing results obtained 21 different insertion sites among 72 flanking sequences from 12 field populations (Table 1 and Additional file 2: Table S1). Two insertion sites were found in all populations (insertion sites 1 and 2). One insertion site was found in nine populations (insertion site 8). One insertion site was found in seven populations (insertion site 7). One insertion site was found in five populations (insertion site 3). Two insertion sites were found in four populations (insertion sites 4 and 14). One insertion site was found in three populations (insertion site 9) and the remaining 13 insertion sites were found in only one or two populations. Almost one half of all populations had a unique insert site, found only in that population (Table 1).

Among these 21 different insertion sites, most of the insertions occurred at a TTAA target site, which is characteristic of the TTAA-specific family of *piggyBac* transposons. Only four target site duplications (TSDs) contained variations. In insertion site 5, the TSD was CTAT; In insertion site 9 and 16, the TSDs were ATAT; In insertion site 10, the TSD was CTAA. The ITR analysis showed that the 13 bp ITRs of CsuPLE1 were conserved in most individuals, with only two ITRs containing slight variations. In insertion site 14, there was a C-A variation in the ITRs; In insertion site 16, there was a G-A variation in the ITRs (Table 2).

CsuPLE1.1 insertion frequency and sequence variations

Flanking PCRs for testing the presence or absence of *CsuPLE1.1* insertions were performed on 45 randomly collected individuals from the 21 populations. The frequencies of individuals with the insertion varied between populations (Table 3). From the 945 individuals tested, 384 were heterozygous for the *CsuPLE1.1*

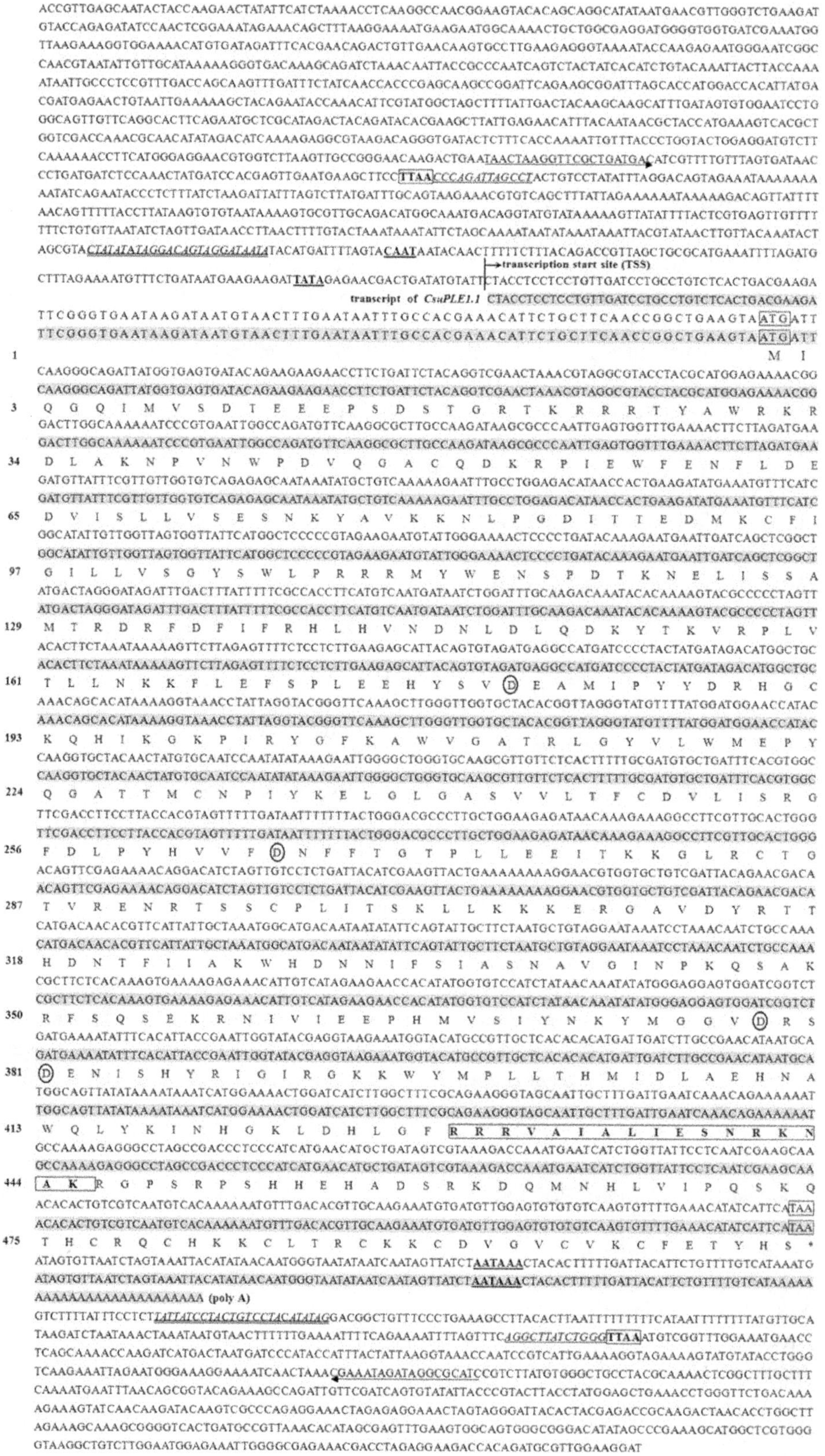

ACCGTTGAGCAATACTACCAAGAACTATATTCATCTAAAACCTCAAGGCCAACGGAAGTACACAGCAGGCATATAATGAACGTTGGGTCTGAAGAT
GTACCAGAGATATCCAACTCGGAAATAGAAACAGCTTTAAGGAAAATGAAGAATGGCAAAACTGCTGGCGAGGATGGGGTGGTGATCGAAATGG
TTAAGAAAGGTGGAAAACATGTGATAGATTTCACGAACAGACTGTTGAACAAGTGCCTTGAAGAGGGTAAAATACCAAGAGAATGGGAATCGGC
CAACGTAATATTGTTGCATAAAAAGGGTGACAAAGCAGATCTAAACAATTACCGCCCAATCAGTCTACTATCACATCTGTACAAATTACTTACCAAA
ATAATTGCCCTCCGTTTGACCAGCAAGTTTGATTTCTATCAACCACCCGAGCAAGCCGGATTCAGAAGCGGATTTAGCACCATGGACCACATTATGA
CGATGAGAACTGTAATTGAAAAAGCTACAGAATACCAAACATTCGTATGGCTAGCTTTTATTGACTACAAGCAAGCATTTGATAGTGTGGAATCCTG
GGCAGTTGTTCAGGCACTTCAGAATGCTCGCATAGACTACAGATACACGAAGCTTATTGAGAACATTTACAATAACGCTACCATGAAAGTCACGCT
GGTCGACCAAACGCAACATATAGACATCAAAAGAGGCGTAAGACAGGGTGATACTCTTTCACCAAAATTGTTTACCCTGGTACTGGAGGATGTCTT
CAAAAAACCTTCATGGGAGGAACGTGGTCTTAAGTTGCCGGGAACAAGACTGAATAACTAAGGTTCGCTGATGACATCGTTTTGTTTAGTGATAAC
CCTGATGATCTCCAAACTATGATCCACGAGTTGAATGAAGCTTCC**TTAA***CCCAGATTAGCCT*ACTGTCCTATATTTAGGACAGTAGAAATAAAAAAA
AATATCAGAATACCCTCTTTATCTAAGATTATTTAGTCTTATGATTTGCAGTAAGAAACGTGTCAGCTTTATTAGAAAAAATAAAAAGACAGTTATTTT
AACAGTTTTTACCTTATAAGTGTGTAATAAAAGTGCGTTGCAGACATGGCAAATGACAGGTATGTATAAAAAGTTATATTTTACTCGTGAGTTGTTTT
TTTCTGTGTTAATATCTAGTTGATAACCTTAACTTTTGTACTAAATAAATATTCTAGCAAAATAATATAAATAAATTACGTATAACTTGTTACAAATACT
AGCGTA*CTATATATAGGACAGTAGGATAATA*TACATGATTTTAGTA**CAAT**AATACAACTTTTTCTTTACAGACCGTTAGCTGCGCATGAAATTTTAGATG
→transcription start site (TSS)
CTTTAGAAAATGTTTCTGATAATGAAGAAGAT**TATA**GAGAACGACTGATATGTATTCTACCTCCTCCTGTTGATCCTGCCTGTCTCACTGACGAAGA
transcript of *CsuPLE1.1* CTACCTCCTCCTGTTGATCCTGCCTGTCTCACTGACGAAGA
TTCGGGTGAATAAGATAATGTAACTTTGAATAATTTGCCACGAAACATTCTGCTTCAACCGGCTGAAGTAATGATT
TTCGGGTGAATAAGATAATGTAACTTTGAATAATTTGCCACGAAACATTCTGCTTCAACCGGCTGAAGTAATGATT
1 M I
CAAGGGCAGATTATGGTGAGTGATACAGAAGAAGAACCTTCTGATTCTACAGGTCGAACTAAACGTAGGCGTACCTACGCATGGAGAAAACGG
CAAGGGCAGATTATGGTGAGTGATACAGAAGAAGAACCTTCTGATTCTACAGGTCGAACTAAACGTAGGCGTACCTACGCATGGAGAAAACGG
3 Q G Q I M V S D T E E E P S D S T G R T K R R R T Y A W R K R
GACTTGGCAAAAAATCCCGTGAATTGGCCAGATGTTCAAGGCGCTTGCCAAGATAAGCGCCCAATTGAGTGGTTTGAAAACTTCTTAGATGAA
GACTTGGCAAAAAATCCCGTGAATTGGCCAGATGTTCAAGGCGCTTGCCAAGATAAGCGCCCAATTGAGTGGTTTGAAAACTTCTTAGATGAA
34 D L A K N P V N W P D V Q G A C Q D K R P I E W F E N F L D E
GATGTTATTTCGTTGTTGGTGTCAGAGAGCAATAAATATGCTGTCAAAAAGAATTTGCCTGGAGACATAACCACTGAAGATATGAAATGTTTCATC
GATGTTATTTCGTTGTTGGTGTCAGAGAGCAATAAATATGCTGTCAAAAAGAATTTGCCTGGAGACATAACCACTGAAGATATGAAATGTTTCATC
65 D V I S L L V S E S N K Y A V K K N L P G D I T T E D M K C F I
GGCATATTGTTGGTTAGTGGTTATTCATGGCTCCCCCGTAGAAGAATGTATTGGGAAAACTCCCCTGATACAAAGAATGAATTGATCAGCTCGGCT
GGCATATTGTTGGTTAGTGGTTATTCATGGCTCCCCCGTAGAAGAATGTATTGGGAAAACTCCCCTGATACAAAGAATGAATTGATCAGCTCGGCT
97 G I L L V S G Y S W L P R R R M Y W E N S P D T K N E L I S S A
ATGACTAGGGATAGATTTGACTTTATTTTTCGCCACCTTCATGTCAATGATAATCTGGATTTGCAAGACAAATACACAAAAGTACGCCCCCTAGTT
ATGACTAGGGATAGATTTGACTTTATTTTTCGCCACCTTCATGTCAATGATAATCTGGATTTGCAAGACAAATACACAAAAGTACGCCCCCTAGTT
129 M T R D R F D F I F R H L H V N D N L D L Q D K Y T K V R P L V
ACACTTCTAAATAAAAAGTTCTTAGAGTTTTCTCCTCTTGAAGAGCATTACAGTGTAGATGAGGCCATGATCCCCTACTATGATAGACATGGCTGC
ACACTTCTAAATAAAAAGTTCTTAGAGTTTTCTCCTCTTGAAGAGCATTACAGTGTAGATGAGGCCATGATCCCCTACTATGATAGACATGGCTGC
161 T L L N K K F L E F S P L E E H Y S V D E A M I P Y Y D R H G C
AAACAGCACATAAAAGGTAAACCTATTAGGTACGGGTTCAAAGCTTGGGTTGGTGCTACACGGTTAGGGTATGTTTTATGGATGGAACCATAC
AAACAGCACATAAAAGGTAAACCTATTAGGTACGGGTTCAAAGCTTGGGTTGGTGCTACACGGTTAGGGTATGTTTTATGGATGGAACCATAC
193 K Q H I K G K P I R Y G F K A W V G A T R L G Y V L W M E P Y
CAAGGTGCTACAACTATGTGCAATCCAATATATAAAGAATTGGGGCTGGGTGCAAGCGTTGTTCTCACTTTTTGCGATGTGCTGATTTCACGTGGC
CAAGGTGCTACAACTATGTGCAATCCAATATATAAAGAATTGGGGCTGGGTGCAAGCGTTGTTCTCACTTTTTGCGATGTGCTGATTTCACGTGGC
224 Q G A T T M C N P I Y K E L G L G A S V V L T F C D V L I S R G
TTCGACCTTCCTTACCACGTAGTTTTTGATAATTTTTTTACTGGGACGCCCTTGCTGGAAGAGATAACAAAGAAAGGCCTTCGTTGCACTGGG
TTCGACCTTCCTTACCACGTAGTTTTTGATAATTTTTTTACTGGGACGCCCTTGCTGGAAGAGATAACAAAGAAAGGCCTTCGTTGCACTGGG
256 F D L P Y H V V F D N F F T G T P L L E E I T K K G L R C T G
ACAGTTCGAGAAAACAGGACATCTAGTTGTCCTCTGATTACATCGAAGTTACTGAAAAAAAAGGAACGTGGTGCTGTCGATTACAGAACGACA
ACAGTTCGAGAAAACAGGACATCTAGTTGTCCTCTGATTACATCGAAGTTACTGAAAAAAAAGGAACGTGGTGCTGTCGATTACAGAACGACA
287 T V R E N R T S S C P L I T S K L L K K K E R G A V D Y R T T
CATGACAACACGTTCATTATTGCTAAATGGCATGACAATAATATATTCAGTATTGCTTCTAATGCTGTAGGAATAAATCCTAAACAATCTGCCAAA
CATGACAACACGTTCATTATTGCTAAATGGCATGACAATAATATATTCAGTATTGCTTCTAATGCTGTAGGAATAAATCCTAAACAATCTGCCAAA
318 H D N T F I I A K W H D N N I F S I A S N A V G I N P K Q S A K
CGCTTCTCACAAAGTGAAAAGAGAAACATTGTCATAGAAGAACCACATATGGTGTCCATCTATAACAAATATATGGGAGGAGTGGATCGGTCT
CGCTTCTCACAAAGTGAAAAGAGAAACATTGTCATAGAAGAACCACATATGGTGTCCATCTATAACAAATATATGGGAGGAGTGGATCGGTCT
350 R F S Q S E K R N I V I E E P H M V S I Y N K Y M G G V D R S
GATGAAAATATTTCACATTACCGAATTGGTATACGAGGTAAGAAATGGTACATGCCGTTGCTCACACACATGATTGATCTTGCCGAACATAATGCA
GATGAAAATATTTCACATTACCGAATTGGTATACGAGGTAAGAAATGGTACATGCCGTTGCTCACACACATGATTGATCTTGCCGAACATAATGCA
381 D E N I S H Y R I G I R G K K W Y M P L L T H M I D L A E H N A
TGGCAGTTATATAAAATAAATCATGGAAAACTGGATCATCTTGGCTTTCGCAGAAGGGTAGCAATTGCTTTGATTGAATCAAACAGAAAAAAT
TGGCAGTTATATAAAATAAATCATGGAAAACTGGATCATCTTGGCTTTCGCAGAAGGGTAGCAATTGCTTTGATTGAATCAAACAGAAAAAAT
413 W Q L Y K I N H G K L D H L G F R R R V A I A L I E S N R K N
GCCAAAAGAGGGCCTAGCCGACCCTCCCATCATGAACATGCTGATAGTCGTAAAGACCAAATGAATCATCTGGTTATTCCTCAATCGAAGCAA
GCCAAAAGAGGGCCTAGCCGACCCTCCCATCATGAACATGCTGATAGTCGTAAAGACCAAATGAATCATCTGGTTATTCCTCAATCGAAGCAA
444 A K R G P S R P S H H E H A D S R K D Q M N H L V I P Q S K Q
ACACACTGTCGTCAATGTCACAAAAAATGTTTGACACGTTGCAAGAAATGTGATGTTGGAGTGTGTGTCAAGTGTTTTGAAACATATCATTCATAA
ACACACTGTCGTCAATGTCACAAAAAATGTTTGACACGTTGCAAGAAATGTGATGTTGGAGTGTGTGTCAAGTGTTTTGAAACATATCATTCATAA
475 T H C R Q C H K K C L T R C K K C D V G V C V K C F E T Y H S *
ATAGTGTTAATCTAGTAAATTACATATAACAATGGGTAATATAATCAATAGTTATCT**AATAAA**CTACACTTTTTGATTACATTCTGTTTTGTCATAAATG
ATAGTGTTAATCTAGTAAATTACATATAACAATGGGTAATATAATCAATAGTTATCT**AATAAA**CTACACTTTTTGATTACATTCTGTTTTGTCATAAAAA
AAAAAAAAAAAAAAAAAAA **(poly A)**
GTCTTTTATTTCCTCT*TATTATCCTACTGTCCTACATATAG*GACGGCTGTTTCCCTGAAAGCCTTACACTTAATTTTTTTTTCATAATTTTTTTTATGTTGCA
TAAGATCTAATAAACTAAATAATGTAACTTTTTTGAAAATTTTCAGAAAATTTTAGTTTC*AGGCTTATCTGGG***TTAA**ATGTCGGTTTGGAAATGAACC
TCAGCAAAACCAAGATCATGACTAATGATCCCATACCATTTACTATTAAGGTAAACCAATCCGTCATTGAAAAGGTAGAAAAGTATGTATACCTGGG
TCAAGAAATTAGAATGGGAAAGGAAAATCAACTAAACGAAATAGATAGGCGCATCCGTCTTATGTGGGCTGCCTACGCAAAACTCGGCTTTGCTTT
CAAAATGAATTTAACAGCGGTACAGAAAGCCAGATTGTTCGATCAGTGTATATTACCCGTACTTACCTATGGAGCTGAAACCTGGGTTCTGACAAA
AGAAAGTATCAACAAGATACAAGTCGCCCAGAGGAAACTAGAGAGGAAACTAGTAGGGATTACACTACGAGACCGCAAGACTAACACCTGGCTT
AGAAAGCAAAGCGGGGTCACTGATGCCGTTAAACACATAGCGAGTTTGAAGTGGCAGTGGGCGGGACATATAGCCCGAAAGCATGGCTCGTGGG
GTAAGGCTGTCTTGGAATGGAGAAATTGGGGCGAGAAACGACCTAGAGGAAGACCACAGATGCGTTGGAAGGAT

Figure 1 (See legend on next page.)

(See figure on previous page.)
Figure 1 Nucleotide sequence, transcript and putative transposase of the *CsuPLE1.1*. The nucleotide sequence in the light gray shadow is the *CsuPLE1.1* transcript. The inverted terminal repeats (ITRs) are underlined and italicized. The sub-terminal inverted repeats are double-underlined and italicized. The 4 bp (TTAA) target site duplications (TSDs) are boxed and in bold. Putative CAAT, TATA and AATAAA polyadenylation signal site sequences are underlined and in bold. The primers used in flanking PCR are indicated by left and right arrows below the corresponding sequences. The "DDD-motif" amino acid residues in the putative transposase are indicated by open circles. The bipartite nuclear localization signal (NLS) is boxed and in bold.

insertion. All remaining individuals did not have a *CsuPLE1.1* insertion.

A total of 84 copies of the *CsuPLE1.1* insertion were cloned from 84 individuals across the 21 field populations. These sequences share high levels of similarity with the exception of the *Flk-PLE1JZ4* copy, which has an approximately 130 bp sequence deletion. There were substitutions, deletions and/or insertions in each copy (Additional file 3: Table S2). Furthermore, in some *CsuPLE1.1* copies, there were indels of three or more bases. We defined this position as special variation position (SVP). Six SVPs were identified (Figure 3). SVP1 from copy *Flk-PLE1GZL4* has an 11 bp deletions at nt ~170. In SVP3, the copies *Flk-PLE1NC2*, *Flk-PLE1NC3*, *Flk-PLE1NC4* and *Flk-PLE1SY4* have 2 bp deletions and 7 bp insertions at nt ~630. In SVP4, the copies *Flk-PLE1JZ3* and *Flk-PLE1LS4* have 9 bp deletions at nt ~842. In SVP5, the copies *Flk-PLE1GX3*, *Flk-PLE1YS3* and *Flk-PLE1GN4* have 7 bp insertions at nt ~1889. In SVP6, the copies *Flk-PLE1YJ2*, *Flk-PLE1YJ3*, *Flk-PLE1YJ4* and *Flk-PLE1LH4* have 9 bp deletions at nt ~2210. Three bases, ACG, were found in only some PLE1 copies particularly at nt ~298 of SVP2.

Among the 84 sequences, 63 sequences have a putative intact ORF. These 63 sequences shared a high degree of sequence similarity. While, position nt ~2310, in the 3′ region outside of the ORF, showed the highest nucleotide variability between *CsuPLE1.1* copies. This site, which we called the variation hotspot, contained a number of indels (Additional file 4: Figure S2).

The variation rate (*Rv*) of the *CsuPLE1.1* copies differed inside and outside the ORF region. The variation rates inside the ORF region and in the PLE 5′ region outside the ORF were significantly lower than the PLE 3′ outside the ORF region (Table 4).

Phylogenetic tree of *C. suppressalis* field populations

A MP phylogenetic tree and a UPGMA phylogenetic tree were constructed independently. In the MP phylogenetic tree, there were many clusters and the phylogenetic relationship between each was ambiguous (Additional file 5: Figure S3). In the UPGMA tree, there were three clusters: JZ, GY and XY formed one clade; YJ, LS, JJ and DY formed a second clade; and TC, SY, HX, GZL and YX formed a third clade (Additional file 6: Figure S4).

Discussion

Since *piggyBac* is one of the most popular transposons used for transgenesis, searching for new active PLEs has attracted lots of attention. However, only a few active PLEs have been reported to date, including *IFP2*, *Uribo2*, *McrPLE* and *AgoPLE1.1* [15,17-19]. Here we identified another potentially active PLE. This PLE has the intact structure of a *piggyBac* transposon, including TTAA insertion sites, 13 bp ITRs, 25 bp subterminal inverted repeats, and a single ORF encoding a transposase of 505 amino acids with a perfect "DDD-motif". The transposase was also shown to be transcriptionally active with a 1748 bp transcript cloned from *C. suppressalis*.

There was a high degree of sequence similarity in *CsuPLE1s* from different field populations. All the tested populations shared two identical insertion sites, and each population also had their own unique insert sites. Since all inactive copies of TE will be fixed or lost in the population over time if they are neutral, insertion sites will become more homogeneous in populations over time [20,21]. Our results therefore suggest that a few *CsuPLE1* copies in *C. suppressalis* may be functional and still moving. Investigators have previously hypothesized that if TEs with high sequence similarities could maintain their original structure in their hosts, then the invasion of the TE was a recent event [15,22,23]. The high sequence similarities between *CsuPLE1* copies found in this study suggest that the invasion of *CsuPLE1* was a recent event.

The transposition activities of intact transposons are often regulated or silenced at the transcriptional, translational, or transpositional level for the survival of transposons and their hosts [6,24-26]. As a result, there are many transposable elements with mutations or variations within their host organisms [27-29]. Sequence mutations may be randomly distributed in transposons. However our results indicate that the 3′ region outside the ORF in the *CsuPLE1* transposons sequence had the highest variation rates. A variation hotspot was also found in this 3′ region. Generally, regions with high numbers of mutation are the result of complex cellular processes including (i) interactions between DNA and mutagens, (ii) repair of premutational lesions, (iii) local reduction in the fidelity of DNA polymerization, and (iv) expression and selection of a protein (RNA) molecule

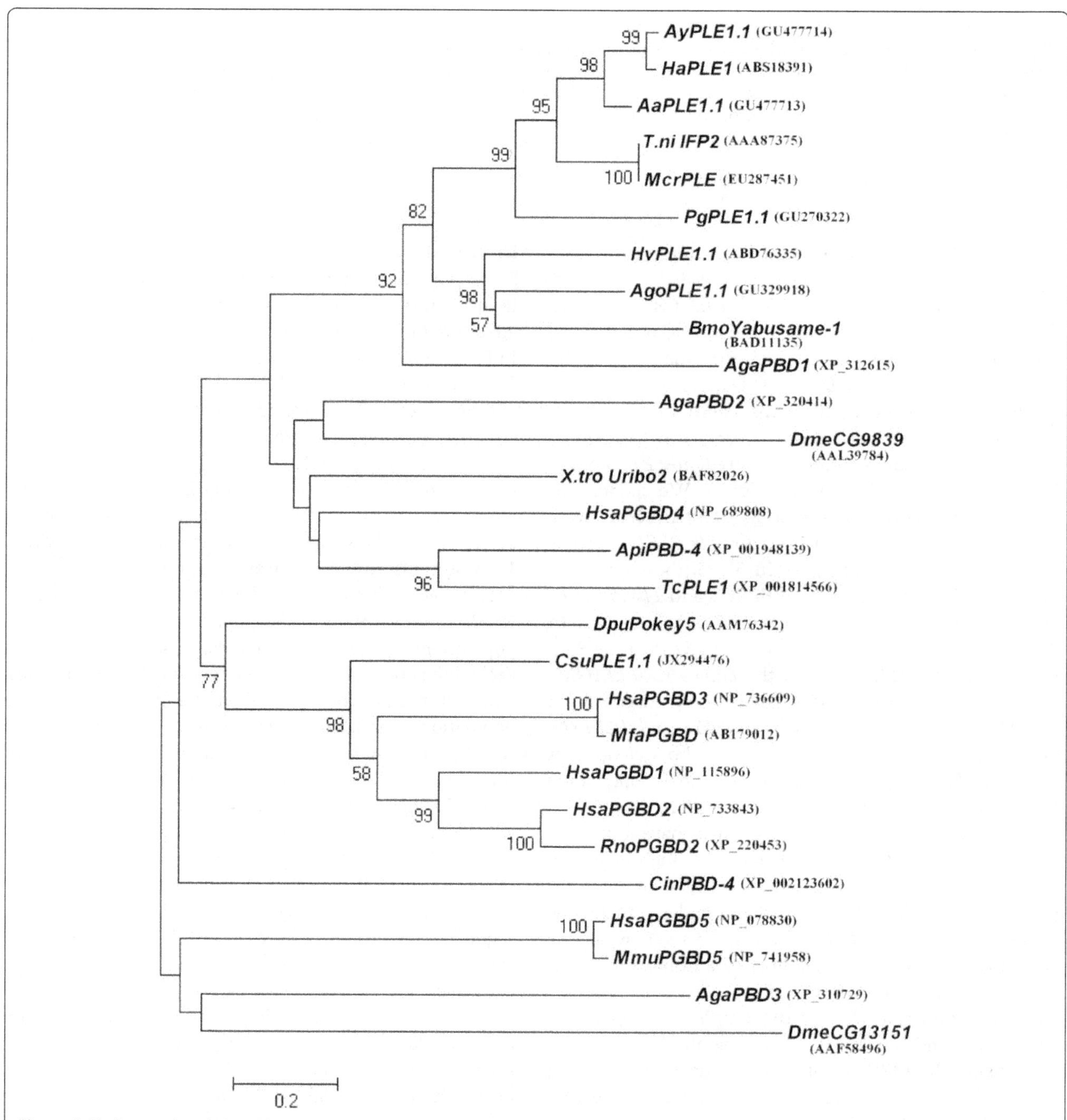

Figure 2 Phylogenetic relationships among *piggyBac*-like element transposase amino acid sequences. The tree was generated by the neighbor-joining method. The numbers at the nodes are the bootstrap values (>50%) for 5000 replications. The GenBank accession numbers are in brackets. Abbreviations: *Aa, Argyrogramma agnata; Aga, Anopheles gambiae; Ago, Aphis gossypii; Api, Acyrthosiphon pisum; Ay, Agrotis ypsilon; Bmo, Bombyx mori; Cin, Ciona intestinalis; Csu, Chilo suppressalis; Dme, Drosophila melanogaster; Dpu, Daphnia pulicaria; Ha, Helicoverpa armigera; Hsa, Homo sapiens; Hv, Heliothis virescens; Mcr, Macdunnoughia crassisigna; Mfa, Macaca fascicularis; Mmu, Mus musculus; Pg, Pectinophora gossypiella; Rno, Rattus norvegicus; Tc, Tribolium castaneum; T.ni, Trichoplusia ni; X.tro, Xenopus tropicalis.*

from which mutations have been detected [30,31]. This region is complex and deserves further study.

Of the 21 insertion sites found, two occurred in all field populations and five occurred in multiple field populations. Thus, these results imply that *CsuPLE1* existed in *C. suppressalis* prior to the expansion of the insect host populations into new regions. Meng et al. stated that *C. suppressalis* had strong population structure with three genetic clusters, i.e. a central China (CC) clade, a northern plus northeastern China (NN) clade and a

Table 1 5′ insertion sites of *CsuPLE1* in *C. suppressalis* genome

5′ Insertion site	Sampling locations											
	GY	JJ	XY	SY	JZ	DY	LS	GZL	HX	TC	YJ	YX
Site 1	√	√	√	√	√	√	√	√	√	√	√	√
Site 2	√	√	√	√	√	√	√	√	√	√	√	√
Site 3	√	√	×	×	×	√	√	×	×	×	√	×
Site 4	√	√	×	×	×	×	×	×	√	×	√	×
Site 5	√	×	×	×	×	×	×	×	×	×	×	×
Site 6	√	×	×	×	×	×	×	×	×	×	×	×
Site 7	√	×	×	×	×	√	√	√	√	√	×	√
Site 8	√	√	√	√	√	√	×	√	√	×	×	√
Site 9	×	√	×	×	×	√	√	×	×	×	×	×
Site 10	×	×	√	×	×	×	×	×	×	×	×	×
Site 11	×	×	×	√	×	×	×	×	×	×	×	×
Site 12	×	×	×	√	×	×	×	×	×	×	×	×
Site 13	×	×	×	√	×	×	×	√	×	×	×	×
Site 14	×	×	×	×	√	×	√	×	×	√	√	×
Site 15	×	×	×	×	×	√	×	×	×	×	×	×
Site 16	×	×	×	×	×	×	√	×	×	×	×	×
Site 17	×	×	×	×	×	×	√	√	×	×	×	×
Site 18	×	×	×	×	×	×	×	√	√	×	×	×
Site 19	×	×	×	×	×	×	×	×	√	×	×	×
Site 20	×	×	×	×	×	×	×	×	×	√	×	×
Site 21	×	×	×	×	×	×	×	×	×	×	√	×

The presence and absence of each insertion site within each population is indicated by a √ and × respectively.

Table 2 The variations of 5′TSDs and 5′ITRs of *CsuPLE1*s

Insertion site	5′ TSDs	5′ ITRs
1	TTAA	5′-CCCAGATTAGCCT
2	TTAA	5′-CCCAGATTAGCCT
3	TTAA	5′-CCCAGATTAGCCT
4	TTAA	5′-CCCAGATTAGCCT
5	*CTAT*	5′-CCCAGATTAGCCT
6	TTAA	5′-CCCAGATTAGCCT
7	TTAA	5′-CCCAGATTAGCCT
8	TTAA	5′-CCCAGATTAGCCT
9	*ATAT*	5′-CCCAGATTAGCCT
10	*C*TAA	5′-CCCAGATTAGCCT
11	TTAA	5′-CCCAGATTAGCCT
12	TTAA	5′-CCCAGATTAGCCT
13	TTAA	5′-CCCAGATTAGCCT
14	TTAA	5′-*A*CCAGATTAGCCT
15	TTAA	5′-CCCAGATTAGCCT
16	*ATAT*	5′-CCCA*A*ATTAGCCT
17	TTAA	5′-CCCAGATTAGCCT
18	TTAA	5′-CCCAGATTAGCCT
19	TTAA	5′-CCCAGATTAGCCT
20	TTAA	5′-CCCAGATTAGCCT
21	TTAA	5′-CCCAGATTAGCCT

The variant nucleotides are in italics.

southwestern China (SW) clade [32]. In these three clades, *C. suppressalis* had arisen from separate refuges and experienced parallel evolution. However, our results suggest that Chinese *C. suppressalis* populations have a common ancestor. The research of retrotransposon Ty3/gypsy in *C. suppressalis* shown that one insertion site of Ty3/gypsy existed in all *C. suppressalis* filed populations [33]. This also supports our conclusion that *C. suppressalis* populations have a common ancestor in China.

The MP phylogenetic tree showed that many small clusters. This was due to the high sequence similarity. However, in the 84 *CsuPLE1.1* copies, there were six SVPs. Based on our results and Meng et al.'s findings [32], we conclude that the *C. suppressalis* populations of SY, GX, YS and GN belong to the central China (CC) clade; and the *C. suppressalis* populations in JZ and LS belong to the southwestern China (SW) clade.

In the UPGMA phylogenetic tree, three clades were found. In the first clade, JZ and XY come from similar geographic areas. However, they are far from GY population. In the second clade, LS, JJ and DY come from the same geographic (the Sichuan Basin). YJ belongs to coastal areas, with a similar temperature, humidity and seasonal temperature difference to the Sichuan Basin. In the third clade, TC and HX is close to each other and have similar environmental conditions. However, SY, GZL and YX are far from each other, and have different environments. This result is not entirely consistent with Meng et al.'s result. This may be due to our small sample size or may be because our choice of method reveals a different phylogenetic relationship between *C. suppressalis* populations. Our research indicated that the insertion sites were a candidate marker for phylogenetic research of *C. suppressalis*.

Rice is the main host plant for *C. suppressalis*. Gene flow in *C. suppressalis* follows a similar pattern to the expansion of rice domestication in China [32]. It has been suggested that the lower reaches of the Yangtze River in China were the first rice farming region, although there are debates about the origin of rice [34-36]. Meng et al. found gene flow of *C. suppressalis* in CC and SW regions tends to move west. In the CC region this is from Ningbo towards regions such as Quzhou, Nangchang, and in the SW region, this is from Liuzhou towards regions such as Guiyang and Yaan. In the NN region gene flow moves northward from Wuhan or Zhumadian to Changchun [32]. These results together suggest that the lower reaches

Table 3 The frequency of *CsuPLE1.1* insertion in each field population

Rank of *CsuPLE1.1* insertion frequency	Sampling locations	Frequency of *CsuPLE1.1* insertion (mean ± SEM)
1	YX	0.8 ± 0.0387
2	HX	0.6667 ± 0.0771
3	GZL	0.578 ± 0.0588
3	SY	0.578 ± 0.089
5	GY	0.511 ± 0.022
5	YZ	0.511 ± 0.097
7	TC	0.489 ± 0.097
8	QC	0.4667 ± 0.0384
9	JZ	0.4443 ± 0.0588
10	GX	0.3557 ± 0.0802
10	LH	0.3557 ± 0.0588
12	DY	0.3333 ± 0.0771
12	NC	0.3333 ± 0.0667
12	MH	0.3333 ± 0.0384
15	JJ	0.311 ± 0.0588
16	FN	0.289 ± 0.0443
16	YS	0.2887 ± 0.0802
18	YJ	0.2667 ± 0.0667
19	XY	0.222 ± 0.0588
19	GN	0.222 ± 0.0588
21	LS	0.1777 ± 0.0447

of the Yangtze River are probably the original settlement of *C. suppressalis* in China. Our results have shown that the *CsuPLE1.1* insertion frequency was the highest in YX populations, located in the lower reaches of the Yangtze River. Moreover, the frequency of *CsuPLE1.1* insertions decreases with increasing distance from YX. If the lower reaches of the Yangtze River are the original settlement of *C. suppressalis* in China, we proposed that the *CsuPLE1.1* invasion event initially occurred at the lower reaches of the Yangtze River. As the transposition of transposons can help host organisms adapt, we suggest that the *CsuPLE1.1* had more transpositional opportunities as *C. suppressalis* expanded into new areas and new environments. The *CsuPLE1.1* insertion frequencies in Clade A (including GZL, GY, YZ and HX populations) were higher than in Clade B (including YJ, MH and LH populations) (Figure 4 and Table 3). This may be due to fewer *C. suppressalis* generations each year in Clade A (2 ~ 3 generations per year) compared to Clade B (3 ~ 4 generations per year) and therefore less opportunity for *CsuPLE1.1* transposition in Clade A. Another reason for differences in insertion frequency could be differences in environmental stress. For example, the *CsuPLE1.1* insertion frequency in FN population was lower than all other nearby populations (Figure 4 and Table 3). In FN, upland rice, which has lower nutrient levels than non-upland rice, was planted. Our previous research showed that the average individual body weight of *C. suppressalis* in FN was lighter than other field population [37]. Also, winters in FN are colder and drier than other nearby populations. *C. suppressalis* in FN therefore faces greater challenges to survive and such stress potentially provides more transpotition opportunity for the *CsuPLE1.1* in this population.

Conclusions

C. suppressalis occurs in all rice-growing areas in China, and they are non-migratory insects. Based on our results, we suggest that *C. suppressalis* populations have a common ancestor in China. The initial invasion of *CsuPLE1* in *C. suppressalis* occurred before *C. suppressalis* populations spread throughout China, and the invasion of *CsuPLE1* transposons was a recent event. Additionally, the lower reaches of the Yangtze River are probably the original settlement of *C. suppressalis* in China. Moreover, the insertion sites of *CsuPLE1s* should be a candidate marker for the phylogenetic research of *C. suppressalis.*

Methods

Sample collection and DNA isolation

The *C. suppressalis* samples were collected from 21 paddy rice field locations in China (Figure 4 and Additional file 7: Table S3), and were kept at −80°C until DNA extraction. Forty-five individual samples were randomly picked from each field population, and genomic DNA (gDNA) was prepared using an AxyPrep DNA Extraction Kit (Axygen Biosciences, Hangzhou, China) by following the protocol provided by the manufacturer.

PCR amplification and sequence analysis

In order to obtain intact sequences of PLEs from *C. suppressalis*, the transcriptome of *C. suppressalis* was surveyed and a putative PLE fragment with the longest sequence (about 1700 bp) was found. Based on this sequence, one pair of specific primers was designed (SPF1: 5′-TGTATTCTACCTCCTCCTGTTG-3′; SPR1: 5′- AAAACACTTGACACACACTCCA-3′). Using these specific primers, a 1616 bp fragment of *CsuPLE1* was amplified from Hexian (HX) individual samples by PCR. The purpose of this PCR was to verify the PLE fragment sequence originating from the transcriptome of *C. suppressalis*. The PCR was performed using *LA Taq* polymerase (TaKaRa Biotechnology, Dalian, China) with the following protocol: 95°C for 3 min; 94°C for 30 s, after which the annealing temperature of the reaction was decreased by 1°C for 30 s every cycle, from 60°C to 50°C, followed by 72°C for 1 min 40 s; 26 cycles of 94°C for 30 s, 53°C for 30 s, and 72°C for 1 min 40 s; and final elongation at 72°C for 10 min. The final PCR volume was 25 μl

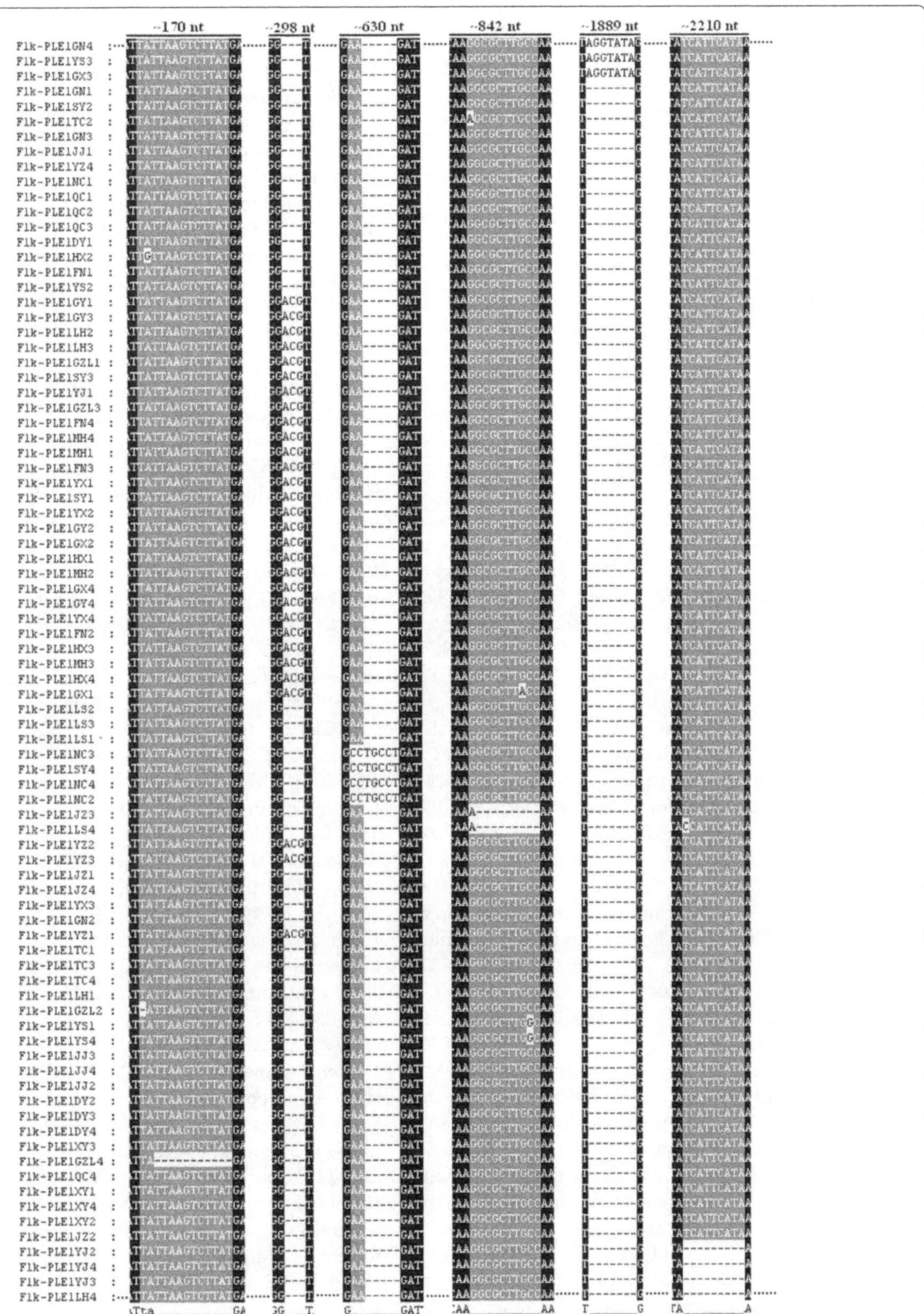

Figure 3 Six special variation positions (SVP). The location of each SVP in the sequence is indicated by Arabic numbers on the top. The omitted sequences are indicated by ellipses. The black short line indicates that there is no base. Flk-PLE1GN is from the Guangning (GN) population. Flk-PLE1YS is from the Yangshuo (YS) population. Flk-PLE1GX is from the Ganxian (GX) population, and so on.

Table 4 The variation rate in different areas of *CsuPLE1.1*

Area	Length (average ± SEM)	No. of position variances* (average ± SEM)	Variation rate* (average ± SEM)
5′ outside the ORF region	613.2 ± 0.3	3.8 ± 0.4	0.0062 ± 0.0007 B
Inside the ORF region	1518 ± 0	8 ± 0.8	0.0053 ± 0.0005 B
3′ outside the ORF region	278 ± 0.1	3.1 ± 0.3	0.0113 ± 0.0011 A

*A certain position contains an/a insertion, deletion, transition or transversion was recorded as one variance position. For the variation rate, were significant variations ($p < 0.01$) between the 3′ outside the ORF region and inside the ORF region, and the 5′ outside the ORF region.

containing approximately 50 ng gDNA, 0.2 mM of each dNTP, 1.5 mM of Mg2+, 0.2 μM of each primer, 2.5 μl of 10 × *LA* PCR buffer (Mg2+ free) and 0.25 μl (5 U/μl) of *LA Taq* polymerase (TaKaRa).

Based on this 1616 bp fragment, two pairs of nested primers for inverse PCR were designed. Inverse PCR was then performed on HX individuals with these two pairs of nested primers to obtain the full-length *CsuPLE1*. The following two pairs of primers were used for the nested inverse PCR: external primer pair (IPS1: 5′-TTGGCT TTCGCAGAAGGGTA-3′ and IPA1: 5′-CGTTTTCTC CATGCGTAGGTA-3′); Internal primer pair (IPS2: 5′-CATGCTGATAGTCGTAAAGACCA-3′ and IPA2: 5′-ATCACTCACCATAATCTGCCCT-3′). The inverse PCR was performed using LA Taq polymerase (TaKaRa) with the following protocol: 95°C for 3 min; 94°C for 30 s, after which the annealing temperature of the reaction was decreased by 1°C for 30 s every cycle, from 63°C to 53°C, followed by 72°C for 4 min; 25 cycles of 94°C for 30 s, 53°C for 30 s, and 72°C for 4 min; and final elongation at 72°C for 10 min. The final PCR volume is 25 μl containing approximately 50 ng gDNA, 0.2 mM of each dNTP, 1.5 mM of Mg2+, 0.2 μM of each primer, 2.5 μl of 10 × LA PCR buffer (Mg2+ free) and 0.25 μl (5 U/μl) of *LA Taq* polymerase (TaKaRa).

All PCR products were purified with an AxyPrep DNA Gel Extraction Kit (Axygen) and directly cloned into the pGEM-T Easy vector (Promega, Madison, WI, USA), and three clones were sequenced by GenScript Biotechnology Co., Ltd. Nanjing, China. The sequencing results were compared with non-redundant databases in the NCBI server using BLASTX and TBLASTX (http://blast.ncbi.nlm.nih.gov/Blast.cgi). CLUSTAL X1.8 [38] was used to align the putative transposase sequence of *CsuPLE1* to 27 PLEs with full-length transposases from other species sourced from Genbank. Phylogenetic analysis of the different PLE transposases was then performed on the aligned sequences in MEGA version 4 using the neighbor-joining method [39].

RNA extraction and first-strand cDNA synthesis for RT-PCR

Chilo suppressalis individuals from our laboratory strain were chosen at random for total RNA extraction. One forth-instar larvae was stored at −80°C for subsequent RNA extraction. Total RNA was isolated using a Promega SV Total RNA Isolation system (Promega) using the manufacturer's protocol.

Approximately 1 μg of total RNA was used as a template for first-strand cDNA synthesis with the PrimeScript RT reagent kit (TaKaRa) in a 20 μl reaction. Reactions were conducted at 37°C for 15 min, followed by 85°C for 5 s, and stopped by cooling on ice for 5 min.

Determination of *CsuPLE1* transcript by RACE

In order to determine the sequence of the intact transcript of *CsuPLE1*, 5′- and 3′-RACE was conducted using the SMART RACE cDNA Amplification Kit (Clontech, Mountain View, CA, USA) using the manufacturer's protocol. The first-strand 5′-RACE-ready cDNA and 3′-RACE-ready cDNA was synthesized from 1 μg of total RNA using SMARTScribe Reverse Transcriptase (Clontech). The synthesized first-stand 5′-RACE-ready cDNA was used as a template to amplify the 5′ end of *CsuPLE1* cDNA using the Universal Primer A Mix (UPM) and the Nested Universal Primer A (NUP) with the two *CsuPLE1*-specific reverse primers (5RACE01: 5′ CTCCAGGCAAATTCTTTTGACAGCA-3′ and 5RACE02: 5′- TCAAACCACTCAATTGGGCGC TTATC-3′). The first round of PCR was performed using *LA Taq* polymerase (TaKaRa) with the following protocol: 95°C for 3 min; 94°C for 30 s, after which the annealing temperature of the reaction was decreased by 2°C for 30 s every cycle, from 65°C to 55°C, followed by 72°C for 1 min; 25 cycles of 94°C for 30 s, 55°C for 30 s, and 72°C for 1 min; and final elongation at 72°C for 10 min. The primers used were UPM and 5RACE01. The final PCR volume was 25 μl containing approximately 50 ng cDNA, 0.2 mM of each dNTP, 1.5 mM of Mg^{2+}, 0.2 μM of each primer, 2.5 μl of 10 × LA PCR buffer (Mg^{2+} free) and 0.25 μl (5 U/μl) of *LA Taq* polymerase (TaKaRa). The second round of PCR, also *LA Taq* polymerase (TaKaRa), was performed on 1 μl of the first round PCR product. The PCR conditions were: 95°C for 3 min; 30 cycles of 94°C for 30 s, 66°C for 30 s, and 72°C for 1 min; and final elongation at 72°C for 10 min. The primers used were NUP and 5RACE02. The final PCR volume was 25 μl and concentrations of all other reagents within the reaction were the same as the first round of PCR. Likewise, the synthesized first-stand 3′-RACE-ready cDNA was employed as a template to amplify the 3′ end of *CsuPLE1* cDNA using UPM and NUP and the two *CsuPLE1*-specific forward primers (3RACE01: 5′-GAAATGGTACATGCCGTTGCTCACAC-3′ and 3RACE02: 5′-TCATCTTGGCTTTCGCAGAAGG

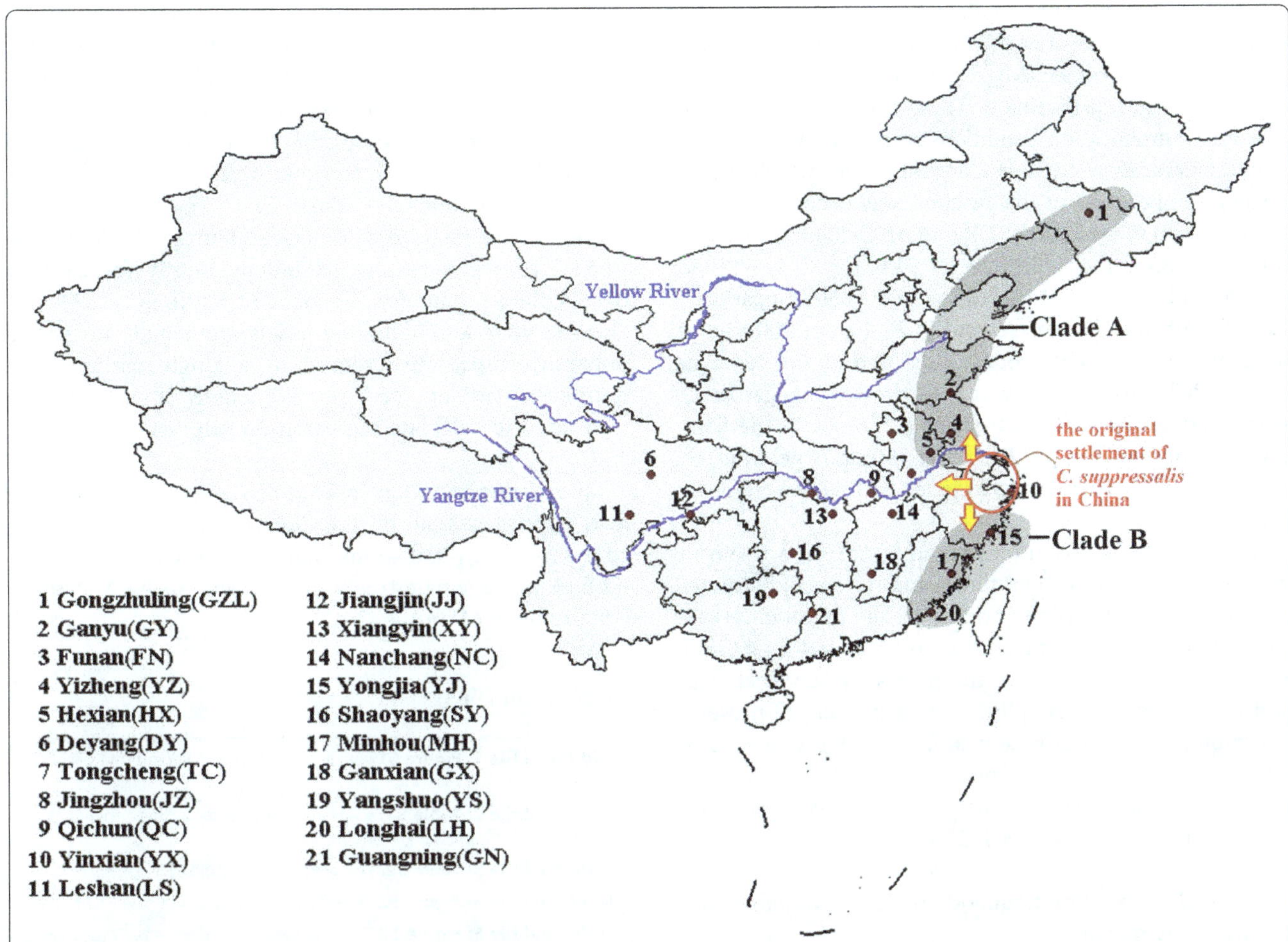

Figure 4 Sampling locations and potential migration directions of *C. suppressalis*. The letters in the brackets are the location abbreviations. The putative original settlement of *C. suppressalis* in China is indicated by red circle. The putative migration directions of *C. suppressalis* are indicated by yellow arrows.

GTAG-3′). The conditions of these two rounds of PCR were the same as those for the amplication of the 5′ end of *CsuPLE1* cDNA, respectively. All the PCR products were subcloned into the pGEM-T Easy vector (Promega) and three clones were sequenced as described above.

Vectorette PCR for TE-display and UPGMA tree construction

Vectorette PCR was used to isolate the *CsuPLE1*'s flanking sequences and to examine insertion site diversity. Vectorette PCR was performed as previously described [40]. Two anchoring bubble linker oligonucleotides were designed to make the vectorette unit for ligation to the *Hind* III digested gDNA.

The vectorette unit was prepared as described in Ko et al. [40]. For this analysis, eight individuals were randomly selected from each of 12 field populations. Approximately 1 μg of gDNA was digested at 37°C for 6 h using *Hind* III (NEB, Ipswich, MA, USA) in a 20 μl reaction. The digested gDNA was ligated with the vectorette unit using T4 DNA ligase (NEB). Two rounds of nested PCR with two pairs of primers were then carried out using the following primers: VPCR1: 5′-CCCTTCTCGAATCG TAACCG-3′ (vectorette external primer), VPCR2: 5′-CGTAACCGTTCGGTCCTCTG-3′ (vectorette internal primer), VP5R1: 5′-TGCCATGTCTGCAACGCACT-3′ (5′ specific external primer), VP5R2: 5′-AGCTGAC ACGTTTCTTACTGC-3′ (5′ specific internal primer). The Vectorette PCR was performed in a 10 μl reaction

```
vect 53 : 5'-CCCTTCTCGAATCGTAACCGTTCGGTCCTCTG-3'
             ||||||||     ||     |  |  ||||||||
vect 57 : 3'-GGGAAGAGAGCAGGCAAGAAATGGCAGGAGACTCGA-5'(HindIII)
```

volume containing approximately 25 ng gDNA, 0.2 mM of each dNTP, 1.5 mM of Mg^{2+}, 0.2 μM of each primer, 2.5 μl of 10 × LA PCR buffer (Mg^{2+} free) and 0.1 μl (5 U/μl) of *LA Taq* polymerase (TaKaRa). These two rounds of Vectorette PCR amplification conditions were 3 min at 95°C for initial denaturation, then 94°C for 30 s, after which the annealing temperature of the reaction was decreased by 1°C for 1 min every cycle, from 65°C to 51°C, followed by 72°C for 2 min 30 s; then 23 cycles of 94°C for 30 s, 52°C for 1 min, and 72°C for 2 min 30 s; and final elongation at 72°C for 10 min. For the second PCR, 1 μl of 100-fold diluted first round PCR product was used as the template. All the PCR products were visualised on a 2% agarose gel with ethidium bromide (EB) staining. To obtain the flanking sequences of *CsuPLE1* in *C. suppressalis* genome, the vectorette PCR products were cloned and sequenced as described above.

Based on the 5′ insertion sites of *CsuPLE1*, a 0 (no insertion) / 1 (with insertion) binary matrix was constructed. Genetic similarities (GS) (Additional file 8: Table S4) between pairs of field populations were measured as $GS(ij) = 2a/(2a + b + c)$ where a is the number of co-existed insertion sites in both samples, b is the number of presence insertion sites in i but absent in j, and c is the number of presence insertion sites in j but absent in i [41,42]. Genetic similarities were used to construct a UPGMA phylogenetic tree in Phylip version 3.695 [43].

The *CsuPLE1.1* insertion frequency, sequence variations and MP tree construction

Inverse PCR identified approximately 1480 bp of flanking sequence from the putative intact copy of *CsuPLE1.1*. The absence or presence of *CsuPLE1.1* in the insertion site (corresponding to the 5′ insertion site 2 in Table 1) in individual *C. suppressalis* was examined by flanking PCR with the primer pairs Flk-F (5′-TAACTAAGGTTCGCT GATGAC-3′) and Flk-R (5′-GATGCGCCTATCTATTT CG-3′). These primers flank the insertion site. For this analysis, a total of 945 individuals were randomly selected from 21 field populations. For each field populations, 45 individuals were selected (three groups, 15 individuals in each group). The absence of *CsuPLE1.1* at the insertion site was indicated by a 264 bp amplicon, and its presence was indicated by an approximately 2670 bp amplicon. The flanking PCR amplification conditions were as follows: initial denaturing at 94°C for 3 min, followed by 30 cycles of 94°C for 30 s, 55°C for 30 s, 72°C for 3 min, and a final extension at 72°C for 10 min. The PCR products were run on a 1.5% agarose gel with EB staining to detect the presence or absence of the *CsuPLE1.1* insertion in each individual. The resulting PCR products were purified, cloned and sequenced as described above.

To compare nucleotide variation inside and outside ORF regions, the *CsuPLE1.1* transposon sequence was divided into three parts: (i) the PLE 5′ outside the ORF region, (ii) inside the ORF region (from the initiator codon to the termination codon) and (iii) the PLE 3′ outside the ORF region. Four individuals containing the *CsuPLE1.1* copy in each of the 21 field populations were randomly selected from those samples mentioned in this section above and the *CsuPLE1.1* copy was purified, cloned and sequenced as described above. The four *CsuPLE1.1* copies were aligned independently for each of the 21 field populations using CLUSTAL X1.8. All nucleotide variation, including indels and single nucleotide polymorphisms, were scored as a single variant. The number of variants (Nv) and the length of each part (L) were used to calculate the variation rate (Rv) where $Rv = Nv/L$.

To examine variation in *CsuPLE1.1* sequences between field populations, all 84 *CsuPLE1.1* copies (4 *CsuPLE1.1* copies from 21 populations) were aligned using CLUSTAL X1.8, A phylogenetic tree was generated using Maximum Parsimony in MEGA 4.

Additional files

Additional file 1: Figure S1. Gel analysis of the 5′ Vectorette PCR of CsuPLE1s in 21 populations.

Additional file 2: Table S1. 5′ insertion sites and flanking sequences of CsuPLE1 in C. suppressalis.

Additional file 3: Table S2. The viorations in CsuPLE1.1 copies.

Additional file 4: Figure S2. The Variation hotspot in CsuPLE1.1 copies.

Additional file 5: Figure S3. Phylogenetic tree constructed based on multiple sequence alignments of CsuPLE1.1 copies.

Additional file 6: Figure S4. Phylogenetic tree constructed based on the 5′ insertion sites of CsuPLE1s.

Additional file 7: Table S3. Longitude and latitude of sampling locations.

Additional file 8: Table S4. The genetic similarities between pairs of field populations.

Abbreviations

TE: Transposable element; PLE: *piggyBac*-like element; ITRs: Inverted terminal repeats; ORF: Open reading frame; TSDs: Target site duplications; NLS: Nuclear localization signal; UPGMA: Unweighted pair group method analysis.

Competing interests

The authors declare that they have no competing interests.

Authors' contributions

G-HL and X-HL performed most of the experiments together. G-HL and J-CF conceived and designed the study together. G-HL, Z-CZ and B-SL collected the insect samples together. G-HL, Z-JH, H-FG and J-CF wrote the paper together. All authors analysed the data. All authors read and approved the final manuscript.

Acknowledgements

We are grateful to Dr. Kostas D. Mathiopoulos for helpful suggestions for improving this manuscript. This work was supported by National Rice Industry Technology System Project grant Cars-001-25, National Key Technology R&D Program grant 2012BAD19B03, National Natural Science Foundation of China grant 31201505 and 31101435, and Jiangsu Agriculture Science and Technology Innovation Fund grant CX(13)3038.

Author details
[1]Institute of Plant Protection, Jiangsu Academy of Agricultural Sciences, Nanjing 210014, China. [2]Education Ministry Key Laboratory of Integrated Management of Crop Diseases and Pests, College of Plant Protection, Nanjing Agricultural University, Nanjing 210095, China. [3]Jiangsu Entry-Exit Inspection and Quarantine Bureau, Nanjing 210001, China.

References

1. Sarkar A, Sim C, Hong YS, Hogan JR, Fraser MJ, Robertson HM, Collins FH: **Molecular evolutionary analysis of the widespread piggyBac transposon family and related "domesticated" sequences.** *Mol Genet Genomics* 2003, **270**(2):173–180.
2. Feschotte C, Pritham EJ: **DNA transposons and the evolution of eukaryotic genomes.** *Annu Rev Genet* 2007, **41**:331–368.
3. Langley CH, Montgomery E, Hudson R, Kaplan N, Charlesworth B: **On the role of unequal exchange in the containment of transposable element copy number.** *Genet Res* 1988, **52**(3):223–235.
4. Kazazian HH Jr: **Mobile elements: drivers of genome evolution.** *Science* 2004, **303**(5664):1626–1632.
5. Oliver KR, Greene WK: **Transposable elements: powerful facilitators of evolution.** *Bioessays* 2009, **31**(7):703–714.
6. Feschotte C: **Transposable elements and the evolution of regulatory networks.** *Nat Rev Genet* 2008, **9**(5):397–405.
7. Finnegan DJ: **Transposable elements.** *Curr Opin Genet Dev* 1992, **2**(6):861–867.
8. Fraser MJ, Smith GE, Summers MD: **Acquisition of host cell DNA sequences by baculoviruses: relationship between host DNA Insertions and FP mutants of autographa californica and galleria mellonella nuclear polyhedrosis viruses.** *J Virol* 1983, **47**(2):287–300.
9. Cary LC, Goebel M, Corsaro BG, Wang HG, Rosen E, Fraser MJ: **Transposon mutagenesis of baculoviruses: analysis of Trichoplusia ni transposon IFP2 insertions within the FP-locus of nuclear polyhedrosis viruses.** *Virology* 1989, **172**(1):156–169.
10. Fraser MJ, Ciszczon T, Elick T, Bauser C: **Precise excision of TTAA-specific lepidopteran transposons piggyBac (IFP2) and tagalong (TFP3) from the baculovirus genome in cell lines from two species of Lepidoptera.** *Insect Mol Biol* 1996, **5**(2):141–151.
11. Zimowska GJ, Handler AM: **Highly conserved piggyBac elements in noctuid species of Lepidoptera.** *Insect Biochem Mol Biol* 2006, **36**(5):421–428.
12. Xu HF, Xia QY, Liu C, Cheng TC, Zhao P, Duan J, Zha XF, Liu SP: **Identification and characterization of piggyBac-like elements in the genome of domesticated silkworm, Bombyx mori.** *Mol Genet Genomics* 2006, **276**(1):31–40.
13. Wang J, Miller ED, Simmons GS, Miller TA, Tabashnik BE, Park Y: **piggyBac-like elements in the pink bollworm, Pectinophora gossypiella.** *Insect Mol Biol* 2010, **19**(2):177–184.
14. Wu M, Sun Z, Luo G, Hu C, Zhang W, Han Z: **Cloning and characterization of piggyBac-like elements in lepidopteran insects.** *Genetica* 2011, **139**(1):149–154.
15. Luo GH, Wu M, Wang XF, Zhang W, Han ZJ: **A new active piggyBac-like element in Aphis gossypii.** *Insect Sci* 2011, **18**(6):652–662.
16. Sheng C, Wang H, Sheng S, Gao L, Xuan W: **Pest status and loss assessment of crop damage caused by the rice borers, Chilo suppressalis and Tryporyza incertulas in China.** *Entomol Knowl* 2003, **40**(4):289–294.
17. Ding S, Wu X, Li G, Han M, Zhuang Y, Xu T: **Efficient transposition of the piggyBac (PB) transposon in mammalian cells and mice.** *Cell* 2005, **122**(3):473–483.
18. Hikosaka A, Kobayashi T, Saito Y, Kawahara A: **Evolution of the Xenopus piggyBac transposon family TxpB: domesticated and untamed strategies of transposon subfamilies.** *Mol Biol Evol* 2007, **24**(12):2648–2656.
19. Wu M, Sun ZC, Hu CL, Zhang GF, Han ZJ: **An active piggyBac-like element in Macdunnoughia crassisigna.** *Insect Sci* 2008, **15**(6):521–528.
20. Deceliere G, Charles S, Biemont C: **The dynamics of transposable elements in structured populations.** *Genetics* 2005, **169**(1):467–474.
21. Wang J, Ren X, Miller TA, Park Y: **piggyBac-like elements in the tobacco budworm, Heliothis virescens (Fabricius).** *Insect Mol Biol* 2006, **15**(4):435–443.
22. Garcia Guerreiro MP, Fontdevila A: **The evolutionary history of Drosophila buzzatii. XXXVI. Molecular structural analysis of Osvaldo retrotransposon insertions in colonizing populations unveils drift effects in founder events.** *Genetics* 2007, **175**(1):301–310.
23. Bui QT, Delauriere L, Casse N, Nicolas V, Laulier M, Chenais B: **Molecular characterization and phylogenetic position of a new mariner-like element in the coastal crab, Pachygrapsus marmoratus.** *Gene* 2007, **396**(2):248–256.
24. Lippman Z, May B, Yordan C, Singer T, Martienssen R: **Distinct mechanisms determine transposon inheritance and methylation via small interfering RNA and histone modification.** *PLoS Biol* 2003, **1**(3):E67.
25. Tran RK, Zilberman D, de Bustos C, Ditt RF, Henikoff JG, Lindroth AM, Delrow J, Boyle T, Kwong S, Bryson TD, Jacobsen SE, Henikoff S: **Chromatin and siRNA pathways cooperate to maintain DNA methylation of small transposable elements in Arabidopsis.** *Genome Biol* 2005, **6**(11):R90.
26. Castro JP, Carareto CM: **Drosophila melanogaster P transposable elements: mechanisms of transposition and regulation.** *Genetica* 2004, **121**(2):107–118.
27. Wang J, Du Y, Wang S, Brown SJ, Park Y: **Large diversity of the piggyBac-like elements in the genome of Tribolium castaneum.** *Insect Biochem Mol Biol* 2008, **38**(4):490–498.
28. Osanai-Futahashi M, Suetsugu Y, Mita K, Fujiwara H: **Genome-wide screening and characterization of transposable elements and their distribution analysis in the silkworm, Bombyx mori.** *Insect Biochem Mol Biol* 2008, **38**(12):1046–1057.
29. Schnable PS, Ware D, Fulton RS, Stein JC, Wei F, Pasternak S, Liang C, Zhang J, Fulton L, Graves TA, Minx P, Reily AD, Courtney L, Kruchowski SS, Tomlinson C, Strong C, Delehaunty K, Fronick C, Courtney B, Rock SM, Belter E, Du F, Kim K, Abbott RM, Cotton M, Levy A, Marchetto P, Ochoa K, Jackson SM, Gillam B, *et al*: **The B73 maize genome: complexity, diversity, and dynamics.** *Science* 2009, **326**(5956):1112–1115.
30. Rogozin I, Kondrashov F, Glazko G: **Use of mutation spectra analysis software.** *Hum Mutat* 2001, **17**(2):83–102.
31. Rogozin IB, Pavlov YI: **Theoretical analysis of mutation hotspots and their DNA sequence context specificity.** *Mutat Res* 2003, **544**(1):65–85.
32. Meng XF, Shi M, Chen XX: **Population genetic structure of Chilo suppressalis (Walker) (Lepidoptera: Crambidae): strong subdivision in China inferred from microsatellite markers and mtDNA gene sequences.** *Mol Ecol* 2008, **17**(12):2880–2897.
33. Li XH, Luo GH, Zhang ZC, Liu BS, Fang JC: **Coling and characterization of Ty3/gypsy retrotransposon in *Chilo suppressalis* (Lepidoptera: Pyralidae).** *Chin J Rice Sci* 2014, **28**(3):314–321.
34. Wang W-M, Ding J-L, Shu J-W, Chen W: **Exploration of early rice farming in China.** *Quat Int* 2010, **227**(1):22–28.
35. Molina J, Sikora M, Garud N, Flowers JM, Rubinstein S, Reynolds A, Huang P, Jackson S, Schaal BA, Bustamante CD, Boyko AR, Purugganan MD: **Molecular evidence for a single evolutionary origin of domesticated rice.** *Proc Natl Acad Sci U S A* 2011, **108**(20):8351–8356.
36. Crawford G: **Early rice exploitation in the lower Yangzi valley: what are we missing?** *The Holocene* 2012, **22**(6):613–621.
37. Luo GH, Zhang ZC, Han GJ, Han ZJ, Fang JC: **Characteristics of overwintering populations of rice stem borers and mutation frequencies of resistance to triazophos.** *Chin J Rice Sci* 2012, **26**(4):481–486.
38. Thompson JD, Gibson TJ, Plewniak F, Jeanmougin F, Higgins DG: **The CLUSTAL_X windows interface: flexible strategies for multiple sequence alignment aided by quality analysis tools.** *Nucleic Acids Res* 1997, **25**(24):4876–4882.
39. Tamura K, Dudley J, Nei M, Kumar S: **MEGA4: Molecular Evolutionary Genetics Analysis (MEGA) software version 4.0.** *Mol Biol Evol* 2007, **24**(8):1596–1599.
40. Ko WY, David RM, Akashi H: **Molecular phylogeny of the Drosophila melanogaster species subgroup.** *J Mol Evol* 2003, **57**(5):562–573.
41. Dice LR: **Measures of the amount of ecologic association between species.** *Ecology* 1945, **26**(3):297–302.
42. Nei M, Li WH: **Mathematical model for studying genetic variation in terms of restriction endonucleases.** *Proc Natl Acad Sci U S A* 1979, **76**(10):5269–5273.
43. Felsenstein J: **PHYLIP-phylogeny inference package (version 3.2).** *Cladistics* 1989, **5**:164–166.

10

Gene expression studies for the analysis of domoic acid production in the marine diatom *Pseudo-nitzschia multiseries*

Katie Rose Boissonneault[1,2*], Brooks M Henningsen[1,3*], Stephen S Bates[4], Deborah L Robertson[5*], Sean Milton[2,6], Jerry Pelletier[7], Deborah A Hogan[8] and David E Housman[2]

Abstract

Background: *Pseudo-nitzschia multiseries* Hasle (Hasle) (*Ps-n*) is distinctive among the ecologically important marine diatoms because it produces the neurotoxin domoic acid. Although the biology of *Ps-n* has been investigated intensely, the characterization of the genes and biochemical pathways leading to domoic acid biosynthesis has been limited. To identify transcripts whose levels correlate with domoic acid production, we analyzed *Ps-n* under conditions of high and low domoic acid production by cDNA microarray technology and reverse-transcription quantitative PCR (RT-qPCR) methods. Our goals included identifying and validating robust reference genes for *Ps-n* RNA expression analysis under these conditions.

Results: Through microarray analysis of exponential- and stationary-phase cultures with low and high domoic acid production, respectively, we identified candidate reference genes whose transcripts did not vary across conditions. We tested eleven potential reference genes for stability using RT-qPCR and GeNorm analyses. Our results indicated that transcripts encoding JmjC, dynein, and histone H3 proteins were the most suitable for normalization of expression data under conditions of silicon-limitation, in late-exponential through stationary phase. The microarray studies identified a number of genes that were up- and down-regulated under toxin-producing conditions. RT-qPCR analysis, using the validated controls, confirmed the up-regulation of transcripts predicted to encode a cycloisomerase, an SLC6 transporter, phosphoenolpyruvate carboxykinase, glutamate dehydrogenase, a small heat shock protein, and an aldo-keto reductase, as well as the down-regulation of a transcript encoding a fucoxanthin-chlorophyll a-c binding protein, under these conditions.

Conclusion: Our results provide a strong basis for further studies of RNA expression levels in *Ps-n*, which will contribute to our understanding of genes involved in the production and release of domoic acid, an important neurotoxin that affects human health as well as ecosystem function.

Keywords: Gene expression, Gene regulation, cDNA microarray, RT-qPCR, Normalization, Reference gene, Domoic acid, *Pseudo-nitzschia multiseries*, Bacillariophyceae, Diatom

* Correspondence: katieroseboissonneault@gmail.com; bmhenningsen@gmail.com; debrobertson@clarku.edu
[1]Department of Biological Sciences, Plymouth State University, MSC 64, 17 High St, Plymouth, NH 03264, USA
[5]Biology Department, Clark University, 950 Main Street, Worcester, MA 01610, USA
Full list of author information is available at the end of the article

Background

The marine diatom *Pseudo-nitzschia multiseries* Hasle (Hasle) (*Ps-n*) produces the neurotoxin domoic acid (DA), which causes amnesic shellfish poisoning (ASP) [1-4]. DA is a neuroexcitatory, water-soluble amino acid that exhibits structural similarity with the neurotransmitter glutamate [5]. DA binds with high affinity to glutamate receptors, leading to excitation and ultimately cell death of neurons exposed to this toxin [6]. Production of DA by *Ps-n*, and at least 14 other members of the genus *Pseudo-nitzschia*, has been verified in oceanic regions throughout the world, primarily in coastal and upwelling zones [7,8]. The documented effects of DA on humans, birds, finfish, cephalopods, and marine mammals, and the economic costs of shellfishery closures due to DA contamination, has generated ongoing interest in understanding the regulation and control of DA production in this genus [7,9-11]. Yet, the biosynthetic pathways leading to DA production and the genes that govern these pathways remain unresolved [12,13].

Numerous studies on *Ps-n* growth dynamics have shown that DA production does not begin until early stationary phase, i.e. toxin is not typically produced in detectable amounts during the exponential growth phase (reviewed in [9]). In other studies that exposed *Ps-n* to conditions that slowed cell division during the mid-exponential phase, cells produced low levels of toxin. Therefore, toxin production appears to be associated with stages in the cell cycle when cell division has slowed or stopped due to some limiting nutrient factor, most notably silicon (Si) [10,14]. In addition, several bacterial isolates have been shown to enhance DA production by *Ps-n* [15-17]. *Ps-n* can produce DA in axenic cultures [2,18], yet, reintroduction of bacteria to axenic cultures results in increased *Ps-n* DA production [15-17].

In this study, we developed a *Ps-n* cDNA library and used it to construct a microarray in order to screen for genes that were differentially expressed under high-toxin-producing versus low-toxin-producing conditions. A total of 5,265 *Ps-n* cDNAs were printed in replicate, and mRNAs from cells that were in late-exponential growth phase were compared to those that were in stationary phase in both axenic and non-axenic cultures. Using these array data, we identified candidate reference and target genes for further study. Eleven reference genes were evaluated for stability in reverse-transcription quantitative PCR (RT-qPCR) analyses of *Ps-n* mRNA from Si-limited cultures. We performed a GeNorm analysis to validate transcripts that did not vary across conditions. Using the validated reference transcripts, we then confirmed the differential regulation of several transcripts whose expression correlates with DA production. These findings will facilitate future work aimed at elucidating the DA biosynthesis pathway and identifying transcriptional biomarkers indicative of DA production.

Results

Pseudo-nitzschia growth and toxin production for microarray studies

Samples for microarray analysis were obtained from three biological experiments using *Ps-n* strain CL-125. These trials included one axenic and two non-axenic cultures, all grown in standard medium f/2. DA production began at the onset of stationary phase and continued to increase over time in all three experiments (Figure 1). Final DA concentrations, expressed on a per mL basis, were ~30 times lower in the axenic growth experiment compared to the non-axenic growth experiments, as expected based on previous studies [2,15-18]. Previous studies also indicated that Si is the limiting nutrient for *Ps-n* cells grown in batch cultures with medium f/2 [9,10,14]; therefore, we presume that the cells in these experiments were Si-limited during stationary phase. Samples were harvested for microarray analysis during the late-exponential and stationary phases to compare gene expression between low-toxin-producing vs. high-toxin-producing cells. These time points are indicated by arrows in Figure 1a, 1b.

Identification and validation of reference transcripts

Our initial goal was the identification of transcripts whose expression levels were stable between late-exponential and stationary phases, which could then be used for normalization of other transcripts' expression levels under these conditions. We selected eleven candidate reference genes to evaluate in RT-qPCR studies based on their stability in the microarray results as well as their biological roles and use as controls in previous studies (Table 1; Additional file 1). These included transcripts encoding: dynein, histone H3, cyclophilin, ubiquitin, elongation factor 1 alpha (EF-1α), phosphoglycerate kinase 1 (PGK), eukaryotic initiation factor 2 (eIF-2), a JmjC-domain containing protein (JmjC), an AAA-domain containing ATPase, glyceraldehyde-3-phosphate dehydrogenase (GAPDH), and 18s rRNA. RT-qPCR primer sets for each candidate reference gene were designed and tested, and exhibited high sequence specificity and PCR efficiency under our assay conditions with an annealing temperature of 60°C (Table 2).

To validate the stability of candidate reference genes, biological triplicates of *Ps-n* strain GGB1 were grown under non-axenic, Si-limited conditions. RNA was harvested at multiple time points during the late-exponential and stationary phases (Figure 1c). The initiation of DA production again corresponded with the onset of stationary phase, which in this study was on day four. Initial silicate concentrations were reduced to 37.2 μM in the culture medium (vs. 107 μM in the standard f/2 medium). The measured

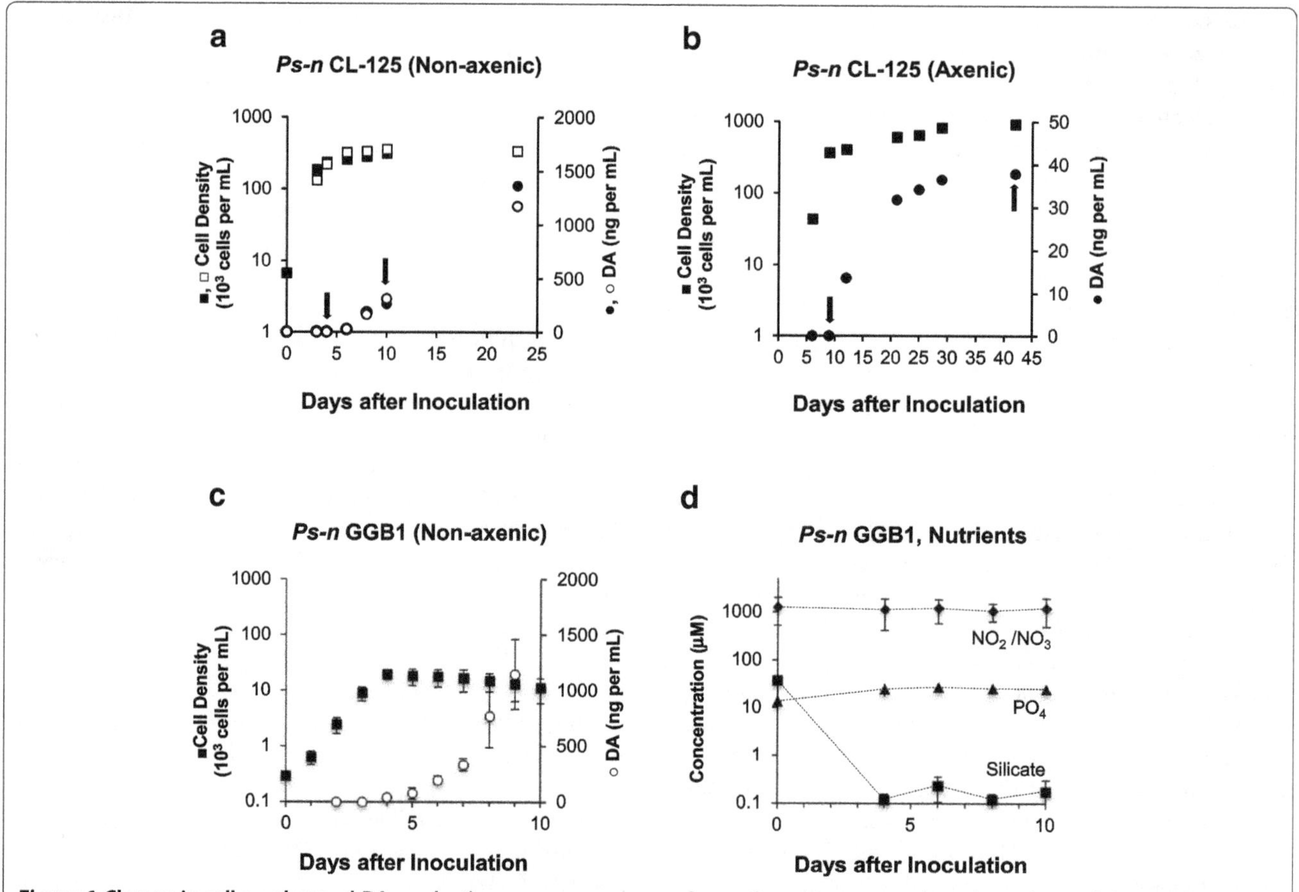

Figure 1 Change in cell number and DA production as a consequence of growth under non-axenic and axenic conditions. a) *Ps-n* strain CL-125, Non-axenic culture experiments 1 (solid) and 2 (open). **b)** *Ps-n* strain CL-125, Axenic culture experiment. Cells were harvested for RNA extraction on the days indicated by arrows. **c)** Increase in cell number (squares) and in DA concentration (circles) of *Ps-n* strain GGB1 in non-axenic, triplicate cultures. Cells were harvested for RNA extraction on days 3–10. **d)** Nutrient concentrations over time in GGB1 cultures (nitrite/nitrate, phosphate, silicate). Data for **(c)** and **(d)** represent the mean change of triplicate samples (± 1 SD).

silicate concentration was below 1.0 µM by day four (Figure 1d), corresponding with the entry into stationary phase. Nitrate and phosphate appear to be present in sufficient quantities throughout the experiment (Figure 1d) [9,10,14], further supporting that growth of these cultures was Si-limited.

In an initial GeNorm analysis, all eleven candidate genes were tested for stability across a subset of samples from the Si-limitation experiment, including replicates from days 3, 4, and 10 (Figure 2a). The four genes with the best stability values (M-value), i.e. JmjC (0.33), dynein (0.33), histone H3 (0.37) and cyclophilin (0.39), along with EF-1α (0.62), were further evaluated for stability using GeNorm analysis across the complete set of expression data for the Si-limitation experiment (Figure 2b). All five genes showed acceptable stability (M-value <0.5) when evaluated across the complete set of data. GeNorm pairwise-variation analysis determined that only two genes (JmjC, dynein) were necessary for subsequent normalization. However, since the JmjC, dynein, and histone H3 genes had equivalent M-values and were matched for 1st rank, we used all three for normalization of the expression data as described below. The expression profiles of the reference genes show the stability of these top-ranked genes, and the slight variability of the EF-1α and cyclophilin genes (Figure 3).

Identification and verification of differentially expressed transcripts

In the microarray study, only those transcripts that were up- or down-regulated in all three trials were considered further (Tables 3 and 4; Additional file 1). Higher transcript levels in stationary (high-toxin-producing) as compared to late-exponential (low-toxin-producing) phase were observed for 12 transcripts, corresponding to 76 cDNA clones printed on the *Ps-n* array (Table 3; Additional file 1); reduced transcript levels under these conditions were observed for six genes, corresponding to 17 cDNA clones printed on the array (Table 4; Additional file 1). In addition to those genes

a

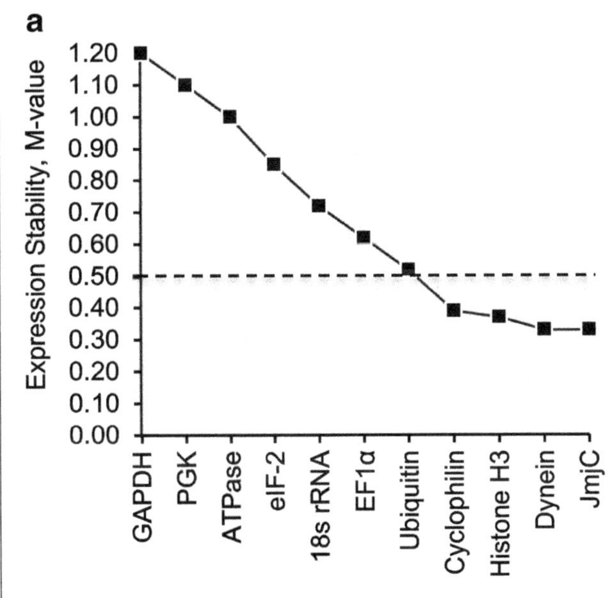

b

Predicted Gene Product	Rank order	M-value
JmjC	1	0.23
Dynein	1	0.23
Histone H3	1	0.23
EF-1α	2	0.34
Cyclophilin	3	0.43

Figure 2 Average expression stability (M-value) of the reference genes determined by GeNorm analysis. An individual reference gene is tested against the other reference genes in a pairwise variation that serially excludes the least stable genes from the analysis. The most stable reference genes exhibit the lowest M-values. The accepted cut-off for stability of reference genes is an M-value of 0.50. **a)** In the initial analysis, which tested a subset of mRNA samples under Si-limited conditions for all eleven of our potential reference genes, four reference genes were determined to be acceptably stable. **b)** In the analysis of the Si-limited growth experiment, these four reference genes and EF1-alpha were tested. All five showed acceptable stability across the mRNAs, and JmjC, Dynein, and Histone H3 were tied for the 1st rank.

genes identified in our study relative to DA production. For example, up-regulation of a gene encoding a putative *Ps-n* cycloisomerase is intriguing as its product may be directly involved in the proposed cyclization step leading to the pyrrolidine ring in DA. Alternatively, the cycloisomerase, similar to other enzymes in the related pfam20282 group, may be involved in converting aromatic compounds into citric acid cycle intermediates, proposed to feed the pathway leading to DA synthesis [34-38]. The identification of a differentially expressed transcript encoding a member of the SLC6 amino acid transporter family is also interesting. The translated *Ps-n* open reading frame aligns most closely with characterized γ-aminobutyric (GABA) neurotransmitter transporters [39], suggesting the hypothesis that the *Ps-n* transporter is involved in movement of DA, or a synthetic precursor, into or out of cells.

Our discovery of the up-regulation of a predicted cycloisomerase belonging to the lactonase/lactonizing family, as well as the SLC6 transporter, entertains the speculation that these gene products are involved in communication between *Ps-n* cells, or *Ps-n* and bacteria. The parallels with GABA in plant signaling pathways [40-42] pose a potential role for DA in *Ps-n* biology, which has not yet been defined [7,9]. For example, GABA produced by wounded plant tissues appears to control the lactone quorum-sensing signal in *Agrobacterium tumefaciens* by regulating the *A. tumefaciens* lactonase gene [41]. Bacterial production of lactones in *Ps-n* cultures is correlated with increased DA production [16,17,43], suggesting a possible relationship between DA and quorum sensing [44]. Characterization of the predicted cycloisomerase's enzymatic properties will be of significant interest in relation to these hypotheses. Similarly, demonstration that the SLC6 transporter is involved in movement of DA into or out of the cell would be a valuable contribution to understanding the role of DA in *Ps-n* biology. While we have taken the perspective that DA may function in signaling pathways, including quorum sensing or pheromone communication [7,8,42], some studies suggest that DA may function as a chelating agent [45-47]. Hence, studying the transport of DA into and out of *Ps-n* cells directly would contribute to describing the role(s) of DA in *Ps-n* biology. A family of four SLC6 transporters was identified in the recently released *Ps-n* draft genome [22,39], so characterization of this entire family should advance our understanding of *Ps-n* biology.

Many of the differentially expressed genes in this study relate to general metabolic pathways. Therefore, further investigation is needed to resolve the role of these genes in relation to both the growth state and DA synthesis in *Ps-n*. For example, the up-regulation of PEPCK, as well as the potential down-regulation of PFK, suggests a change in energy metabolizing pathways as *Ps-n* cells transition from exponential to stationary phase, consistent with a shift to gluconeogenesis and carbon metabolism through the citric acid cycle [48,49]. Similarly, glutamate dehydrogenase, which catalyzes the reversible conversion of glutamate and the citric acid cycle intermediate α-ketoglutarate, is a key enzyme involved in nitrogen and energy metabolism [50,51]. In addition, the differential expression of a predicted acyl-CoA synthetase (Table 3) suggests the possibility that lipids and fatty acids are being broken down, and while this may be a physiological response to growth-limiting conditions, the products

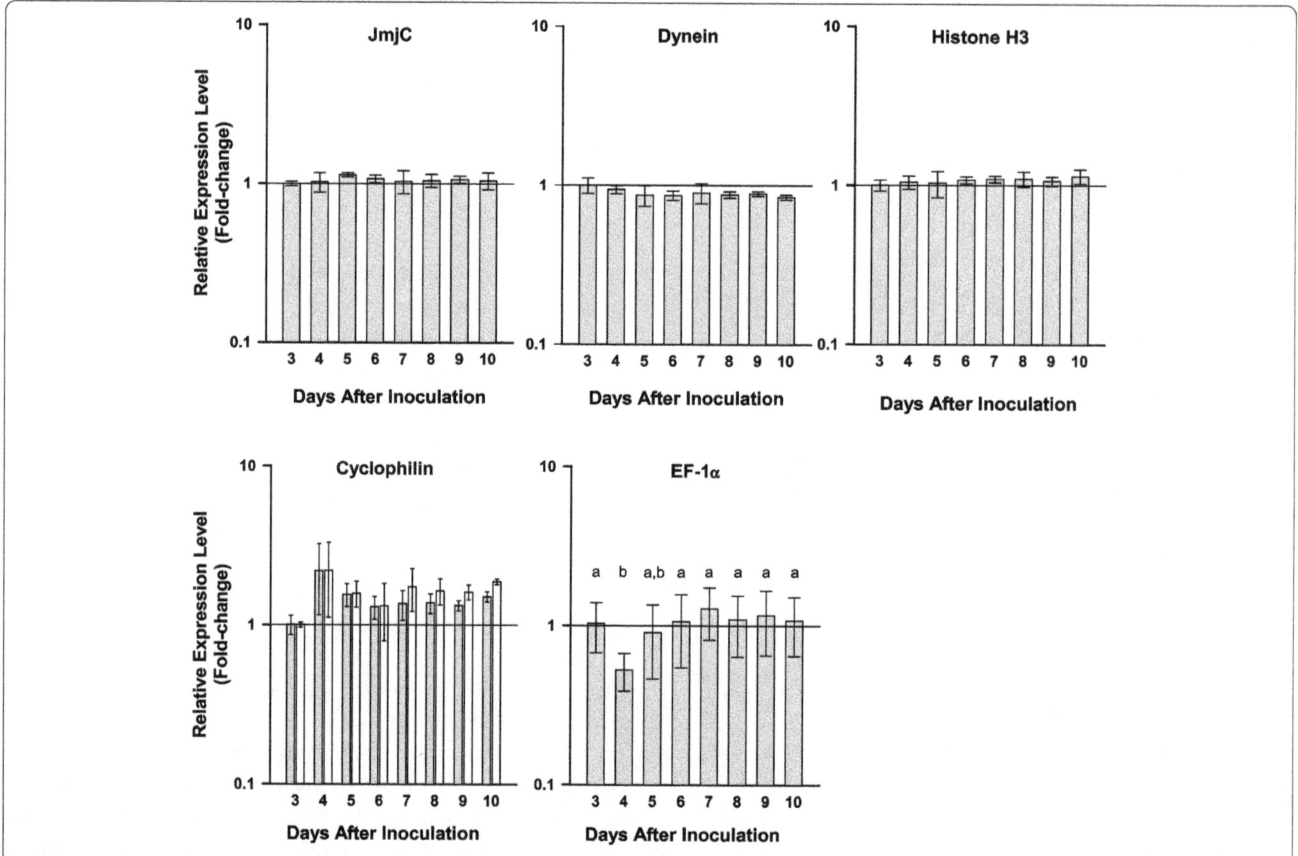

Figure 3 RT-qPCR analysis of candidate reference genes for normalization from *Pseudo-nitzschia multiseries*. Bars represent the mean change (± 1 SD) in expression relative to Day 3. Open bars represent measurements using primers that spanned an exon-exon junction; grey bars represent measurements using standard primers that did not span an exon-exon junction. Means were not significantly different ($p < 0.05$), except EF-1α. Means with different letters were significantly different ($p < 0.05$). Statistical analyses were performed using a general linear model ANOVA with Bonferroni *post-hoc* test, 95% confidence intervals.

could then be channeled as precursors into DA synthesis. Previous studies have shown that *Ps-n* lipid content decreases in response to Si deficiency during stationary phase [9,52]. Acyl-CoA synthetases are also involved in amino acid acylation, so could be directly involved in the condensation of the glutamate and isoprenoid-like moieties [13,53,54].

A small heat shock protein gene was most highly up-regulated later in the stationary phase as determined by RT-qPCR, suggestive of its expression relative to physiological stress. The aldo-keto reductase transcript levels showed a step-wise progression from the exponential into the stationary phase, with the highest expression levels later in stationary phase, as well. The expression patterns of these genes may be useful for monitoring the physiological state of *Ps-n* cells. The aldo-keto reductase may also have a functional role in DA synthesis, as the labeling studies indicated that the C7' in DA is selectively oxidized to a carboxyl group [12]. Several of the genes that were identified as being up-regulated in this study have not been previously characterized from diatoms and represent potential targets for further studies of DA synthesis.

The enhancement of DA production by co-existing microbes is a complex and fascinating aspect of DA biology [7]. A limited number of genes in our study indicated significantly different expression patterns between the non-axenic vs. axenic growth experiments. For example, a subtilisin-like gene, predicted to encode a secreted protease, was up-regulated in the non-axenic cultures relative to the axenic culture (Additional file 2). In addition, microbes may influence the metabolic pathways predicted to be involved in fatty acid production. Ramsey et al. [12] suggested that the principal pathway to the isoprenoid portion of DA is via an alternative glyceraldehyde 3-phosphate (G3P)-independent route, and it is interesting to note that glyceraldehyde-3-phosphate dehydrogenase (GAPDH) was up-regulated only in the axenic growth experiment (Table 3), suggesting that bacteria may influence this pathway. Future studies will focus on the specific roles that co-existing microbes play in the regulation of *Ps-n* genes and domoic acid production.

Table 3 Transcripts at higher levels in stationary (high-toxin-producing) as compared to late-exponential (low-toxin-producing) phase in *Pseudo-nitzschia multiseries* (*Ps-n*) as determined by cDNA microarray analysis

		Fold change[a]			
		Stationary versus late-exponential phase			
Ps-n NR Identifier	JGI *Ps-n* Genome hit *Scaffold:Start-End*	Non-axenic Expt. 1	Non-axenic Expt. 2	Axenic Expt.	Predicted gene product
PSN0011	*481:47438-49830*	3.99 ± 0.96	3.46 ± 1.90	2.19 ± 0.37	Cycloisomerase (pfam10282 lactonase/lactonizing enzyme), COG2706 (3-carboxymuconate cyclase)
PSN0072	*269:13929-16593*	3.40 ± 0.38	3.39 ± 0.32	2.01 ± 0.24	SLC6, Sodium and Chloride-dependent amino acid transporter
PSN0014	*2396:2127-4845*	4.64 ± 0.89	4.10 ± 0.89	2.14 ± 0.27	Acyl-CoA synthetase with transit peptide
PSN0016	*37:106113-109070*	3.78 ± 0.27	3.01 ± 0.29	3.11 ± 0.50	Phosphoenolpyruvate carboxykinase, ATP-dependent with transit peptide (PEPCK)
PSN0025	*155:75271-76268*	6.65 ± 1.56	7.07 ± 1.75	4.18 ± 0.57	Small heat shock protein with alpha-crystallin domain, chloroplastic (sHSP)
PSN0052	*70:256654-258177*	3.89 ± 0.59	2.61 ± 0.21	1.57 ± 0.01	Mitochondrial carrier protein
PSN0015	*66:296511-298386*	3.19 ± 0.57	3.15 ± 0.43	1.86 ± 0.17	Aldo-keto reductase with signal peptide
PSN0042	*21:457033-458813*	5.47 ± 0.75	7.52 ± 1.69	3.37 ± 0.50	Predicted protein with signal peptide
6H1	*117:15135-16102*	5.45 ± 0.55	3.76 ± 0.00	2.57 ± 0.04	Predicted protein with signal or transit peptide
73D12	*1312:10048-11278*	3.45 ± 0.11	4.01 ± 0.03	1.63 ± 0.07	*Ps-n* specific, no hits in NR or Swissprot
46A5	*447:56474-57752*	4.36 ± 0.00	4.61 ± 0.11	1.73 ± 0.02	*Ps-n* specific, no hits in NR or Swissprot
17F11	*303:41858-44529*	5.42 ± 0.18	6.02 ± 0.85	2.07 ± 0.06	Predicted protein with glycosyltransferase domain
PSN1428[b]	*95:293619-297233*	2.22 ± 0.26	1.86 ± 0.20	1.81 ± 0.37	NAD-specific glutamate dehydrogenase (GDH)

[a]The fold-change data presented are the average of all of the cDNA clones that were printed on the array for each transcript. The number of cDNA clones for each transcript was: PSN0011 (21), PSN0072 (3), PSN0014 (14), PSN0016 (14), PSN0025 (5), PSN0052 (4), PSN0015 (7), PSN0042 (4), 6H1 (1), 73D12 (1), 46A5 (1), 17F11 (1), PSN1428 (2). Each clone was printed on the array twice.

[b]PSN1428 was included as a gene of interest in RT-qPCR analysis, below, although it did not meet our statistical criteria for the original microarray analysis. Please see Methods for statistical analysis, and Additional file 1 for FDR and LDFR data.

Table 4 Transcripts at lower levels in stationary (high-toxin-producing) as compared to late-exponential (low-toxin-producing) phase in *Pseudo-nitzschia multiseries* (*Ps-n*) as determined by cDNA microarray analysis

		Fold change[a]			
		Stationary versus late-exponential phase			
Ps-n NR Identifier	JGI *Ps-n* Genome Hit *Scaffold:Start-End*	Non-axenic Expt. 1	Non-axenic Expt. 2	Axenic Expt.	Predicted gene product
PSN0100	*32:397085-398987*	0.34 ± 0.01	0.33 ± 0.05	0.39 ± 0.01	Pyrophosphate-dependent phosphofructokinase (PFK)
PSN0060	*133:16659-18505*	0.20± 0.02	0.16 ± 0.03	0.44 ± 0.04	Predicted protein with signal or transit peptide
PSN0048	*188:179178-180986*	0.36± 0.08	0.25 ± 0.06	0.52 ± 0.05	Predicted protein with signal or transit peptide
PSN0080	*461:123066-124152*	0.32 ± 0.02	0.33 ± 0.03	0.57 ± 0.07	Predicted protein with mitochondrial transit peptide
135E4	*1441:17433-18891*	0.17 ± 0.00	0.19 ± 0.02	0.45 ± 0.02	Predicted protein
165G9	*8:175639-176910*	0.37 ± 0.02	0.37 ± 0.02	0.62 ± 0.04	Tetratricopeptide repeat protein
135H6[b]	*214:78592-7971*	0.41 ± 0.00	0.22 ± 0.02	0.59 ± 0.00	Fucoxanthin-chlorophyll a-c binding protein, chloroplastic (FCP)

[a]The fold-change data presented are the average of all of the cDNA clones that were printed on the array for each transcript. The number of cDNA clones for each transcript was: PSN0100 (2), PSN0060 (5), PSN0048 (5), PSN0080 (3), 135E4 (1), 165G9 (1), 135H6 (1), and each clone was printed twice.

[b]135H6 was included as a gene of interest in RT-qPCR analysis, below, although it did not meet our fold-change cut-off for the original FDR microarray analysis (yet, all LFDRs were <10%). Please see Methods for statistical analysis, and Additional file 1 for FDR and LDFR data.

Table 5 RT-qPCR target gene primer sequences and characteristics

Predicted gene product	Primer sequence	GC (%)	Tm (°C)[a]	Amplicon (bp)	Ex-Ex Spanning [b]	Efficiency (%)	R^2
Cycloisomerase	F: TCATAGGTGGCGTCAAGAACGTGT	50.0	60.3	127	No	99.5	0.995
	R: TCAGCTTGTCGTGCCGAAATTGTG	50.0	60.3				
SLC6	F: TCGGACACTACGGAGACTACG	57.1	57.1	73	No	104.2	0.997
	R: ACCAAGGTGAAGGCGACG	61.1	58.0				
SLC6 Ex-Ex	F: CATGCACGATACTGTCTATTTCG	43.5	53.6	122	Yes	100.0	0.998
	R: CGTCCAACCAAAATAAGCCAGC	50.0	57.0				
Aldo-keto reductase	F: GAATGGGCTACGGAGAGACG	60.0	57.3	114	No	99.5	0.998
	R: GTACAGGCGTGAATTTGGTAGC	50.0	56.2				
sHSP	F: GACGAAGGATTCATCACCGTCG	54.5	57.7	141	No	102.9	0.998
	R: GACACCGTTGTCGAGGGTAG	60.0	57.4				
PEPCK	F: GCATTGCTCTGCAAACGTCG	55.0	57.7	107	No	100.0	0.997
	R: CAATCAAGGCTCGGTGAGGATC	54.5	57.7				
GDH	F: CAATGCCATCAACGCCATCAAGGA	50.0	60.2	128	No	98.4	0.998
	R: CAAAGCCGAGGTTGGCAAGAGTTT	50.0	60.3				
PFK	F: CGAGGTGGCATCCAAACGATTGC	56.5	61.1	84	No	105.1	0.998
	R: GCAGCCTGTGTATTGGTATCGTCG	54.1	59.7				
PFK Ex-Ex	F: GGAGAAAATCCGCTCGAGGTG	57.1	57.9	111	Yes	99.4	0.997
	R: CTTTGAGAGAACCGCAGCCTG	57.1	58.4				
FCP	F: CGTCTCATACCACGGCAC	61.1	55.7	184	No	97.8	0.997
	R: CTTGGATTGATGGTCCACGAG	52.3	55.6				

[a]The annealing temperature for all standard curve analyses was performed at 60°C to demonstrate the efficiency of the primer sets under our assay conditions; the calculated Tm values are provided for reference.

[b]'Ex-Ex spanning' refers to primers that span an exon-exon junction; these primer sets did not yield a product using gDNA as a template.

The *Ps-n* genome is predicted to include 19,703 genes [22]; thus, the estimated 3,675 non-redundant transcripts monitored via this microarray represent ~20% of the genome. Future studies using RNA sequencing methods will determine if other transcripts related to those highlighted here are also differentially expressed in correlation with DA production.

Conclusions

Our study identified a number of significantly up- and down-regulated genes that provide the basis for future studies on DA production, growth state, stress, and amino acid transport in *Ps-n*. The identified transcripts may be particularly useful as early indicators of toxin production and the switch of *Ps-n* cells to an alternative growth state. The reliability of RT-qPCR data will be enhanced by use of the validated internal reference genes presented in this study. To our knowledge, this is the first identification and validation of reference genes for RT-qPCR studies in *Ps-n*.

Methods

Pseudo-nitzschia multiseries strains and culture conditions

Ps-n strain CL-125 was isolated by Claude Léger (Fisheries and Oceans Canada, Gulf Fisheries Centre, Moncton, New Brunswick, Canada) from a sample collected on September 23, 2000, in Mill River (a brackish water estuary), Prince Edward Island, Canada. Cultures for the microarray studies were grown in 0.2 μm-filtered, autoclaved seawater (from Woods Hole, MA) enriched with f/2 nutrients [55] and amended with 10^{-8} μM Se. These batch cultures were grown in 15 L of f/2 medium in 19 L borosilicate carboys, and incubated at 20°C. The irradiance was maintained at 100 μmol photons m^{-2} s^{-1}, with a 14:10-h light:dark (L:D) cycle for the cDNA library cultures, and continuous light for the experimental cultures. The cultures were aerated using aquarium pumps with sterile cotton and activated carbon filters and were constantly mixed with magnetic stirrers. An axenic culture of CL-125 was obtained by antibiotic treatment for 72 h, using 1.6:0.8 mg mL^{-1} penicillin:streptomycin [2]. These were tested for culturable bacteria by incubation in Bacto-peptone broth (Difco Laboratories, Detroit, MI, USA; 1 g L^{-1} seawater) and 2216 Marine Agar (Difco) at ~20°C for at least 20 d.

Ps-n strain GGB1 was isolated by Michael Carlson and Kyle Frishkorn (University of Washington, Seattle, WA, USA) in July 2010, from Puget Sound, WA, USA. Cultures for the RT-qPCR study were grown in 0.45 μm-

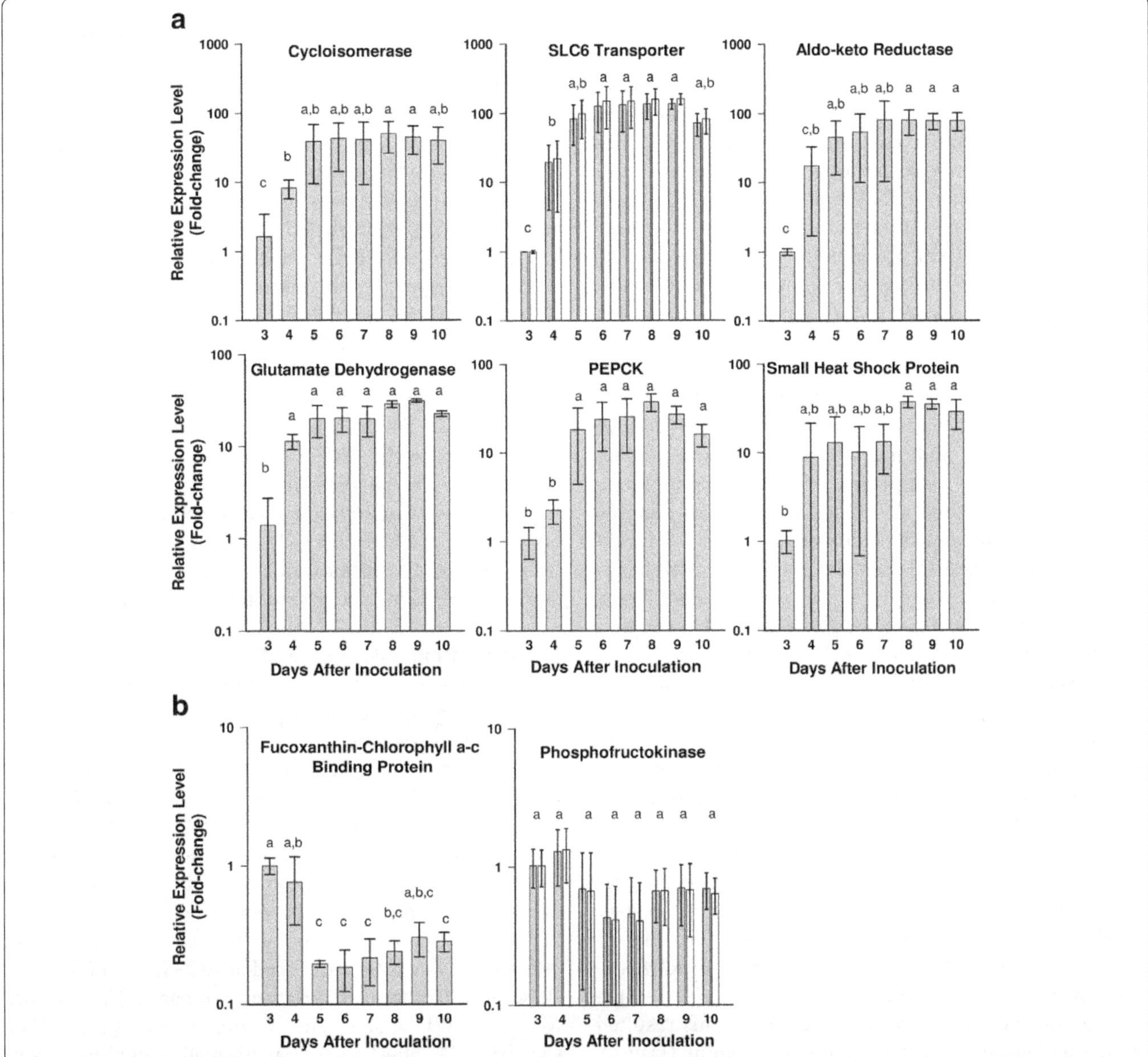

Figure 4 RT-qPCR analysis of *Pseudo-nitzschia multiseries* genes whose expression was up-regulated (a) or down-regulated (b) in microarray analysis. Bars represent the mean change (± 1 SD) in expression relative to Day 3. Open bars represent measurements using primers that spanned an exon-exon junction; grey bars represent measurements using standard primers that did not span an exon-exon junction. Means with different letters were significantly different ($p < 0.05$). Statistical analyses were performed using a general linear model ANOVA with Bonferroni *post-hoc* test, 95% confidence intervals.

filtered, autoclaved seawater (from Portsmouth Harbor, Newcastle, NH, USA) enriched with f/2 nutrients [55], except that the initial Si was lowered to 37.2 μM and ferric sequestrene was replaced with $Na_2EDTA \cdot 2H_2O$ and $FeCl_3 \cdot 6H_2O$ (Provasoli-Guillard National Center for Marine Algae and Microbiota). Before inoculation of experimental cultures, cells were maintained in exponential growth for at least two preceding transfers. Triplicate GGB1 experimental cultures were grown in 2.6 L of f/2 medium in 3-L polycarbonate baffled flasks, and incubated at 15°C. The irradiance was maintained at ~100 μmol photons m^{-2} s^{-1}, with a 16:8-h L:D cycle. Flasks were aerated by constant mixing supplied by magnetic stirrers.

Sampling, toxin, and nutrient analysis

The microarray study included three biological replicates: two non-axenic cultures and one axenic culture. Samples were taken every two to three days for cell counts and domoic acid (DA) analysis. Cell concentrations were estimated by averaging the number of cells enumerated by light microscopy, using a Neubauer hemacytometer chamber in triplicate counts of individual samples preserved in Lugol's iodine. DA was analyzed in whole-culture samples (cells plus medium [1]), using HPLC of the FMOC (fluorenylmethoxycarbonyl) derivative [56]. The lower limit of detection was 15 ng mL^{-1} for the first non-axenic experiment, 3 ng mL^{-1} for the second non-axenic experiment, and 7.5 ng mL^{-1} for the axenic experiment. RNA was prepared from cells harvested during an initial time point from the late-exponential (low-toxin-producing) growth phase, and a final time point during the stationary (high-toxin-producing) phase (Figure 1a, 1b; RNA extraction protocol outlined below).

For RT-qPCR evaluation, three non-axenic biological replicate cultures were sampled daily, from the time of inoculation until day 10 of growth, for cell counts, whole-culture DA and nutrients. Cell count samples were taken by preserving 5 mL of culture with 250 μL of formalin and stored at 4°C until cells were counted (400 cells or the entire slide) on a Sedgwick-Rafter slide. Whole-culture DA samples were taken by freezing 15 mL of culture at −20°C; samples were sonicated at 50% power on ice for 2 min and filtered through a 0.2-μm filter prior to analysis. DA samples were analyzed using the Abraxis Domoic Acid ELISA kit ([57], Warminster, PA, USA). The limit of detection was 0.06 ng mL^{-1}. Filtered samples were stored at −80°C for nutrient analyses. Silicate was measured using the molybdate method [58-60]; phosphate was measured by the ascorbic acid-molybdate method [61,62]. Nitrite and nitrate were measured on an auto-analyzer (Lachat Instruments, Loveland, CO, USA) using a copper-cadmium reduction and colorimetric assay [61,63,64]. Total RNA was extracted daily from each flask beginning on day three of growth (Figures 1c, 3, 4; RNA extraction protocol outlined below). One RNA sample was lost on the initial day of extraction, so this resulted in two biological replicates for this time point (Day 3). The remaining RNA samples for the profile through to Day 10 included three biological replicates for each time point.

Microarray and cDNA library construction

Ps-n strain CL-125 cells from non-axenic cultures were harvested during the late-exponential through mid-stationary phases, under predominantly toxin-producing conditions. Cultures were split into 250–500 mL aliquots that were centrifuged for 15 min at 1000 *g*. The loose pellets were pooled, and centrifuged again briefly to remove any residual culture medium. Total RNA was extracted immediately by homogenizing the cells in TRIzol® (Invitrogen Corporation, Carlsbad, CA, USA). Insoluble material was removed by low-speed centrifugation of the samples, which increased quality and yield of the resulting total RNA. Precipitating twice with salt and ethanol also contributed to high-quality total RNA, as indicated by both 260/280 O.D. ratios and gel electrophoresis. Poly (A)+ RNA was then isolated from total RNA using a biotin-labeled oligo(dT)20-streptavidin kit (Roche Molecular Biochemicals, Indianapolis, IN, USA) following the manufacturer's instructions.

First-strand cDNA was prepared from 5 μg poly (A)+ RNA using Superscript II (Invitrogen, Grand Island, NY, USA), NC-p7 (an RNA chaperone), and oligo pd(TZ) (an oligo-dT primer with some of the internal thymidine residues replaced with 3-nitropyrrole to minimize mispriming to internal A-rich sequences). Double-stranded cDNA was generated using RNase H, *E. coli* DNA polymerase I, and *E. coli* ligase. The ends of the cDNAs were polished with T4 DNA polymerase, and BstXI adaptors were ligated to the cDNA ends. The cDNAs were then fractionated on sucrose gradients, ligated into pMD1 (a pUC-based vector) and transformed, by electroporation, into *E. coli* DH10B cells [65]. Following an initial library plating, 19,200 individual colonies were picked and stored at −80°C in 15% glycerol for further analysis.

EST sequencing, assembly, and annotation

2,220 cDNA inserts were sequenced to verify the quality of the library and to begin gene discovery. Many of the cDNAs were sequenced more than once in the 5′- and 3′- direction, yielding a set of 3,533 *Ps-n* ESTs. These sequences are deposited in the NCBI dbEST database [GenBank accession numbers FD476666-FD480212]. (Note: Of the sequenced cDNAs, 1,889 were a subset of the 5,265 *Ps-n* cDNAs printed on the microarray (see below)). Sequence reactions were run on an automated DNA sequencer (ABI 3700 with dye terminators); and selected cDNAs were sequenced at ACGT, Inc. (Wheeling, IL, USA). ESTs were edited using Seqman (DNAStar, Inc., Madison, WI, USA), and ContigExpress (VectorNTI, Carlsbad, CA, USA), as well as manually edited to remove low quality data, poly (A) tails, and vector sequence. The *Ps-n* ESTs were assembled into consensus or contig sequences, using a criterion of 95% identity over more than 50 nucleotides (Seqman, DNAStar). Average sequence length of the individual reads was 639 bp (after editing). The ESTs were assembled into 1,550 non-redundant (NR) sequences, indicating a redundancy of ~43% within our *Ps-n* library. From this, we estimate that approximately 3,675 unique genes were printed on the *Ps-n* microarray since 5,265 *Ps-n* cDNAs from the *Ps-n* library were printed on the microarray. The ESTs, final assembled NR sequences, and annotations are provided in Additional files 2, 3, 4 and 5.

Assembled sequences were compared against NCBI's NR and Swissprot databases using the Basic Local Alignment Search tools, blastx and tblastn, via theBlast2Go application [66-68]. The *Ps-n* sequences from this study were also compared against the unpublished *Pseudo-nitzschia multiseries* CLN-47 genome sequence, Assembly v1 (October 2011), sequenced by the US Department of Energy Joint Genome Institute (JGI) [22]. The reference and target genes specifically discussed in this paper were further annotated using NCBI's ORF Finder [69] and BlastP [66,67], and the Center for Biological Sequence Analysis' (Technical University of Denmark) ChloroP 1.1, TargetP 1.1, and SignalP 4.1 [70], as well as searching for the conserved diatom AFAP chloroplast targeting domain [71-73].

Microarray construction

cDNA preparation: 5,265 clones from the *Ps-n* cDNA library were grown overnight in Luria broth with carbenicillin (50 μg mL^{-1}) at 37°C on a shaker table. Three "vector-only" and twelve *Homo sapiens* cDNA control clones (J. Pelletier) were also grown under these conditions. Ten microliters of bacterial culture were used in 100-μL PCR reactions with primers T7 forward (TAA TACGACTCACTATAGGG) and M13 reverse (CAG GAAACAGCTATGAC), which flanked the cloning site of the pMD1 vector. PCR amplification was performed using HiFI Taq polymerase (Invitrogen Corporation). An initial DNA denaturation step at 94°C for 2 min was followed by 35 amplification cycles (0:30 melting at 94°C, 0:30 annealing at 55°C, 1:00 extension at 68°C). PCR products were purified using MultiScreen size-exclusion filter plates (Millipore, Billerica, MA, USA). The DNA was then resuspended in 100 μL of nuclease-free de-ionized water and transferred to clean plates using a mechanical pipetting station. DNA quality was verified by 1% agarose gel electrophoresis for eight samples per 96-well plate; DNA concentration was determined by PicoGreen fluorescent staining [63]. Fifty μL of each PCR product was dried by vacuum centrifugation and then resuspended in 10 μL of 1.5 M Betain /3X SSC print buffer, yielding an average final concentration of 600 ng μL^{-1}.

Ps-n cDNA probes were printed onto CMT-GAPS slides (Corning, Corning, NY, USA), using a MicroGrid 610 TAS array printer (Biorobotics, Woburn, MA, USA) with quill pins. A total of 5,169 *Ps-n* cDNAs were printed in duplicate, and 96 were printed in quadruplicate; in addition, 3 "vector-only" cDNAs, 12 *H. sapiens* control cDNAs, and 10 control cDNAs from the SpotReport Alien Array Validation System (Stratagene, La Jolla, CA, USA) were printed in duplicate, resulting in a final chip that included 10,772 features. Spots were printed with a 32 print-tip head, producing a lay-out represented by 8 × 4 grids. Each grid was sub-divided into two sections, representing replicate spots. Individual features were 13 μm in diameter and were separated by 130 μm (from one spot to the next). 0.005 μL of ~600 ng μL^{-1} DNA (2–3 ng) was transferred to each spot. Final *Ps-n* arrays displayed a strong signal-to-noise ratio, with virtually no background, as demonstrated visually (Figure 5). Experimental hybridization results also confirmed the high degree of reproducibility between replicate spots on the *Ps-n* chip (See Additional files 1 and 2; and, the corresponding Gene Expression Omnibus (GEO) [74] file, accession number GSE46845).

RNA preparation and microarray hybridizations

RNA was prepared from cells harvested during both the late-exponential (low-toxin-producing) and stationary (high-toxin-producing) phases for all three biological replicates. Eight liters of culture were harvested at an initial time point during the mid- to late-exponential growth phase and the remaining 7 L were harvested at a final time point during the stationary phase (Figure 1a, 1b). Cell suspensions were centrifuged in 0.5-L aliquots for 15 min at 1,000 *g*, which resulted in loose pellets that were pooled, split among 2–4, 50-mL conical tubes and spun again briefly to remove any remaining liquid. Ten to 20 mL of TRIzol (depending on cell pellet volume) were added to the conical tubes, and the pellets were homogenized for 60 s, frozen in liquid N, and stored at −80°C until RNA extraction. Total RNA was extracted, as above, cleaned with RNeasy columns (Qiagen, Valencia, CA, USA) and run on formaldehyde agarose gels to confirm the quality of the RNA.

Ten micrograms of *Ps-n* RNA from each harvest were spiked with mRNA from the SpotReport Alien Array Validation System, incubated for 10 min at 65°C with oligo-dT and then cooled at 25°C for 5 min. Four microliters of 1 mM Cy3- or Cy5-conjugated dUTPs were added to each RNA sample and the mixtures were incubated at 42°C for 2 min. A master mix, including 4.5 μL of 0.2 M DTT, 18 μL of 5X 1st strand buffer, 1.8 μL of 25 mM dATP, dGTP and dCTP, 1.8 μL of 10 mM dTTP, and 2 μL of Superscript II reverse transcriptase, was added to each RNA mixture and incubated for 1 h at 42°C. After 1 h, an additional 1 μL of Superscript II was added to each and the reactions were incubated at 42°C for another hour. Starting RNA was degraded by addition of stop solution (3 μL of 0.5 M EDTA, pH 8; 3 μL of 1 N NaOH) and incubated for 30 min at 60°C. Labeled cDNA was cleaned using RNeasy columns (Qiagen); Cy3-labeled cDNA and the corresponding Cy5-labeled cDNA that were to be compared were combined and loaded onto the same column. The labeled target cDNA pools were then hybridized to the probe cDNAs on the *Ps-n* cDNA microarrays (construction described above). *Ps-n* microarrays were processed before hybridization by holding them face-down over a steaming water bath for a few seconds, and then snap-drying them on a 95°C heat block. The DNA was immobilized onto the slides by UV cross-linking

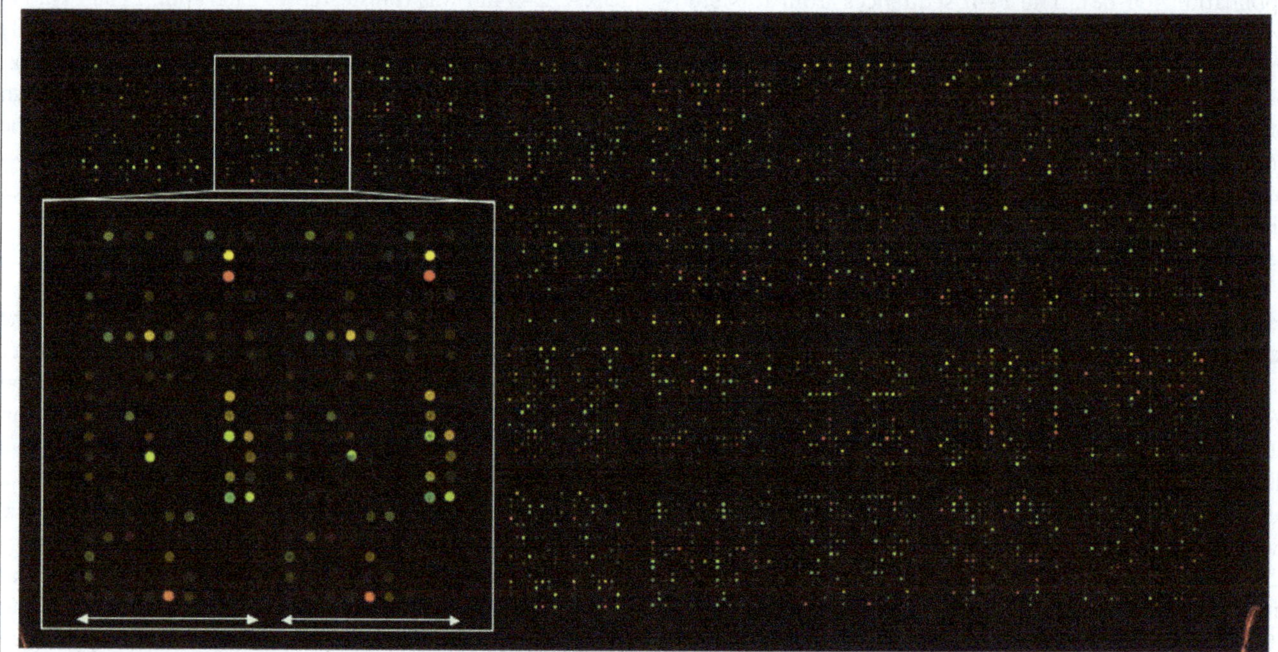

Figure 5 Scanning fluorescence image of the *Pseudo-nitzschia multiseries* (*Ps-n*) cDNA microarray hybridized with Cy3- and Cy5-labeled cDNAs from non-toxin-producing vs. toxin-producing cells. 5,169 individual *Ps-n* cDNAs and 25 control cDNAs were printed in replicate, and 96 *Ps-n* cDNAs were printed in quadruplicate, yielding a final chip including 10,772 features. A representative grid is enlarged to illustrate the subdivision of each 8 × 4 grid into two replicate sections, differentiated by the arrowed lines.

at 65 mJoules. Cross-linked slides were soaked for 15 min in freshly prepared succinic anhydride/sodium borate solution with gentle agitation, soaked for 2 min in boiling nuclease-free, de-ionized water and finally, rinsed in 95% ethanol, and spun dry. Processed microarrays were prehybridized at room temperature for 1 h. Pre-hybridization solution was composed of 50% formamide, 5X SSC, 0.1% SDS, and 1% BSA. Hybridization buffer was composed of 50% formamide, 10X SSC, 0.2% SDS, and 0.26% salmon sperm. Labeled cDNA was denatured prior to hybridization by heating for 2 min at 80°C, while the cassette and microarray were pre-warmed at 42°C. The cDNA was then loaded onto the array, and arrays were hybridized for 16 h at 42°C in humidified chambers. Hybridized arrays were washed successively in 1X SSC, 0.03% SDS, 0.1X SSC, 0.01% SDS, and 0.1X SSC, and dried by brief centrifugation.

Replicate hybridizations were repeated within each biological experiment and dye-swapped to account for differences in dye labeling and detection efficiencies. Non-axenic experiments 1 and 2 each included six technical replicates, while the axenic experiment included four technical replicates.

Microarray image analysis and normalization

Dual-channel arrays were scanned at 595 nm (Cy3) and 685 nm (Cy5) on ArrayWoRx scanners (Applied Precision, Inc., Issaquah, WA, USA). The scanning system converts signal from fluors to "pixel" values, which allows the data to be saved as tiff files. DigitalGenome software (MolecularWare, Cambridge, MA, USA) was then used to integrate annotated chip information with the tiff files and to visualize, edit and export the data for normalization. A loess algorithm was applied to the spot mean intensity values across replicate arrays within each biological experiment to correct for systematic biases using S+ArrayAnalyzer software (Insightful Corp., Seattle, WA, USA) [75,76]. Quality control included analyzing final intensity ratios for the control set of data after normalization. The normalized intensity data for each control spike that corresponded to time zero (T0) and time final (TF) experimental mRNAs in the labeling reactions were analyzed using linear regression analysis to verify that the mean integrated intensity across the control spots was equal (slope ≈ 1). The slope of the linear regression of T0 to TF control intensity values averaged across arrays approached 1 for all three biological experiments: Non-axenic experiment 1 = 0.95, R^2 = 0.97; Non-axenic experiment 2 = 0.91, R^2 = 0.95; Axenic experiment = 0.91, R^2 = 0.98. Any negative values, outliers (defined as two standard deviations away from the mean for individual spots), and any spots that did not include data for at least three replicate arrays within each dataset, were removed from further analysis. These parameters resulted in the removal of data for <1.5% of the original 10,772 features printed on array (final datasets: Axenic experiment

1 = 10,723; Axenic experiment 2 = 10,614; Non-axenic experiment = 10,630).

Microarray statistics

Significance analysis of gene expression was performed using a t-test algorithm modified for multiple tests: Significance Analysis of Microarrays (SAM) [77] [http://www-stat.stanford.edu/~tibs/SAM/]. SAM reports those genes with statistically significant differences between treatments based on an overall false discovery rate (FDR). A score d(i) is assigned to each gene based on changes in gene expression relative to the standard deviation of repeated measurements. The FDR is an estimate of the percentage of genes identified by chance that would have an observed relative difference d(i) greater than the expected relative difference dE(i) set by an adjustable threshold, delta. An FDR of 1% estimates that for every 100 genes called significant, less than one would be identified incorrectly. The FDR may be adjusted by changing the delta and fold-change thresholds. While SAM does not report individual p-values, each gene is assigned its own "local FDR" (LFDR), which is the comparable statistical measurement to identify individual genes with changes in expression. LFDR can be used to review the data beyond the defined set of differentially expressed genes based on FDR. Hong *et al.* [78] demonstrated an LFDR of 10% as a reliable cut-off to successfully identify changes in expression for specific genes. While the FDR is considered the most reliable measure of the statistically accurate gene list within an experiment, the LFDR offers a second method for reviewing the statistical likelihood of changes in expression for a particular gene. In our study, we used the overall FDR to define the initial set of differentially expressed genes. We used the LFDR to confirm the overall change in gene expression for those transcripts that had multiple cDNAs printed on the microarray.

Initially, each dataset in our study was analyzed independently. Non-axenic experiments 1 and 2 were analyzed for statistical significance using a relatively stringent fold-change cut-off of 2.5 to target genes that were substantially up- or down-regulated during the transition to stationary phase, when toxin was produced. A delta value of 0.275 resulted in overall FDRs that were <1% in both of these experiments. Expression levels were consistently lower in the axenic experiment as compared to the non-axenic experiments. For example, the cDNAs with positive fold-change differences averaged 4.07 ± 0.97 in Non-axenic experiment 1, 3.85 ± 1.17 Non-axenic experiment 2, and 1.92 ± 0.54 in the Axenic growth experiment. Therefore, a lower fold-change cut-off of 1.5, and a delta value of 0.275, resulted in a comparable FDR that was <2.5% for the Axenic experiment. Only those transcripts that were determined to be significantly up- or down-regulated in all three biological experiments were further analyzed.

Two layers of replicates (replicate cDNAs on the array and replicate hybridizations) were accounted for by first analyzing the replicates spots as uncollapsed, independent data points using the normalization and statistical analysis described above. Then, replicate spots were averaged and collapsed, accordingly, depending on whether or not they fell into a greater contig. For singletons, both replicate spots were required to be statistically significant in all three biological experiments to be further considered. In this case, replicate spots were averaged for a final expression ratio and standard deviation. For cDNA features that fell into a larger contig, 90% of the cDNAs that fell within the contig were required to be significantly differentially expressed (based on initial SAM analysis or LFDR) to further consider the overall contig as up- or down-regulated. In this case, the replicates were collapsed by averaging the mean ratios of all cDNAs for a final fold-change ratio and standard deviation for the overall contig. The individual cDNAs within contigs served as additional replicates for these transcripts, and the results confirmed the consistent change in gene expression (see Additional file 1). The final fold-change values for those transcripts that were statistically higher or lower in stationary (toxin-producing) as compared to late-exponential (low-toxin-producing) growth phase are presented in Tables 3 and 4. Additional data files complying with MIAME format [79] were deposited at the GEO [74] data repository, accession number GSE46845.

Primer design and validation for RT-qPCR

Primers were designed manually to be within a length of 18–26 nucleotides with a GC content between 50-65%. These values resulted in high sequence specificity and melting temperatures (Tm) that worked well under our assay conditions using an annealing temperature of 60°C. JmjC forward, ATPase reverse and SLC6 Ex-Ex forward had lower GC contents than the original specifications, but still worked efficiently under these assay conditions. All primer sets were designed for PCR amplicons of 50–200 bp in length. Primers were synthesized by Integrated DNA Technologies, Inc. (Coralville, IA, USA) and purified by standard desalting. Efficiencies of amplification were initially determined for each primer set by running standard curves with 5-fold serial dilutions of *Ps-n* cDNA derived from stationary phase cultures, as well as genomic DNA. PCR conditions are described, below. Primer sequences and information can be found in Tables 2 and 5. Reported efficiencies in the tables correspond with the initial cDNA standard curve analyses. Standard curves were also run using 2-fold serial dilutions of pooled cDNA from the experimental samples combined in equal amounts.

The primer sets used in the Si-limitation experimental analyses showed efficiencies >95%, with R^2 values >0.99. Primers that span an exon-exon junction were designed for one reference and two of the target genes; these primer sets did not yield a product using gDNA as a template.

RNA isolation, cDNA synthesis and RT-qPCRs

Total RNA was extracted daily from each flask beginning on day three of growth. Cells were collected from 250 mL of culture by filtering onto a 5.0-µm, 47-mm membrane filter (MF-Millipore mixed cellulose ester). Filters were transferred to 50 mL conical tubes and 3 mL of TRIzol were added. Cells were washed quickly and gently from the filter and homogenized at full speed with a Polytron homogenizer (Kinematica, Inc, Bohemia, NY) for 90 s. Samples were incubated at room temperature for 5 min following homogenization and centrifuged at 3000 *g* for 10 min at 4°C in order to pellet cellular debris. 200 µL of chloroform was added for every 1 mL of homogenate and samples were incubated for 3 min with periodic shaking at room temperature. Samples were centrifuged at 12,000 *g* for 20 min at 4°C in order to separate the aqueous and organic phases. 75-80% of the aqueous phase was transferred to fresh tubes and an equal volume of 70% EtOH was added. The RNA-EtOH mixture was cleaned using RNeasy mini columns with on-column RNAase-free DNAase digestion (Qiagen). Clean RNA samples were eluted in 100 µL of DEPC-treated water, and stored at −80°C until analyzed. RNA concentrations were analyzed using a Nanodrop 2000 spectrophotometer (Wilmington, DE, USA), and RNA quality was verified by gel electrophoresis using Lonza 1.2% RNA cassettes (Walkersville, MD, USA). RNA samples were diluted to 20 ng μL^{-1}. 600 ng of total RNA from each sample was reverse transcribed using the iscript cDNA synthesis kit in 50 µL volume reactions, using both poly A and random hexamer primers (Bio-Rad Laboratories, Inc., Hercules, CA, USA). The reverse transcription reaction was carried out by incubating at 25°C for 5 min, followed by 42°C for 30 min. The enzyme was deactivated by heating to 85°C for 5 min and samples were held at 4°C until retrieved and stored at −20°C.

RT-qPCR reactions were set up as follows: 10 µL of SYBR Green PCR mix (Bio-Rad Laboratories, Inc.), 0.75 µL of cDNA, 0.2 µM of forward primer, 0.2 µM of reverse primer, and nuclease-free water to a final volume of 20 µL. An exception was that the β-tubulin Ex-Ex efficiency was within the 95-105% range using 0.4 µM for both forward and reverse primers. Each experimental cDNA was amplified in triplicate for each primer set using the following cycling parameters: 1) 95°C for 3 min; 2) 95°C for 10 s; 3) 60°C for 15 s; 4) 72°C for 30 s (plate read); 5) repeat 39 more cycles of steps 2–4; 6) 72°C for 10 min; 8) melting curve analysis from 65-95°C in 0.5°C increments every 5 s; and 9) hold at 4°C. Cq values were determined for each reaction at 150 relative fluorescent units.

Evaluation of DNA contamination in RT-qPCRs

qPCR reactions were run on 1 µL of each RNA from the Si-limitation growth experiment to test for amplification due to contaminating DNA. The absence of DNA contamination was further confirmed by using both standard and exon-exon spanning primer sets in parallel for one control and two target genes in the RT-qPCR reactions (Figures 3 and 4).

Analysis of candidate reference gene expression stability

Cq values were inputted into the GeNorm plus algorithm [20,80] and a stability value (M-value) was calculated for each gene (Figure 2A, 2B). Genes with the lowest M-values are considered the most stable; M-values <0.5 are ideal for use in normalization of qPCR data. The optimum number of reference genes to use for normalization is determined by calculating the geometric average of the two, three, four, and five most stable genes. The pairwise-variation between subsequent normalization factors is calculated and when the variation is below 0.15, the amount of change caused by the addition of the new control gene is considered negligible, and therefore unnecessary to include in subsequent normalization calculations.

Normalization, quantification, and statistics of RT-qPCR analysis

The arithmetic mean of triplicate technical replicates was calculated and used for subsequent calculations. The ΔCq values were calculated by the difference between each sample and the average Cq of the chosen reference point, which was the first time point (T3). Relative quantities (RQ) were calculated by exponentiation of the ΔCqs ($2^{\Delta Cq}$). RQ values of the target genes, for each sample, were divided by the geometric average of the chosen reference genes' RQ values, resulting in a normalized relative quantity (NRQ) [19]. The arithmetic mean and standard deviation of biological replicate NRQs were then calculated and plotted (Figures 3 and 4). NRQ values were log transformed (Log_2 NRQ) into Cq' values for statistical analysis [81]. Statistically significant changes in gene expression were determined using a general linear model analysis of variance (ANOVA) with Bonferroni *post-hoc* test. Statistical analyses were performed with 95% confidence intervals ($p < 0.05$) using Minitab16 statistical software (State College, PA, USA). Additional files 6 and 7 contain the complete set of RT-qPCR data and statistical results. RT-qPCR data analyses and reporting were in accordance with MIQE guidelines [82,83].

Additional files

Additional file 1: Fold-change data and statistics for cDNA replicates on the *Ps-n* microarray for each of the transcripts discussed in this paper.

Additional file 2: Excel file with the annotation and array data for the entire set of cDNA clones printed on the *Ps-n* microarray.

Additional file 3: Fasta file with all of the *Ps-n* ESTs from this study.

Additional file 4: Fasta file with all of the assembled sequences from this study.

Additional file 5: Contig alignments in .ace format. A number of the alignments (46A5, 53B6, 73D12, 135H6, 165G9, 177F1, PSN0011, PSN0014, PSN0016, PSN0019, PSN0032, PSN0042, PSN0060, PSN0072, PSN0080, PSN0100, PSN0332, PSN0547, PSN0918, and PSN1327) include sequences from the JGI *Pseudo-nitzschia* genome project [22] for comparison; these sequences are designated within the contig by the JGI modeled gene name or genome location. Note that contigs corresponding with three transcripts discussed in the manuscript, PSN0014, PSN0016, and PSN0052, showed splice variants. Expression of the individual cDNAs within these contigs were not significantly different in the microarray analysis (see Additional file 1).

Additional file 6: RT-qPCR data.

Additional file 7: RT-qPCR statistical results.

Abbreviations

ASP: Amnesic shellfish poisoning; cDNA: Complementary DNA; DA: Domoic acid; EF-1α: Elongation factor 1-alpha, eIF-2, Elongation initiation factor 2; ESTs: Expressed sequence tags; FCP: Fucoxanthin-chlorophyll a-c binding protein; FDR: False discovery rate; FMOC: Fluorenylmethoxycarbonyl; GABA: γ-aminobutyric; GAPDH: Glyceraldehyde-3-phosphate dehydrogenase; GDH: Glutamate dehydrogenase; HPLC: High performance liquid chromatography; LFDR: Local false discovery rate; MIAME: Minimum information about a microarray experiment; PCR: Polymerase chain reaction; PEPCK: Phosphoenolpyruvate carboxykinase; PGK: Phosphoglycerate kinase; PFK: Phosphofructokinase; *Ps-n*: *Pseudo-nitzschia multiseries*; RT-qPCR: Reverse-transcription quantitative PCR; SAM: Significance analysis of microarrays; SLC6: Solute carrier family 6.

Competing interests

The authors declare that they have no competing interests.

Authors' contributions

KRB contributed to the design, completion, and analysis of the RT-qPCR and microarray studies, and wrote the manuscript. BMH contributed to the design, completion, and analysis of the RT-qPCR experiments. KRB and JP constructed the cDNA library and sequencing was completed in JP's lab at McGill University (or ACGT, Inc., for a small subset of samples). KRB and SM completed the microarray construction and data analysis. DLR participated in data review and in revising the manuscript. SSB participated in the design of the array studies, data review, and in drafting the manuscript. HPLC assays were completed in SSB's lab in Moncton, NB (Fisheries and Oceans Canada). DAH contributed to the design of the RT-qPCR studies, ELISA domoic acid analyses, data review and in revising the manuscript. DEH contributed to the design of the RT-qPCR and microarray studies, data review and analyses, and drafting the manuscript. Microarray studies were completed in DEH's lab at MIT; RT-qPCR studies were completed in KRB's lab at PSU. All authors read and approved the final manuscript.

Acknowledgements

We thank Claude Léger for running HPLC domoic acid analyses for the microarray studies; we also thank Michael Quilliam for running HPLC and mass spectrometry to confirm validity of the ELISA DA analysis in the RT-qPCR studies. We thank Nhi Nguyen and Isabelle Harvey for assistance in library construction and sequencing, and Xuan Shirley Li for assistance during microarray construction. We thank Charlie Whittaker and Sebastian Hoersch for assistance with bioinformatics analysis. We thank Vittoria Roncalli for providing an introduction to the Blast2Go application. We thank Petra Lenz and Brad Jones for generously providing space and computer support for KRB to work on data analysis during Jan 2012. We thank Christopher Wilk and Megan Cooper for completing cell counts in the strain GGB1 growth experiment. We thank Katherine Lozano and Lauren Oakes for running ELISA assays. We thank Nichole DiLuzio for technical lab assistance. We thank Micaela Parker (University of Washington) and the JGI *Pseudo-nitzschia* genome team for access to the genome data before public release, and permission to include JGI genome project sequences in our contig alignments for comparison. We also thank Michael Carlson and Kyle Frishkorn (University of Washington) for providing the *Ps-n* GGB1 culture. We thank Thomas Boucher for consultation on statistical analysis of RT-qPCR data, including a tutorial of the Minitab 16 software. We thank Brian Howes, Sara Sampieri Horvet, and Jennifer Benson at the Coastal Systems Program, University of Massachusetts, Dartmouth, for running the silicate, nitrite/nitrate, and phosphate analyses. We thank Don Anderson, Mark Hahn, and Senjie Lin for their advice on the microarray studies. We thank Jefferson Turner and Junne Kamihara for helpful discussions throughout the course of this work and in preparation of the manuscript. And, we thank Ron Taylor, Steve Fiering, Chuck Wise, and George Tuthill for all of their support and encouragement of this project. Financial support for this research was provided by the Woods Hole Oceanographic Institution Academic Programs Office, the Plymouth State University Graduate Programs Office, and the New Hampshire IDeA Network of Biological Research Excellence (NH-INBRE), with grants from the National Center for Research Resources (5P20RR030360-03) and the National Institute of General Medical Sciences (8P20GM103506-03), National Institutes of Health.

Author details

[1]Department of Biological Sciences, Plymouth State University, MSC 64, 17 High St., Plymouth, NH 03264, USA. [2]Koch Institute, Massachusetts Institute of Technology, 76-553, 77 Massachusetts Avenue, Cambridge, MA 02139, USA. [3]Present address: Mascoma Corporation, 67 Etna Road Suite 300, Lebanon, NH 03766, USA. [4]Fisheries and Oceans Canada, Gulf Fisheries Centre, P.O. Box 5030, Moncton, New Brunswick E1C 9B6, Canada. [5]Biology Department, Clark University, 950 Main Street, Worcester, MA 01610, USA. [6]Present address: Vertex Pharmaceuticals, 130 Waverly Street, Cambridge, MA 02139, USA. [7]Department of Biochemistry, McGill University, 3655 Promenade Sir William Osler, Montreal, Quebec H3G 1Y6, Canada. [8]Department of Microbiology and Immunology, Vail Building Room 208, Dartmouth Medical School, Hanover, NH 03755, USA.

References

1. Bates SS, Bird CJ, de Freitas ASW, Foxall R, Gilgan M, Hanic LA, Johnson GR, McCulloch AW, Odense P, Pocklington R, Quilliam MA, Sim PG, Smith JC, Subba Rao DV, Todd ECD, Walter JA, Wright JLC: **Pennate diatom *Nitzschia pungens* as the primary source of domoic acid, a toxin in shellfish from eastern Prince Edward Island, Canada.** *Can J Fish Aquat Sci* 1989, **46:**1203–1215.
2. Douglas DJ, Bates SS: **Production of domoic acid, a neurotoxic amino acid, by an axenic culture of the marine diatom *Nitzschia pungens* f. *multiseries* Hasle.** *Can J Fish Aquat Sci* 1992, **49:**85–90.
3. Douglas DJ, Ramsey UP, Walter JA, Wright JLC: **Biosynthesis of the neurotoxin domoic acid by the marine diatom *Nitzschia pungens* forma**

multiseries, determined with [13C]-labelled precursors and nuclear magnetic resonance. *J Chem Soc Chem Commun* 1992, **1992**:714–7156.

4. Wright JLC, Boyd RK, de Freitas ASW, Falk M, Foxall RA, Jamieson WD, Laycock MV, McCulloch AW, McInnes AG, Odense P, Pathak VP, Quilliam MA, Ragan MA, Sim PG, Thibault P, Walter JA, Gilgan M, Richard DJA, Dewar D: **Identification of domoic acid, a neuroexcitatory amino acid, in toxic mussels from eastern Prince Edward Island.** *Can J Chem* 1989, **67**:714–716.
5. Takemoto T, Daigo K: **Constituents of *Chondria armata*.** *Chem Pharm Bull* 1958, **6**:578–580.
6. Ramsdell JS: **The molecular and integrative basis to domoic acid toxicity.** In *Phycotoxins: Chemistry and Biochemistry.* Edited by Botana L. Cambridge, MA: Blackwell Publishing Professional; 2007:223–250.
7. Lelong A, Hégaret H, Soudant P, Bates SS: ***Pseudo-nitzschia* (Bacillariophyceae) species, domoic acid and amnesic shellfish poisoning: revisiting previous paradigms.** *Phycologia* 2012, **51**:168–216.
8. Trainer VL, Bates SS, Lundholm N, Thessen AE, Cochlan WP, Adams NG, Trick CG: ***Pseudo-nitzschia* physiological ecology, phylogeny, toxicity, monitoring and impacts on ecosystem health.** *Harmful Algae* 2012, **14**:271–300.
9. Bates SS: **Ecophysiology and metabolism of ASP toxin production.** In *Physiological Ecology of Harmful Algal Blooms.* Edited by Anderson DM, Cembella AD, Hallegraeff GM. Heidelberg: Springer-Verlag; 1998:405–426.
10. Bates SS, Garrison DL, Horner RA: **Bloom dynamics and physiology of domoic-acid-producing *Pseudo-nitzschia* species.** In *Physiological Ecology of Harmful Algal Blooms.* Edited by Anderson DM, Cembella AD, Hallegraeff GM, Anderson DM, Cembella AD, Hallegraeff GM. Heidelberg: Springer-Verlag; 1998:267–292.
11. Bates SS, Trainer VL: **The ecology of harmful diatoms.** In *Ecology of Harmful Algae Ecological Studies. Volume 189.* Edited by Granéli E, Turner J. Heidleberg: Springer-Verlag; 2006:81–93.
12. Ramsey UP, Douglas DJ, Walter JA, Wright JLC: **Biosynthesis of domoic acid by the diatom *Pseudo-nitzschia multiseries*.** *Nat Toxins* 1998, **6**:137–146.
13. Savage TJ, Smith GJ, Clark AT, Saucedo PN: **Condensation of the isoprenoid and amino precursors in the biosynthesis of domoic acid.** *Toxicon* 2012, **59**:25–33.
14. Pan Y, Bates SS, Cembella AD: **Environmental stress and domoic acid production by *Pseudo-nitzschia*: a physiological perspective.** *Nat Toxins* 1998, **6**:127–135.
15. Bates SS, Douglas DJ, Doucette GJ, Léger C: **Enhancement of domoic acid production by reintroducing bacteria to axenic cultures of the diatom *Pseudo-nitzschia multiseries*.** *Nat Toxins* 1995, **3**:428–435.
16. Osada M, Stewart JE: **Gluconic acid/gluconolactone: physiological influences on domoic acid production by bacteria associated with *Pseudo-nitzschia multiseries*.** *Aquat Microb Ecol* 1997, **12**:203–209.
17. Stewart JE: **Bacterial involvement in determining domoic acid levels apparent in *Pseudo-nitzschia multiseries* cultures.** *Aquat Microb Ecol* 2008, **50**:135–144.
18. Kotaki Y, Koike K, Yoshida M, Thuoc CV, Huyen NTM, Hoi NC, Fukuyo Y, Kodama M: **Domoic acid production in *Nitzschia* sp. isolated from a shrimp-culture pond in Do Son, Vietnam.** *J Phycol* 2000, **36**:1057–1060.
19. Hellemans J, Mortier G, De Paepe A, Speleman F, Vandesompele J: **qBase relative quantification framework and software for management and automated analysis of real-time quantitative PCR data.** *Genome Biol* 2007, **8**:R19.
20. Vandesompele J, De Preter K, Pattyn F, Poppe B, Van Roy N, De Paepe A, Speleman F: **Accurate normalization of real-time quantitative RT-PCR data by geometric averaging of multiple internal control genes.** *Genome Biol* 2002, **3**:1–12.
21. Gunjan A, Paik J, Verreault A: **Regulation of histone synthesis and nucleosome assembly.** *Biochimie* 2005, **87**:625–635.
22. US Department of Energy Joint Genome Institute, Armbrust EV, Parker MS, Rocap G, Jenkins B, Bates SS: ***Pseudo-nitzschia multiseries* CLN-47 genome sequence, Assembly v1.** 2011. [http://genome.jgi.doe.gov/Psemu1/Psemu1.home.html]
23. Cui B, Liu Y, Gorovsky M: **Deposition and function of histone H3 variants in *Tetrahymena thermophila*.** *Mol Cell Biol* 2006, **26**:7719–7730.
24. Elsaesser S, Goldberg A, Allis C: **New functions for an old variant: no substitute for histone H3.3.** *Curr Opin Genet Dev* 2010, **20**:110–117.
25. Jin J, Cai Y, Li B, Conaway R, Workman J, Conaway J, Kusch T: **In and out: histone variant exchange in chromatin.** *Trends Biochem Sci* 2005, **30**:680–687.
26. Yu L, Gorovsky M: **Constitutive expression, not a particular primary sequence, is the important feature of the H3 replacement variant hv2 in *Tetrahymena thermophila*.** *Mol Cell Biol* 1997, **17**:6303–6310.
27. Anju V, Kapros T, Waterborg J: **Identification of a replication-independent replacement histone H3 in the basidiomycete *Ustilago maydis*.** *J Biol Chem* 2011, **286**:25790–25800.
28. Siaut M, Heijde M, Mangogna M, Montsant A, Coesel S, Allen A, Manfredonia A, Falciatore A, Bowler C: **Molecular toolbox for studying diatom biology in *Phaeodactylum tricornutum*.** *Gene* 2007, **406**:23–35.
29. Pan Y, Subba Rao DV, Mann KH, Brown RG, Pocklington R: **Effects of silicate limitation on production of domoic acid, a neurotoxin, by the diatom *Pseudo-nitzschia multiseries* (Hasle). I. Batch culture studies.** *Mar Ecol Prog Ser* 1996, **131**:225–233.
30. Valenzuela J, Mazurie A, Carlson RP, Gerlach R, Cooksey KE, Peyton BM, Fields MW: **Potential role of multiple carbon fixation pathways during lipid accumulation in *Phaeodactylum tricornutum*.** *Biotechnol Biofuels* 2012, **5**:40.
31. Corstjens PLAM, González EL: **Effects of nitrogen and phosphorus availability on the expression of the coccolith-vesicle V-ATPase (subunit c) of *Pleurochrysis* (Haptophyta).** *J Phycol* 2004, **40**:82–87.
32. Dyhrman ST, Haley ST, Birkeland SR, Wurch LL, Cipriano MJ, McArthur AG: **Long serial analysis of gene expression for gene discovery and transcriptome profiling in the widespread marine coccolithophore *Emiliania huxleyi*.** *Appl Environ Microbiol* 2006, **72**:252–260.
33. Reinbothe S, Reinbothe C: **The regulation of enzymes involved in chlorophyll biosynthesis.** *Eur J Biochem* 1996, **237**:323–343.
34. Kajander T, Merckel M, Thompson A, Deacon A, Mazur P, Kozarich J, Goldman A: **The structure of *Neurospora crassa* 3-carboxy-cis, cis-muconate lactonizing enzyme, a beta propeller cycloisomerase.** *Structure* 2002, **10**:483–492.
35. Mazur P, Henzel W, Mattoo S, Kozarich J: **3-Carboxy-cis, cis-muconate lactonizing enzyme from *Neurospora crassa:* an alternate cycloisomerase motif.** *J Bacteriol* 1994, **176**:18–28.
36. Mazur P, Pieken W, Budihas S, Williams S, Wong S, Kozarich J: **Cis, cis-muconate lactonizing enzyme from *Trichosporon cutaneum:* evidence for a novel class of cycloisomerases in eucaryotes.** *Biochemistry* 1994, **33**:1961–1970.
37. Thomason L, Court D, Datta A, Khanna R, Rosner J: **Identification of the *Escherichia coli* K-12 *ybh*E gene as *pgl*, encoding 6-phosphogluconolactonase.** *J Bacteriol* 2004, **186**:8248–8253.
38. Zimenkov D, Gulevich A, Skorokhodova A, Biriukova I, Kozlov Y, Mashko S: ***Escherichia coli* ORF *ybh*E is *pgl* gene encoding 6-phosphogluconolactonase (EC 3.1.1.31) that has no homology with known 6PGLs from other organisms.** *FEMS Microbiol Lett* 2005, **224**:275–280.
39. Henningsen B: **Bioinformatic and Gene Expression Analysis of an SLC6 Homolog in the Toxin-Producing Marine Diatom *Pseudo-nitzschia multiseries*. MS thesis.** Plymouth State University, Biological Sciences; 2012.
40. Beuve N, Rispail N, Laine P, Cliquet J-B, Ourry A, Le Deunff E: **Putative role of gamma-aminobutyric acid (GABA) as a long-distance signal in up-regulation of nitrate uptake in *Brassica napus* L.** *Plant Cell Environ* 2004, **27**:1035–1046.
41. Chevrot R, Rosen R, Haudecoeur E, Cirou A, Shelp B, Ron E, Faure D: **GABA controls the level of quorum-sensing signal in *Agrobacterium tumefaciens*.** *Proc Natl Acad Sci USA* 2006, **103**:7460–7464.
42. Palanivelu R, Brass L, Edlund AF, Preuss D: **Pollen tube growth and guidance is regulated by POP2, an *Arabidopsis* gene that controls GABA levels.** *Cell* 2003, **114**:47–59.
43. Johnston M, Gallacher S, Smith EA, Glover LA: **Detection of N-acyl homoserine lactones in marine bacteria associated with production and biotransformation of sodium channel blocking toxins and the microflora of toxin-producing phytoplankton.** In *Harmful Algal Blooms.* Edited by Hallegraeff GM, Blackburn SI, Bolch CJ, Lewis RJ. Paris: Intergovernmental Oceanographic Commission of UNESCO; 2001:375–378.
44. Amin S, Parker M, Armbrust E: **Interactions between diatoms and bacteria.** *Microbiol Mol Biol R* 2012, **76**:667–684.
45. Maldonado MT, Hughes MP, Rue EL, Wells ML: **The effect of Fe and Cu on growth and domoic acid production by *Pseudo-nitzschia multiseries* and *Pseudo-nitzschia australis*.** *Limnol Oceanogr* 2002, **47**:515–526.
46. Rue E, Bruland K: **Domoic acid binds iron and copper: a possible role for the toxin produced by the marine diatom *Pseudo-nitzschia*.** *Mar Chem* 2001, **76**:127–134.
47. Wells ML, Trick CG, Cochlan WP, Hughes MP, Trainer VL: **Domoic acid: the synergy of iron, copper, and the toxicity of diatoms.** *Limnol Oceanogr* 2005, **50**:1908–1917.

48. Carnal N, Black C: **Phosphofructokinase activities in photosynthetic organisms: the occurrence of pyrophosphate-dependent 6-phosphofructokinase in plants and algae.** *Plant Physiol* 1983, **71**:150–155.
49. Lea PJ, Chen ZH, Leegood RC, Walker RP: **Does phosphoenolpyruvate carboxykinase have a role in both amino acid and carbohydrate metabolism?** *Amino Acids* 2001, **20**:225–241.
50. Lehmann T, Ratajczak L: **The pivotal role of glutamate dehydrogenase (GDH) in the mobilization of N and C from storage material to asparagine in germinating seeds of yellow lupine.** *J Plant Physiol* 2008, **165**:149–158.
51. Qui X, Wie W, Lian X, Zhang Q: **Molecular analyses of the rice glutamate dehydronase gene family and their response to nitrogen and phosphorous deprivation.** *Plant Cell Rep* 2009, **28**:1115–1126.
52. Parrish CC, de Freitas ASW, Bodennec G, MacPherson EJ, Ackman RG: **Lipid composition of the toxic marine diatom, *Nitzschia pungens*.** *Phytochemistry* 1991, **30**:113–116.
53. Black P, DiRusso C: **Yeast acyl-CoA synthetases at the crossroads of fatty acid metabolism and regulation.** *Biochim Biophys Acta* 2007, **1771**:286–298.
54. Shockey J, Browse J: **Genome-level and biochemical diversity of the acyl-activating enzyme superfamily in plants.** *Plant J* 2011, **66**:143–160.
55. Guillard RRL, Ryther JH: **Studies of marine planktonic diatoms. I. *Cyclotella nana* Hustedt, and *Detonula confervacea* (Cleve) Gran.** *Can J Microbiol* 1962, **8**:229–239.
56. Pocklington R, Milley JE, Bates SS, Bird CJ, de Freitas ASW, Quilliam MA: **Trace determination of domoic acid in seawater and phytoplankton by high-performance liquid chromatography of the fluorenylmethoxycarbonyl (FMOC) derivative.** *Internat J Environ Anal Chem* 1990, **38**:351–368.
57. Garthwaite I, Ross K, Miles C, Hansen R, Foster D, Wilkins A, Towers N: **Polyclonal antibodies to domoic acid, and their use in immunoassays for domoic acid in sea water and shellfish.** *Nat Toxins* 1998, **6**:93–104.
58. Clesceri LS, Greenberg AE, Trussell RR (Eds): *Standard Methods for the Examination of Water and Wastewater.* 17th edition. Washington, D.C: American Public Health Association; 1989.
59. Mullin JB, Riley JP: **The colorimetric determination of silicate with special reference to sea and natural waters.** *Anal Chim Acta* 1955, **12**:162.
60. Strickland JDH, Parsons TR: **A Manual of Seawater Analysis.** In *Fisheries Research Board of Canada.* ; 1965.
61. Eaton AD, Clesceri LS, Greenberg AE (Ed): *Standard Methods for the Examination of Water and Wastewater.* 19th edition. Washington, D.C: American Public Health Association; 1995.
62. Murphy J, Riley JP: **A modified single solution method for determination of phosphate in natural waters.** *Anal Chim Acta* 1962, **27**:31–36.
63. Bendschneider K, Robinson R: **A new spectrophotometic method for the determination of nitrite in seawater.** *J Mar Res* 1952, **11**:87–96.
64. Wood E, Armstrong F, Richards F: **Determination of nitrate in sea water by cadmium copper reduction to nitrite.** *J Mar Biol Assoc UK* 1967, **47**:23–31.
65. Das M, Harvey H, Chu LL, Sinha M, Pelletier J: **Full-length cDNAs: more than just reaching the ends.** *Physiol Genomics* 2001, **6**:57–80.
66. Altschul SF, Madden TL, Schaffer AA, Zhang J, Zhang Z, Miller W, Lipman DJ: **Gapped BLAST and PSI-BLAST: a new generation of protein database search programs.** *Nucleic Acids Res* 1997, **25**:3389–3402.
67. Altschul SF, Gish W, Miller W, Myers EW, Lipman DJ: **Basic local alignment search tool.** *J Mol Biol* 1990, **215**:403–410.
68. Conesa A, Götz S, García-Gómez J, Terol J, Talón M, Robles M: **Blast2GO: a universal tool for annotation, visualization and analysis in functional genomics research.** *Bioinformatics (Oxford, England)* 2005, **21**:3674–3676.
69. *The Open Reading Frame Finder.* http://www.ncbi.nlm.nih.gov/gorf/gorf.html.
70. Emanuelsson O, Brunak S, Von Heijne G, Nielsen H: **Locating proteins in the cell using TargetP, SignalP and related tools.** *Nat Protoc* 2007, **2**:953–971.
71. Apt KE, Zaslavkaia L, Lippmeier JC, Lang LC, Kilian O, Wetherbee R, Grossman AR, Kroth PG: ***In vivo* characterization of diatom multipartite plastid targeting signals.** *J Cell Sci* 2002, **115**:4061–4069.
72. Gruber A, Vugrinec S, Hempel F, Gould SB, Maier U-G, Kroth PG: **Protein targeting into complex diatom plastids: functional characterisation of a specific targeting motif.** *Plant Mol Biol* 2007, **64**:519–530.
73. Kilian O, Kroth PG: **Identification and characterization of a new conserved motif within the presequence of proteins targeted into complex diatom plastids.** *Plant J* 2005, **41**:175–183.
74. Edgar R, Barrett T: **NCBI GEO standards and services for microarray data.** *Nat Biotechnol* 2006, **24**:1471–1472.
75. Park T, Yi SG, Kang SH, Lee S, Lee YS, Simon R: **Evaluation of normalization methods for microarray data.** *BMC Bioinformatics* 2003, **4**:33.
76. Quackenbush J: **Microarray data normalization and transformation.** *Nat Genet* 2002, **32**(Suppl):496–501.
77. Tusher VG, Tibshirani R, Chu G: **Significance analysis of microarrays applied to the ionizing radiation response.** *Proc Natl Acad Sci USA* 2001, **98**:5116–5121.
78. Hong W, Tibshirani R, Chu G: **Local false discovery rate facilitates comparison of different microarray experiments.** *Nucleic Acids Res* 2009, **37**:7483–7497.
79. Brazma A, Hingamp P, Quackenbush J, Sherlock G, Spellman P, Stoeckert C, Aach J, Ansorge W, CA B, Causton H, *et al*: **Minimum information about a microarray experiment (MIAME)-toward standards for microarray data.** *Nat Genet* 2001, **4**:365–371.
80. Vandesompele J, Kubista M, Pfaffl MW: **Reference gene validation software for improved normalization.** In *Real-time PCR: Current Technology and Application.* Edited by Logan J, Edwards K, Saunders N. Norfolk: Caister Academic Press; 2009:47–64.
81. Rieu I, Powers SJ: **Real-time quantitative RT-PCR: design, calculations, and statistics.** *Plant Cell* 2009, **21**:1031–1033.
82. Bustin S, Beaulieu J-F, Huggett J, Jaggi R, Kibenge F, Olsvik P, Penning L, Toegel S: **MIQE precis: Practical implementation of minimum standard guidelines for fluorescence-based quantitative real-time PCR experiments.** *BMC Molec Biol* 2010, **11**:74.
83. Bustin SA, Benes V, Garson JA, Hellemans J, Huggett J, Kubista M, Mueller R, Nolan T, Pfaffl MW, Shipley GL, Vandesompele J, Wittwer CT: **The MIQE guidelines: minimum information for publication of quantitative real-time PCR experiments.** *Clin Chem* 2009, **55**:611–622.

11

APOBEC3 inhibits DEAD-END function to regulate microRNA activity

Sara Ali[1], Namrata Karki[3], Chitralekha Bhattacharya[1], Rui Zhu[1], Donna A MacDuff[2], Mark D Stenglein[2], April J Schumacher[2], Zachary L Demorest[2], Reuben S Harris[2], Angabin Matin[1*] and Sita Aggarwal[3*]

Abstract

The RNA binding protein DEAD-END (DND1) is one of the few proteins known to regulate microRNA (miRNA) activity at the level of miRNA-mRNA interaction. DND1 blocks miRNA interaction with the 3'-untranslated region (3'-UTR) of specific mRNAs and restores protein expression. Previously, we showed that the DNA cytosine deaminase, APOBEC3 (apolipoprotein B mRNA-editing enzyme, catalytic polypeptide like 3), interacts with DND1. APOBEC3 has been primarily studied for its role in restricting and inactivating retroviruses and retroelements. In this report, we examine the significance of DND1-APOBEC3 interaction. We found that while human DND1 inhibits miRNA-mediated inhibition of *P27*, human APOBEC3G is able to counteract this repression and restore miRNA activity. APOBEC3G, by itself, does not affect the 3'-UTR of *P27*. We found that APOBEC3G also blocks DND1 function to restore miR-372 and miR-206 inhibition through the 3'-UTRs of *LATS2* and *CX43*, respectively. In corollary experiments, we tested whether DND1 affects the viral restriction function or mutator activity of APOBEC3. We found that DND1 does not affect APOBEC3 inhibition of infectivity of exogenous retrovirus HIV (ΔVif) or retrotransposition of MusD. In addition, examination of *Ter/Ter;Apobec3−/−* mice, lead us to conclude that DND1 does not regulate the mutator activity of APOBEC3 in germ cells. In summary, our results show that APOBEC3 is able to modulate DND1 function to regulate miRNA mediated translational regulation in cells but DND1 does not affect known APOBEC3 function.

Keywords: DND1, APOBEC3G, APOBEC3, microRNA, P27

Background

The RNA binding protein DEAD-END (DND1) is essential for germ cell viability [1,2]. When *Dnd1* is functionally inactivated, as in the *Ter* mutant mouse strain, this results in death of germ cells, sterility [2], and in some cases development of testicular germ cell tumors [2,3].

DND1 encodes canonical RNA recognition motifs [1,4] through which it interacts with the 3′-UTRs of mRNAs. For example, DND1 inhibits miR-221 function from the 3′-UTR of *P27* resulting in increased P27 protein expression [4,5]. Two U-rich DND1 binding sites have been mapped adjacent to two miR-221 binding sites in the 3′-UTR of *P27* [4]. DND1 has also been shown to inhibit miR-372 from the 3′-UTRs of *LATS2* (serine/threonine-protein kinase, large tumor suppressor, homolog 2) and inhibit miR-1 and miR-206 from the 3′-UTRs of *CX43* (connexin-43) [4]. However, DND1 binding sites have not been mapped within the 3′-UTRs of *LATS2* or *CX43*.

miRNA association with mRNA usually results in translation inhibition or degradation of mRNA. It is thought that DND1 binds to mRNA and prevents miRNAs and miRISC (miRNA-induced silencing complexes) from binding. miRISCs are composed of ribonucleoproteins that assemble with the miRNA and mediate either translational repression or degradation of mRNA [6-8]. Alternately, DND1 may bind and sequester mRNAs away from miRNA access.

Although DND1 was initially identified for its role in germ cells and germ cell tumors, emerging evidence indicates a wider role for DND1 in mammalian tissues, especially in cancers. For example, over-expression of DND1 is detected in some histological sub-types of human testicular cancers, leukemia, lung and ovarian cancers (ONCOMINE and NCBI Geo Profiles). A recent study detected DND1 in human tongue squamous cell

* Correspondence: amatin@mdanderson.org; sita.aggarwal@pbrc.edu
[1]Department of Genetics, University of Texas, MD Anderson Cancer Center, 1515 Holcombe Blvd, Houston, TX 77030, USA
[3]Pennington Biomedical Research Center, 6400 Perkins Road, Baton Rouge, LA 70808, USA
Full list of author information is available at the end of the article

carcinoma (TSCC) and found that miR-24 directly targets *DND1* mRNA [9]. Up-regulation of miR-24 decreased DND1 expression resulting in lower P27 levels and increased proliferation and reduced apoptosis in TSCC cells. Another study showed that *Ras* transformed keratinocytes down regulate DND1 which results in increased miR-21 mediated inhibition of MSH2 [10].

Work in our laboratory and others show that DND1 interacts with a broad range of mRNA targets [11,12]. The targets include transcripts encoding cell cycle regulators (*P27, TP53, LATS2*), pluripotency factors (*OCT4, SOX2, NANOG*) and pro- and anti-apoptotic factors (*BAX* and *BCLX*). Expression of these genes is required at specific developmental stages in germ cells such as during active proliferation or quiescence. Because DND1 interacts with a range of mRNAs, this raises the question as to the factors which might serve to modulate DND1 interaction with physiologically appropriate targets. Therefore one important goal is to determine how DND1 function is regulated in cells.

In a related study, we found that DND1 interacts with APOBEC3 [13]. We showed that mouse DND1 immunoprecipitated with mouse APOBEC3 in mammalian cells, including in germ cells. In addition, fluorescent tagged DND1 and APOBEC3 co-localized at peri-nuclear regions in mammalian cells.

One well-studied function of mouse Apobec3 and its human counterpart, APOBEC3G (apolipoprotein B mRNA-editing enzyme, catalytic polypeptide-like 3G, A3G) is contributing to innate immunity through retrovirus and retrotransposon restriction [14-16]. Restriction occurs through a well-established cDNA cytosine deamination mechanism and by a less well-characterized deamination-independent mechanism [17-20]. Human APOBEC3G and mouse APOBEC3, each possess two zinc-binding motifs [21,22]. The active domain is responsible for deaminase activity and the pseudo-active domain contributes most of the RNA/ssDNA binding affinity [23-27]. Although both human and mouse proteins have such a division of labor, the domain organization is opposite with N-terminal of mouse APOBEC3 and the C-terminal domain of human APOBEC3G active for deamination [28-30].

Other studies indicate that overexpression of specific members of the human APOBEC3 family (such as the single cytidine deaminase domain containing, APOBEC3A) can hypermutate the cellular genome or mitochondrial DNA [31,32]. Thus APOBEC3 family members are potentially powerful mutators [33] and very likely cells possess mechanisms to keep the latent deleterious activity of APOBEC3 in check. One way that cells protect their genomes from APOBEC3 is that mouse APOBEC3 and most human APOBEC3 proteins are localized to the cytoplasm [14,29,34].

In this report, we examined the significance of APOBEC3 interaction with DND1. Our results show that APOBEC3 can oppose DND1 function to restore miRNA-mediated inhibition of translation. We therefore propose that interaction of APOBEC3 with DND1 may be one way in which DND1 activity is regulated in cells.

Methods

Transient transfections

Human DND1 with HA tag in C-terminus was cloned into pCDNA3.1 nV5-DEST (Invitrogen) expression vector (DND1-HA). Human APOBEC3G with myc tag in C-terminus was cloned into pcDNA3.1(+)(APOBEC3G-myc). Transient transfections were performed using the 293 T cell line as this cell line has previously been used for testing DND1 function [4] and the results using 293 T were similar to that using other cell lines such as MCF-7 and Tera1. On advantage is that 293 T cells take up transfected DNA efficiently to give reproducible results. 293 T cells were cultured in DMEM supplemented with 10% fetal bovine serum in 5% CO_2 at 37°C. The cells were transiently transfected using SuperFect transfection reagent (QIAGEN) with 1 ng pGL3-P27-3′UTR [4] together with constructs encoding miRVec-221 (50 ng), DND1-HA (10 ng) and/or APOBEC3G-myc (1 ng to 25 ng range). *LacZ* expression constructs (4 ng) were co-transfected into all cells. Equivalent amounts of DNA were introduced into all cells with pGEM DNA being used to equalize for DNA levels used for transfections. After 48 h the cells were washed and treated with cell culture lysis buffer (Promega). 5 uL of the lysates were used for luciferin assays. All transfection experiments were performed in triplicates. Results shown are the mean and standard error from three independent experiments. Similar transfections also tested the effect of DND1, APOBEC3G and miR-372 (mirVec-372) on pGL3 3′UTR LATS2, and miR-206 (miR vec-206) on pGL3 Cx43 3′UTR and pGL3-control vector. Mutant P27 vectors used were pGL3 3′UTR min mut1 (m1 or mut1, in which both DND1 binding sites are mutated) and pGL3-p27mut-3′-UTR (m3; in which both miRNA binding sites are mutated) [4].

Statistical analysis

Data are expressed as mean ± standard deviation/or standard error. Statistical analyses were performed using GraphPad Prism (software version 5.0. VA). Differences were determined by Student's t test. A P value of < 0.05 was considered significant.

Luciferase assays

The assays were performed using Luciferase assay kit (Pomega) according to manufacturer's directions. β-galactosidase assay results were used to normalize the

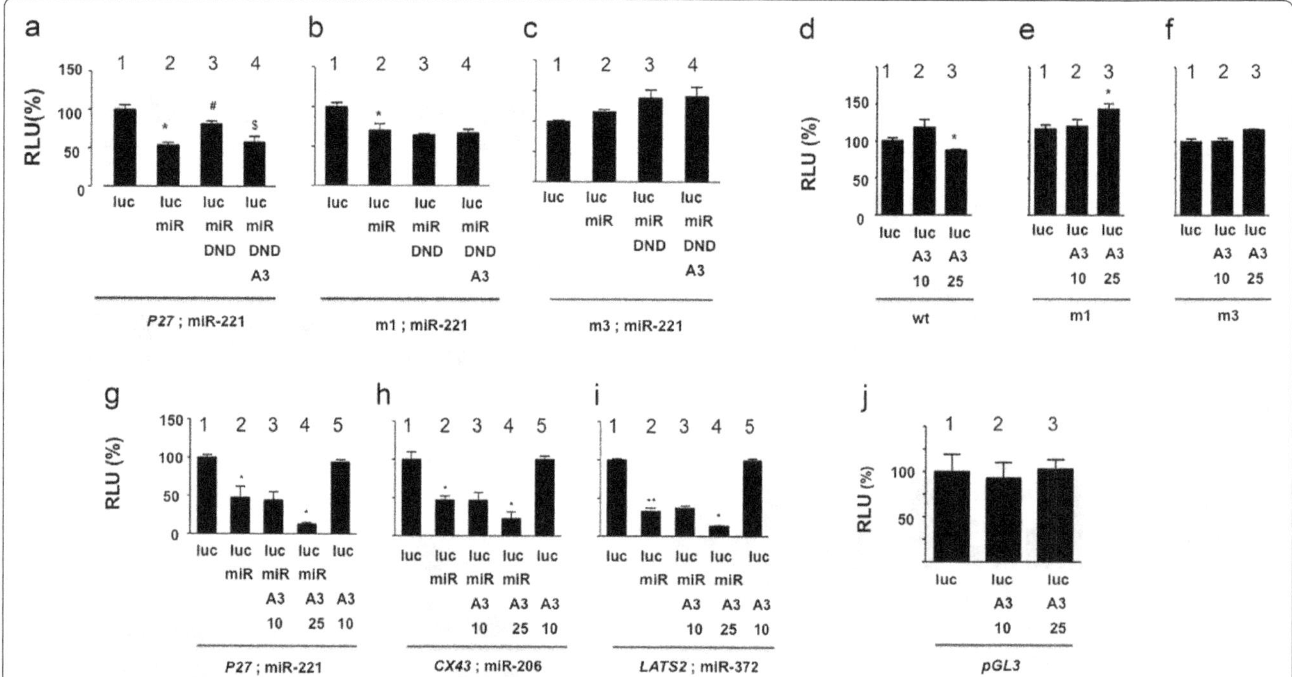

Figure 2 APOBEC3G likely functions through DND1 to restore miRNA activity. (a) APOBEC3G blocks DND1 to restore miR-221 inhibition of luciferase from wild-type *P27*-3'-UTR (lane 4). (*) miR-221 inhibited luc-P27 expression ($P = 0.0116$) (lane 2), (#) DND1 rescued the inhibition ($P= 0.0138$) (lane 3). ($) APOBEC3G opposed the function of DND ($P = 0.0467$) (lane 4). **(b)** APOBEC3G does not restore miR-221 inhibition from mutated *P27* 3'-UTR in m1 (or luc-m1; in which both DND1 binding sites mutated) [4] (lane 4). (*) miR-221 inhibited luc-m1 expression ($P= 0.0460$). DND1 does not rescue miR-221inhibition ($P=0.2665$) (lane 3) and A3 does not affect DND1 function ($P = 0.3211$). **(c)** miR-221 does not inhibit m3 (or luc-m3, both miRNA binding sites mutated [4]) ($P= 0.2665$) (lane 2). DND1 (P= 0.1184) (lane 3) or A3 (P= 0.4569) (lane 4) also do not affect luc-m3 activity. **(d)** Effect of increasing levels of APOBEC3G on wild-type pGL3-P27-3'-UTR. APOBEC3G was cotransfected at two concentrations (10 ng and 25 ng). Higher APOBEC3G (A3) (25 ng) inhibits (*) pGL3-P27-3'-UTR (lane 3) ($P=0.023$). **(e)** Higher A3 (25 ng) slightly enhances luc-m1 expression (*) (lane 3) (both DND1 binding sites mutated) ($P=0.0432$). **(f)** Higher A3 (25 ng) also enhances luc-m3 expression (*) (lane 3) (both miR-221 binding sites mutated) ($P = 0.0141$). However, lower A3 levels (10 ng) have no effect on pGL3-P27-3'-UTR, luc-m1, luc-m3 expression (lane 2 in **d**, **e** and **f**: $P= 0.0798$, 0.3567 and 0.4544 respectively). **(g)** Effect of increasing APOBEC3G on miR function. miR-221 inhibits *luc-P27* ($P = 0.0153$) (lane 2). A3 (10 ng), together with miR-221, does not further inhibit *luc-P27* ($P = 0.4553$). However, A3 at 25 ng (lane 4), together with miR-221, further inhibits *luc-P27* ($P= 0.0488$). **(h)** miR-206 inhibits luc-*CX43* ($P=0.0105$) (lane 2). A3 (10 ng), together with miR-221 does not further inhibit *luc-CX43* ($P = 0.4765$). However, 25 ng A3 (*) (lane 4), together with miR-221, further inhibits luc-*CX43* ($P= 0.0385$). **(i)** miR-372 inhibits luc-*LATS2* ($P= 0.0011$) (lane 2). A3 (10 ng), together with miR-221 does not further inhibit luc-*LATS2* ($P= 0.1673$). However, 25 ng A3 (*) (lane 4), together with miR-221, further inhibits luc-*LATS2* ($P = 0.013$).
(j) Increasing levels of APOBEC3G have no effect on luciferase translation. Control luciferase reporter, pGL3-luciferase (1 ng) was cotranfected with increasing concentrations (10 ng and 25 ng) of APOBEC3G expression constructs ($P= 0.4021$ and $P = 0.4573$ respectively).

We also tested the direct effect of APOBEC3G on pGL3-control vector. Increasing levels of APOBEC3G did not affect the luciferase activity from pGL3-control vector implying that APOBEC3G did not affect translation of luciferase (Figure 2j). The pGL3 vector does not have an extensive 3′-UTR with miRNA or DND1 binding sites unlike the 3′-UTRs of *P27, CX43* or *LATS2,* and therefore is not affected by higher levels of APOBEC3G.

In summary, these experiments lead us to conclude that APOBEC3G is able to block DND1 function and restore miRNA-mediated translation repression. APOBEC3G does not directly affect the 3′-UTRs of *P27, CX43* or *LATS2* or miRNA interaction with mRNAs. However, higher levels of APOBEC3G appear to have additional independent functions on the 3′-UTRs of genes.

DND1 does not affect APOBEC3 function

The above results show that APOBEC3G can block DND1 function to regulate miRNA activity. However, mouse APOBEC3 and human APOBEC3G have been widely studied for their role in inhibiting viral infectivity [14-16]. Therefore, we tested whether DND1 affects APOBEC3 function of inhibiting retroviral infectivity [28,39]. Single cycle infectivity assay shows that, as expected, mouse APOBEC3 severely reduces HIV(ΔVif) viral infectivity in 293 T cells (Figure 3a, lane 3). However, presence of mouse DND1 together with APOBEC3 did not change ability of APOBEC3 to reduce HIV(ΔVif) infectivity (Figure 3a, lane 4). DND1 alone also had no effect on HIV(ΔVif) viral infectivity (Figure 3a, lane 2). Empty vector was used to equalize the amount of construct used in each assay lane [28,39].

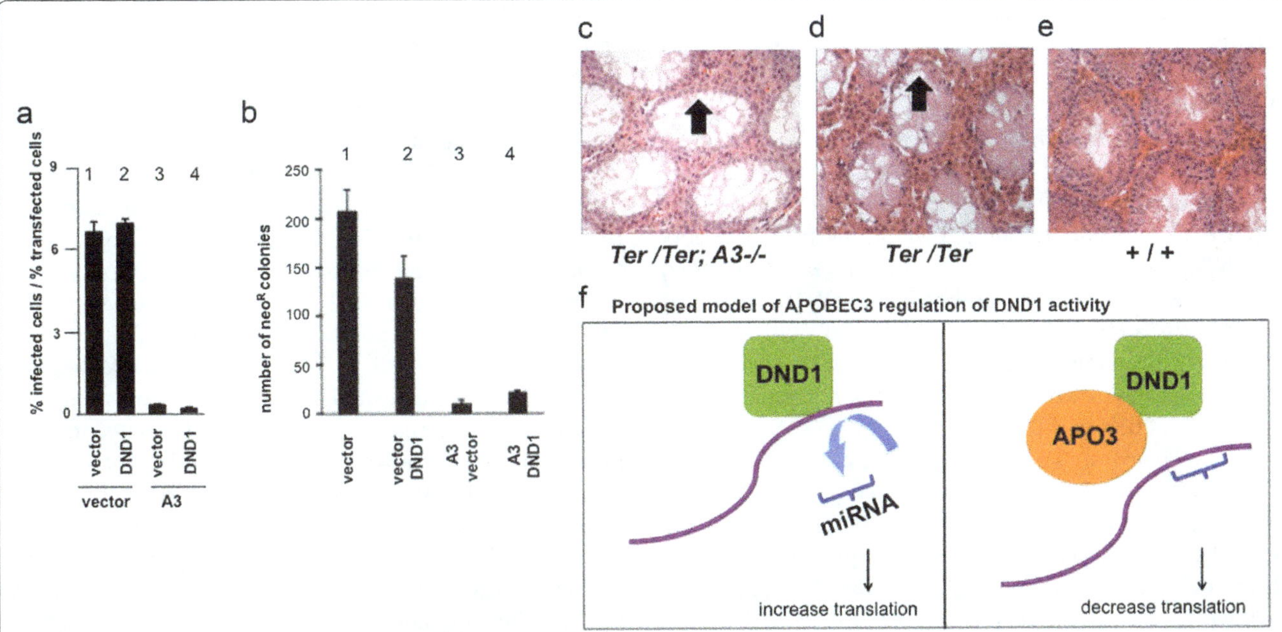

Figure 3 DND1 does not affect APOBEC3 function. (a) DND1 does not affect APOBEC3 antiretroviral activity. Infectivity of HIV-GFP produced in the presence of control vector (vector) (lane 1) or vector encoding mouse DND1 (lane 2) ($P = 0.2082$). Infectivity of HIV-GFP produced in the presence of mouse APOBEC3 (A3) (lane 3) and APOBEC3 plus DND1 (lane 4) ($P = 0.1464$). Empty vector was used to equalize the amount of construct used in each assay lane. Results from two independent experiments were averaged. Error bars indicate the difference in infectivity observed between the two experiments. **(b)** DND1 does not affect MusD restriction by APOBEC3. Effect of control vector (lane 1), mouse DND1 (lane 2), mouse APOBEC3 (lane 3), both mouse DND1 and APOBEC3 (lane 4) on MusD retrotransposition, relative to the vector control. Transposition was monitored by the number of G418-resistant colonies. Results from two independent experiments were averaged. *P* value comparing lanes 3 and 4 is 0.10980. **(c)** Histology section through testes of double homozygous male *Ter/Ter ;A3–/– (or Ter/Ter; Apobec3–/–)*, **(d)** *Ter/Ter* and **(e)** wild-type (+/+) mice. Arrow points to lumen in seminiferous tubules showing lack of germ cells persist in *Ter/Ter ; A3–/–* testis similar to that in testis of *Ter/Ter* mice. Testes of *Apobec3–/– (A3–/–)* mice have normal wild-type germ cell histology similar to +/+ (not shown). **(f)** Proposed model how APOBEC3G regulates DND1 function. DND1 blocks miRNA activity (left panel). We observe decreased protein translation (as measured by luciferase activity) in the presence of APOBECG3 suggesting that APOBEC3G blocks DND1 function. When DND1 binding sites on P27-3'-UTR are inactivated, both DND1 and APOBEC3G fail to affect luciferase activity. This suggests that APOBEC3G functions through DND1, may be by removing or sequestering DND1 to restore miRNA access to the 3'-UTR of *P27* (right panel).

In a second set of experiments, we tested whether DND1 affects the ability of APOBEC3 to inhibit MusD retrotransposition [35]. APOBEC3, by itself, drastically inhibits MusD retrotransposition (as indicated by decrease in neo^R colonies by Apo3 + vector) (Figure 3b, lane 3). However, DND1, when combined with APOBEC3, did not significantly affect APOBEC3 function of inhibiting MusD retrotransposition (A3 + DND1, lane 4). DND1, by itself, also had no effect on MusD retrotransposition (vector + DND1, lane 2). The results from these two experiments indicate that DND1 does not modulate the viral restriction function of APOBEC3.

To further explore the functional relationship between *Dnd1* and *Apobec3*, we asked whether the germ cell phenotype of mice lacking wild-type *Dnd1* (*Ter* mutant mice) is dependent on *Apobec3*. It is known that APOBEC3 family members are potentially powerful mutators [31-33]. We reasoned that interaction of DND1 with APOBEC3 in germ cells may be one mechanism to keep the latent deleterious activity of APOBEC3 in check. Thus it is formally possible that depletion of germ cells in mice lacking normal *Dnd1* (as observed in *Ter* mutant mice) is due to uncontrolled Apobec3 genomic mutating activity. We therefore tested whether removing *Apobec3* from mice in which *Dnd1* is inactivated, would restore normal germ cells. Mice lacking wild-type *Dnd1* (in *Ter* mice) have no germ cells [2] but mice lacking *Apobec3* are normal and fertile [37]. Thus, we generated double mutant mice lacking both wild-type *Dnd1* and *Apobec3* (*Ter/Ter;Apobec3–/–* or *Ter/Ter;A3–/–*) mice. However, examination of newborn or adult testes of *Ter/Ter;A3–/–* mice indicated that they have no germ cells and are sterile (Figure 3c) and are indistinguishable from *Ter/Ter;A3+/+* animals (Figure 3d). Thus this genetic assay suggests that DND1 does not regulate the mutator function of APOBEC3 in germ cells and germ cell loss.

in *Ter/Ter* mice is likely not due to unregulated activity of APOBEC3. Together, results from the restriction assays and the genetic crosses are consistent with the idea that DND1 does not modulate mouse APOBEC3 activity.

Discussion

Our results demonstrate that APOBEC3G is able to block DND1 function and restore miRNA mediated inhibition of translational repression. This function of APOBEC3G appears to apply to multiple mRNA targets of DND1 as APOBEC3G has a similar effect on *P27, LATS2* and *CX43*. On the other hand, DND1 does not appear to affect the viral restriction function of APOBEC3. In addition, our genetic crosses suggest that DND1 interaction with APOBEC3 does not regulate the mutator function of APOBEC3 in germ cells.

The mechanism of how APOBEC3G blocks DND1 remains to be determined. At least, three possible mechanisms can be proposed. The first possibility is that APOBEC3G may bind to DND1 and sequester it away from mRNAs and miRNAs (Figure 3f). In support of this, we have observed that mouse APOBEC3 and DND1 co-immunoprecipitate and co-localize in cells [13].

The second possibility is that APOBEC3G may bind mRNAs (maybe together with DND1) and subsequently interact with components of the miRISC to activate translation repression or interact with translation initiation factors to inhibit them. Indeed, mass spectrometric analysis show that a large number of cellular RNA-binding proteins associate with APOBEC3G [40-42]. Some of these are known components of the miRISC such as ARGONAUTE 1(Agol), ARGONAUTE 2 (Ago2), GW182, MOV10, YB-1, DCP1A and RCK/P54, and are involved in post-transcriptional silencing of gene expression [38,41-43]. These interactions of APOBEC3G with RNA binding proteins were found to be either direct protein-protein interactions or mediated by RNA. In addition, confocal microscopy experiments showed that APOBEC3G co-localized with many of the miRISC RNA-binding proteins to mRNA processing, P-bodies [40,41]. Thus it is conceivable that interaction of APOBEC3 with specific miRISC proteins may override the effect of DND1 to enhance miRNA activity.

The third possibility is that the cytidine deaminase activity of APOBEC3 may allow it to edit the 3′-UTR sequences of *P27, LATS2* and *CX43* to inhibit DND1 binding. Interestingly, indirect support of this hypothesis comes from recent reports by a number of groups that analysed deep sequencing data and found greater than expected incidence of editing present in the mammalian transcriptome [44,45]. As to which of the three possible mechanisms apply to APOBEC3G blocking DND1 function is currently under investigation.

APOBEC3 proteins have been studied as factors that restrict viruses and retrotransposons. However, we found that DND1 has no effect on the viral restriction function of APOBEC3. APOBEC3G can bind both cellular RNAs and RNA binding proteins [27,40,41] and the RNA binding activity of the N-terminal cytidine deaminase domain of APOBEC3G is essential for viral restriction [23,24]. The significance of APOBEC3G interactions with cellular proteins and RNAs is not clear. However, during viral infection of cells, such as HIV-1 (Vif) infection of T lymphocytes, APOBEC3G gains access to viral particles through a ribonucleoprotein interaction and thus APOBEC3 binding to RNA is a critical for antiviral function [23]. The incorporation of APOBEC3G into new viral particles allows it to be released into infected cells where APOBEC3G can deaminate the replicating viral cDNAs to effect reduction of viral infectivity. Interestingly, it has been shown that other P-body proteins, such as MOV10 [46-48] are also involved in reducing the infectivity of exogenous retroviruses and retrotransposons. Thus, P-body protein components such as MOV10, and as we report here, APOBEC3G, participate both in miRNA silencing and viral restriction processes.

An earlier study on APOBEC3G function showed that APOBEC3G can, by itself, inhibit miRNA activity [38]. The previous study used luciferase constructs encoding only miRNA binding sites or encoding multiple miRNA binding sites in tandem. In contrast, our assays used the 3′-UTR of endogenous genes to test APOBEC3G activity and we also tested how APOBEC3G blocks DND1 function on these 3′-UTRs. Moreover, the amount of APOBEC3G expression vectors transfected into cells were considerably higher in the study [38] and we also found that higher APOBEC3G levels may have alternate effects on the 3′-UTRs and miRNA function. Another possibility is that APOBEC3G may have different effects on different miRNAs and transcripts, inhibiting some miRNAs while activating others.

One caveat of our studies is that the miRNA studies were performed using human DND1 and APOBEC3G whereas the infectivity assays and genetic studies were performed using mouse factors. We did this because human DND1 function was previously characterized using human *P27* 3′-UTR and DND1 binding sites have been mapped in the 3′UTR of *P27* [4]. Humans encode multiple members of the APOBEC3 family proteins (APOBEC3A, 3B, 3C, etc.) [21]. We selected human APOBEC3G for our miRNA studies which is functionally most closely related to mouse APOBEC3. The role of the other human APOBEC3 factors in inhibiting DND1 activity remains to be determined.

APOBEC3 proteins are expressed in germ cells [13,21,49]. But although APOBEC3 proteins inhibit retrotransposition [17,50,51], to date, this has not been demonstrated in germ cells. In fact, *Apobec3* null mice are normal and fertile [37]. We found that double null *Ter/Ter;Apobec3−/−* mice have similar phenotype as

Ter/Ter mice and lack of *Apobec3* does not rescue the *Ter/Ter* (*Dnd1−/−*) phenotype to restore germ cells. This supports the idea that DND1 does not regulate the mutator activity of APOBEC3 in germ cells, and deregulated APOBEC3 mutator activity is not responsible for germ cell loss in *Ter/Ter* (*Dnd1−/−*) mice.

In summary, our studies focus on a novel aspect of APOBEC3 function in that we show APOBEC3G regulates DND1 function and in this way affects miRNA activity. Viral restriction and miRNA mediated gene silencing are evolutionarily related processes utilizing similar protein complexes, which localizes to cytoplasmic RNA granules. Our data provides compelling evidence that APOBEC3G may be involved in both these processes.

Conclusion

We present our novel finding that RNA binding protein DEAD END (DND1) blocks miRNA function to permit translation, and APOBEC3G (apolipoprotein B mRNA-editing enzyme, catalytic polypeptide like 3) antagonizes DND1 to reduce translation. Not much is known about how microRNA target interactions are regulated by RNA binding proteins, thus our result advances an area of importance to help understand how microRNA activity can be modulated in response to signals.

Abbreviations

3'-UTR: 3'-untranslated region; Mouse APOBEC3 and human APOBEC3G: Apolipoprotein B mRNA-editing enzyme, catalytic polypeptide like 3; LATS2: Serine/threonine-protein kinase, large tumor suppressor, homolog 2; CX43: Connexin-43; miRNA: microRNA; mRNA: Messenger RNA; P27, LATS2: Human genes and transcripts are in italics and capitals; P27, LATS2, APOBEC3G: Human and mouse proteins are in capitals; *Apobec3*: Mouse genes and transcripts are in italics and lowercase.

Competing interests

The authors declare that they have no competing interests.

Authors' contributions

AM and SA conceived and designed the experiments, performed analysis and interpretation of data and wrote the manuscript. NK performed the transfection assays. CB carried out the genetic crosses and mouse histology. RZ, S Ali and SA performed the experiments. DAM, MDS, AJS, ZLD and RH designed, performed and analyzed viral infectivity and MusD transposition assay data. All authors read and approved the final manuscript.

Acknowledgements

This project was funded by Texas ARP. We thank Q. Wang, M. Zhang, K. Luo, and G. Seawood for technical assistance. We thank R. Agami for the generous gift of the luciferase and mirVec constructs. We thank R. Harris for the human APOBEC3G-myc construct and M.S. Neuberger and C. Rada for Apobec3−/− mice.

Author details

[1]Department of Genetics, University of Texas, MD Anderson Cancer Center, 1515 Holcombe Blvd, Houston, TX 77030, USA. [2]Department of Biochemistry, Molecular Biology, and Biophysics, University of Minnesota, 321 Church Street SE, Minneapolis, MN 55455, USA. [3]Pennington Biomedical Research Center, 6400 Perkins Road, Baton Rouge, LA 70808, USA.

References

1. Weidinger G, Stebler J, Slanchev K, Dumstrei K, Wise C, Lovell-Badge R, Thisse C, Thisse B, Raz E: **Dead end, a novel vertebrate germ plasm component, is required for zebrafish primordial germ cell migration and survival.** *Curr Biol* 2003, **13**(16):1429–1434.
2. Youngren KK, Coveney D, Peng X, Bhattacharya C, Schmidt LS, Nickerson ML, Lamb BT, Deng JM, Behringer RR, Capel B, *et al*: **The Ter mutation in the dead end gene causes germ cell loss and testicular germ cell tumours.** *Nat* 2005, **435**:360–364.
3. Noguchi T, Noguchi M: **A recessive mutation (ter) causing germ cell deficiency and a high incidence of congenital testicular teratomas in 129/Sv-ter mice.** *J Natl Cancer Inst* 1985, **75**(2):385–392.
4. Kedde M, Strasser MJ, Boldajipour B, Vrielink JA, Slanchev K, le Sage C, Nagel R, Voorhoeve PM, van Duijse J, Orom UA, *et al*: **RNA-binding protein Dnd1 inhibits microRNA access to target mRNA.** *Cell* 2007, **131**:1273–1286.
5. Ketting RF: **A dead end for microRNAs.** *Cell* 2007, **131**:1226–1227.
6. Bushati N, Cohen SM: **microRNA functions.** *Ann Rev Cell Dev Biol* 2007, **23**:175–205.
7. Peters L, Meister G: **Argonaute poteins: mediators of RNA silencing.** *Mol Cell* 2007, **26**:611–623.
8. Filipowicz W, Bhattacharya SN, Sonenberg N: **Mechanisms of post-transcriptional regulation by microRNAs: are the answers in sight?** *Nat Rev Genet* 2008, **9**:102–114.
9. Liu X, Wang A, Heidbreder CE, Jiang L, Yu J, Kolokythas A, Huang L, Dai Y, Zhou X: **MicroRNA-24 targeting RNA-binding protein DND1 in tongue squamous cell carcinoma.** *FEBS Lett* 2010, **584**:4115–4120.
10. Bhandari A, Gordon W, Dizon D, Hopkin AS, Gordon E, Yu Z, Andersen B: **The Grainyhead transcription factor Grhl3/Get1 suppresses miR-21 expression and tumorigenesis in skin: modulation of the miR-21 target MSH2 by RNA-binding protein DND1.** *Oncogene* 2012. doi:10.1038/onc.2012.168:1–11.
11. Cook MS, Munger SC, Nadeau JH, Capel B: **Regulation of male germ cell cycle arrest and differentiation by DND1 is modulated by genetic background.** *Dev* 2011, **138**:23–32.
12. Zhu R, Iacovino M, Mahen E, Kyba M, Matin A: **Transcripts that associate with the RNA binding protein, DEAD-END (DND1), in embryonic stem (ES) cells.** *BMC Mol Biol* 2011, **12**(1):37.
13. Bhattacharya C, Aggarwal S, Kumar M, Ali A, Matin A: **Mouse apolipoprotein B editing complex 3 (APOBEC3) is expressed in germ cells and interacts with dead-end (DND1).** *PLoS One* 2008, **3**(5):e2315.
14. Mangeat B, Turelli P, Caron G, Friedli M, Perrin L, Trono D: **Broad antiretroviral defence by human APOBEC3G through lethal editing of nascent reverse transcripts.** *Nat* 2003, **424**:99–103.
15. Shindo K, Takaori-Kondo A, Kobayashi M, Abudu A, Fukunaga K, Uchiyama T: **The enzymatic activity of CEM15/Apobec-3G is essential for the regulation of the infectivity of HIV-1 virion but not a sole determinant of its antiviral activity.** *J Biol Chem* 2003, **278**:44412–44416.
16. Wissing S, NLK G, Greene WC: **HIV-1 Vif versus the APOBEC3 cytidine deaminases: an intracellular duel between pathogen and host restriction factors.** *Mol Aspects Med* 2010, **31**:383–397.
17. Esnault C, Heidmann O, Delebecque F, Dewannieux M, Ribet D, Hance AJ, Heidmann T, Schwartz O: **APOBEC3G cytidine deaminase inhibits retrotransposition of endogenous retroviruses.** *Nat* 2005, **433**:430–433.
18. Muckenfuss H, Hamdorf M, Held U, Perkovic M, Löwer J, Cichutek K, Flory E, Schumann GG, Münk C: **APOBEC3 proteins inhibit human LINE-1 retrotransposition.** *J Biol Chem* 2006, **281**:22161–22172.
19. Bishop KN, Holmes RK, Malim MH: **Antiviral potency of APOBEC proteins does not correlate with cytidine deamination.** *J Virol* 2006, **80**:8450–8458.
20. Iwatani Y, Chan DS, Wang F, Maynard KS, Sugiura W, Gronenborn AM, Rouzina I, Williams MC, Musier-Forsyth K, Levin JG: **Deaminase independent inhibition of HIV-1 reverse transcription by APOBEC3G.** *Nucleic Acids Res* 2007, **35**:7096–7108.
21. Jarmuz A, Chester A, Bayliss J, Gisbourne J, Dunham I, Scott J, Navaratnam N: **An anthropoid-specific locus of orphan C to U RNA-editing enzymes on Chromosome 22.** *Genomics* 2002, **79**:285–296.
22. Conticello SG, Thomas CJ, Petersen-Mahrt SK, Neuberger MS: **Evolution of the AID/APOBEC family of polynucleotide (deoxy)cytidine deaminases.** *Mol Biol Evol* 2005, **22**:367–377.
23. Friew YN, Boyko V, Hu WS, Pathak VK: **Intracellular interactions between APOBEC3G, RNA, and HIV-1 Gag: APOBEC3G multimerization is dependent on its association with RNA.** *Retrovirol* 2009, **6**:56.

24. Gooch BD, Cullen BR: **Functional domain organization of human APOBEC3G.** *Virol* 2008, **379:**118–124.
25. Iwatani Y, Takeuchi H, Strebel K, Levin JG: **Biochemical activities of highly purified, catalytically active human APOBEC3G: correlation with antiviral effect.** *J Virol* 2006, **80:**5992–6002.
26. Li J, Potash MJ, Volsky DJ: **Functional domains of APOBEC3G required for antiviral activity.** *J Cell Biochem* 2004, **92:**560–572.
27. Navarro F, Bollman B, Chen H, Konig R, Yu Q, Chiles K, Landau NR: **Complementary function of the two catalytic domains of APOBEC3G.** *Virol* 2005, **333:**374–386.
28. Hache G, Liddament MT, Harris RS: **The retroviral hypermutation specificity of APOBEC3F and APOBEC3G is governed by the C-terminal DNA cytosine deaminase domain.** *J Biol Chem* 2005, **280:**10920–10924.
29. Jónsson SR, Haché G, Stenglein MD, Fahrenkrug SC, Andrésdóttir V, Harris RS: **Evolutionarily conserved and non-conserved retrovirus restriction activities of artiodactyl APOBEC3F proteins.** *Nucleic Acids Res* 2006, **34:**5683–5694.
30. Hakata Y, Landau NR: **Reversed functional organization of mouse and human APOBEC3 cytidine deaminase domains.** *J Biol Chem* 2006, **281:**36624–36631.
31. Suspène R, Aynaud M-M, Guétard D, Henry M, Eckhoff G, Marchio A, Pineau P, Dejean A, Vartanian J-P, Wain-Hobson S: **Somatic hypermutation of human mitochondrial and nuclear DNA by APOBEC3 cytidine deaminases, a pathway for DNA catabolism.** *Proc Natl Acad Sci USA* 2011, **17:**2–7.
32. Landry S, Narvaiza I, Linfesty DC, Weitzman MD: **APOBEC3A can activate the DNA damage response and cause cell-cycle arrest.** *EMBO Rep* 2011, **12:**444–450.
33. Petit V, Vartanian J-P, Wain-Hobson S: **Powerful mutators lurking in the genome.** *Phil Trans R Soc B* 2009, **364:**705–715.
34. Stenglein MD, Harris RS: **APOBEC3B and APOBEC3F inhibit L1 retrotransposition by a DNA deamination-independent mechanism.** *J Biol Chem* 2006, **281:**16837–16841.
35. Schumacher AJ, Hache G, MacDuff DA, Brown WL, Harris RS: **The DNA deaminase activity of human APOBEC3G is required for Ty1, MusD, and human immunodeficiency virus type 1 restriction.** *J Virol* 2008, **82:**2652–2660.
36. Bhattacharya C, Aggarwal S, Zhu R, Kumar M, Zhao M, Meistrich ML, Matin A: **The mouse dead-end gene isoform alpha is necessary for germ cell and embryonic viability.** *Biochem Biophys Res Commun* 2007, **355:**194–199.
37. Mikl MC, Watt IN, Lu M, Reik W, Davies SL, Neuberger MS, Rada C: **Mice deficient in APOBEC2 and APOBEC3.** *Mol Cell Biol* 2005, **25:**7270–7277.
38. Huang J, Liang Z, Yang B, Tian H, Ma J, Zhang H: **Derepression of microRNA-mediated protein translation inhibition by apolipoprotein B mRNA-editing enzyme catalytic polypeptide-like 3G (APOBEC3G) and its family member.** *J Biol Chem* 2007, **282:**33632–33640.
39. Schumacher AJ, Nissley DV, Harris RS: **APOBEC3G hypermutates genomic DNA and inhibits Ty1 retrotransposition in yeast.** *Proc Natl Acad Sci USA* 2005, **102:**9854–9859.
40. Kozak SL, Marin M, Rose KM, Bystrom C, Kabat D: **The anti-HIV-1 editing enzyme APOBEC3G binds HIV-1 RNA and messenger RNAs that shuttle between polysomes and stress granules.** *J Biol Chem* 2006, **281:**29105–29119.
41. Gallois-Montbrun S, Kramer B, Swanson CM, Byers H, Lynham S, Ward M, Malim MH: **Antiviral protein APOBEC3G localizes to ribonucleoprotein complexes found in P bodies and stress granules.** *J Virol* 2007, **81:**2165–2178.
42. Gallois-Montbrun S, Holmes RK, Swanson CM, Fernandez-Ocana M, Byers HL, Ward MA, Malim MH: **Comparison of cellular ribonucleoprotein complexes associated with the APOBEC3F and APOBEC3G antiviral proteins.** *J Virol* 2008, **11:**5636–5642.
43. Wichroski MJ, Robb GB, Rana TM: **Human retroviral host restriction factors APOBEC3G and APOBEC3F localize to mRNA processing bodies.** *PLoS Pathog* 2006, **2:**374–383.
44. Zaranek AW, Levanon EY, Zecharia T, Clegg T, Church GM: **A survey of genomic traces reveals a common sequencing error, RNA editing, and DNA editing.** *PLoS Genet* 2010, **6:**e1000954.
45. Li M, Wang IX, Li Y, Bruzel A, Richards AL, Toung JM, Cheung VG: **Widespread RNA and DNA sequence differences in the human transcriptome.** *Sci* 2011, **333:**53–58.
46. Nathans R, Chu C, Serquina AK, Lu C-C, Cao H, Rana TM: **Cellular microRNA and P bodies modulate host-HIV-1 interactions.** *Mol Cell* 2009, **34:**696–709.
47. Chable-Bessia C, Meziane O, Latreille D, Triboulet R, Zamborlini A, Wagschal A, Jacquet J-M, Reynes J, Levy Y, Saib A, *et al*: **Suppression of HIV-1 replication by microRNA effectors.** *Retrovirol* 2009, **6:**26.
48. Burdick R, Smith JL, Chaipan C, Friew Y, Chen J, Venkatachari NJ, Delviks-Frankenberry KA, Hu W-S, Pathak VK: **P body-associated protein Mov10 inhibits HIV-1 replication at multiple stages.** *J Virol* 2010, **84:**10241–10253.
49. Koning FA, Newman ENC, Kim E-Y, Kunstman KJ, Wolinsky SM, Malim MH: **Defining APOBEC3 expression patterns in human tissues and hematopoietic cell subsets.** *J Virol* 2009, **83:**9474–9485.
50. Feinberg AP, Ohlsson R, Henikoff S: **The epigenetic progenitor origin of human cancer.** *Nature Rev Genet* 2006, **7:**21–33.
51. Bogerd HP, Wiegand HL, Doehle BP, Lueders KK, Cullen BR: **APOBEC3A and APOBEC3B are potent inhibitors of LTR-retrotransposon function in human cells.** *Nucleic Acids Res* 2006, **34:**89–95.

12

Fluorescence-based monitoring of ribosome assembly landscapes

Rainer Nikolay[1,3*], Renate Schloemer[1], Silke Mueller[2] and Elke Deuerling[1*]

Abstract

Background: Ribosomes and functional complexes of them have been analyzed at the atomic level. Far less is known about the dynamic assembly and degradation events that define the half-life of ribosomes and guarantee their quality control.

Results: We developed a system that allows visualization of intact ribosomal subunits and assembly intermediates (i.e. assembly landscapes) by convenient fluorescence-based analysis. To this end, we labeled the early assembly ribosomal proteins L1 and S15 with the fluorescent proteins mAzami green and mCherry, respectively, using chromosomal gene insertion. The reporter strain harbors fluorescently labeled ribosomal subunits that operate wild type-like, as shown by biochemical and growth assays. Using genetic and chemical perturbations by depleting genes encoding the ribosomal proteins L3 and S17, respectively, or using ribosome-targeting antibiotics, we provoked ribosomal subunit assembly defects. These defects were readily identified by fluorometric analysis after sucrose density centrifugation in unprecedented resolution.

Conclusion: This strategy is useful to monitor and characterize subunit specific assembly defects caused by ribosome-targeting drugs that are currently used and to characterize new molecules that affect ribosome assembly and thereby constitute new classes of antibacterial agents.

Keywords: Ribosome assembly, Ribosome biogenesis, Fluorescent proteins, Antimicrobials, Knock out, λ-red recombineering, High throughput screening

Background

The bacterial 70S ribosome is formed by a small 30S- and a large 50S subunit. While the small subunit consists of one 16S ribosomal RNA (rRNA) and 21 ribosomal proteins (r-proteins), the large subunit contains two rRNAs (23S and 5S rRNA) and 33 r-proteins [1]. Reconstitution of intact ribosomal subunits in the test tube is possible using components derived from purified ribosomes, but requires non-physiological conditions, such as high Mg^{2+} concentration and incubation temperatures of up to 50°C [2,3]. *In vivo*, this process critically depends on biogenesis factors, which are proteins that process, modify and chaperone rRNA or r-proteins [4,5]. Ribosome assembly is characterized by a highly coordinated sequence of events consisting of rRNA synthesis and r-protein uptake. Since assembly takes place co-transcriptionally (i.e. during rRNA synthesis) there is a hierarchical order of binding events with early- and late assembly r-proteins [5,6]. Each r-protein gene is present in a single copy per genome, whereas rRNAs are encoded by multiple *rrn* operons (seven ones in *E. coli*). One 16S, 23S and 5S rRNA (and several tRNAs) are contained in one primary transcript that is processed by site specific RNases [4]. Due to this genetic organization ribosomal subunits are consequently produced in equal stoichiometric amounts. Furthermore, synthesis of rRNA and r-proteins are synchronized [7] with the consequence that free cytosolic pools of r-proteins are close to zero under optimal conditions [8-10]. In addition, ribosome assembly is a fast process taking place within a couple of minutes at 37°C [11]. It follows that r-proteins upon synthesis are rapidly taken up by nascent ribosomal subunits. If selected r-proteins were fluorescence labeled, it further follows that the fluorescences signal would represent subunit precursors and whole subunits rather than free cytosolic pools of r-proteins.

Treatment of cells with chemical agents or occurrence of mutations (e.g. affecting genes encoding r-proteins or

* Correspondence: rainer.nikolay@Charite.de; elke.deuerling@uni-konstanz.de
[1]Molecular Microbiology, University of Konstanz, Constance 78457, Germany
Full list of author information is available at the end of the article

biogenesis factors) can lead to ribosome assembly defects [12]. One possible fate of defective assembly intermediates is a selective clearance by RNase based control mechanisms [13,14]. Alternatively, assembly intermediates can accumulate and possibly mature into intact subunits, as soon as the source of defect is eliminated [5,12,15-18].

Analyses of protein and RNA content of ribosomal assembly intermediates are possible using quantitative mass spectrometry approaches and cryo-electron microscopy [19-25]. Analyses of sucrose gradient fractions by agarose and two-dimensional gel electrophoresis with subsequent quantitation of rRNA and r-proteins are established methods but time-consuming and insensitive to small differences in quantity and quality.

Therefore, we set out to establish a convenient fluorescence-based method to assess amounts and assembly-states of ribosomal particles.

In our previous approach [26] late assembly r-proteins were labeled with fluorescent proteins (FPs) to monitor and compare the intact portions of both ribosomal subunits. In this study intact ribosomal subunits and assembly intermediates of all maturation states are detectable by labeling early assembly r-proteins. A reporter strain harboring the fusion proteins L1-mAzami green (a coral-derived monomeric green fluorescent protein [27], hereafter mAzami) and S15-mCherry was constructed and exhibited normal growth, indicating an intact translation apparatus. We used synthetic gene knock out of *rplC* (encoding L3) and *rpsQ* (encoding S17), respectively, or ribosome directed antibiotics to induce subunit assembly defects. A_{254} and fluorescence analysis of sucrose gradient centrifugates allowed *in vitro* analysis of ribosomal subunits and all of their assembly intermediates in unprecedented resolution.

Results

Rationale

In order to generate a reporter strain suitable for monitoring ribosome assembly landscapes, we selected ribosomal protein candidates from each subunit according to the following criteria [28,29]: The candidates should be i) distant from functional sites, ii) accessible to C-terminal tagging with fluorescent proteins, iii) early assembly proteins [10] and iv) subject of feedback regulation. The ribosomal proteins S15 and L1 fulfill all these criteria: Their surface exposed C-termini (Additional file 1A) allow convenient tagging (with mCherry and mAzami). Although these proteins are not essential [30] deletion strains have exaggerated generation times at 37°C. In addition, absence of S15 results in severe cold sensitivity [31] and ribosomes lacking L1 show 50% reduced translation activity *in vitro* [32]. Therefore, growth would be severely hampered if the fusion proteins do not fully complement the wild type protein's function. According to *in vivo* ribosome assembly maps (Additional file 2), both are early assembly proteins and consequently present in ribosomal particles of each state of maturation. Finally, feedback regulation by autogenous control [33,34] ensures that they are not produced in excess.

Phenotypic and biochemical characterization of the engineered strains

To generate *E. coli* strains harboring modified genes coding for S15-mCherry (MCr*) and L1-mAzami (MCg*) fusion proteins (Additional file 1B), we used the technique of lambda red recombineering [35,36]. The final reporter strain MCrg* producing both S15-mCherry and L1-mAzami fusion proteins was constructed, using P1 phage transduction.

To exclude that tagging of r-proteins with FPs interferes with regular cell functions and growth, we analyzed reporter strains in more detail. Spot tests revealed that growth of the genetically engineered strains did not differ from that of the wild type strain at various temperatures (Figure 1A). To analyze possible growth differences more precisely, all strains were grown to stationary phase at different temperatures and their growth rates were calculated (Figures 1B). It turned out that the growth rate of MCrg* at 37°C was 5-10% less than the wild type strain.

Next, the protein content of MCr*, MCg* and MCrg*-derived ribosomes was analyzed by SDS-PAGE and immunoblotting (Figure 1C, D). While MCr* and MCg* ribosomes contained one fusion protein (migrating at 37 and 57 kDa, respectively), two fusion proteins were observed in MCrg* ribosomes.

Collectively, the data indicate that growth behavior and functional competence of the ribosomes of MCrg* are similar to those of the parental strain.

Generation of ribosome subunit specific assembly defects and *in vitro* analysis

To induce assembly defects in the small or the large ribosomal subunit, conditional gene knock outs of *rpsQ* (encoding S17) and *rplC* (encoding L3), respectively, were generated in the reporter strain background (Additional file 1A). It has been shown previously that defects in each of these genes caused ribosome assembly defects that were supposed to be subunit specific [26,37,38].

The resulting strains (MCrg*ΔsQ and MCrg*ΔlC) carried plasmids containing wild type copies of the genes deleted from the chromosome under control of an IPTG inducible promoter. The withdrawal of IPTG in liquid cultures should result in impaired growth and in subunit specific assembly defects [39] as soon as the number of intact ribosomes becomes limiting.

To this end, we grew MCrg*, MCrg*ΔsQ and -ΔlC cells in the absence of IPTG to mid-logarithmic phase and examined the ribosomes by sucrose gradient ultracentrifugation and

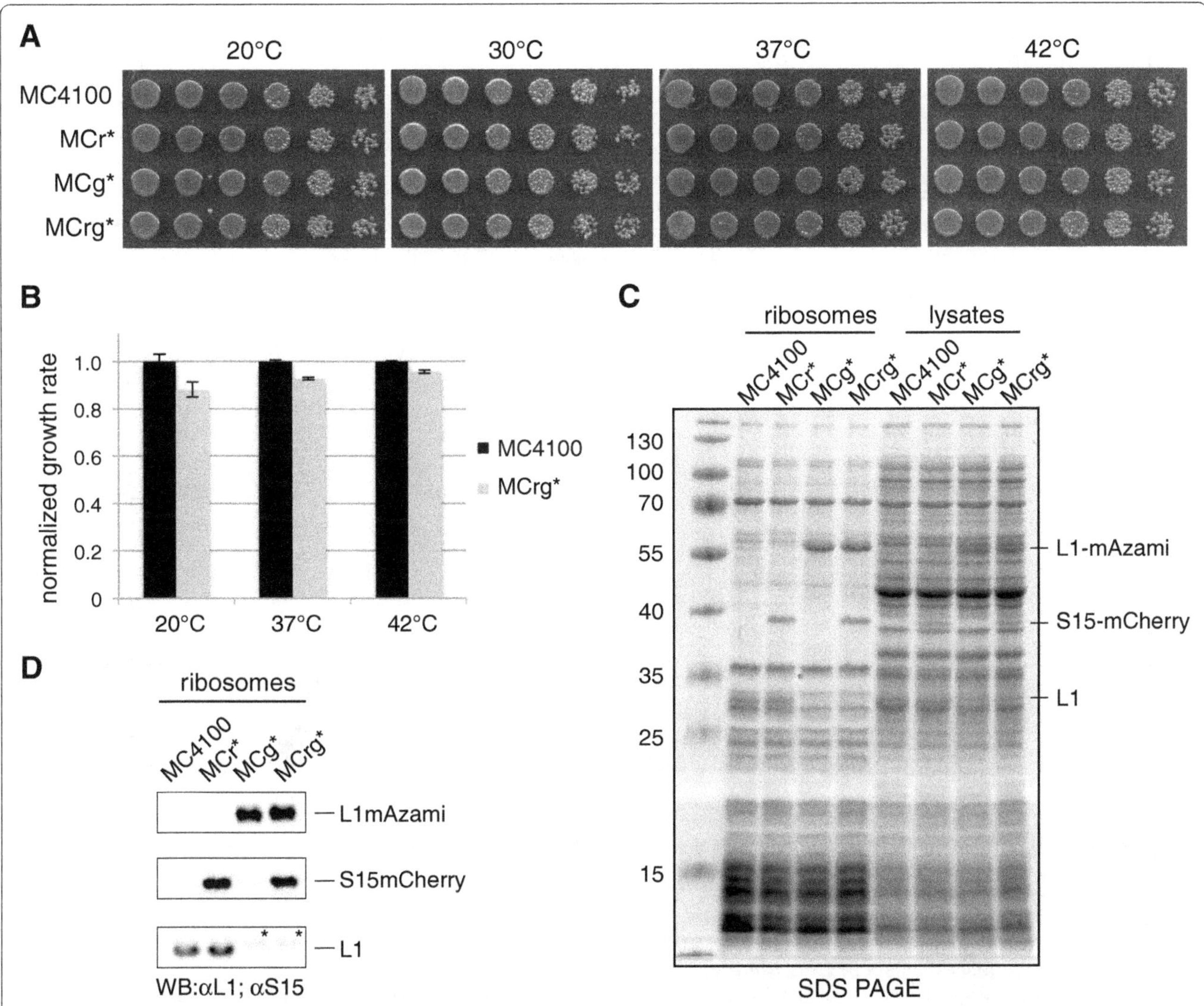

Figure 1 Physiological and biochemical characterization. (A) Growth comparison on solid medium: Cells of the indicated strains were spotted onto LB agar in a serial dilution and incubated at the given temperatures. **(B)** Cells as indicated were grown at 20, 37 and 42°C to stationary phase. Growth rates were calculated and normalized values are given for each strain at each incubation temperature. Data were obtained from three independent experiments. Ribosomes from the indicated strains were isolated by sucrose cushion centrifugation and subjected to SDS-PAGE **(C)** and western blot analysis **(D)**. For immunodetection, S15 and L1 specific antisera were used. Note that S15 wild type protein was not resolved by SDS-PAGE and immunoblotting. Asterisks denote unspecific protein bands.

polysome profile analysis (Figures 2A-C). MCrg* ribosomes showed the expected pattern consisting of 30S-, 50S-, 70S-, and polysome peaks (Figure 2A), whereas depletion of *rpsQ* (Figure 2B) led to dramatically reduced amounts of 70S ribosomes and polysomes, increased amounts of 50S subunits and a broad peak of particles in the region of the 30S subunits. Likewise, depletion of *rplC* reduced the amount of 70S ribosomes and led to a defined peak of 30S subunits and a large and broad peak of particles migrating between mature 50S and 30S subunits (Figure 2C). This was expected because both absence of *rpsQ* and *rplC* should result in defective small and large ribosomal subunits, respectively. Consequently, the reduced number of functional subunits limited the amount of monosomes and polysomes.

Fluorometric analysis of the sucrose fractions provided fluorescence profiles of MCrg*, MCrg*ΔsQ and -ΔlC derived ribosomes (Figures 2D-F). Comparing A_{254} and fluorescence profiles of MCrg* ribosomes (Figure 2G) revealed reasonable coincidence of the individual peaks. For completeness the entire profile, also including early low molecular weight fractions, is depicted as insert (Figure 2G, insert).

When analyzing A_{254} and fluorescence profiles of MCrg*ΔsQ ribosomes by overlay (Figure 2H) several aspects attracted attention: The largest peak of red

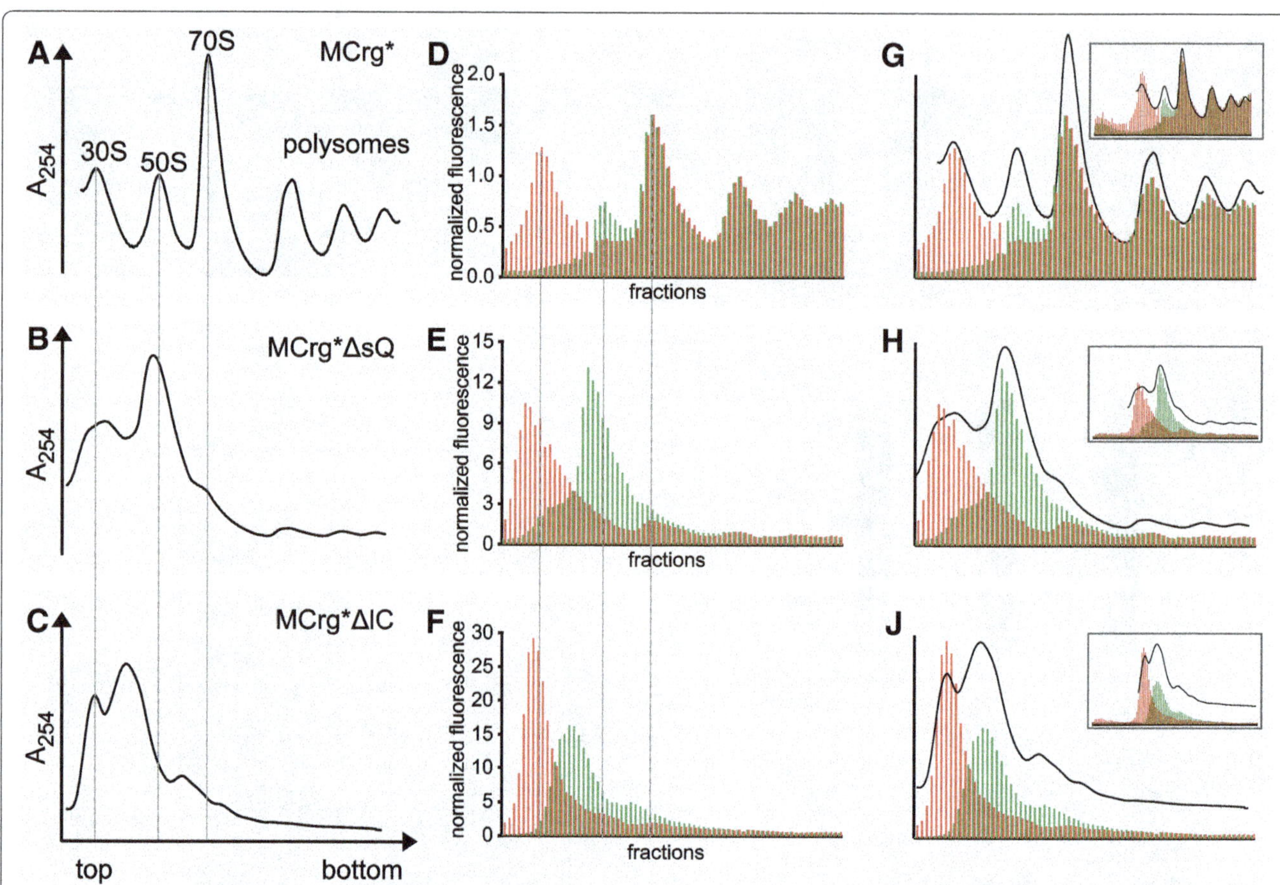

Figure 2 Polysome analysis and fluorescence detection of sucrose fractions. Cells were grown in M9 medium at 37°C to $OD_{600} = 0.4$ and harvested. Lysates were subjected to sucrose gradient centrifugation. Centrifugates were analyzed by A_{254} detection and fractionated. Polysome profiles derived from: **(A)** MCrg*, **(B)** MCrg*ΔsQ, **(C)** MCrg*ΔlC. Sucrose gradient fractions of samples **A-C** were analyzed for mAzami- and mCherry specific fluorescence and normalized results are given in bar charts for: **(D)** MCrg* **(E)** MCrg*ΔsQ **(F)** MCrg*ΔlC. Superposition of A_{254} profiles and corresponding fluorescence bar charts: **(G)** MCrg*, **(H)** MCrg*ΔsQ, **(J)** MCrg*ΔlC. The inserts show fluorescence analysis of all available fractions from each sucrose gradient run. Red bars: normalized mCherry fluorescence; Green bars: normalized mAzami fluorescence. Fluorescence was normalized to the first polysome peak ("disome") where subunits are supposed to be present in 1:1 ratio. Experiments were done in duplicates, representative profiles are shown.

fluorescence -representing the small subunit- was decreased in intensity relative to largest peak of green fluorescence -representing the large subunit- (in comparison with Figure 2G). Moreover, the red fluorescence peak was in addition left shifted due to absence of *rpsQ*. The largest peak of green fluorescence was slightly left shifted and showed a shoulder at the lower left side overlapping with the red peak, indicating defective large ribosomal subunits. This indicates that a selective assembly defect of the small subunit is also associated with an accumulation of 50S assembly intermediates. Finally, as shown in the insert there was no increased fluorescence in the low molecular weight fractions, indicating proper autogenous control of S15-mCherry and L1-mAzami.

Combined analysis of A_{254} and fluorescence profiles of MCrg*ΔlC ribosomes (Figure 2J) revealed a decrease in the green fluorescence peak relative to the red fluorescence peak, in comparison with Figure 2H. In addition, the green fluorescence peak was clearly left shifted, due to assembly defects in the absence of *rplC*. In the A_{254} profile the peak of the large subunit appeared higher than the peak of the small subunit. This is presumably the case because peaks of both subunits are overlapping each other (Figure 2F), thereby producing a dominant broad peak in the region of the large subunit. Investigation of the low molecular weight fractions in the insert showed strict feedback regulation of S15-mCherry and L1-mAzami.

In summary, assembly defects of the small and large ribosomal subunit could be provoked and were readily detectable by fluorescence analysis of sucrose gradient centrifugates.

In vivo analysis of subunit specific assembly defects

Next, we asked whether subunit assembly defects could be detected by fluorescence readout *in vivo* using MCrg*?

Fluorescent labeling of the early assembly r-proteins L1 and S15 results in fluorescent subunits of all stages of maturation. Since both subunits are systemically produced in equal amounts, any shift in the fluorescence ratios is expected to be a consequence of subunit specific turnover (facilitated by ribonucleases and proteases). Increased turnover of the large subunit should reduce the amount of green fluorescence and consequently lower the normalized fluorescence emission ratio of mAzami/mCherry, while increased turnover of the small subunit in turn should increase the ratio.

MCrg*, MCrg*ΔsQ and MCrg*ΔlC cells were transferred to 384-well plates and incubated at 37°C for 10 hours in M9 medium. Fully automated sample handling was possible, using a robotic platform equipped with incubator, microplate reader and robotic arm. Both A_{650} values and fluorescence intensities were measured (Figure 3) in one-hour intervals. From the latter, normalized fluorescence ratios were calculated. While MCrg* grew unperturbed, MCrg*ΔsQ and MCrg*ΔlC cells showed impaired growth and reached lower cell densities after 10 hours (Figure 3A). The background corrected and normalized fluorescence ratios of MCrg*ΔlC reached a minimum of 0.8 after 6 hours, revealing a defect of the 50S assembly, whereas the ratios of MCrg*ΔsQ increased instead reaching a maximum of 1.3 after 9 hours indicating a defect in the 30S assembly (Figure 3B).

We conclude that depletion of *rplC* and *rpsQ*, respectively, causes severe assembly defects of ribosomal subunits. Moreover, the changes in fluorescence ratios suggest that there is turnover of defective subunits.

Probing MCrg* with ribosome-targeting antibiotics

It has been shown that translation inhibitors such as chloramphenicol [16,17,40,41], erythromycin [16,17,42,43] and neomycin [18,44,45] cause directly or indirectly assembly defects of both ribosomal subunits. The mechanistic interpretation of the antibiotic mediated effects is controversial [15-18]. We tested all of the before mentioned antibiotics and included a fourth one (kanamycin) that, to our knowledge, was not investigated so far for its potential to cause ribosome assembly defects. Using MCrg*, we set out to clarify, whether treatment of cells with ribosome targeting antibiotics results in assembly defects of one or both of the ribosomal subunits. The fact that early assembly proteins are fluorescently labeled should allow detailed analysis of the assembly landscapes upon antibiotic treatment. In addition, cell based assays were used to elicit whether there are indications for subunit specific turnover.

MCrg* cells were grown in M9 medium for seven hours in the absence or presence of chloramphenicol (7 μg/ml), erythromycin (100 μg/ml), kanamycin (7 μg/ml) or neomycin (7 μg/ml). While each antibiotic led to impaired cell growth (Figure 4A), chloramphenicol caused strongest growth defects. Erythromycin, kanamycin and neomycin treatment did not show a significant change in fluorescence ratios (Figure 4B). Treatment with chloramphenicol, by contrast, led to an increased fluorescence ratio, with a maximum of about 1.20 after 7 hours. This suggests that treatment of cells with chloramphenicol might decrease the relative amounts of the small subunit.

This hypothesis was tested by analyzing ribosome profiles obtained from MCrg* cells that grew in the presence of the antibiotics or without (Figure 4C-G).

Figure 4C shows the A_{254} profile of ribosomes derived from non-treated cells. The 70S peak and the polysomes (not shown) were reduced in intensity since no chloramphenicol was added prior to harvesting. In our previous study, it turned out that addition of chloramphenicol (immediately before harvest) to antibiotic-treated cells caused extremely broadened 70S peaks. Ribosomes derived from chloramphenicol (D) or erythromycin (E)

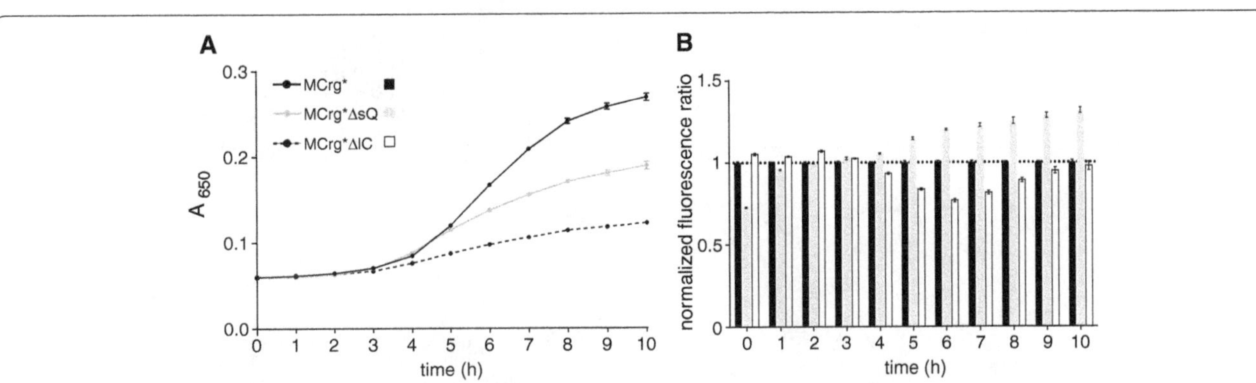

Figure 3 Growth in 384-well pates and fully automated fluorescence analysis of reporter strains. Aliquots of MCrg*, MCrg*ΔsQ and MCrg*ΔlC cultures were transferred into 384- well plates in quadruplicates. Cells were grown in M9 medium at 37°C for 10 hours. Measurements were made in one-hour intervals. **(A)** A_{650} values were determined and **(B)** mAzami and mCherry fluorescence emission were detected and ratios were calculated for MCrg*, MCrg*ΔsQ and MCrg*ΔlC. Fluorescence ratios of MCrg* were normalized to 1. Data points given in the growth curves and fluorescence ratios are mean values from four independent experiments; error bars show standard deviation.

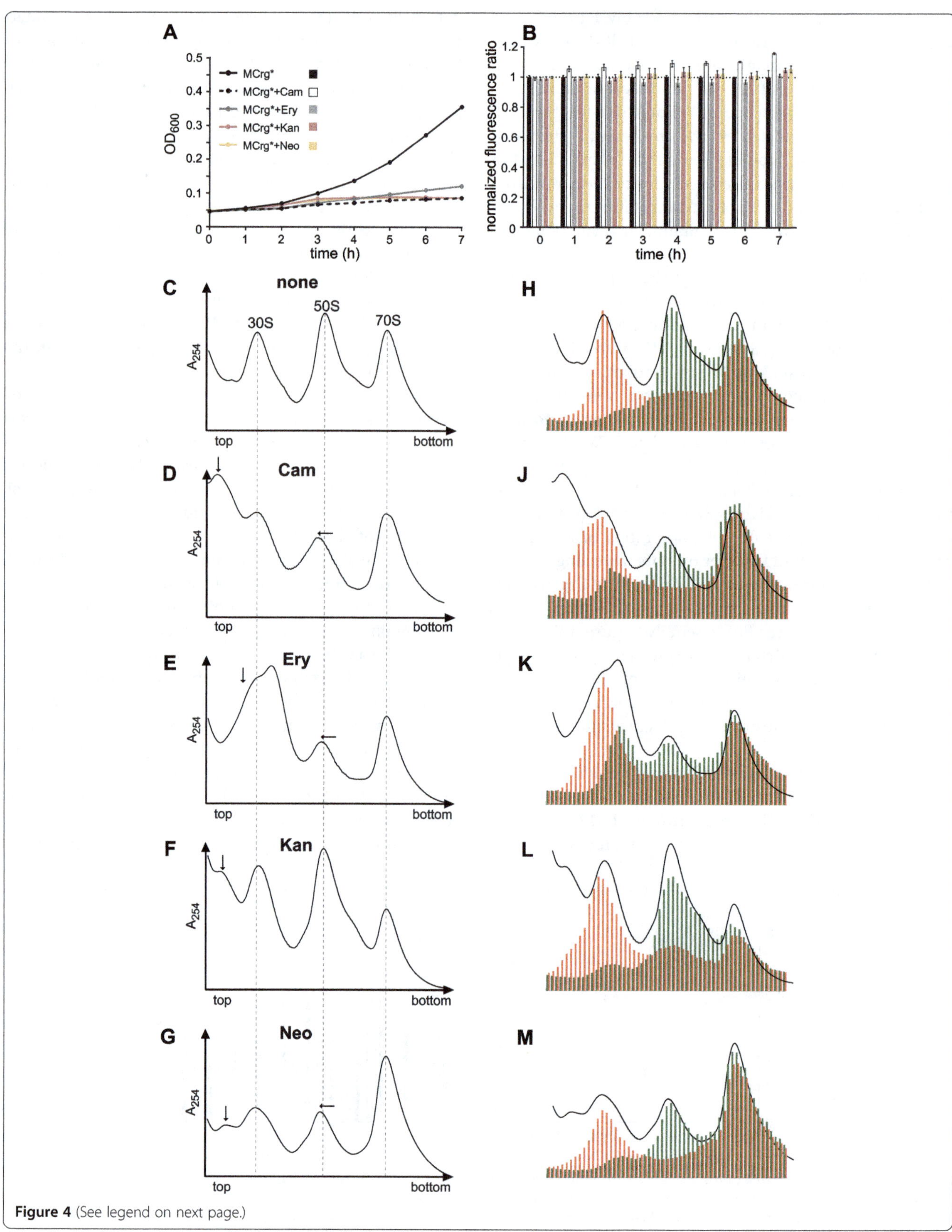

Figure 4 (See legend on next page.)

(See figure on previous page.)
Figure 4 Testing MCrg* with inhibitors of translation. Cell-based assay: MCrg* cells were cultured in Erlenmeyer flasks at 25°C in M9 medium for 7 hours in the absence and presence of antibiotics, as indicated. Samples were taken every hour. **(A)** OD_{600} values were determined and **(B)** mAzami and mCherry fluorescence emission were detected and ratios were calculated. Fluorescence ratios of MCrg* were normalized to 1. Exemplary growth curves are given and fluorescence ratios are means from three independent experiments; error bars show standard deviation. **Analyses of isolated ribosomal particles**: Sucrose density gradient (10-25%) centrifugation profiles from **(C)** control cells with no antibiotic (none), **(D)** chloramphenicol (Cam), **(E)** erythromycin (Ery), **(F)** kanamycin (Kan) and **(G)** neomycin (Neo) treated cells. Sucrose gradient fractions from **(C)** to **(G)** were analyzed for fluorescence by a microplate reader. A_{254} profiles and fluorescence bar charts were superimposed for **(H)** control cells with no antibiotic (none), **(J)** chloramphenicol (Cam), **(K)** erythromycin (Ery), **(L)** kanamycin (Kan) and **(M)** neomycin (Neo) treated cells. Cells in presence and absence of antibiotics were cultured in LB medium at 25°C for 3 hours before subsequent polysome analysis. Left shifted peaks of the large subunit are indicated by horizontal arrows, abnormal portions of the small subunit by vertical arrows. Red bars: normalized mCherry fluorescence; Green bars: normalized mAzami fluorescence. Fluorescence was normalized to the first polysome peak ("disome") where subunits are present in 1:1 ratio. Experiments were done in duplicates, representative profiles are shown.

treated cells showed reduced and left shifted 50S peaks. Shifted peaks of the large subunit are indicated with horizontal arrows, additional peaks and shoulders left-lateral of the 30S peak are marked with vertical arrows. Chloramphenicol treatment caused an additional peak (left-lateral of 30S), while after erythromycin treatment the 30S peak had a shoulder on the left. Treatment with kanamycin had no apparent influence on the 50S peak but provoked an additional peak left-lateral of the 30S peak (F). Neomycin treatment caused a slight reduction and left shift of the 50S peak and an additional peak on the left of the 30S peak (G).

Overlay diagrams (combining A_{254} and fluorescence read outs) (Figure 4H-M) show that ribosome profiles from non-treated cells are congruent (Figure 4H). Profiles from chloramphenicol and in particular erythromycin treated cells revealed an extra portion of green fluorescence within the region of the 30S peak (Figure 4J-K). Ribosome profiles derived from kanamycin and neomycin treated cells possessed additional green fluorescent peaks of very weak intensity but at similar positions within the profile (Figure 4L-M). The particles within the green peaks migrating slower than that of the 50S subunits presumably represent defective assembly intermediates of the large subunit that are caused by all the antibiotics used in this study, but to different extent. For a more detailed analysis the green fluorescence profiles were aligned and the one derived from non-treated cells was compared with all the others obtained from antibiotic treated cells (Additional file 3), illustrating the presence of assembly intermediates of the large subunit. Comparison of the red fluorescent peaks (Figure 4H-M) indicates that treatment with all antibiotics produced a more or less pronounced shoulder on the left side. To analyze potential defects of the small subunit in more detail, the red fluorescence profiles derived from antibiotic treated cells were compared with the one obtained from non-treated cells (Additional file 4). It turned out that treatment with all four antibiotics caused distinct left-sided shoulders of the red fluorescence peak, indicating small subunit assembly defects. This is either due to accumulation of the 21S precursor of the 30S subunit, or assembly dead-ends of this subunit. A more thorough analysis, which includes quantitation of the 16S and 23S rRNA within the sucrose fractions is given in the supplementary material (Additional file 5), confirms all hypothesized assembly intermediates. Taken together, the administration of a representative collection of ribosome-targeting antibiotics caused assembly defects of both subunits in each case, but to different extents.

In summary, our analyses demonstrated that ribosome assembly defects are detectable by fluorometric analysis of fractions collected after sucrose gradient ultracentrifugation of ribosomal preparations. Assembly defects caused by gene depletions (of *rplC* and *rpsQ*) or by treatment of reporter cells with four different ribosome-targeting antibiotics revealed the presence of defective assembly intermediates of both ribosomal subunits.

Discussion

Here, we report construction and characterization of the strain MCrg* harboring large and small ribosomal subunits labeled with mAzami and mCherry proteins, respectively. This strain has growth properties similar to the parental strain. We have provided evidence that the reporter strain reveals assembly defects of ribosomal subunits by fluorescence-based readouts of sucrose density centrifugates and should report differences in subunit specific turnover when subjected to fluorescence-based *in vivo* analysis.

In our previous study, we labeled the late assembly r-proteins L19 and S2 with FPs to identify and compare the portion of intact ribosomal subunits. A detailed characterization of physiological and biochemical properties of the reporter strain demonstrated that there were no substantial limitations of the translation apparatus resulting from labeling of two ribosomal proteins with two different FPs. When ribosome assembly had been perturbed genetically or chemically fluorescence-based read out allowed identification of these assembly defects [26].

In the current study, we labeled the early assembly r-proteins L1 and S15 with FPs. Again, we made sure that tagging of the r-proteins with FPs did not compromise cell physiology and ribosome composition, as confirmed by growth assays and SDS-PAGE analysis of purified ribosomes (Figure 1). In addition, we provided evidence that MCrg* allows for *in vitro* analysis of isolated ribosomes to monitor fully matured ribosomal subunits and their assembly intermediates (Figure 2). Moreover, MCrg* is a strong tool for cell-based assays. It allows fluorometric *in vivo* determination of the ratio between the two ribosomal subunits, including all assembly states. In untreated reporter cells this ratio is expected to be one, due to equal production of ribosomal subunits. A change in the fluorescence ratio is therefore interpreted as consequence of unequal turnover of ribosomal subunits.

Generation of subunit specific assembly defects by gene depletion

Subunit specific ribosome assembly defects were induced by depletion of r-protein genes encoding the early assembly proteins S17 and L3, respectively (Figure 2). Depletion of *rpsQ* (encoding S17) caused an assembly defect of the small subunit (Figure 3E) and increased the fluorescence ratio by 30% (Figure 3B), indicating increased turnover of the small subunit. Depletion of *rplC* (encoding L3) caused an assembly defect of the large subunit (Figure 2F) and decreased the fluorescence ratio by 20% (Figure 3B), indicating increased turnover of the large subunit. It seems that assembly defects on one subunit disturb assembly of the other subunit. It is possible that assembly defects reduce the pool of active 70S ribosomes, thereby reducing levels of translation, which in turn induces in assembly defects of both subunits (also see below).

Treatment of MCrg* with ribosome targeting antibiotics

Chloramphenicol and erythromycin target the large ribosomal subunit, while kanamycin and neomycin interact predominantly with the small ribosomal subunit. Chloramphenicol binds to the peptidyl transferase center (PTC) and inhibits fixation of the CCA-aminoacyl end of an aminoacyl tRNA at the A-site region of the PTC [46], while the macrolide erythromycin binds in the upper region of the ribosomal exit tunnel and hinders protein synthesis by blocking a full occupation of the exit tunnel [47]. Kanamycin and neomycin belong to the aminoglycoside and 2-deoxystreptamine aminoglycoside families of antibiotics, respectively. They both target the 30S subunit and interact with the internal loop 44 of the decoding center [48]. For neomycin an additional binding site within helix 69 of the 23S rRNA was reported [49,50]. Both antibiotics cause translational misreading by stabilizing and binding of near cognate tRNAs to the mRNA [48].

Antibiotics used in this study induce assembly defects in both ribosomal subunits (Figure 4). This raises the question, whether a perturbation of the translation apparatus always results in mutual assembly defects? According to previous studies, inhibition of translation by ribosome targeting antibiotics reduces protein biosynthesis and provokes overproduction of rRNA [16,51,52]. Consequently, there is a stoichiometric imbalance between rRNA and r-proteins, which promotes accumulation of defective precursor particles of both ribosomal subunits [5,16]. Siibak et al. provided evidence that the presence of sublethal concentrations of chloramphenicol or erythromycin resulted in reduced and unbalanced protein biosynthesis [16,17]. In particular ribosomal proteins were produced in amounts that correlated with their presence in assembly intermediates [17]. In that sense assembly defects reflect an indirect consequence of impaired translation.

In our study defective subunits were monitored by fluorescence-based *in vitro* analyses in great detail. Notably, the defects observed in subunit assembly upon treatment with antibiotics were most reveling (Figure 4, Additional files 3 and 4). Ribosome profiles clearly differed both in A_{254} and in fluorescence analysis. Detailed analyses of 30S subunit formation *in vitro* [53] and evidence from *in vivo* experiments [18] suggest that there are parallel routes for ribosomal subunit formation. Therefore, assembly intermediates described here (and in literature) may reflect a heterogeneous collection of similar sizes and shapes [5]. This could indicate that the used antibiotics have intrinsic properties that cause specific assembly defects with particles slightly differing in size and composition. Champney and coworkers have provided evidence, using different bacterial species and dozens of different antibiotics, that certain ribosome targeting drugs bind to ribosomal precursor particles and thereby presumably hinder further maturation to intact subunits [15,54].

Even though mechanistic concepts and causative explanations for ribosome assembly defects caused by current translation inhibitors differ, attempts to identify primary inhibitors of ribosome assembly are important and worthwhile for several reasons:

1. The availability of small molecules that selectively inhibit the assembly process of one subunit at a defined stage would clearly be highly appreciated by researchers in the ribosome field to obtain a more detailed understanding of subunit specific assembly *in vivo*.
2. Such compounds could be the basis for the development of drugs with a defined target leading to subunit specific assembly breakdown. Molecules can be envisioned that exclusively target either assembly of the large or the small ribosomal subunit.
3. If the latter was possible, these compounds could be combined in one preparation to achieve defined and

sustained assembly defects of both ribosomal subunits simultaneously.

Conclusions

The described here fluorescence-based assays allow for monitoring of ribosome assembly landscapes in hitherto unmatched resolution. Assembly intermediates can be precisely detected and unambiguously allocated to either the large or small ribosomal subunit. Rather than replacing quantitative mass spectrometry approaches, it is supposed to be a valuable method to select the optimal fractions for subsequent mass spectrometric or structural analyses. Therefore, we believe that MCrg* significantly enriches the tool kit for a more thorough investigation of ribosome assembly and characterization of new classes of antimicrobials. The methodology described here might also be applied to eukaryotic systems from yeast up to mammalian cells, using CRISPR Cas 9 based knock-in techniques [55-57] to screen for small molecule inhibitors of eukaryotic ribosome assembly. The high metabolic activity of tumor cells goes along with high rates of protein production, which is expected to make them more susceptible towards inhibition of the translation apparatus than normal cells [58]. Therefore, molecules found in such screenings could be the basis for the development of new anti-cancer drugs.

Methods

Media, buffers, antibodies and antibiotics

LB medium (5 g yeast extract, 10 g trypton, 5 g NaCl/ l); M9 medium (64 g $Na_2HPO_4 7H_2O$, 15 g KH_2PO_4, 2.5 g NaCl, 5 g NH_4Cl, 2 mM $MgSO_4$, 0.1 mM $CaCl_2$, 0.4% glucose/l); PBS (137 mM NaCl, 2.7 mM KCl, 10 mM Na_2HPO_4, 1.8 mM KH_2PO_4, pH 7.4); S15 and L1 specific antisera, raised in sheep were obtained from Dr. Nierhaus. Horseradish peroxidase (HRP)-conjugated rabbit anti-sheep (CodeNo: 313-035-003; LotNo: 106383) and donkey anti-rabbit (CodeNo: 711-035-152; LotNo: 103871) secondary antibodies were from Jackson ImmunoResearch. HRP-substrate: for detection a mixture of 1 ml solution A + 100 µl solution B + 1 µl solution C was freshly prepared (solution A: 0.1 mM TRIS (pH8.6), 25 mg Luminol, 100 ml distilled H_20; solution B: 11 mg *p*-hydroxycoumaric acid, in 10 ml DMSO; solution C: H_2O_2 (30%). Antibiotics were used in concentrations as indicated: Ampicillin (Applichem-A0839,0100), chloramphenicol (Sigma-C0378), erythromycin (Sigma-E6367), kanamycin (Roth-T832.4) and neomycin (Sigma-N1876).

Plasmids and bacterial strains

rpsQ and *rplC* were amplified from genomic *E. coli* DNA using specific primers with *Sac*I and *Xba*I restriction sites, respectively. Digested inserts were ligated with an opened pTRC99a vector (*lacIq*, trc promoter, *bla*-gene for ampicillin resistance) [59] using *Sac*I/ *Xba*I restriction sites generating pTRC-*rplQ* and pTRC-*rplC*, respectively. Plasmids were brought into DH5α-Z1 by chemical transformation for amplification and were isolated using Qiagen mini-prep kit.

MC4100 (F^- *[araD139]*$_{B/r}$ *Δ(argF-lac)169 lambda*$^-$ *e14-flhD5301 Δ(fruK-yeiR)725 (fruA25) relA1 rpsL150(strR) rbsR22 Δ(fimB-fimE)632(::IS1) deoC1*); **DY330** (W3110 Δ *lacU169 gal490 λcI857 Δ (cro-bioA)*)[36], **DH5α-Z1** (F *endA1 hsdR17*(r_k m_{k+}) *supE44 thi-1 recA1 gyrA relA1* Δ (*lacZYA-argF*)U169 *deoR* Φ80 *lacZ*Δ M15 *Lac*R *Tet*R and *Sp*^{r}) [60]

λ-red recombineering

Coding sequences of mAzami green [27] (hereafter mAzami) and mCherry (fused with kanamycin resistance cassettes (kanR) derived from plasmid pKD4 [35]) with flanking homologous regions (40–50 nucleotides) for 3'prime genomic insertion in frame with *rplA* and *rpsO*, respectively, were amplified using Phusion DNA-Polymerase. PCR products of the expected size were purified and brought into competent DY330 cells via electroporation. Successful genomic integration was verified by colony PCR and DNA-sequencing. Genetic modifications were transferred to strains of interest using P1-phage transduction. Resistance cassettes were eliminated by transforming strains of interest with pCP20 [61] encoding FLP-recombinase from *Saccharomyces cerevisiae*, which eliminates kanR flanked by FRT-sites. Strains were cured from pCP20 by incubation at 42°C for 15 hours. Gene deletions of *rpsQ* and *rplC* were achieved as described previously [26].

Cell growth analysis

Growth on LB agar plates: Stationary *E. coli* cells were diluted in LB medium to an initial cell density of OD_{600} = 0.025. 1:5 serial dilutions were prepared and transferred to LB agar plates with a plating stamp. Agar plates were incubated at 20, 30, 37 and 42°C until visible single colonies had formed.

Growth in liquid media: Stationary *E. coli* cells as indicated (cultured in LB medium) were diluted in LB medium to an initial cell density of OD_{600} = 0.05 (for incubation at 20°C) or 0.025 (for incubation at 37 and 42°C). Alternatively, stationary pre-cultures of the indicated strains were washed and diluted in M9 medium to an initial cell density of OD_{600} = 0.05. Cell suspensions were cultured in baffled flasks in a water bath incubator with a shaking frequency of 200 rpm until stationary phase was reached or a maximum of 10 hours had passed. Cell density was determined using a photometer (Amersham ultrospec 3110 pro). Growth rates were calculated for periods of exponential growth.

SDS-PAGE and immunoblot analysis of purified ribosomes

Proteins from cell lysates (20 µg total protein) and purified ribosomes (15 pmol) from the indicated strains were resolved by a 13% SDS gel. Proteins were stained using Coomassie brilliant blue 250G.

For immunoblot analysis proteins from cell lysates (3 µg total protein) and purified ribosomes (3 pmol) were resolved by a 13% SDS gel and blotted to nitrocellulose membranes, which were decorated with S15 (raised in rabbit) and L1 (raised in sheep) specific anti-sera (1:10000 and 1:15000 in TBS +3% milk powder). HRP-conjugated donkey anti-rabbit and rabbit anti-sheep secondary antibody (1:20000 and 1:10000 in TBS + 3% milk powder) were used in combination with HRP-substrate to allow immunodetection. Chemiluminescence was monitored using LAS 3000 imager (Fuji Film).

Purification of ribosomes by sucrose cushion centrifugation

E. coli cells were cultured in LB medium at 37°C to cell densities as indicated, harvested by centrifugation and processed as described previously [26].

Sucrose gradient centrifugation and ribosome analysis

Stationary pre-cultures of the individual strains were washed and diluted in M9 medium to $OD_{600} = 0.05$ and cultured at 37°C to cell densities as indicated. Chloramphenicol (250 µg/ml) was added 5 min before cells were harvested by centrifugation, flash-frozen, and stored at −80°C. Frozen cell pellets were resuspended in buffer IV (10 mM TRIS, 10 mM $MgCl_2$, 100 mM NH_4Cl, 250 µM chloramphenicol, pH7.5) and cells were lyzed using Fastprep-24. Cleared lysates (0.5 ml of a solution with $A_{260} = 15$ or 20) were loaded on 10-40% sucrose gradients (in buffer V: 10 mM TRIS, 10 mM $MgCl_2$, 100 mM NH_4Cl, 0.5 mM DTT, 1xTM Complete, pH7.5) and centrifuged in a Sorvall TH-641 rotor for 2:40 h at 41 krpm. A_{254} profiles of sucrose centrifugates were obtained using a Teledyne Isco gradient reader. Fractions of the sucrose gradient were collected in 96-well plates (5 drops per well) for further fluorometric analysis.

For testing MCrg* with antibiotics, cells were washed and diluted in M9 medium to $OD_{600} = 0.1$ and grown to $OD_{600} = 0.15$ before chloramphenicol (7 µg/ml), erythromycin (100 µg/ml), kanamycin (7 µg/ml) or neomycin (7 µg/ml) were added. Cells were cultured at 25°C. After 3 hours of incubation cells were harvested and processed as described above with two exceptions: No Chloramphenicol was added 5 min before harvesting and 10-25% sucrose gradients were used for separation of ribosomal particles.

Agarose gel electrophoresis

20 µl of sucrose gradient fractions were mixed with 6xDNA sample buffer (Thermo Scientific) and loaded on a 1% agarose gel. The RNA content was separated at 150 V for 1 hour. The gel was stained in an ethidium bromide solution for 15 min and RNA bands were visualized at 302 nm on an UV-transilluminator (UVP) by optical inspection.

Fluorometric analyses

Manual measurements using Fluorospectrometer (Jasco FP-6500): 1 ml aliquots of cell suspension of various strains were transferred to quartz cuvettes and mAzami- (excitation 480 nm/ emission 510 nm ± 5 nm band width) and mCherry specific fluorescence intensities (excitation 580 nm/ emission 610 nm ± 5 nm band width) were determined. Fluorescence ratios were calculated by dividing mAzami by mCherry fluorescence intensities. Ratios were normalized to the reporter strain (MCrg*) and plotted in bar charts.

For the fully automated in *vivo assay* a Freedom EVO® 200 robotic platform (Tecan) was used, as described before [26]. Briefly, stationary cells of various strains were washed and diluted in M9 medium to a cell density of $OD_{600} = 0.05$. 80 µl aliquots of each strain were transferred to a 384-well plate in quadruplicates. Cells were incubated for 10 hours at 37°C in a monitored incubator (MIO2™) with 8.5 Hz shaking frequency. The samples were analyzed in a microplate reader in one-hour intervals for mAzami- and mCherry specific fluorescence, using filter combinations 485/535 nm and 535/612 nm, respectively. Fluorescence ratios were calculated by dividing mAzami by mCherry fluorescence intensities. Background corrected ratios were normalized to the reporter strain (MCrg*) and plotted in bar charts. Cell densities were determined simultaneously by detecting the absorbance at 650 nm ± 5 nm (A_{650}). Obtained values were plotted in a spread chart. Calculations and diagrams were made using Magellan 7 (Tecan) and Graph Pad Prism v6 (Graph Pad) software packages.

Sucrose fraction were analyzed using an Infinite F500 (Tecan) fluorescence microplate reader. Sucrose gradient fractions collected in 96-well plates (5 drops per well) were analyzed for mAzami- and mCherry specific fluorescence using filter combinations 485/535 nm and 535/612 nm, respectively. Fluorescence intensities were normalized to a fraction containing the first polysome ("disome") peak, as indicated. In Figure 4 and Figures based on it, the polysome peaks are not shown. Nevertheless, in each case the first polysome peak was used for normalization.

Additional files

Additional file 1: Overview constructed strains and 70S ribosome structure. (A) Surface representation of a T. termophilus 70S ribosome crystal structure. The 16S rRNA is colored in light gray proteins of the small subunit in yellow. 23S and 5S rRNA are shown in dark gray, proteins of the large subunit in cyan. S15 is highlighted in red, L1 in green. Their surface exposed C-termini are shown in purple. The figure was generated with pymol, based on PDB files 4KCZ and 4KCY (29). (B) Given are the names of the constructed strains as used in this study (lab nomenclature in brackets), relevant genotype and fluorescent fusion proteins produced. pTRC-rpsQ, pTRC-rplC: complementation plasmids with copies of the chromosomally deleted genes. Endogenous genes are shown as gray arrows, genes encoding fluorescent proteins as colored boxes. Genes to be deleted were replaced by kanamycin resistance cassettes (KanR). mCherry gene and protein portions are shown in red, mAzami accordingly in green.

Additional file 2: *in vivo* 30S and 50S ribosome assembly maps adapted from [10]. (A) 30S assembly map reflecting order and interdependency of r-protein interaction with the nascent small ribosomal subunit. Early interacting r-proteins are shaded in dark gray, late interacting ones in light gray. Boxed proteins are contained in the 21S precursor of the 30S subunit. S15 is circled in red. (B) 50S assembly map reflecting order and interdependency of r-protein interaction with the nascent large ribosomal subunit. Early interacting r-proteins are shaded in dark gray, late interacting ones in light gray. Proteins contained in the 32S or in the 43S precursor of the large ribosomal subunit are boxed or circled in black, respectively. L1 is circled in green.

Additional file 3: Alignment of mAzami specific fluorescence intensities from samples analyzed in Figure 4. mAzami fluorescence profiles from untreated cells in combination with mAzami profiles derived from (A) chloramphenicol (Cam), (B) erythromycin (Ery), (C) kanamycin (Kan) and (D) neomycin (Neo) treated cells. The diagrams show normalized mAzami fluorescence intensities from sucrose fractions derived from untreated cells (gray bars) in direct comparison with the ones from antibiotic treated cells (green bars). 70S peaks were used for normalization.

Additional file 4: Alignment of mCherry specific fluorescence intensities from samples analyzed in Figure 4. mCherry fluorescence profiles from untreated cells in combination with mCherry profiles derived from (A) chloramphenicol (Cam), (B) erythromycin (Ery), (C) kanamycin (Kan) and (D) neomycin (Neo) treated cells. The diagrams show normalized mCherry fluorescence intensities from sucrose fractions derived from untreated cells (gray bars) in direct comparison with the ones from antibiotic treated cells (red bars). 70S peaks were used for normalization.

Additional file 5: A254 detection and fluorescence analysis of sucrose density gradient fractions. Sucrose density gradient (10-25%) centrifugation profiles derived from (A) control cells with no antibiotic (none), (B) chloramphenicol (Cam), (C) erythromycin (Ery), (D) kanamycin (Kan) and (E) neomycin (Neo) treated cells. Sucrose fractions were collected and analyzed for mAzami and mCherry specific fluorescence (green and red bars). A254 profiles and fluorescence bar charts were superimposed and sucrose fractions were analyzed for presence of 16S and 23S rRNA by agarose gelelectrophoresis and subsequent optical inspection. Open circles: No rRNA, gray circles: fractions with low-intermediate amounts of rRNA, black circles: fractions with high amounts of rRNA.

Abbreviations

mAzami green: A monomeric green fluorescent protein; mCherry: A monomeric red fluorescent protein; r-protein: ribosomal protein; r-RNA: ribosomal RNA; LB: Luria Broth; PCR: Polymerase chain reaction; SDS-PAGE: Sodium dodecyl sulphate polyacrylamide gel electrophoresis; IPTG: Isopropyl β-D-1-thiogalacto-pyranoside; UV: Ultraviolet.

Competing interests

The authors declare competing financial interests. Engineered bacterial strains described in this study are part of a patent application (application No: 13175775.9-1410) that is pending.

Authors' contributions

RN developed the idea of the project, designed and performed experiments, evaluated results, and prepared the figures. RS performed most of the experiments. SM designed, performed and evaluated the fully automated *in vivo* assay. RN and ED conceived research and wrote the manuscript. All authors read and approved the final manuscript.

Acknowledgements

We would like to thank K. H. Nierhaus for critical reading of the manuscript and helpful suggestions. We are grateful to W. Boos and M. Koch for proof reading of the manuscript and critical discussions. This work was supported by a fellowship of the Zukunftskolleg from the University of Konstanz (RN), German Science Foundation (DE 783/3-1, SFB 969/A01) (ED), Human Frontier in Science Program (RGP0025/2012) (ED).

Author details

[1]Molecular Microbiology, University of Konstanz, Constance 78457, Germany. [2]Screening Center Konstanz, University of Konstanz, Constance 78457, Germany. [3]Current address: Institute of Medical Physics and Biophysics, Charité-Universitaetsmedizin Berlin, Berlin 10117, Germany.

References

1. de Narvaez CC, Schaup HW. In vivo transcriptionally coupled assembly of Escherichia coli ribosomal subunits. J Mol Biol. 1979;134(1):1–22.
2. Traub P, Nomura M. Structure and function of E. coli ribosomes. V. Reconstitution of functionally active 30S ribosomal particles from RNA and proteins. Proc Natl Acad Sci U S A. 1968;59(3):777–84.
3. Nierhaus KH, Dohme F. Total reconstitution of functionally active 50S ribosomal subunits from Escherichia coli. Proc Natl Acad Sci U S A. 1974;71(12):4713–7.
4. Kaczanowska M, Ryden-Aulin M. Ribosome biogenesis and the translation process in Escherichia coli. Microbiol Mol Biol Rev. 2007;71(3):477–94.
5. Shajani Z, Sykes MT, Williamson JR. Assembly of bacterial ribosomes. Annu Rev Biochem. 2011;80:501–26.
6. Nierhaus KH. The assembly of prokaryotic ribosomes. Biochimie. 1991;73(6):739–55.
7. Nomura M, Gourse R, Baughman G. Regulation of the synthesis of ribosomes and ribosomal components. Annu Rev Biochem. 1984;53:75–117.
8. Marvaldi J, Pichon J, Delaage M, Marchis-Mouren G. Individual ribosomal protein pool size and turnover rate in Escherichia coli. J Mol Biol. 1974;84(1):83–96.
9. Ulbrich B, Nierhaus KH. Pools of ribosomal proteins in Escherichia coli. Studies on the exchange of proteins between pools and ribosomes. Eur J Biochem. 1975;57(1):49–54.
10. Chen SS, Williamson JR. Characterization of the ribosome biogenesis landscape in E. coli using quantitative mass spectrometry. J Mol Biol. 2013;425(4):767–79.
11. Lindahl L. Intermediates and time kinetics of the in vivo assembly of Escherichia coli ribosomes. J Mol Biol. 1975;92(1):15–37.
12. Maguire BA. Inhibition of bacterial ribosome assembly: a suitable drug target? Microbiol Mol Biol Rev. 2009;73(1):22–35.
13. Basturea GN, Zundel MA, Deutscher MP. Degradation of ribosomal RNA during starvation: comparison to quality control during steady-state growth and a role for RNase PH. RNA. 2011;17(2):338–45.
14. Deutscher MP. Maturation and degradation of ribosomal RNA in bacteria. Prog Mol Biol Transl Sci. 2009;85:369–91.
15. Champney WS. The other target for ribosomal antibiotics: inhibition of bacterial ribosomal subunit formation. Infect Disord Drug Targets. 2006;6(4):377–90.
16. Siibak T, Peil L, Xiong L, Mankin A, Remme J, Tenson T. Erythromycin- and chloramphenicol-induced ribosomal assembly defects are secondary effects of protein synthesis inhibition. Antimicrob Agents Chemother. 2009;53(2):563–71.
17. Siibak T, Peil L, Donhofer A, Tats A, Remm M, Wilson DN, et al. Antibiotic-induced ribosomal assembly defects result from changes in the synthesis of ribosomal proteins. Mol Microbiol. 2011;80(1):54–67.
18. Sykes MT, Shajani Z, Sperling E, Beck AH, Williamson JR. Quantitative proteomic analysis of ribosome assembly and turnover in vivo. J Mol Biol. 2010;403(3):331–45.
19. Talkington MW, Siuzdak G, Williamson JR. An assembly landscape for the 30S ribosomal subunit. Nature. 2005;438(7068):628–32.
20. Popova AM, Williamson JR. Quantitative analysis of rRNA modifications using stable isotope labeling and mass spectrometry. J Am Chem Soc. 2014;136(5):2058–69.
21. Siibak T, Remme J. Subribosomal particle analysis reveals the stages of bacterial ribosome assembly at which rRNA nucleotides are modified. RNA. 2010;16(10):2023–32.

22. Guo Q, Goto S, Chen Y, Feng B, Xu Y, Muto A, et al. Dissecting the in vivo assembly of the 30S ribosomal subunit reveals the role of RimM and general features of the assembly process. Nucleic Acids Res. 2013;41(4):2609–20.
23. Li N, Chen Y, Guo Q, Zhang Y, Yuan Y, Ma C, et al. Cryo-EM structures of the late-stage assembly intermediates of the bacterial 50S ribosomal subunit. Nucleic Acids Res. 2013;41(14):7073–83.
24. Clatterbuck Soper SF, Dator RP, Limbach PA, Woodson SA. In vivo X-ray footprinting of pre-30S ribosomes reveals chaperone-dependent remodeling of late assembly intermediates. Mol Cell. 2013;52(4):506–16.
25. Jomaa A, Jain N, Davis JH, Williamson JR, Britton RA, Ortega J. Functional domains of the 50S subunit mature late in the assembly process. Nucleic Acids Res. 2014;42(5):3419–35.
26. Nikolay R, Schloemer R, Schmidt S, Mueller S, Heubach A, Deuerling E. Validation of a fluorescence-based screening concept to identify ribosome assembly defects in Escherichia coli. Nucleic Acids Res. 2014;42(12):e100.
27. Karasawa S, Araki T, Yamamoto-Hino M, Miyawaki A. A green-emitting fluorescent protein from Galaxeidae coral and its monomeric version for use in fluorescent labeling. J Biol Chem. 2003;278(36):34167–71.
28. Schuwirth BS, Borovinskaya MA, Hau CW, Zhang W, Vila-Sanjurjo A, Holton JM, et al. Structures of the bacterial ribosome at 3.5 A resolution. Science. 2005;310(5749):827–34.
29. Zhou J, Lancaster L, Donohue JP, Noller HF. Crystal structures of EF-G-ribosome complexes trapped in intermediate states of translocation. Science. 2013;340(6140):12360861–9.
30. Baba T, Ara T, Hasegawa M, Takai Y, Okumura Y, Baba M, et al. Construction of Escherichia coli K-12 in-frame, single-gene knockout mutants: the Keio collection. Mol Syst Biol. 2006;2:1–11.
31. Philippe C, Eyermann F, Benard L, Portier C, Ehresmann B, Ehresmann C. Ribosomal protein S15 from Escherichia coli modulates its own translation by trapping the ribosome on the mRNA initiation loading site. Proc Natl Acad Sci U S A. 1993;90(10):4394–8.
32. Subramanian AR, Dabbs ER. Functional studies on ribosomes lacking protein L1 from mutant Escherichia coli. Eur J Biochem. 1980;112(2):425–30.
33. Portier C, Dondon L, Grunberg-Manago M. Translational autocontrol of the Escherichia coli ribosomal protein S15. J Mol Biol. 1990;211(2):407–14.
34. Yates JL, Nomura M. Feedback regulation of ribosomal protein synthesis in E. coli: localization of the mRNA target sites for repressor action of ribosomal protein L1. Cell. 1981;24(1):243–9.
35. Datsenko KA, Wanner BL. One-step inactivation of chromosomal genes in Escherichia coli K-12 using PCR products. Proc Natl Acad Sci U S A. 2000;97(12):6640–5.
36. Yu D, Ellis HM, Lee EC, Jenkins NA, Copeland NG, Court DL. An efficient recombination system for chromosome engineering in Escherichia coli. Proc Natl Acad Sci U S A. 2000;97(11):5978–83.
37. Lhoest J, Colson C. Cold-sensitive ribosome assembly in an Escherichia coli mutant lacking a single methyl group in ribosomal protein L3. Eur J Biochem. 1981;121(1):33–7.
38. Herzog A, Yaguchi M, Cabezon T, Corchuelo MC, Petre J, Bollen A. A missense mutation in the gene coding for ribosomal protein S17 (rpsQ) leading to ribosomal assembly defectivity in Escherichia coli. Mol Gen Genet. 1979;171(1):15–22.
39. Maguire BA, Wild DG. The roles of proteins L28 and L33 in the assembly and function of Escherichia coli ribosomes in vivo. Mol Microbiol. 1997;23(2):237–45.
40. Osawa S, Otaka E, Itoh T, Fukui T. Biosynthesis of 50 s ribosomal subunit in Escherichia coli. J Mol Biol. 1969;40(3):321–51.
41. Dodd J, Kolb JM, Nomura M. Lack of complete cooperativity of ribosome assembly in vitro and its possible relevance to in vivo ribosome assembly and the regulation of ribosomal gene expression. Biochimie. 1991;73(6):757–67.
42. Chittum HS, Champney WS. Erythromycin inhibits the assembly of the large ribosomal subunit in growing Escherichia coli cells. Curr Microbiol. 1995;30(5):273–9.
43. Usary J, Champney WS. Erythromycin inhibition of 50S ribosomal subunit formation in Escherichia coli cells. Mol Microbiol. 2001;40(4):951–62.
44. Mehta R, Champney WS. 30S ribosomal subunit assembly is a target for inhibition by aminoglycosides in Escherichia coli. Antimicrob Agents Chemother. 2002;46(5):1546–9.
45. Mehta R, Champney WS. Neomycin and paromomycin inhibit 30S ribosomal subunit assembly in Staphylococcus aureus. Curr Microbiol. 2003;47(3):237–43.
46. Ulbrich B, Mertens G, Nierhaus KH. Cooperative binding of 3'-fragments of transfer ribonucleic acid to the peptidyltransferase center of Escherichia coli ribosomes. Arch Biochem Biophys. 1978;190(1):149–54.
47. Tenson T, Lovmar M, Ehrenberg M. The mechanism of action of macrolides, lincosamides and streptogramin B reveals the nascent peptide exit path in the ribosome. J Mol Biol. 2003;330(5):1005–14.
48. Wilson DN. Ribosome-targeting antibiotics and mechanisms of bacterial resistance. Nat Rev Microbiol. 2014;12(1):35–48.
49. Borovinskaya MA, Pai RD, Zhang W, Schuwirth BS, Holton JM, Hirokawa G, et al. Structural basis for aminoglycoside inhibition of bacterial ribosome recycling. Nat Struct Mol Biol. 2007;14(8):727–32.
50. Wang L, Pulk A, Wasserman MR, Feldman MB, Altman RB, Cate JH, et al. Allosteric control of the ribosome by small-molecule antibiotics. Nat Struct Mol Biol. 2012;19(9):957–63.
51. Hosokawa K, Nomura M. Incomplete ribosomes produced in chloramphenicol- and Puromycin-inhibited Escherichia Coli. J Mol Biol. 1965;12:225–41.
52. Cortay JC, Cozzone AJ. Effects of aminoglycoside antibiotics on the coupling of protein and RNA syntheses in Escherichia coli. Biochem Biophys Res Commun. 1983;112(3):801–8.
53. Mulder AM, Yoshioka C, Beck AH, Bunner AE, Milligan RA, Potter CS, et al. Visualizing ribosome biogenesis: parallel assembly pathways for the 30S subunit. Science. 2010;330(6004):673–7.
54. Champney WS. Bacterial ribosomal subunit assembly is an antibiotic target. Curr Top Med Chem. 2003;3(9):929–47.
55. Jinek M, Chylinski K, Fonfara I, Hauer M, Doudna JA, Charpentier E. A programmable dual-RNA-guided DNA endonuclease in adaptive bacterial immunity. Science. 2012;337(6096):816–21.
56. Mali P, Yang L, Esvelt KM, Aach J, Guell M, DiCarlo JE, et al. RNA-guided human genome engineering via Cas9. Science. 2013;339(6121):823–6.
57. Cong L, Ran FA, Cox D, Lin S, Barretto R, Habib N, et al. Multiplex genome engineering using CRISPR/Cas systems. Science. 2013;339(6121):819–23.
58. Loreni F, Mancino M, Biffo S. Translation factors and ribosomal proteins control tumor onset and progression: how? Oncogene. 2014;33(17):2145–56.
59. Amann E, Ochs B, Abel KJ. Tightly regulated tac promoter vectors useful for the expression of unfused and fused proteins in Escherichia coli. Gene. 1988;69(2):301–15.
60. Hanahan D. Studies on transformation of Escherichia coli with plasmids. J Mol Biol. 1983;166(4):557–80.
61. Cherepanov PP, Wackernagel W. Gene disruption in Escherichia coli: TcR and KmR cassettes with the option of Flp-catalyzed excision of the antibiotic-resistance determinant. Gene. 1995;158(1):9–14.

13

Differential regulation of the rainbow trout (*Oncorhynchus mykiss*) MT-A gene by nuclear factor interleukin-6 and activator protein-1

Peter Kling[2†], Carina Modig[1†], Huthayfa Mujahed[1], Hazem Khalaf[1], Jonas von Hofsten[3] and Per-Erik Olsson[1*]

Abstract

Background: Previously we have identified a distal region of the rainbow trout (*Oncorhynchus mykiss*) metallothionein-A (rtMT-A) enhancer region, being essential for free radical activation of the rtMT-A gene. The distal promoter region included four activator protein 1 (AP1) cis-acting elements and a single nuclear factor interleukin-6 (NF-IL6) element. In the present study we used the rainbow trout hepatoma (RTH-149) cell line to further examine the involvement of NF-IL6 and AP1 in rtMT-A gene expression following exposure to oxidative stress and tumour promotion.

Results: Using enhancer deletion studies we observed strong paraquat (PQ)-induced rtMT-A activation via NF-IL6 while the AP1 cis-elements showed a weak but significant activation. In contrast to mammals the metal responsive elements were not activated by oxidative stress. Electrophoretic mobility shift assay (EMSA) mutation analysis revealed that the two most proximal AP1 elements, $AP1_{1,2}$, exhibited strong binding to the AP1 consensus sequence, while the more distal AP1 elements, $AP1_{3,4}$ were ineffective. Phorbol-12-myristate-13-acetate (PMA), a known tumor promoter, resulted in a robust induction of rtMT-A via the AP1 elements alone. To determine the conservation of regulatory functions we transfected human Hep G2 cells with the rtMT-A enhancer constructs and were able to demonstrate that the cis-elements were functionally conserved. The importance of NF-IL6 in regulation of teleost MT is supported by the conservation of these elements in MT genes from different teleosts. In addition, PMA and PQ injection of rainbow trout resulted in increased hepatic rtMT-A mRNA levels.

Conclusions: These studies suggest that AP1 primarily is involved in PMA regulation of the rtMT-A gene while NF-IL6 is involved in free radical regulation. Taken together this study demonstrates the functionality of the NF-IL6 and AP-1 elements and suggests an involvement of MT in protection during pathological processes such as inflammation and cancer.

Keywords: Rainbow trout, Metallothionein-A promoter, Nuclear factor interleukin-6, Activator protein-1, Oxidative stress

Background

Reactive oxygen species (ROS) are continuously being generated during oxygen dependent events in living organisms. ROS production is highly correlated to pathological responses both at the cellular and organismal level [1]. These responses include events such as cancer, cell death and aging. At the cellular level the oxidative stress response results in activation and up regulation of several antioxidant enzymes, such as superoxide dismutase (SOD), catalase and glutathione-s-transferase (GST) as well as non-enzymatic antioxidant compounds, such as ascorbic acid, β-carotene and reduced glutathione, GSH [2]. The role of metallothionein (MT) as a free radical scavenger has been well documented in vitro [3], in cell lines [4], and at the organism level [5,6]. Furthermore, it has been shown that MT is induced by ROS inducing agents such as paraquat (PQ), hydrogen peroxide (H_2O_2) and glutathione depleting agents such as diethylmaleate [1,4,7]. In addition, Phorbol-12-myristate-13-acetate (PMA)

* Correspondence: per-erik.olsson@oru.se
†Equal contributors
[1]Örebro Life Science Center, School of Science and Technology, Örebro University, Örebro SE-701 82, Sweden
Full list of author information is available at the end of the article

is a potent tumor promoter shown to induce MT expression via activator protein 1 (AP1) cis-acting elements in mammals [8-10].

Sequencing of several teleost MT enhancer regions have revealed the presence of distally located nuclear factor interleukin 6 (NF-IL6) and AP1 elements, inferring that these elements are involved in a conserved function [7,11-13]. The transcription factor NF-IL6 is activated by the cytokine IL-6 in response to NF-κB activation. NF-IL6 is suppressed in normal tissues, but is rapidly and drastically induced by inflammation [14].

The composite transcription factor AP1 was first identified as a factor mediating optimal basal level expression of the human MT2A (hMT2A) enhancer region [15-17]. Sequencing of the Fugu genome allowed determination of gene similarity between human and teleost genomes [18]. It has been shown that there are more AP1 genes in Fugu than in mammals and that they share high homology in the DNA binding and dimerisation domains [19]. The sequence data thus indicate that the functionality of the AP1 proteins may be highly similar in both teleosts and mammals.

MT-I induction by ROS in mouse has shown to be mediated by metal responsive elements (MREs) and antioxidant responsive elements (ARE)/upstream stimulatory factor (USF) cis-acting elements [20]. The USF/ARE composite transcription factor has also been identified in a number of other terrestrial vertebrates including chicken [21]. The USF cis-acting element has been shown to participate in basal level transcription of the mouse MT-I gene [22]. The ARE cis-acting element has also been identified and characterized in enhancer regions from metabolizing enzymes participating in the phase II drug response and is activated by electrophilic xenobiotics and H_2O_2 [23]. Oxidative stress has been hypothesized to be the main driving force for gene activation via ARE [24,25]. However, it has been indicated that ARE driven gene expression can occur in the absence of oxidative stress through the transcription factor nuclear factor (erythroid-derived 2)-like 2 (Nrf-2) [26]. The AP1 cis-acting element share high homology to the ARE but is not a component of the protein complex that binds the USF /ARE site on the mouse MT I enhancer region [27]. However, AP1 has been shown to bind to ARE in the NADP (H): quinone oxidoreductase hNQO1 gene enhancer resulting in activation of this gene [28,29].

Functional analysis of the rtMT-A promoter has indicated that free radicals regulate the rtMT-A gene via a region containing multiple copies of AP1 elements and one NF-IL6 element [7]. In the rtMT-A gene there is a distal region of the promoter that contain AP1 elements with a potential to regulate MT gene expression. Upstream of the AP1 elements there is one distinct NF-IL6 element that has not been previously analyzed for its involvement in MT regulation. In studies of teleosts there is no clear correlation between the presence of AP1 elements and free radical responsiveness of the MT genes [30,31]. However, a study on the interaction between MT and free radicals indicate that rtMT-A and sea mussel (*Mytilus galloprovincialis*) MT10 has a higher ROS scavenging capacity than rabbit MT [32]. Furthermore, in an open sea study on cod, a strong correlation was observed between hepatic MT levels and the total ROS scavenging capacity of the liver [33]. Thus, while AP1 has not been clearly correlated with MT induction in teleosts it remains that one of the functions of MT is ROS regulation. To further explore the understanding of MT regulation in teleosts by ROS and tumor promoting agents we have analyzed the relative contribution of AP1 and NF-IL6 elements in the rtMT-A enhancer region. In order to study the conservation of the identified cis-elements we tested the regulatory potential using both homologous as well as heterologous systems. Functional analysis of the rtMT-A promoter suggest that the NF-IL-6 element is instrumental to MT induction by oxidative stress while the AP-1 elements exhibited a strong activation in response to tumor promoting agents such as PMA. The AP-1 elements appeared only to a minor extent contribute to free radical inducibility of the rtMT-A gene. Moreover, hepatic expression of rtMT-A mRNA was substantially increased in response to both oxidative stress and tumor promotion, suggesting that MT may be involved in the protection against pathological processes such as inflammation and cancer.

Results

Basal level expression of the rtMT-A promoter

Transfection of the indicated deletion constructs (Figure 1) in both homologous (RTH-149) and heterologous (Hep G2) systems show that the full-length rtMT-A promoter is required for maximal basal level expression (Figure 2). While single pairs ($AP1_{1,2}$ and $AP1_{3,4}$) of AP1 elements did not initiate transcription, the region containing the complete cluster of AP1 elements ($AP1_{tot}$) enhanced basal level activity 2-fold and 20-fold in the RTH-149 and Hep G2 cell lines respectively. The MRE construct (−793, complete set of 6 MREs) was observed to be important for basal level activity of the rtMT-A promoter in both cell lines. Transfection with the MRE-AP1 (−939, complete set of 6 MREs and 4 AP1 elements) construct resulted in further enhancement of basal level expression.

rtMT-A activity following PQ and H_{2O2} exposure

RTH-149 and Hep G2 cells were transfected with the MRE-pGL3 vector (−793), the MRE-AP1-pGL3 vector (−939), and the AP1-pGL3 vector ($AP1_{tot}$). Treatment of transiently transfected RTH-149 cells with 10 μM PQ (Figure 3A) or 100 μM H_2O_2 (data not shown) did not

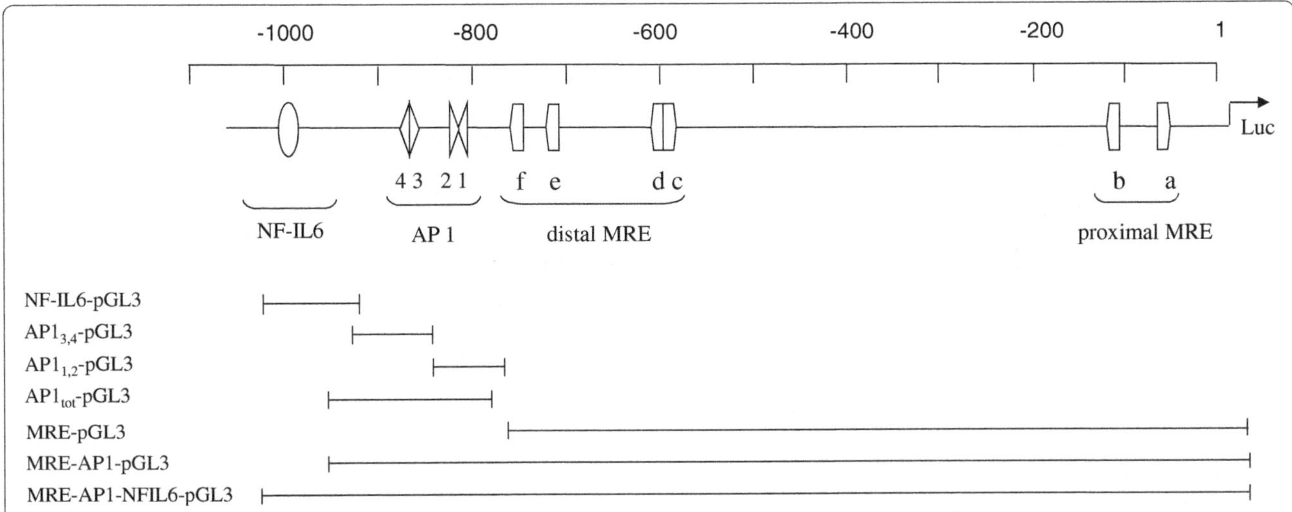

Figure 1 Cis-acting elements of the rtMT-A promoter. The location and orientation of the NF-IL6, AP1 and MRE elements on the rtMT-A enhancer region is shown. There are 6 MREs (MRE), 4 AP1 ($AP1_{tot}$) and 1 NF-IL6 (NF-IL6) element located in the first 1100 bp upstream of the transcriptional start site in the rtMT-A gene. The constructs shown are −1042/-917 that contain the NF-IL6 element alone; -939/-812 that contain $AP1_{3,4}$; -834/-774 that contain $AP1_{1,2}$; -939/-774 ($AP1_{tot}$); -793/+23 (MRE); -939/+23 (MRE-AP1); -1042/+23 (MRE-AP1-NFIL6). The TATA box found in rtMT-A is included in the reverse primers of the distally located constructs. The constructs were cloned into the pGL3-basic vector.

result in a significant increase in luciferase activity. However, transfection of Hep G2 cells followed by 100 μM H_2O_2 exposure resulted in ~1.5-fold induction with both constructs containing the 4 AP1 elements (Figure 3B). Transfection with the MRE-pGL3 vector did not result in up-regulation of the luciferase expression. These data suggest that the AP1 elements of the rtMT-A promoter alone confer free radical inducibility.

Mutation analysis and EMSA of rtMT-A AP1 elements

EMSAs were performed to identify relative binding affinity of the four AP1 elements (the sequences were tested pairwise) of the rtMT-A promoter to the AP1 consensus sequence. A set of normal and mutated AP1 oligonucleotides were used in EMSA competition assays (native and mutated (*) AP1 sequences are presented in Table 1). A strong gel-shift was observed following incubation with labeled AP1 consensus oligonucleotide with HeLa nuclear extracts. This shift was completely abolished by competing with 400-fold excess of either unlabeled AP1 consensus or $AP1_{1,2}$ oligonucleotides (Figure 4A). Mutation of both proximal AP1 elements ($AP1_{1*,2*}$) abolished AP1 binding while separate mutation of the individual elements ($AP1_{1*,2}$ and $AP1_{1,2*}$) resulted in reduced AP1 binding. Thus, both $AP1_1$ and $AP1_2$ were observed to be important for interaction with the AP1 protein complex, with $AP1_2$ exhibiting strongest affinity. The distally located AP1 elements, $AP1_{3,4}$ showed low binding affinity to the AP1 consensus sequence. However, using a 1000 × molar excess of competitor a slight reduction in the intensity of the shift could be observed (Figure 4B). Dose–response competition of native and mutated $AP1_{1,2*}$ oligonucleotides (40, 400 and 1000 × excess) indicate that a 40 fold excess of consensus AP1 element completely competed away the labeled AP1 element. The $AP1_{1,2}$ oligonucleotide was less effective and requires 400 fold excess for complete competition. The two single mutated sequences ($AP1_{1*,2}$ and $AP1_{1,2*}$), also showed dose-dependent competition although requiring higher concentrations of competitor. Mutation of both the proximally located AP1 elements ($AP1_{1,2*}$) resulted in abolished competition with the

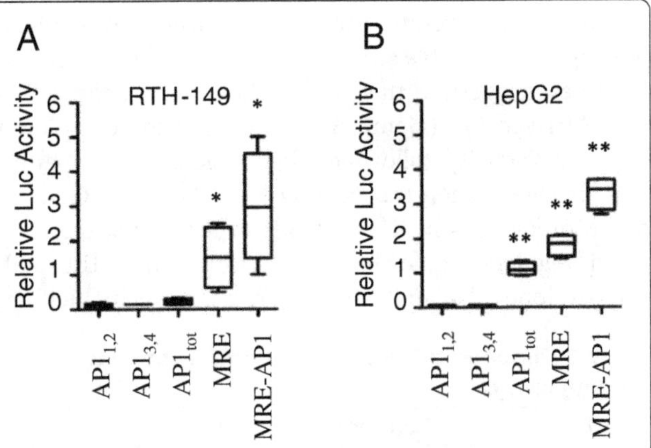

Figure 2 Basal level expression of rtMT-A cis-acting elements. Basal level luciferase activity in RTH-149 **(A)** and Hep G2 **(B)** cells transfected with the indicated rtMT-A promoter constructs. Luciferase activity is given as relative luminescence. All activities were normalized using either the dual luciferase system or β-galactosidase activity. Luciferase activities were analyzed 24 hours post transfection. Results are presented as mean ± SE (n = 4). All other activities were adjusted accordingly. Statistically significant differences from control levels are indicated by *(p < 0.05); ** (p < 0.01).

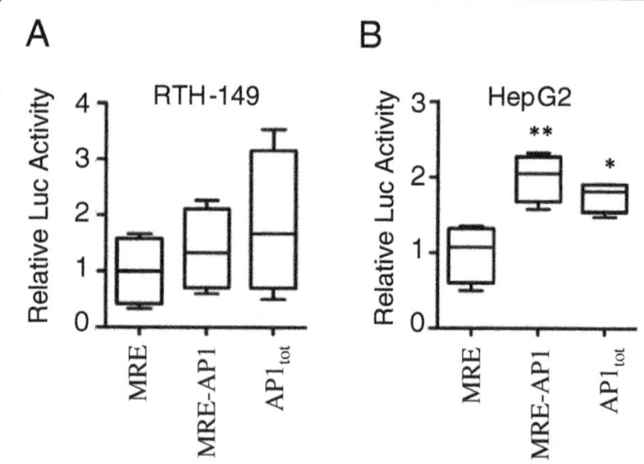

Figure 3 Activation of rtMT-A cis-acting elements following exposure to oxidative stress. Luciferase activity in RTH-149 **(A)** and Hep G2 **(B)** cells transfected with the indicated rtMT-A enhancer vectors. Cells were exposed to 10 μM PQ (RTH-149) or 100 μM H_2O_2 (Hep G2). Activities are given as fold induction. All activities were normalized using either the dual luciferase system or β-galactosidase activity. Luciferase activities were analyzed 24 hours post transfection. Results are presented as mean ± SE (n = 4). Statistically significant differences from control levels are indicated by *($p < 0.05$); ** ($p < 0.01$).

AP1 consensus oligonucleotide. The present functional analysis of the rtMT-A AP1 elements demonstrates that the proximally located $AP1_1$ and $AP1_2$ elements show the highest competition with the AP1consensus sequence.

In vitro inhibition of H_2O_2 induced rtMT-A gene activity

Transfection of Hep G2 cells with AP1-pGL3 and MRE-AP1-pGL3 vectors resulted in a 2-fold, increase in luciferase activity following H_2O_2 exposure. The H_2O_2 induced luciferase activity in cells transfected with AP1 containing constructs was significantly reduced when co-incubated with synthetic double stranded $AP1_{1,2}$ and $AP1_{3,4}$ oligonucleotides, suggesting that the H_2O_2 induced gene activity was AP1-specific (Figure 5). The observed reduction in luciferase activity following oligonucleotide competition was strongest when transfecting with the AP1-pGL3 vector. Moreover, as in the previous experiments, there was no up regulation following transfection of the MRE-pGL3 vector alone (data not shown). A decrease in luciferase activity was also observed in control cells co-incubated with competitor. However this decrease was small compared to the observed decrease in cells exposed to H_2O_2.

Table 1 Oligonucleotides used for EMSA competition binding assays

Primer	Sequence
AP1-1,2	TGG TAT GAC ACA GCT C AA TTA CTC AAG CAG
AP1-1*,2	TGG TAT GAC ACA GCT C AA TTA ATT AAG CAG
AP1-1,2*	TGG TAT AAT ACA GCT CAA TTA CTC AAG CAG
AP1-1*,2*	TGG TAT AAT ACA GCT CAA TTA ATT AAG CAG
AP1-3,4	CTG GTA CTG TCA GTG ACT ATT T

The different $AP1_{1,2}$ and $AP1_{3,4}$ oligonucleotides corresponds to position (−814 to −785) and (−867 to −836) of the rtMT-A promoter respectively. The asterisk (*) indicates a mutated cis-element.

Activity of AP1 and NF-IL6 elements in response to PQ and PMA

Exposure of RTH-149 cells to 10 μM PQ resulted in a 6-fold increase in gene activity following transfection with the MRE-AP1-NF-IL6-pGL3 vector containing the distally located NF-IL-6 element (Figure 6). Transfection with the MRE-AP1-pGL3 vector resulted in a modest increase in luciferase activity while the MRE-pGL3 vector did not elicit an increased luciferase activity. The response to 10 μM PQ was similar following transfection with the NF-IL6-pGL3 vector (Figure 7A), indicating that the NF-IL6 enhancer mediated the observed PQ inducibility. In contrast, transfection with the AP1-pGL3 vector and exposure to 10 μM PQ did not result in an increased luciferase response, suggesting that the AP1 elements are not primarily involved in the PQ response. On the other hand, exposure to 162 nM PMA resulted in a robust induction (5-fold) of luciferase activity following transfection with the AP1-pGL3 vector. The isolated NF-IL-6 element was unresponsive to PMA (Figure 7B). Thus, both the AP1 and the NF-IL6 elements were functional in rainbow trout cells.

In vivo induction of MT-A mRNA by PQ and PMA

Rainbow trout were injected with 10 mg/kg PQ and 10.3 μg/kg PMA in order to determine the effect on endogenous hepatic MT-A gene expression. A 2.5- and a 2-fold induction of hepatic MT-A mRNA was observed following exposure to PQ and PMA respectively (Figure 8), confirming that both inducers result in up regulation of rtMT-A mRNA in vivo.

Discussion

As a consequence of organism utilization of oxygen, deleterious reactive oxygen species are produced. These oxygen species may lead to lipid peroxidation, DNA strand breaks and cell death. However, several antioxidant systems, such as GSH, SOD, catalase and MT have evolved to protect the organism from oxidative stress. A wide variety of stresses, ranging from physical injury to oxidative stress, induce MT in animals [27,34]. Although numerous studies have been performed to confirm MTs antioxidative function, few have focused on the link between regulation of MT and a role during oxidative stress. In the present study we aimed to characterize the regulatory role of the rtMT-A enhancer region in response to oxidative stress and tumor promotion. Previous identification and functional analysis of distal elements on the rtMT-A promoter have revealed that a cluster including 4 AP1 elements and a single NF-IL-6 element may be involved in free radical

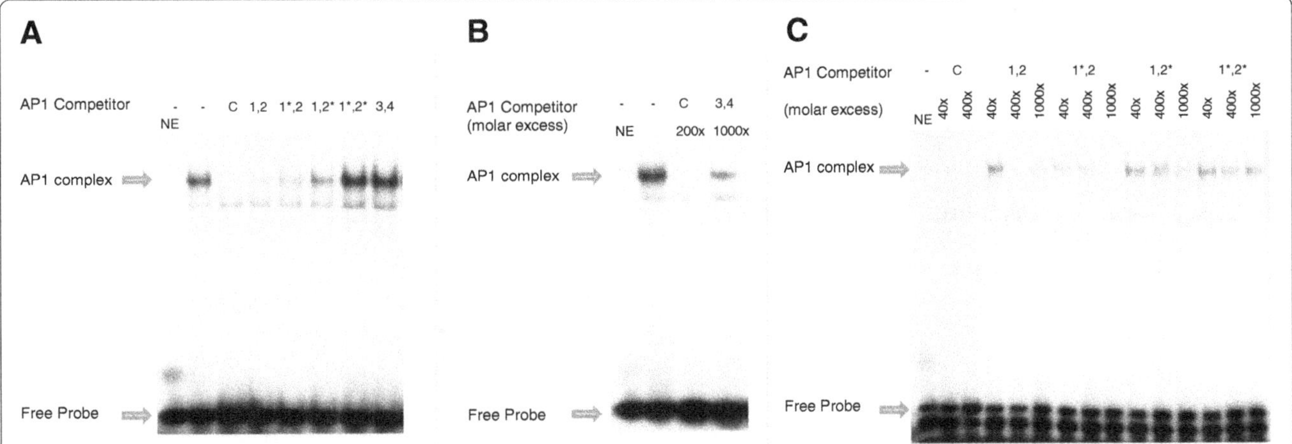

Figure 4 AP1 binding affinity to AP1 consensus sequence. Electrophoretic mobility shift assay (EMSA) on HeLa nuclear extract (10 μg). The AP1 consensus oligonucleotide was labelled with γ-32P ATP. Synthetic oligonucleotides, corresponding to the rtMT-A promoter were used as competitors. The constructs used as competitors were: Consensus AP1 **(C)**; $AP1_{1,2}$ (1,2); $AP1_{1*,2}$ (1*,2); $AP1_{1,2*}$ (1,2*); $AP1_{1*,2*}$ (1*,2*) and $AP1_{3,4}$ (3,4). Mutated AP1 elements are indicated by an asterisk (*). NE, no extract. The location of the AP1 protein complex and free unbound probe is indicated by arrows. **(A)** Competitor oligonucleotides were added in 400 x molar excess. **(B)** The AP1 consensus and $AP1_{3,4}$ were added in 200 x and 10,000 x molar excess respectively. **(C)** Dose–response analysis of AP1 oligonucleotides. Synthetic oligonucleotides, corresponding to the AP1 elements in the rtMT-A promoter were used as competitors. Competitor oligonucleotides were added in 40 x, 400 x and 1000 x molar excess.

inducibility [7]. While free radical regulation of mammalian MT genes seem to be mediated via USF/ARE and MRE cis-acting elements [20], teleost MT genes may be regulated via conserved clusters of cis-acting elements sharing high homology to the NF-IL6 and the AP1 consensus core sequences [7,11,12]. Functional analysis, from the present study, of the rtMT-A promoter suggest that the NF-IL-6 element is instrumental to MT induction by oxidative stress (PQ), and the AP-1 elements may to a minor but significant extent contribute to free radical inducibility.

The observed absence of ROS induced MRE activation in the present study demonstrates that MT regulation in rainbow trout differ from that in mammals where the MRE binding transcription factor MTF-1 is activated following oxidative stress [35]. Since the MRE elements identified on the rtMT-A promoter were not observed to contribute to the oxidative response this suggests that teleost MTF-1 is not activated by oxidative stress. Characterization of MTF-1, from zebrafish and rainbow trout has revealed a high conservation with regard to binding specificity and properties [36]. However, this study points out that there might be different mechanisms that regulate MT gene expression during oxidative stress in different species.

While USF/ARE and MTF-1 mediate oxidative MT expression in mammals, the AP1 cis-acting elements identified on the rtMTA promoter, sharing high homology to the ARE core sequence, showed weak activation in response to oxidative stress. The AP1 cis-acting element was originally discovered on the hMTIIa gene mediating optimal basal level expression of MT [37]. Functional

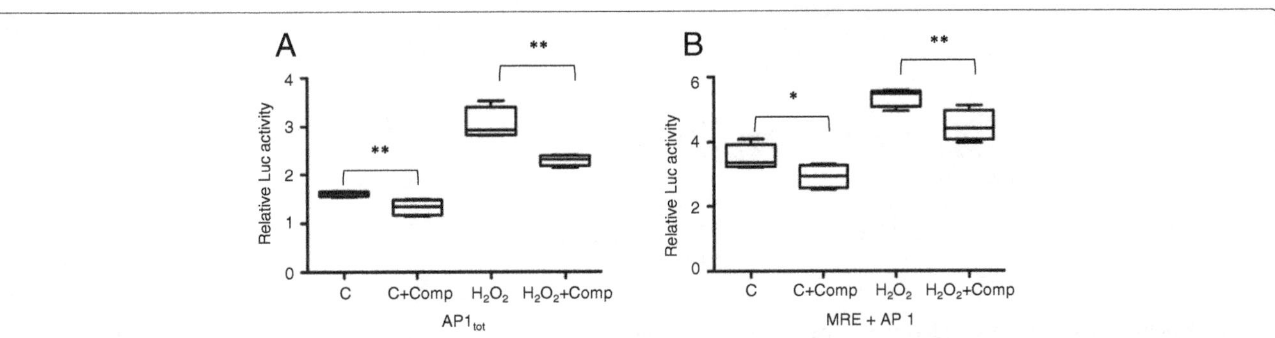

Figure 5 ***In vitro*** **competition assay.** Hep G2 cells were transfected with the 4AP1-pGL3 vector ($AP1_{tot}$) **(A)** or 6MRE-4AP1-pGL3 vector (MRE-AP1) **(B)** alone or together with 50 x molar excess of the $AP1_{1-2}$ and $AP1_{3-4}$ oligonucleotides. Control groups incubated without exposure (Control vs. Control-Comp), while experimental cells were exposed to 200 μM H_2O_2 (H_2O_2 vs. H_2O_2 Comp). Results are presented as mean ± SE (n = 4). Statistically significant differences between groups are indicated by *(p < 0.05); **(p < 0.01).

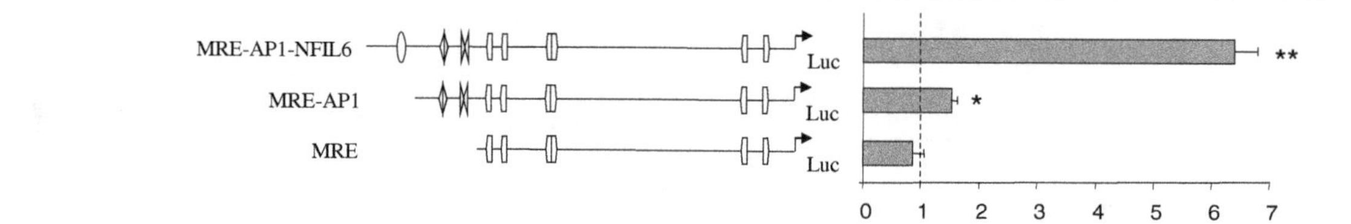

Figure 6 Activation of rtMT-A deletion constructs following exposure to PQ. RTH-149 cells were transfected with 0.5 ug of MRE-AP1-NFIL6-pGL3, MRE-AP1-pGL3 and MRE-pGL3 vectors and 0.3 μg of pRL-CMV vector. Following transfection cells were exposed to 10 μM PQ. The activities are given as fold induction. All activities were normalized to the expression level of the pGL3-basic vector. Luciferase activities were analyzed 24 hours post transfection. Results are presented as mean ± SE (n = 4). Statistically significant differences from control levels are indicated by *($p < 0.05$); **($p < 0.01$).

analysis of the identified rtMT-A elements strongly indicated that the AP1 elements were required for maximal basal level expression in both fish (RTH-149) and mammalian cell (Hep G2) systems. Further analysis of binding affinity for the AP1 consensus sequence indicate that the proximal AP1 pair exhibited highest binding affinity, while the binding activity of the distal AP1 elements was at the border of detection limit. In addition, mutational analysis indicated that $AP1_2$ showed highest binding of the proximally located pair. Hence, functional analysis of the identified AP1 elements suggests that $AP1_{1,2}$ is functional with respect to both AP1 protein complex interaction and transactivation. However, there have been conflicting reports on the involvement of AP1 in the free radical regulation of MT in teleosts [30,31]. While a deletion construct containing MRE and AP1 responded to ROS in common carp [30], the zebrafish MT gene promoter that contains both AP1 and MRE elements did not respond to ROS [30,31]. In the present study co-transfection of Hep G2 cells with rtMT-A AP1 containing constructs with rtMT-A AP1 oligonucleotides abolished H_2O_2 induced promoter activity. These data suggest that the identified AP1 elements on the rtMT-A promoter in rainbow trout specifically mediate free radical MT inducibility. The primary role of the AP1 protein complex is as a modulator of cell proliferation and differentiation. It has been indicated that there is a link between proliferating human hepatic cells and high expression of MT protein [38]. Exposure of RTH-149 cells to the tumor promoter PMA resulted in a strong activation of AP1 cis-elements in the present study. In vivo injection of rainbow trout with PMA strongly induced hepatic rtMT-A mRNA levels, confirming the functionality of the AP1 elements on the rtMT-A promoter. These data strengthen the view of MT as a modulator of cell proliferation and differentiation. In the teleost CHSE-214 cell line MT becomes progressively methylated during prolonged subculturing, coinciding with a decrease in MT expression and reduced cell proliferation [39]. It has been

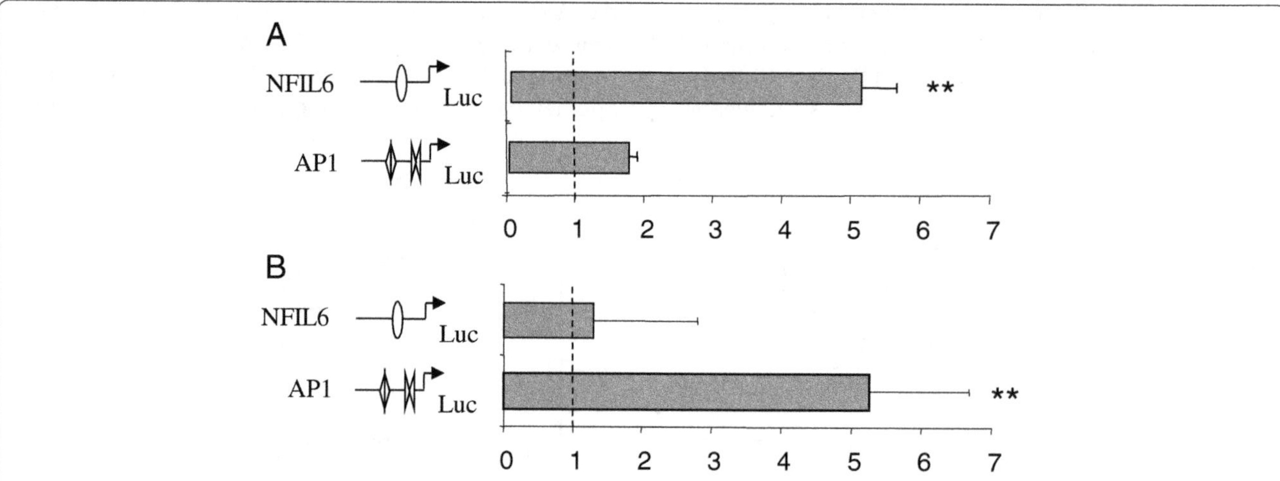

Figure 7 Activation of rtMT-A deletion constructs following exposure to PQ and PMA. RTH-149 cells were transfected with 0.5 μg of either the NFIL6-pGL3 or the AP1-pGL3 (AP_{tot}) vectors and 0.3 μg of pRL-CMV vector (Renilla Luciferase control reporter vector). Following transfection cells were exposed to 10 μM PQ **(A)** or to 162 nM PMA **(B)**. The activities are given as fold induction. All activities were normalized to the expression level of the pGL3-basic vector. Luciferase activities were analyzed 24 hours post transfection. Results are presented as mean ± SE (n = 4). Statistically significant differences from control levels are indicated by **($p < 0.01$).

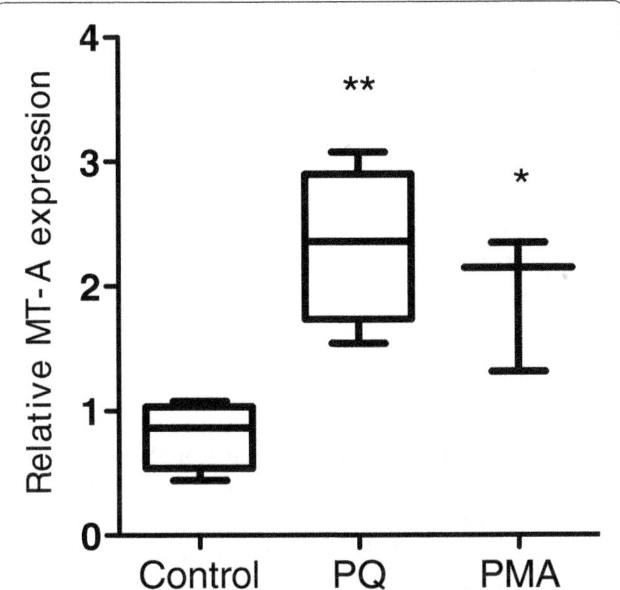

Figure 8 Hepatic MT mRNA expression following injection with PQ or PMA. Real-time quantitative PCR was used to determine the relative gene expression levels of the rtMT-A gene. Control fish were injected with 1% DMSO, while experimental fish were injected with 10 μM PQ or PMA and exposed for 24 hours. Results are presented as mean ± SE (n = 5). Expression levels were normalized against EF1α using ΔΔCT method. Statistically significant differences from control levels are indicated by *(p < 0.05); ** (p < 0.01).

suggested that antioxidants such as ascorbate, α-tocopherol and β-carotene are inhibitory to differentiation [40]. MT has also been suggested to alter the cellular redox state [41], indicating a role for MT during development and differentiation.

Cellular responses following stress, such as injury, promote a transient activation of NF-IL6 and AP1 protein that both play key roles in the initiation of inflammatory responses. NF-IL6 is rapidly activated in response to cytokines and oxidative stress [42]. This is in support with the present study in the RTH-149 cell line demonstrating that the NF-IL6 element was substantially activated in response to PQ but not to PMA. NF-IL6 sites are also present in the MT enhancer of other teleosts such as the crucian carp [12] suggesting a conserved function of MT in response to inflammation where NF-IL6 may mediate ROS inducibility. Studies have demonstrated that NF-IL6 activity can be modulated following protein-protein interaction with the AP1 protein complex [43] and NF-κB [14]. The observed minor increase in AP1 element activity following PQ exposure may enhance MT expression to protect from oxidative stress in fish. Rainbow trout cells respond to ROS exposure by induction of MT, which result in cellular protection from ROS toxicity [3,4,6]. Furthermore, it has been shown in a recent study on cod that there is a close correlation between the hepatic levels of ROS and MT [33]. IL6 and CXCL8 are two of the first inflammatory mediators expressed and contain cis-acting sites for both AP1 and NF-IL6, which indicate their crucial role in acute phase responses [44,45]. Kanekiyo and colleagues [46] demonstrated that MT gene activation by zinc regulates macrophage colony stimulating factor (M-CSF), which in turn recruit and stimulate cytokine production by monocytes. Mice with genetic deletions in the MT proteins have a significant reduction in inflammatory mediators, including TNFα, IL6 and IL1α, compared to wild type mice [47]. Furthermore, growth hormone (GH) was found to induce the expression of rtMT-A) [48]. GH treatment results in phosphorylation of NF-IL6 and increases its transcriptional activity. This suggests that GH may be able to modulate MT regulation through NF-IL6 signaling.

Conclusions

The present study demonstrates that the NF-IL6 and AP1 cis-acting elements of the rtMT-A promoter are functionally active. While NF-IL6 was instrumental for MT induction by oxidative stress, the AP1 elements was primarily and substantially activated in response to tumor promotion. Since NF-IL6 is a key component of initiation of inflammation it appears that MT may regulate this process by free radical scavenging. These data strengthens the idea of MT as an important regulator of proliferation and differentiation. In addition, there appears to be a complex interplay between NF-IL6 and AP1 that needs to be addressed in future studies to understand the link between MT expression, inflammation, development and differentiation.

Methods

Construction of luciferase plasmids

PCR was used to construct the desired rtMT-A promoter luciferase plasmids. The previously sequenced 5′flanking region of the rtMT-A enhancer region was used as a template for all PCR reactions. Forward and reverse primers, containing a Kpn I and Hind III site respectively were used to create the desired luciferase constructs (Table 2). The binding sites for the PCR primers used to create the different rtMT-A enhancer region constructs are shown in Figure 1. The PCR products were resolved on agarose gels and purified using the Qiagen PCR purification kit. Purified PCR fragments were subsequently cloned into the Kpn I –Hind III site of the luciferase reporter vector pGL3-basic (Promega, USA). The size of each insert was confirmed by restriction digest using the Kpn I and Hind III restriction enzymes. Large preparations of each construct were then performed using the Qiagen Midi kit. Each plasmid construct was resolved on agarose gels to test the purity.

The fish handling procedures were approved by the Swedish Ethical Committee in Umea (Permit A75-01).

Table 2 Oligonucleotides used to create rtMT-A gene promoter deletion mutants

Primer	Location	Sequence*
NF-IL6 forward	(−1042 to −1020)	GCG **GGT ACC** TAT GTT CGA TTG GAC TAT GAT TC
AP1 forward 1	(−939 to −917)	GCG **GGT ACC** TGA TAG ACT ATC CTT GTT GTA GG
AP1 forward 2	(−834 to −815)	GCG **GGT ACC** ATA ACA TTG CAC AAT GTT TG
MRE forward	(−793 to −773)	CGG **GGT ACC** TCA AGC AGG AGA TTC TGG AA
NF-IL6 reverse	(917 to −939)	GCG **AAG CTT** *TTT ATA* TCG CTA CAA TTA ATT ACA AAC GAC CG
AP1 reverse 1	(−774 to −790)	GCG **AAG CTT** *TTT ATA* TCG CCA GAA TCT CCT GCT TG
AP1 reverse 2	(−812 to −829)	GCG **AAG CTT** *TTT ATA* TCG CCA CAA ACA TTG TGC AAT G
MRE reverse	(+23 to +5)	GCG **AAG CTT** CAG TGG TGT GTT GTC AGC G

*The restriction enzyme sites are shown in bold (the 3′ primers contain a Hind III site and the 5′ primers contain a Kpn I site) and TATAA box sequences are shown in italics. The MRE contains the natural TATAA box found in rtMT-A.

Cell culture conditions

The rainbow trout hepatoma (RTH-149) cells were propagated at 18°C and the human hepatoblastoma (Hep G2) cells were propagated at 37°C. Both cell lines were grown in minimum essential medium with Earle's salts (GIBCO, Life Technologies) and supplemented with 10% fetal calf serum (FCS) and 1% L-glutamine in an atmosphere of 5% CO_2

Transfection

The cells were co-transfected with 0.5 μg of enhancer coupled pGL3-basic vector or the empty pGL3-basic vector and 0.3 g of pRL-CMV vector (Renilla Luciferase control reporter vector, Promega, USA), in serum free medium using Lipofectamin 2000 (Invitrogen, USA). The cells were transfected under serum free conditions for 15 hours. Hep G2 cells were grown in 250 ml tissue culture bottles to semi confluence and were then seeded in 3 cm-diameter culture dishes, at a density of 15,600 cells/cm^2, 24 hours prior to transfection. The cells were transfected with 1 μg of enhancer coupled pGL3-basic vectors and co-transfected with 1 μg of the pSV-β-galactosidase plasmid containing the simian virus 40 early promoters. In the competition experiments 50 molar excess of synthetic $AP1_{1,2}$ and $AP1_{3,\ 4}$ (for description of oligonucleotides see Table 1) was co-transfected with the AP1-pGL3 vector or the MRE-AP1-pGL3 vector. The total time of transfection in the competition experiment was 14 hours. It was observed that longer transfection times resulted in reduced effect of the oligonucleotides and was probably due to oligonucleotide degradation.

Reporter gene assays and exposure to PQ and PMA

Following transfection cells were re-fed with fresh media and allowed to recover for 2–4 hours. Fresh media alone or media containing 10 μM PQ, 162 nM PMA or H_2O_2 (100 and 200 μM) was added to the cells. After 24 hours exposure, the cells were washed with phosphate buffered saline (PBS) and harvested using cell lysis buffer (Promega, USA). The lysed Hep G2 cells were centrifuged and the supernatant stored at −20°C until β-galactosidase (control) and luciferase assays were performed. Dual-Luciferase reporter assay system (Promega, USA) was used in RTH-149 cell experiments, using the Renilla luciferase vector (pRL-CMV) as a control. For luciferase assays cell lysate were mixed with luciferase substrate and the luciferase activity of the construct was immediately measured in a luminometer (Turner Designs). The cells co-transfected with pRL-CMV was thereafter measured for Renilla luciferase activity. For β-galactosidase assays cell lysate and substrate buffer were mixed and incubated for 30 minutes or until a faint yellow color appeared. The absorbance was then measured at 420 nm. The luciferase activities in Hep G2 cells were normalized for β-galactosidase activity and expressed as relative luciferase activity or luciferase arbitrary units. The activity in RTH-149 cells was calculated by the quotient of the construct/Renilla luminescence.

Electrophoretic mobility shift assay (EMSA)

EMSAs were performed using commercially available HeLa (human) nuclear extracts (Promega, USA). Synthetic normal and mutated AP1 oligonucleotides were synthesized (DNA technology, Denmark). The oligonucleotides that were used are described in Table 1. Complementary strands were denatured for 5 minutes at 80°C and allowed to anneal by slow cooling to room temperature. Both strands were labelled with [−32P] ATP and T4 polynucleotide kinase. EMSA mixtures contained 40,000 cpm (~0.2 ng) and 10 μg of nuclear protein in a final volume of 15 μl of 4% glycerol; 1 mM $MgCl_2$; 0.5 mM DTT; 50 mM NaCl; 10 mM Tris–HCl (pH 7.5) with 1 μg of poly (dI-dC) as nonspecific competitor. The binding reaction with labeled AP1 consensus oligonucleotide was incubated for 20 minutes at room temperature. Incubation of the competitor DNA fragment, in molar excess, with nuclear protein and binding buffer for 10 minutes at room temperature, was performed to initiate the completion binding experiments. The labeled probe was then added, and incubation was allowed to proceed for another

20 minutes at room temperature. 1.5 μl of 10 × gel-shift loading buffer was added to each sample. The reaction mixtures were then loaded onto 4% non-denaturing polyacrylamide (37.5:1) gels at 250 V for approximately 2 hours. The gels were subsequently dried and autoradiographed at –70°C, using an intensifying screen.

In vivo PQ and PMA exposure

Fish (100 g) were kept in aquaria for 1 week prior to experimental start. Fish were injected with either PQ (10 mg/kg) or PMA (10.3 μg/kg). All fish received a total volume of 200 μl/100 g fish containing 1.5% DMSO dissolved in PBS (n = 5) Controls received 1.5% DMSO alone. Following injection the fish were kept for 4 days. Fish were killed with a blow to the head and the livers were removed and frozen in liquid nitrogen and stored at –80°C until analyzed.

Real time qPCR

Total RNA was isolated from liver samples using NucleoSpin RNAII kit (Macherey-Nagel, Germany) and quantified by Nano-vue (GE Healthcare, USA). cDNA was prepared using qScript cDNA synthesis kit (Quanta Biosciences, USA). The qPCR was performed using KAPA SYBR FAST qPCR kit (Kapabiosystems, USA) according to manufacturer's recommendations. Obtained Ct values were normalized against elongation factor EF1 alpha (EF1α). Relative gene-expression was determined by using the ΔΔCt method [49]. The following primer sequences was used for EF1α forward (5′GCATCAAG CAGTGGTCGAGTGA′3), EF1α reverse (5′TTGAAA GAGCCCTTGCCCATCTCA′3), rtMT-A forward, (5′ ACACCACTGACACCCAGACAAACT′3) and rtMT-A reverse (5′AGCTGGTATCACAAGTCTTGCCCT′3).

Statistical analysis

Statistical differences was determined using the two-tailed Student *t*-Test. Statistical significance level was determined at the *P < 0.05 and **P < 0.01 level.

Competing interests

The authors declare that they have no competing interests.

Authors' contribution

PEO obtained the fundings. PK, CM and PEO designed the study. PK, CM, HM, HK and JvH carried out the experimental work. All authors drafted, revised and approved the final manuscript.

Acknowledgements

This work was supported financially by the Knowledge Foundation, Kempes' foundation, Hierta Retzius' foundation, Wallenbergs' foundation and Örebro University.

Author details

[1]Örebro Life Science Center, School of Science and Technology, Örebro University, Örebro SE-701 82, Sweden. [2]Department of Zoology, Göteborg University, Göteborg SE-405 30, Sweden. [3]Department of Molecular Biology, Umeå University, Umeå SE-901 87, Sweden.

References

1. Halliwell B, Gutteridge JM: **Free radicals and antioxidant protection: mechanisms and significance in toxicology and disease.** *Hum Toxicol* 1988, **7:**7–13.
2. Jones DP: **Radical-free biology of oxidative stress.** *Am J Physiol Cell Physiol* 2008, **295:**C849–C868.
3. Thornalley PJ, Vasak M: **Possible role for metallothionein in protection against radiation-induced oxidative stress. Kinetics and mechanism of its reaction with superoxide and hydroxyl radicals.** *Biochim Biophys Acta* 1985, **827:**36–44.
4. Kling PG, Olsson P: **Involvement of differential metallothionein expression in free radical sensitivity of RTG-2 and CHSE-214 cells.** *Free Radic Biol Med* 2000, **28:**1628–1637.
5. Liu J, Liu Y, Hartley D, Klaassen CD, Shehin-Johnson SE, Lucas A, Cohen SD: **Metallothionein-I/II knockout mice are sensitive to acetaminophen-induced hepatotoxicity.** *J Pharmacol Exp Ther* 1999, **289:**580–586.
6. Zheng H, Liu J, Liu Y, Klaassen CD: **Hepatocytes from metallothionein-I and II knock-out mice are sensitive to cadmium- and tert-butylhydroperoxide-induced cytotoxicity.** *Toxicol Lett* 1996, **87:**139–145.
7. Olsson PE, Kling P, Erkell LJ, Kille P: **Structural and functional analysis of the rainbow trout (*Oncorhyncus mykiss*) metallothionein-A gene.** *Eur J Biochem* 1995, **230:**344–349.
8. Ebinu JO, Stang SL, Teixeira C, Bottorff DA, Hooton J, Blumberg PM, Barry M, Bleakley RC, Ostergaard HL, Stone JC: **Ras GRP links T-cell receptor signaling to Ras.** *Blood* 2000, **95:**3199–3203.
9. Angel P, Poting A, Mallick U, Rahmsdorf HJ, Schorpp M, Herrlich P: **Induction of metallothionein and other mRNA species by carcinogens and tumor promoters in primary human skin fibroblasts.** *Mol Cell Biol* 1986, **6:**1760–1766.
10. Imbra RJ, Karin M: **Metallothionein gene expression is regulated by serum factors and activators of protein kinase C.** *Mol Cell Biol* 1987, **7:**1358–1363.
11. He P, Xu M, Ren H: **Cloning and functional characterization of 5′-upstream region of metallothionein-I gene from crucian carp (*Carassius cuvieri*).** *Int J Biochem Cell Biol* 2007, **39:**832–841.
12. Ren H, Xu M, He P, Muto N, Itoh N, Tanaka K, Xing J, Chu M: **Cloning of crucian carp (*Carassius cuvieri*) metallothionein-II gene and characterization of its gene promoter region.** *Biochem Biophys Res Commun* 2006, **342:**1297–1304.
13. Samson SL, Paramchuk WJ, Gedamu L: **The rainbow trout metallothionein-B gene promoter: contributions of distal promoter elements to metal and oxidant regulation.** *Biochim Biophys Acta* 2001, **1517:**202–211.
14. Matsusaka T, Fujikawa K, Nishio Y, Mukaida N, Matsushima K, Kishimoto T, Akira S: **Transcription factors NF-IL6 and NF-kappa B synergistically activate transcription of the inflammatory cytokines, interleukin 6 and interleukin 8.** *Proc Natl Acad Sci USA* 1993, **90:**10193–10197.
15. Haslinger A, Karin M: **Upstream promoter element of the human metallothionein-IIA gene can act like an enhancer element.** *Proc Natl Acad Sci U S A* 1985, **82:**8572–8576.
16. Karin M, Haslinger A, Heguy A, Dietlin T, Imbra R: **Transcriptional control mechanisms which regulate the expression of human metallothionein genes.** *Experientia Suppl* 1987, **52:**401–405.
17. Scholer H, Haslinger A, Heguy A, Holtgreve H, Karin M: **In vivo competition between a metallothionein regulatory element and the SV40 enhancer.** *Science* 1986, **232:**76–80.
18. Aparicio S, Chapman J, Stupka E, Putnam N, Chia JM, Dehal P, Christoffels A, Rash S, Hoon S, Smit A, *et al*: **Whole-genome shotgun assembly and analysis of the genome of *Fugu rubripes*.** *Science* 2002, **297:**1301–1310.
19. Cottage AJ, Edwards YJ, Elgar G: **AP1 genes in Fugu indicate a divergent transcriptional control to that of mammals.** *Mamm Genome* 2003, **14:**514–525.
20. Dalton T, Palmiter RD, Andrews GK: **Transcriptional induction of the mouse metallothionein-I gene in hydrogen peroxide-treated Hepa cells involves a composite major late transcription factor/antioxidant response element and metal response promoter elements.** *Nucleic Acids Res* 1994, **22:**5016–5023.
21. Fernando LP, Andrews GK: **Cloning and expression of an avian metallothionein-encoding gene.** *Gene* 1994, **81:**177–183.

22. Gregor PD, Sawadogo M, Roeder RG: **The adenovirus major late transcription factor USF is a member of the helix-loop-helix group of regulatory proteins and binds to DNA as a dimer.** *Genes Dev* 1990, **4:**1730–1740.
23. Jaiswal AK: **Antioxidant response element.** *Biochem Pharmacol* 1994, **48:**439–444.
24. Ishii T, Itoh K, Takahashi S, Sato H, Yanagawa T, Katoh Y, Bannai S, Yamamoto M: **Transcription factor Nrf2 coordinately regulates a group of oxidative stress-inducible genes in macrophages.** *J Biol Chem* 2000, **275:**16023–16029.
25. Rushmore TH, Morton MR, Pickett CB: **The antioxidant responsive element. Activation by oxidative stress and identification of the DNA consensus sequence required for functional activity.** *J Biol Chem* 1991, **266:**11632–11639.
26. Lee JM, Moehlenkamp JD, Hanson JM, Johnson JA: **Nrf2-dependent activation of the antioxidant responsive element by tert-butylhydroquinone is independent of oxidative stress in IMR-32 human neuroblastoma cells.** *Biochem Biophys Res Commun* 2001, **280:**286–292.
27. Andrews GK: **Regulation of metallothionein gene expression by oxidative stress and metal ions.** *Biochem Pharmacol* 2000, **59:**95–104.
28. Venugopal R, Jaiswal AK: **Nrf2 and Nrf1 in association with Jun proteins regulate antioxidant response element-mediated expression and coordinated induction of genes encoding detoxifying enzymes.** *Oncogene* 1998, **17:**3145–3156.
29. Venugopal R, Jaiswal AK: **Nrf1 and Nrf2 positively and c-Fos and Fra1 negatively regulate the human antioxidant response element-mediated expression of NAD(P)H:quinone oxidoreductas1 gene.** *Proc Natl Acad Sci USA* 1996, **93:**14960–14965.
30. Chan PC, Shiu CK, Wong FW, Wong JK, Lam KL, Chan KM: **Common carp metallothionein-1 gene: cDNA cloning, gene structure and expression studies.** *Biochim Biophys Acta* 2004, **1676:**162–171.
31. Yan CH, Chan KM: **Cloning of zebrafish metallothionein gene and characterization of its gene promoter region in HepG2 cell line.** *Biochim Biophys Acta* 2004, **1679:**47–58.
32. Buico A, Cassino C, Dondero F, Vergani L, Osella D: **Radical scavenging abilities of fish MT-A and mussel MT-10 metallothionein isoforms: an ESR study.** *J Inorg Biochem* 2008, **102:**921–927.
33. Chesman BS, O'Hara S, Burt GR, Langston WJ: **Hepatic metallothionein and total oxyradical scavenging capacity in Atlantic cod, *Gadus morhua* caged in open sea contamination gradients.** *Aquat Toxicol* 2007, **84:**310–320.
34. Oh SH, Deagen JT, Whanger PD, Weswig PH: **Biological function of metallothionein. V. Its induction in rats by various stresses.** *Am J Physiol* 1978, **234:**E282–E285.
35. Dalton TP, Li Q, Bittel D, Liang L, Andrews GK: **Oxidative stress activates metal-responsive transcription factor-1 binding activity. Occupancy in vivo of metal response elements in the metallothionein-I gene promoter.** *J Biol Chem* 1996, **271:**26233–26241.
36. Dalton TP, Solis WA, Nebert DW, Carvan MJ 3rd: **Characterization of the MTF-1 transcription factor from zebrafish and trout cells.** *Comp Biochem Physiol B* 2000, **126:**325–335.
37. Lee W, Haslinger A, Karin M, Tijan R: **Activation of transcription by two factors that bind promoter and enhancer sequences of the human metallothionein gene and SV40.** *Nature* 1987, **325:**368–372.
38. Studer R, Vogt CP, Cavigelli M, Hunziker PE, Kägi JH: **Metallothionein accretion in human hepatic cells is linked to cellular proliferation.** *Biochem J* 1997, **328:**63–67.
39. Kling P: *Metallothionein regulation and function in teleosts during metal- and free radical exposure.* Umea University, Umea, Sweden: Doctoral thesis; 2001.
40. Allen RG, Venkatraj VS: **Oxidants and antioxidants in development and differentiation.** *J Nutr* 1992, **122:**631–635.
41. Ye B, Maret W, Vallee BL: **Zinc metallothionein imported into liver mitochondria modulates respiration.** *Proc Natl Acad Sci USA* 2001, **98:**2317–2322.
42. Klampfer L, Lee TH, Hsu W, Vilcek J, Chen-Kiang S: **NF-IL6 and AP-1 cooperatively modulate the activation of the TSG-6 gene by tumor necrosis factor alpha and interleukin-1.** *Mol Cell Biol* 1994, **14:**6561–6569.
43. Hsu W, Kerppola TK, Chen PL, Curran T, Chen-Kiang S: **Fos and Jun repress transcription activation by NF-IL6 through association at the basic zipper region.** *Mol Cell Biol* 1994, **14:**268–276.
44. Kang SS, Woo SS, Im J, Yang JS, Yun CH, Ju HR, Son CG, Moon EY, Han SH: **Human placenta promotes IL-8 expression through activation of JNK/SAPK and transcription factors NF-kappaB and AP-1 in PMA-differentiated THP-1 cells.** *Int Immunopharmacol* 2007, **7:**1488–1495.
45. Ondrey FG, Dong G, Sunwoo J, Chen Z, Wolf JS, Crowl-Bancroft CV, Mukaida N, Van Waes C: **Constitutive activation of transcription factors NF-(kappa)B, AP-1, and NF-IL6 in human head and neck squamous cell carcinoma cell lines that express pro-inflammatory and pro-angiogenic cytokines.** *Mol Carcinog* 1999, **26:**119–129.
46. Kanekiyo M, Itoh N, Kawasaki A, Matsuda K, Nakanishi T, Tanaka K: **Metallothionein is required for zinc-induced expression of the macrophage colony stimulating factor gene.** *J Cell Biochem* 2002, **86:**145–153.
47. Kanekiyo M, Itoh N, Kawasaki A, Matsuyama A, Matsuda K, Nakanishi T, Tanaka K: **Metallothionein modulates lipopolysaccharide-stimulated tumour necrosis factor expression in mouse peritoneal macrophages.** *Biochem J* 2002, **361:**363–369.
48. Vergani L, Lanza C, Borghi C, Scarabelli L, Panfoli I, Burlando B, Dondero F, Viarengo A, Gallo G: **Efects of growth hormone and cadmium on the transcription regulation of two metallothionein isoforms.** *Mol Cell Endocrinol* 2007, **263:**29–37.
49. Schmittgen TD, Livak KJ: **Analyzing real-time PCR data by the comparative C(T) method.** *Nat Protoc* 2008, **3:**1101–1108.

14

A plasmid toolkit for cloning chimeric cDNAs encoding customized fusion proteins into any Gateway destination expression vector

Raquel Buj[1,3], Noa Iglesias[1], Anna M Planas[1,2] and Tomàs Santalucía[1,2*]

Abstract

Background: Valuable clone collections encoding the complete ORFeomes for some model organisms have been constructed following the completion of their genome sequencing projects. These libraries are based on Gateway cloning technology, which facilitates the study of protein function by simplifying the subcloning of open reading frames (ORF) into any suitable destination vector. The expression of proteins of interest as fusions with functional modules is a frequent approach in their initial functional characterization. A limited number of Gateway destination expression vectors allow the construction of fusion proteins from ORFeome-derived sequences, but they are restricted to the possibilities offered by their inbuilt functional modules and their pre-defined model organism-specificity. Thus, the availability of cloning systems that overcome these limitations would be highly advantageous.

Results: We present a versatile cloning toolkit for constructing fully-customizable three-part fusion proteins based on the MultiSite Gateway cloning system. The fusion protein components are encoded in the three plasmids integral to the kit. These can recombine with any purposely-engineered destination vector that uses a heterologous promoter external to the Gateway cassette, leading to the in-frame cloning of an ORF of interest flanked by two functional modules. In contrast to previous systems, a third part becomes available for peptide-encoding as it no longer needs to contain a promoter, resulting in an increased number of possible fusion combinations. We have constructed the kit's component plasmids and demonstrate its functionality by providing proof-of-principle data on the expression of prototype fluorescent fusions in transiently-transfected cells.

Conclusions: We have developed a toolkit for creating fusion proteins with customized N- and C-term modules from Gateway entry clones encoding ORFs of interest. Importantly, our method allows entry clones obtained from ORFeome collections to be used without prior modifications. Using this technology, any existing Gateway destination expression vector with its model-specific properties could be easily adapted for expressing fusion proteins.

Keywords: Gateway cloning, Fusion protein, Combinatorial, Fluorescent protein, Epitope tag

Background

The availability of genomic data from an ever increasing number of species, arising from the completion of multiple genome projects and the metagenomic analysis of natural samples, has uncovered a potential treasure trove of proteins with as-yet undiscovered properties that could be exploited to solve both existing and future biotechnology challenges. The huge amount of sequence data currently being generated calls not only for bioinformatic prediction of open-reading frames (ORFs) and their putative functions [1], but also for novel screening techniques (functional metagenomics) that will help identifying gene products with potential technological interest [2]. Metagenomics is already revealing a wealth of new protein families with unknown functions, as well as distant relatives of other proteins with well-characterised functions, which will be very valuable for evolutionary and structural studies [3]. These studies will likely require the expression of newly-described ORFs in heterologous hosts

* Correspondence: tomas.santalucia@iibb.csic.es
[1]Department of Brain Ischemia and Neurodegeneration, Institut d'Investigacions Biomèdiques de Barcelona (IIBB)-Consejo Superior de Investigaciones Científicas (CSIC), Barcelona, Spain
[2]Institut d'Investigacions Biomèdiques August Pi i Sunyer (IDIBAPS), Barcelona, Spain
Full list of author information is available at the end of the article

for their initial characterisation, and expressing them as protein fusions with standard tags might be advantageous. One of the main hurdles in metagenomic screening derives from the existence of profound differences in the way that different taxonomic groups of organisms translate ORFs (i.e. codon bias and use of different initiation codons) [2]. These differential features demand for the creation of flexible cloning and expression systems allowing protein expression in multiple hosts [4]. Furthermore, the prediction of the whole ORF complement of an organism (the so called ORFeome) based on the analysis of genomic data has facilitated the creation of representative clone collections that contain a large proportion of such ORFs [5]. The functional characterization of putative ortholog genes identified among ORFeome clone collections might involve monitoring their intracellular behaviour, a task facilitated by the generation of fusions with genetically-encoded fluorescent proteins [6]. Controlling for possible artefacts caused by steric interference of the fluorescent module with important functional domains in the protein of interest should ideally involve comparing the intracellular behaviour when the fluorescent module is attached to either the N- or the C-termini, as well as with that of the protein over-expressed on its own [7], as it may explain unexpected behaviours. It might also be necessary to exchange the fluorescent module so to find the one with the spectral properties best suited to a particular experiment [7]. Meeting those demands may lead to time-consuming subcloning strategies, which would be further complicated by the intrinsically limited availability of in frame-, pre-inserted functional modules in expression vector backbones. The simplification of subcloning strategies, at least allowing the simultaneous insertion of the ORFs of interest into a panel of different expression vectors, would be advantageous [5]. One crucial step in this direction has been the implementation of recombination-based cloning systems such as the Gateway kits commercialized by Invitrogen (Life Technologies). Gateway cloning, which has been adopted for the creation of many ORFeome clone collections [5], bypasses the constraints imposed by restriction and ligation strategies and, in its basic configuration, is based on the recombinase-mediated shuttling of a DNA fragment (e.g. an ORF) from an entry clone into a destination vector. DNA fragments are transferred between vectors regardless of their sequence thanks to flanking recognition signals (the *att* sites) targeted by recombinases that have been borrowed from the λ phage's lysogeny cycle machinery [8]. The creation of an entry clone typically (but not exclusively) involves the PCR-mediated attachment of flanking *attB* sites to the DNA of interest, and its recombination-mediated cloning into an *attP*-containing plasmid. Recombination of this entry clone with an appropriate destination vector containing compatible *att* sequences will lead to an expression vector encoding an ORF, provided that the destination vector is furnished with a suitable promoter. The expression of an ORF of interest as a fusion with a fluorescent protein can be achieved with the help of a limited number of commercially-available destination vectors that contain an in-frame pre-inserted module.

Our aim was to develop a system that could meet the above requirements while avoiding complex subcloning strategies. Owing to the advantages of Gateway cloning, we generated a customizable Gateway-based plasmid toolkit for constructing fusion proteins from any ORF available as a standard entry clone. ORFs can be optionally expressed on their own or as in-frame fusions with a collection of standardized functional modules, in a combinatorial way, and from any Gateway destination expression vector. We describe the design and components of the toolkit, and we demonstrate its feasibility by presenting proof-of-principle expression data obtained in transient transfection experiments with prototype vectors encoding fluorescent fusion proteins.

Results and discussion

Conceptual design of the cloning toolkit

The construction of a vector for the expression of fusion proteins requires the subcloning of the cDNA of interest into a vector's backbone, at a site next to one or more pre-existing in-frame modules that provide the functions to be monitored. Hence, the design of the preassembled backbone vector conditions the nature of the fusion and the relative position of its constituent modules. This limits the possibilities for screening the best modular arrangement and performance. Subcloning of fusion proteins could be streamlined by releasing the functional modules from a fixed position in the backbone vector and regarding them as exchangeable modules to be cloned at the N- and C-termini simultaneously with the ORF of interest (Figure 1A). Ideally, such an approach would be based on recombination-mediated cloning since this should allow standardisation of the procedures and of the peptides linking the modules, facilitating the adoption of high-throughput technology, as described in earlier reports [9,10]. Expressing a protein optionally on its own or as a fusion with two flanking modules from a single vector backbone would be an additional asset (Figure 1A). Thus, when expressed under uniform conditions, comparing a protein's behaviour in either circumstance would facilitate the identification of potential artefacts occurring as a consequence of the fusion with functional modules [6]. It would also allow a large number of fusions based on a given ORF to be cloned simultaneously into an array of model-specific expression vectors in combination with a collection of functional modules (Figure 1B). A cloning toolkit with such features would

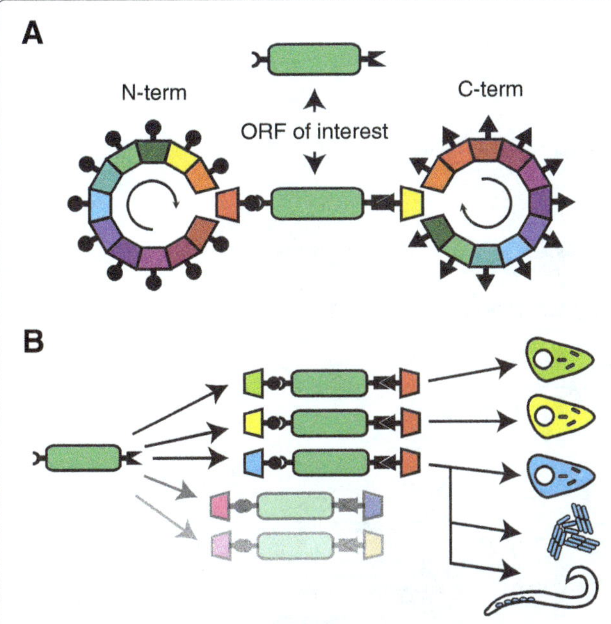

Figure 1 A combinatorial strategy for the expression of proteins as fusions with functional modules. A. Our toolkit's design allows any ORF of interest to be easily expressed either on its own (above) or as a fusion (below) with N- and C-terminal modules picked from among a collection of parts (represented as multicoloured wheels). The fusion's components are linked by standardized short peptide arms (black spheres and triangles) introduced as a result of the cloning procedure. The collection is constructed so that a given functional module can be attached either at the N-terminal or the C-terminal side of the protein of interest, which is helpful for finding the optimal arrangement in each case. **B**. Multiple fusions can potentially be generated by combining an ORF of interest with different modules from the collection. Specific fusions aimed at particular applications will be defined by the choice of modules used. For example, an ORF of interest N-terminally tagged with different fluorescent modules (green, yellow, blue) can be used to monitor the fusion protein's intracellular localization under conditions best suited for each experiment. Furthermore, our versatile design allows a particular fusion protein arrangement to be cloned into vectors specific for expression in different model systems in the context of comparative studies.

represent an advance over previous systems constructed according to similar design principles [10]. Furthermore, our toolkit would ideally tap into the wealth of ORFeome resources produced by the efforts of different laboratories leading the mass-production of parts for systems biology [5], a feature shared with previous designs [9,10]. We chose a cloning toolkit developed with the MultiSite Gateway three-fragment vector construction kit from Invitrogen (See Additional file 1) as it would make it meet the requirements above. A recombination reaction involving three separate entry clones based on the pDONR vectors from the kit (pDONRP4-P1R, pDONR221 and pDONRP2R-P3) and containing DNA fragments flanked by specific variants of the *att* sites, results in their ordered subcloning into the kit's promoterless destination vector pDEST-R4-R3 (See Additional file 1). Interestingly, the DNA fragment that occupies the central position is flanked by *attL1* and *attL2* sites in its entry vector (pDONR221), just as in the entry vectors available from public or commercial cDNA resources (e.g. ORF shuttle clones available through the ORFeome Collaboration [11,12]). Furthermore, new entry clones constructed by PCR-amplification of cDNAs and their BP clonase-mediated recombination into vectors containing *attP1/attP2* sites used for standard Gateway cloning are also compatible with the MultiSite Gateway kit (Figure 2, panel 1) (Also see Additional file 1). The preservation of the translation frame of the three fragments that are pieced together was built into the original MultiSite cloning system's design [9]. Therefore, by cloning the N-term- and C-term-functional modules proposed in the toolkit's design into pDONR-P4-P1R and pDONR-P2R-P3, respectively, they could join an entry vector holding the ORF for a protein of interest in a MultiSite cloning reaction that would create a fusion protein's chimeric ORF (Figure 2, panel 2). However, pDEST-R4-R3 is a promoterless destination vector, meaning that further engineering of this vector would be required in order to insert a promoter upstream of the chimeric ORF. Instead, replacing the original single-fragment Gateway cassette in any destination expression vector furnished with a heterologous promoter, with the R4-R3 cassette from pDEST-R4-R3 seemed a more powerful strategy (Figure 2, panel 3). This would allow taking advantage of numerous existing Gateway destination vectors that would become in this way available for MultiSite Gateway cloning (see also ref. [13]), increasing the functional flexibility of the toolkit. This way, the protein of interest could be expressed on its own or as a fusion protein in the context of the same vector backbone, by choosing for the LR-recombination between the unmodified vector and that with the engineered Gateway cassette, used in combination with the clones encoding functional modules. Furthermore, all of the elements in the proposed collections of flanking N-terminal and C-terminal functional modules would partake of the same construction principles. Thus, they could be used in a combinatorial fashion in MultiSite Gateway LR Clonase-mediated recombination reactions with any existing Gateway entry clone containing an *attL1/attL2*-flanked ORF, where the components of the fusion proteins would be linked by the translated remaining *attB1* and *attB2* sequences acting as universal connectors (Figure 2, panel 4). The open architecture of such toolkit would allow for easily updating any successful fusion arrangement to the latest versions of the functional modules in use (e.g. fluorescent proteins). This would be achieved by simply incorporating such versions as new modules in the collection, so they could be used instead of the superseded modules in a replay of the recombination reaction that led to the fusion in question.

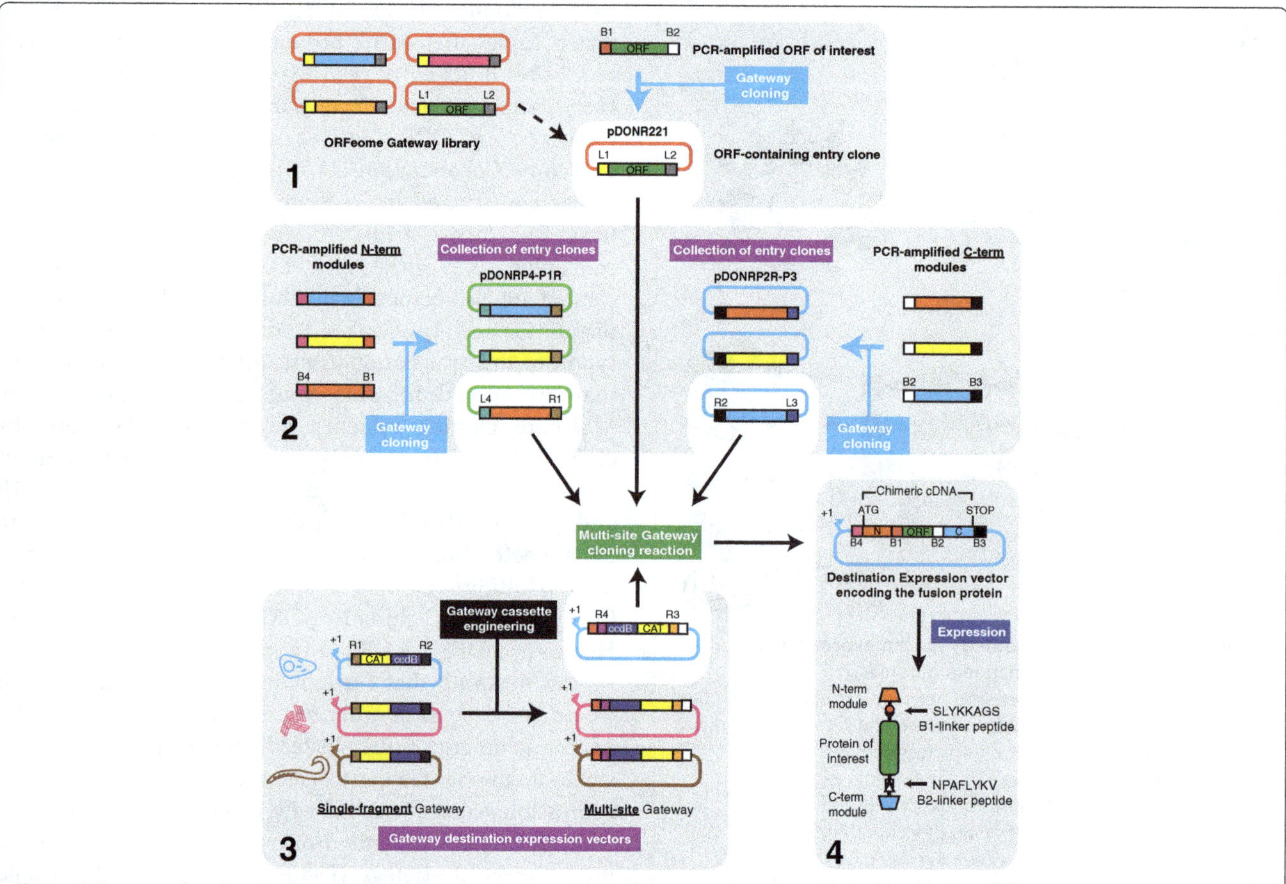

Figure 2 Strategy for the cloning of fusion proteins. Cloning is based on the Multi-site Gateway kit from Invitrogen. The plasmid encoding the ORF of interest can either be an existing entry clone from a Gateway-based ORFeome library, or can be constructed by Gateway-cloning the PCR-amplified ORF into pDONR221 with BP clonase **(panel 1)**. A collection of entry clones encoding the functional modules to be attached to the ORF of interest is constructed similarly by Gateway cloning the PCR-amplified modules into the pDONR P4-P1R (N-term modules) and pDONR P2RP3 (C-term modules) plasmids from the MultiSite Gateway kit (**panel 2**, see main text for details). All of the inserts in the entry vectors are flanked by appropriate *att* sites provided by the PCR-amplification primers, and are indicated throughout this figure with a key (e.g. the *attB1* site is indicated as *B1*). Single-fragment Gateway destination expression vectors designed for expression in different model systems are adapted by engineering of the Gateway cassette to allow MultiSite Gateway cloning reactions **(panel 3)**. Such reactions involve one adapted destination vector, as well as the plasmids encoding the ORF of interest and the chosen N-term and C-term modules (white background), whose matching *att* sites (R4xL4, R1xL1, L2xR2, L3xR3) recombine to produce a destination expression vector encoding the fusion protein in the form of a chimeric cDNA **(panel 4)**. Expression of the cDNA results in the production of a fusion protein whose three constituent parts are linked by short peptides resulting from the translation of the *attB1* and *attB2* sites remaining after the LR recombination (panel 4, see main text for details).

Engineering of a destination expression vector to enable three-fragment cloning

The substitution of the single site- by the MultiSite *attR4-attR3* Gateway cassette in an existing destination expression vector was achieved by modifying a method first reported by Magnani and colleagues [13]. This consists of an LR-recombination reaction between a vector that contains the *attR4-attR3* Gateway cassette from pDEST-R4-R3, flanked by *attL1/attL2* sites (pDONR221-R4-R3), and the destination expression vector in its standard form, furnished with *attR1/attR2* sites [13]. In this way, a successful recombination reaction would place the R4-R3 Gateway cassette between *attB1* and *attB2* sites (resulting from the LR recombination between *attL1/attL2* and *attR1/attR2* sites). However, such a procedure raised a technical issue, common to all reactions where standard Gateway cassettes are replaced with versions containing different flanking recombination sites, i.e. how to easily distinguish, after bacterial transformation, between *E. coli* colonies that contain the original destination vector from those containing the vector with the recombined R4-R3 Gateway cassette, since they would be identical in terms of their selection markers. To solve this issue, we took advantage of the fact that Gateway vectors typically possess two antibiotic resistance genes, one of them being the chloramphenicol-acetyl transferase (*CAT*) gene that lies inside the Gateway cassette, and another one that lies elsewhere in the vector. The ccdB-resistant *E. coli* bacteria transformed with such vectors acquire resistance to both chloramphenicol and the antibiotic (ampicillin, kanamycin, etc.) specified by the other resistance gene. Our strategy consisted in engineering the vector before the recombination reaction, so that an inactivating mutation was introduced by site-directed mutagenesis into the *CAT* gene in the existing Gateway cassette, while the rest of the vector's functionality was preserved. *E. coli* cells transformed with this vector were sensitive to the presence of chloramphenicol in the culture medium (See Additional file 2). On the other hand, the PCR-amplified MultiSite Gateway cassette, which was to replace the original one after recombination, had a functional *CAT* gene. Thus, successful recombination would restore the resistance to chloramphenicol in the engineered vector and allow the growth of colonies of bacteria in the presence of chloramphenicol and ampicillin, while negatively selecting bacteria transformed with the non-recombined destination vector. Since all Gateway vectors contain the same sequence in the *CAT* gene that is targeted by our mutagenesis reaction, this strategy could potentially be used for tailoring the Gateway cassette in any vector. We initially tested it by assembling the pDONR221-R4-R3 vector (Figure 3A). This involved a BP-recombination reaction between a PCR-amplified R4-R3 Gateway cassette furnished with a wild-type *CAT* gene, and a *CAT*-mutant version of the pDONR221 vector (Figure 3A). The screening strategy proved to be successful, and we were able to isolate bacterial clones containing recombined pDONR221-R4-R3, as confirmed by sequencing, which were again resistant to chloramphenicol and kanamycin. We subsequently used the same strategy in the adaptation of a destination expression vector for MultiSite Gateway cloning, as required by the design of our cloning toolkit. Thus, an LR-recombination reaction was set up between pDONR221-R4-R3 and the *CAT*-mutant version of pEF5/FRT/V5-DEST (Figure 3B). After transformation of the reaction and plating on LB medium supplemented with ampicillin and chloramphenicol, several colonies resistant to both antibiotics were isolated. Sequencing of the plasmid prepared from these colonies evidenced the presence of an R4-R3 Gateway cassette flanked by *attB1/attB2* sites in the context of pEF5/FRT/V5-DEST, showing the successful substitution of the single site- by the MultiSite Gateway cassette. Thus, we designated this plasmid pEF5/FRT/V5-DEST-R4-R3 and used it in the construction of the fusion protein prototypes discussed below. Importantly, the mutation of the destination vectors' *CAT* gene required by our method will not only be useful for enabling MultiSite recombination cloning, but could also be used subsequently for other adaptation reactions involving different Gateway cassettes flanked by other types of recombination sites (different *att* sequences, FRT or *loxP* recombination sites), which could be screened with the same procedure.

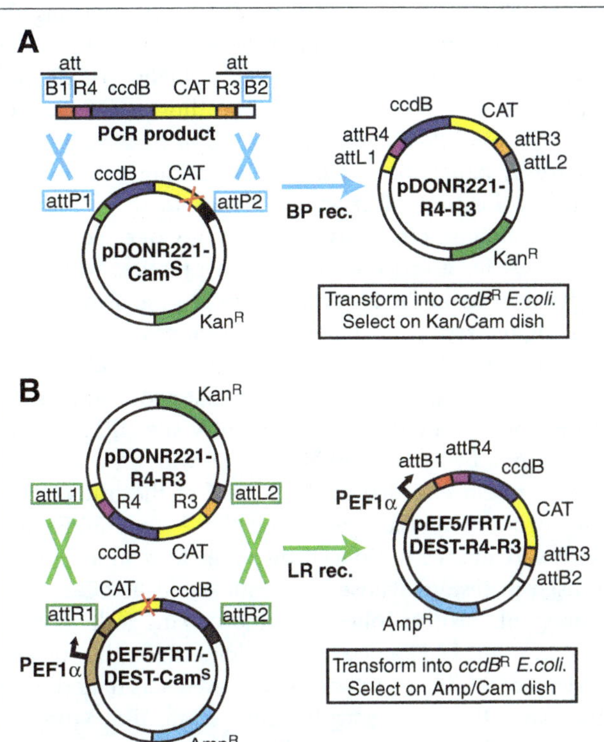

Figure 3 Adaptation of existing Gateway Destination vectors for the expression of fusion proteins. A. Construction of adapter vector pDONR221-R4-R3. A PCR product corresponding to the *attR4-attR3*-containing Gateway cassette in pDEST-R4-R3 was amplified with primers that provided flanking *attB1* and *attB2* sites. An inactivating mutation was introduced into the *CAT* gene (red X) in pDONR221 by site-directed mutagenesis so that ccdB-resistant *E. coli* transformed with this plasmid were sensitive to chloramphenicol (Cam^S). A BP-recombination reaction was set up between the amplified *attR4-attR3* Gateway cassette and pDONR221_Cam^S to create pDONR221-R4-R3 (see details under methods). The *att* sites involved in the reaction are marked with blue squares. Selection with LB medium containing kanamycin and chloramphenicol (*Kan/Cam*) yielded colonies containing the recombined plasmid. **B**. Adaptation of destination expression vector pEF5/FRT/V5-DEST for MultiSite Gateway recombination reactions. The same inactivating mutation in the *CAT* gene as above (red X) was introduced into the commercially available single-site destination expression vector pEF5/FRT/V5-DEST to generate a Cam-sensitive version (pEF5/FRT/DEST-Cam^S). The mutant plasmid was subjected to an LR-recombination reaction with pDONR221-R4-R3 in order to replace its original Gateway cassette with the *attR4-attR3* MultiSite recombination cassette in pDONR221-R4-R3 (endowed with a wild type *CAT* gene). The *att* sites involved in the reaction are marked with green squares. Selection of the colonies containing recombined pEF5/FRT/V5-DEST (pEF5/FRT/-DEST-R4-R3) was carried out with LB medium containing ampicillin and chloramphenicol (*Amp/Cam*). Only the features in the pEF5/FRT/V5-DEST plasmid map that are relevant for the reaction are shown.

Creation of a collection of functional modules

We cloned a collection of functional DNA modules into vectors pDONR-P4-P1R and pDONR-P2R-P3 so they could be fused to the N- or C-termini of the protein of interest, respectively. This was carried out by BP-recombination between the *attB*-flanked PCR products encompassing the functional modules and either of the pDONR vectors. PCR products that were to be cloned into pDONR-P4-P1R were flanked by *attB4/attB1* sites, while those to be cloned into pDONRP2R-P3 were flanked by *attB2/attB3* sites (Figure 2, panel 2). A current list of the components in our collection of modules is provided as Table 1. At the moment, they mostly consist of clones containing the cDNA for a series of fluorescent proteins, which can be fused either at the N-terminal or C-terminal end of the ORF of interest. The fluorescent proteins available are ECFP (cyan), EGFP (green), EYFP (yellow) and mKate2 (far red). There is also a 3′-module that allows expressing ECFP from an internal ribosome entry site (IRES) in the context of a bicistronic mRNA shared with a two-module fusion protein encoded at the 5′-half of the mRNA. We will introduce modules encoding the A206K EGFP and EYFP variants, with reduced oligomer formation [14], to be used in fusions whose behaviour could be affected by fluorescent protein oligomerization. This will not be necessary for ECFP-encoding modules since this protein carries a constitutive mutation that prevents its dimerization [15]. Since the three entry vectors from the MultiSite Gateway cloning kit must be present for successful recombination, we included a non peptide-encoding 3′-module with a stop codon and a SV40 early polyadenylation signal cassette to allow translational termination of chimeric ORFs when fusion of a module at the C-terminus is not necessary or convenient. Please note that the parental pEF5/FRT/V5-DEST vector contains the BGH polyadenylation signal downstream of the Gateway cassette and this is preserved in pEF5/FRT/V5-DEST-R4-R3. Thus, when a 3′ peptide-encoding module is typically used, the BGH polyA signal will allow proper processing of the chimeric transcript, with the 3′ module providing the STOP codon at the end of the encoded peptide. Because an N-terminal tag might interfere with functionally important modifications of the protein (e.g. myristoylation), we will also obtain a neutral 5′-module containing an intronic sequence that can serve a similar function at the N-terminus. We have also constructed modules containing the well-characterised V5 and 6xHis epitope tags. Other possible modules would be those encoding for protein domains that target proteins to an organelle (e.g. NLS and a mitochondrial targeting sequence), an application previously demonstrated [9]. The *attB*-flanked PCR products were generated in two sequential PCRs, with two sets of partially overlapping primers (see Methods section). The sequence of primers used for creating the current collection of functional modules is provided as Additional file 3: Table S1. Since all modules in each category (N- or C-terminal) were cloned the same way into pDONR-P4-P1R or pDONR-P2R-P3, they were fully exchangeable and conferred the desired combinatorial character to the toolkit.

Table 1 Plasmids encoding functional modules currently available on the toolkit

5′ module	3′ module
pDONR P4-P1R-**mKate2**	pDONR P2R-P3-**V5-6xHis**
pDONR P4-P1R-**V5-6xHis**	pDONR P2R-P3-**mKate2**
pDONR P4-P1R-**EGFP**	pDONR P2R-P3-**IRES_ECFP**
pDONR P4-P1R-**EYFP**	pDONR P2R-P3-**SV40 polyA signal**
pDONR P4-P1R-**ECFP**	pDONR P2R-P3-**EGFP**
	pDONR P2R-P3-**EYFP**
	pDONR P2R-P3-**ECFP**

Plasmids based on the pDONR P4-P1R vector contain inserts that will be placed at the 5′ flank of the cDNA and will encode the N-terminal module of the fusion protein. On the other hand, those based on pDONR P2R-P3 will transfer their insert to a position flanking the 3′-end of the cDNA, and will add a C-terminal module to the fusion protein in case they encode a peptide.

Construction and expression of the fusion protein prototypes

We constructed a panel of vectors for expressing fusion proteins in order to test the feasibility of our cloning toolkit. These vectors were obtained through MultiSite Gateway recombination reactions between plasmids encoding the N-terminal modules, the ORFs of interest, the C-terminal modules, and the pEF5/FRT/V5/DEST-R4-R3 destination expression vector described above. LR clonase-mediated recombination between compatible *att* sites on the participating plasmids produced a chimeric ORF that substituted the engineered Gateway cassette on plasmid pEF5/FRT/V5/DEST-R4-R3. The *attB1/attB2* sites flanking the *attR4/attR3* Gateway cassette were not targeted by the LR-recombinase [13]. Expression of the ORFs produced fusion proteins where the three component parts were separated by peptide linker arms resulting from the translation of the *attB1* and *attB2* sites flanking the protein of interest (see above). Since the upstream *attB1* and *attB4* recombination sites are located between the EF1α promoter and the translation initiation codon of the N-terminal module, and the *attB3* and downstream *attB2* sites lay beyond the stop codon in the C-terminal module, none of them would participate in the translated product (Figure 2, panel 4). Even though there seems to have been no systematic analysis of the possible functional interference of the residual *attB1/attB2* sites in the expression of fusion proteins [10], individual studies found those sites to be either neutral [16] or detrimental [17] for protein expression, hence the need to assess their convenience for the intended applications. The fusion proteins obtained with this approach are listed on Table 2 and are grouped in sets that illustrate some of the possibilities of the cloning toolkit. The first set highlights the combinatorial aspect of the collection of functional modules,

Table 2 Panel of expression vectors encoding fusion proteins

Vector	5′-module	Central module	3′-module
Set 1			
pEF5/FRT-DEST-R4-R3_V5-6xHis: PAR2:mKate2	V5-6xHis	PAR-2	mKate2
pEF5/FRT-DEST-R4-R3_V5-6xHis: PAR2: EGFP	V5-6xHis	PAR-2	EGFP
Set 2			
pEF5/FRT-DEST-R4-R3_mKate2: SIRT1	mKate2	SIRT1	*SV40 polyA*
pEF5/FRT-DEST-R4-R3_mKate2: ΔNtermSIRT1[a]	mKate2	ΔNtermSIRT1	*SV40 polyA*
Set 3			
pEF5/FRT-DEST-R4-R3_mkate2: RelA:V5-6xHis	mKate2	p65	V5-6xHis
pEF5/FRT-DEST-R4-R3_V5-6xHis:p65:mKate2	V5-6xHis	p65	mKate2

MultiSite Gateway recombination reactions were carried out between the adapted destination expression vector pEF5/FRT-DEST-R4-R3, and plasmids encoding 5′-module, the ORF of interest (central module) and the 3′-module as described. The resulting expression plasmids are listed on the table, with an indication of the functional modules used for making each fusion protein, and are organised in sets, according to the toolkit properties intended to illustrate (see main text). The central module encoding wt SIRT1 is an entry clone obtained from an ORFeome library, used in the MultiSite cloning reaction with no prior modifications. The SV40 polyA module (italics) does not encode for a peptide but contains a polyA signal that precedes the BGH polyA signal provided by the pEF5/FRT-DEST-R4-R3 vector.

[a]The "Δ" symbol stands for "deleted".

since it is composed by two fusion proteins that share the N-terminal and central (i.e. the protein of interest) modules but that are fused to different fluorescent proteins at the C-terminus. This would be helpful for optimising fluorescent fusion proteins since different combinations can be produced up front and tested for the existence of steric hindrance, poor fusion stability and functional interference by oligomerization of the fluorescent module [18]. Furthermore, it would also provide some flexibility for designing coexpression experiments, because fusion proteins with different emission spectra would be available to suit each case (See panel A, Additional file 4). This set comprises two fusion proteins containing the Proteinase-activated receptor PAR2 (V5-6xHis: PAR-2: mKate2 and V5-6xHis: PAR-2: EGFP, Figure 4A). The second set represents the possibility of generating mutations in the ORF of interest while still in the entry clone. In this way, the construction of one mutational library would be sufficient for preparing a range of expression vector libraries to be screened on different model systems (mammalian cells, yeast, etc.) or with different reporters. This would avoid the need to go through independent mutagenesis reactions for creating libraries based on different expression vectors (See panel B, Additional file 4), and should be helpful for studying the effects of point mutations or deletions on the function of the protein of interest (e.g. intracellular trafficking). This set comprises the wild type (wt) form as well as an N-terminal deletion of the mouse sirtuin SIRT1 (*ΔN-t-SIRT1,* Figure 4A), both N-terminally fused to mKate2. This deletion affects the first nuclear location signal (N1) and may produce alterations on the intracellular distribution of the protein [19]. Finally, the third set represents the two possible arrangements of a fusion protein with functional modules flanking the ORF of interest (See panel C, Additional file 4). The ability to generate the two proteins up front would facilitate the identification of the order that causes the least interference on the protein's function. Our example consists of two fusion proteins where the far-red fluorescent protein mKate2 [20] and a V5-6xHis epitope tag cassette were placed either N-terminal or C-terminal relative to the human p65/RelA subunit of NFkB (mKate2: p65: V5-6xHis and V5-6xHis: p65: mKate2, Figure 4A).

The constructs described above were transiently transfected into cells in order to test the feasibility of our approach (Figure 4B, panels a-j). In HeLa cells, C-terminal fusions of PAR-2 with mKate2 (Figure 4B, panels a, b) or with EGFP (panels c, d) were expressed separately in transient transfection experiments. Cells under control conditions showed localisation of PAR-2 fluorescent fusions mostly at a perinuclear compartment [21], with a weaker signal located near the plasma membrane (a, c). Treatment of the cells with the PAR-2-specific agonist AC55541 caused a slight redistribution of the fluorescent signal from either fusion protein that concentrated on a smaller area around the nucleus (b, d). Also in HeLa cells, expression of wt SIRT1 fused to mKate2 showed nuclear localisation (Figure 4B, panel e), but the loss of an N-terminal fragment of mSIRT1 encompassing N1, caused the fusion protein to be localised in the cytoplasm (Figure 4B, panel f), in agreement with previous data [19]. In Raw264.7 monocytes, both the mKate2: p65: V5-6xHis and V5-6xHis:p65:mKate2 fusions (Figure 4A, set 3) demonstrated cytoplasmic localisation under control conditions (Figure 4B, panels g, i), as expected [22]. LPS treatment of the cells induced nuclear translocation of both fusion proteins (Figure 4B, panels h, j). The existence of slight differences in the intracellular distribution of fluorescence between different fusions of the same ORF (compare panels a and c, or g and i) suggest there might

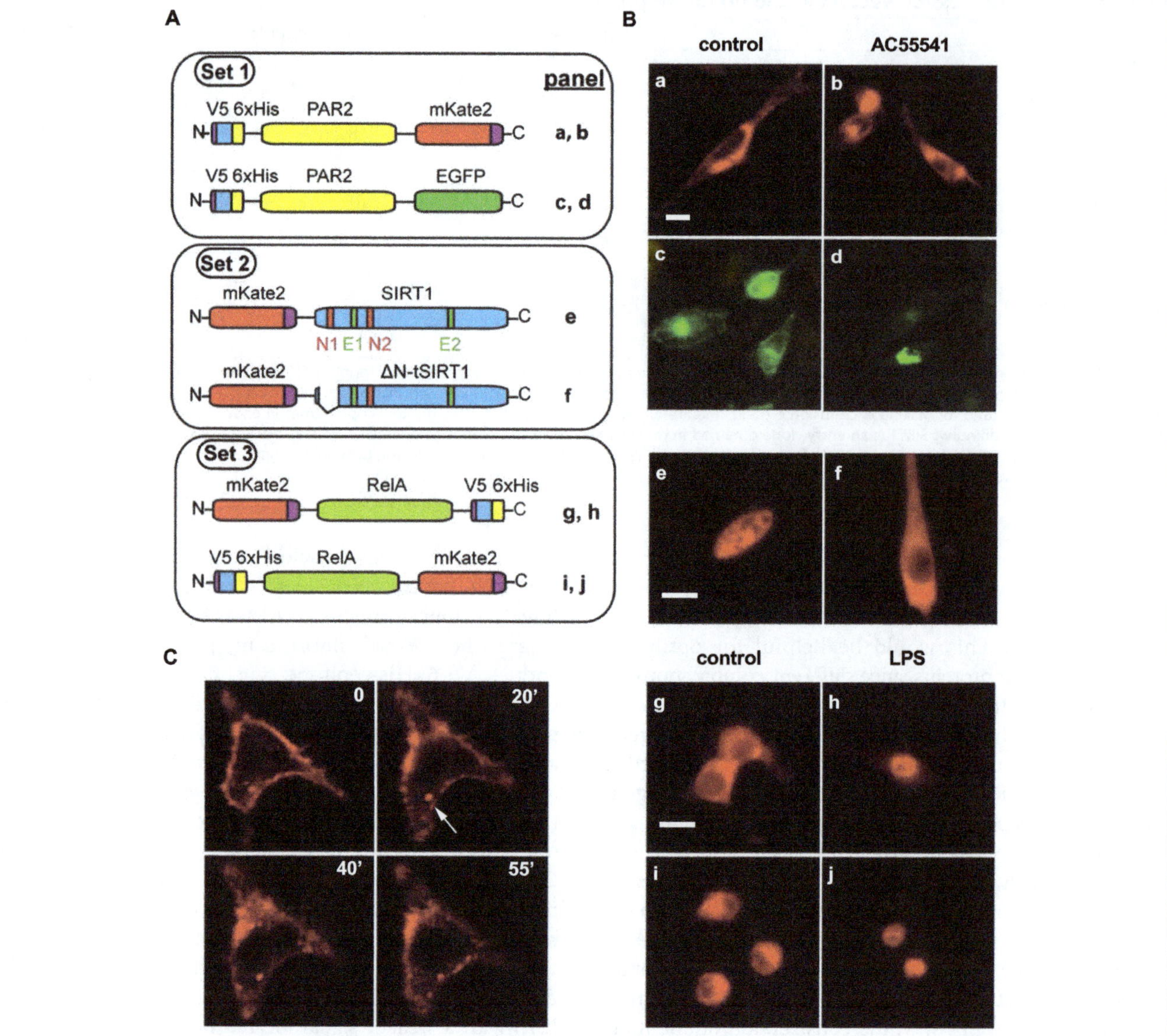

Figure 4 Expression of fluorescent fusion proteins. A. Schematic representation of the fusion proteins produced in order to test the feasibility of our toolkit, indicating the micrograph panels in Figure 4B where each protein's expression is shown. The relative positions of the nuclear location signals (N1, N2) and nuclear export signals (E1, E2) in murine wild type (wt) SIRT1 is shown. *ΔN-tSIRT1*, N-terminal-deleted murine SIRT1 (see main text); the portion deleted from the wt isoform is indicated. Fusion proteins are clustered in functional sets, as described in the main text. **B**. Detection of fluorescent protein fusions under an epifluorescence microscope. Panels a-d: expression of PAR-2 fused to mKate2 **(a,b)** or EGFP **(c,d)** in transiently-transfected HeLa cells under control conditions **(a,c)** or after treatment with the PAR2-specific agonist AC55541 (5 μM, 1 h; panels **b**, **d**). White bar: 10 μm. Panels **e**, **f**: Differential localisation of wt- **(e)** or *ΔN-tSIRT1* **(f)** proteins fused to mKate2 in transiently-transfected HeLa cells, under control conditions. White bar: 10 μm. Panels **(g-j)**: expression of two p65/RelA-based fusions with mKate2 after transient trasfection of their expression vectors into Raw264.7 monocytes, under control **(g,i)** or LPS-stimulated conditions (100 ng/ml, 1 h; panels **h,j**). White Bar: 10 μm. All micrographs were taken at 200X magnification. **C.** Frames from a time-lapse movie of a Raw264.7 monocyte transfected with the expression vector for the PAR2-mKate2 fusion protein and imaged on a confocal microscope (63X objective lens) after adding 5 μM AC55541 in order to monitor cellular trafficking of the receptor. The complete time-lapse experiment is provided as Additional file 5: Movie S1. Time elapsed after adding the agonist is indicated in each frame. The white arrow in the 20 min-panel points at some PAR2-mKate2-labelled intracellular vesicles that are seen trafficking in Additional file 5: Movie S1.

be effects related to the fusion arrangements that would deserve further investigation. If confirmed, they would illustrate the usefulness of our cloning strategy. As we pointed out in the background section, the study of the intracellular behaviour of a newly-described protein may help delineate its possible functions. Our cloning toolkit would be useful for the generation of fusions to be specifically used in such studies. This is exemplified by Additional file 5: Movie S1, where the fusion of the monomeric fluorescent protein mKate2 to the C-terminus of PAR-2 was expressed in a Raw264.7 monocyte by transient transfection of the pEF5/FRT-DEST-R4-R3_V5-6xHis: PAR2:mKate2 expression vector, and observed under a confocal microscope in a time-lapse experiment aimed at tracking the protein's behaviour following treatment with the PAR-2 specific agonist AC55541. Figure 4C provides a series of still frames extracted from the time-lapse recording at the indicated times after adding the agonist. These show the agonist-induced clustering and internalization of the plasma membrane-associated PAR-2: mKate2 fusion, as well as the presence of intracellular vesicles containing the fluorescent protein (Figure 4C, arrow), as previously described [21]. Active perinuclear trafficking of the vesicles can be appreciated in the complete time-lapse movie (Additional file 5: Movie S1). Additional file 6 Movie S2 corresponds to a shorter time-lapse sequence of the same cell recorded under control conditions, just before adding the agonist, and shows expression of the fusion protein in continuous association with the plasma membrane, as well as in some intracellular vesicles that display a more modest trafficking behaviour.

All the fusion proteins described above were expressed from pEF5/FRT/V5/DEST-R4-R3, a plasmid derived from pEF5/FRT/V5/DEST (see above and Methods section). This is a vector compatible with the FLP-In system (Life Technologies), which allows the generation of isogenic clones with stable plasmid integration by using a selection of cell lines that contain a single FRT recombination sequence in their genome (FLP-In cell lines). Thus, our system should facilitate the assembly of expression vector libraries consisting of fusions between selected functional modules and a library of randomized mutations of a cDNA, so they could be stably-transfected into an FLP-In cell line and screened for a mutation conferring the properties sought after, in the absence of clonal variegation caused by genome-positional effects. Furthermore, this would not be restricted to randomized mutation libraries, but could also be applied to libraries consisting of collections of cDNAs (e.g. the Gateway-based Human Kinase Open Reading Frame Collection from the Centre for Cancer Systems Biology (DFCI)/ Broad Institute (MIT) [23]). Our system would also be useful for the generation of chimeric proteins. Thus, a protein coding sequence could be divided into three fragments at appropriate sites that did not compromise the main structure or function of the protein, with each fragment cloned into one of the three pDONR vectors in the toolkit. Different isoforms in a family of proteins could be cloned in this way so that homologous domains cloned in the same type of pDONR vector could be shuffled between the isoforms in order to analyse the impact on the proteins' function. Similarly, a collection of mutants from a single domain could be cloned with the remainder of the wild type ORF in order to screen for residues involved in a specific function. Although we have not addressed these specific applications, pioneering work on the yeast STE2 receptor by Cheo and colleagues with an analogous set up [9] suggests that our design could be used in such studies.

Interestingly, some of the clones encoding functional modules in our toolkit could be exchanged with those from similar platforms built with the same MultiSite Gateway cloning kit as ours, expanding the number of possible fusion protein combinations. One example is the platform known as Tol2Kit [24]. It consists of a series of 5′-, middle- and 3′-clones arranged in the context of several destination vectors specific for zebra fish transgenesis, and was created to facilitate the obtention of strains expressing a variety of minigene constructs consisting of *promoter_coding sequence_3′tag* arrangements [24]. Another example is the pTransgenesis system [25], which consists of collections of entry vectors initially developed to facilitate the construction of minigenes to be expressed in transgenic *Xenopus* strains, but are also compatible with transgenesis in other model organisms such as *Drosophila*, zebrafish and mammalian cell models, thus facilitating the study of the evolutionarily conserved properties of biological systems [25]. Finally, Nagels-Durand and colleagues have recently reported a novel set of destination vectors specific for expression in *Saccharomyces cerevisae* that allow MultiSite recombination cloning of promoters, ORFs and epitope tags in combination with a choice of auxotrophy markers and replication mechanisms in this model organism [26]. Some of the functional modules described in our work overlap with those from these other platforms (Table 1 and [24-26]), and have been successfully used for the applications described therein. Fluorescent protein-encoding modules from our toolkit as well as from the other platforms mentioned above could also be used as reporters in new projects aimed at characterizing ORFeome-inspired promoteromes, such as in the pioneering work of Denis Dupuy and collaborators [27], who constructed a library of intergenic regions comprising proximal transcriptional regulatory elements plus the 5′UTRs cloned in the pDONR P4-P1R vector, and used it in combination with ORFs derived from the nematode's ORFeome in Multisite Gateway cloning reactions that generated vectors with a *promoter:ORF:reporter* arrangement. These were used subsequently in studies of the regulatory mechanisms controlling gene transcription and protein localization at genome-level [27,28].

Despite the availability of a more recent version of the MultiSite Gateway cloning kit (*MultiSite Gateway Pro*) that allows using any single-fragment destination vector in two-, three-, or four-fragment recombination reactions without further modification [29], we chose the earlier version of the kit because this would allow the use of entry vectors containing the *attL1/attL2*-flanked ORF of interest as obtained from public sources, without further subcloning. This is precluded in the Pro version of the kit, since those *att* sites ought to lie on separate entry vectors because of the kit's configuration. Thus, with a *MultiSite Gateway Pro*-based cloning platform, any *attL1/attL2* flanked-ORF of interest obtained from a public repository would still have to be PCR-amplified and subcloned into at least one pDONR vector appropriate for the Pro version of the kit prior to the MultiSite recombination reaction, blunting the advantage conferred by the ready availability of the entry clone. In our approach, engineering is instead restricted to the destination vector of choice, which once adapted for MultiSite Gateway recombination can be used to generate multiple fusion proteins with the unmodified entry clones from the library. Nevertheless, cloning platforms that use the MultiSite Gateway Pro kit are perfectly possible and one such has been developed [30], which allows the versatile creation of minigenes to be used in *Drosophila* transgenesis for the tissue-specific expression of proteins, including fluorescent reporters. Platforms reported in earlier publications, such as the *Drosophila* Gateway vector collection devised by the Murphy lab at the Carnegie Institution [31], and the *Saccharomyces cerevisae* collection of Gateway destination vectors constructed at the Lindquist lab [32] also allow directly producing fusion proteins with both custom made-, or ORFeome-derived entry clones, since their input consists of plasmids containing ORFs flanked by *attL1/attL2* sites, which are used in single-fragment LR recombinations to generate the expression vectors. In these platforms, though, variability lies with the destination vectors, which contain functional modules preinserted at 5′ or 3′ relative to the standard Gateway cassette, which will be expressed in-frame with the Gateway-shuttled ORF of interest. Since the functional modules lie beyond the *att* sites, construction of these platforms required complex cloning of Gateway cassette-derivatives into a large number of expression vectors in order to achieve the desired number of module combinations.

Although toolkits for fusion protein construction based on sequence-directed cloning methods provide versatility and ease of use, the possible functional impact of peptide "scars" linking the modules (resulting from the translation of intervening residual cloning sites, see above), needs to be taken into account [33]. In view of this, other more economical, non-sequence-directed, "seamless" recombinational cloning methods for joining two or more DNA fragments without the operation of extraneous sequences would represent an attractive alternative [34,35]. Nevertheless, in the most widely used seamless methods, DNA fragments to be assembled ought to be amplified by PCR with primers that provide some sequence overlap at their ends. The distribution of these sequence overlaps among the fragments ensures they are stitched together in the required order. In other instances, "stitching" oligonucleotides can be used that act as a bridge encompassing the ends of the unrelated sequences to be joined [35]. In either case, the generation of a combinatorial cloning platform based on such methods would involve prior knowledge of the fusions that would be of interest in order to provide necessary tools in advance. This would somewhat curtail the implementation of high-throughput strategies based on the use of ORFeome-derived clones, as it would require a case-based design, which is not necessary when universal external connectors (e.g. *att* sites) are used. Furthermore, PCR-amplification of modules for constructing specific fusions would require checking every resulting chimeric cDNAs for the presence of unwanted mutations, while such controls are only required at the time of subcloning if the same modules are perpetuated as inserts of a plasmid in the context of a Gateway-based platform. In the absence of a perfect recombination-based solution, one ought to consider what is best for a given application, whether lowering the risk of encountering scar-derived functional effects, or using high throughput applications. Nevertheless, this is no obstacle to using both approaches in a complementary way since, for instance, the assembly of composite modules derived from the sequence of two or more proteins could be achieved by using seamless recombinational cloning, and the new modules could then be incorporated into Gateway-based cloning platforms for their fusion to other peptides. Interestingly, non-recombination-based cloning platforms such as Golden Gate, Golden Braid, and MoClo [33,36,37], which have shown great potential for scalability in the construction of multipartite expression vectors, could also be adapted for the seamless cloning of fusion proteins with the possibility of using peptide-encoding modules in high-throughput applications. Careful design of the module-cloning strategy may allow the construction of fusion proteins with a minimal contribution of scar sequence between the modules [33,36]. The downsides are that these systems require that cloned DNA fragments are free from the infrequent target sites for the type IIS restriction enzymes they are based upon (this may require the fragment's sequence to be modified prior to cloning), and that the ORFs of interest cannot be used straight from Gateway-based ORF collections but would require PCR-amplification with primers providing the appropriate restriction enzyme sites. All in all, the above considerations should be a guide for the future development of cloning strategies with the potential to use peptide-encoding modules across different platforms.

Conclusions

We describe the blueprint for a system that streamlines the cloning of fusion proteins from Gateway-based, ORF-containing entry clones. It is based on sets of plasmids that encode customized functional modules to be fused to an ORF of interest, plus an adaptor plasmid that allows existing destination expression vectors to participate in MultiSite cloning reactions leading to the cloning of fusion proteins. Multiple fusions can thus be potentially assembled from a single ORF in a combinatorial way, and expressed in diverse cellular models. Expression vectors are easy to construct and update, since cloning is based on Gateway technology. We believe this approach will be widely useful for the scientific community as it should extend the range of possible studies based on fusion proteins beyond those currently afforded by cloning vectors with preinserted modules. Plasmids will be made available through Addgene.

Methods

Molecular biology

PCR reactions were carried out with AccuPrime Pfx SuperMix (Life Technologies) as indicated by the manufacturer. The pDONR plasmids and cloning vector pDEST-R4-R3 were obtained as part of the MultiSite Gateway cloning kit (Life Technologies catalog # 12537023). Destination expression vector pEF5/FRT/V5-DEST was obtained from Life Technologies (catalog # V6020-20). This is a vector for cloning and expressing proteins in FLP-In™ isogenic cell lines. The expression of cloned proteins is driven by the human elongation factor 1α promoter located upstream of the Gateway cassette. The vector has an FRT recombination site that mediates FLP recombinase-directed integration of the vector into a unique homologous FRT site in the genome of FLP-In™ cell lines, and the hygromycin resistance gene acts as a selectable marker for integration. Nevertheless, it behaves just as any other expression vector in transient transfection experiments. BP- and LR-recombination reactions were carried out with BP clonase II and LR Clonase II Plus enzyme mixes, respectively (Life Technologies), following the manufacturer's instructions. Reactions were stopped by addition of Proteinase K and incubation at 37°C for 10 min. All competent *E. coli* strains were obtained from Life Technologies and used in one-shot format for plasmid transformation. LB medium was supplemented where indicated with selection antibiotics at the following concentrations: ampicillin (100 μg/mL), kanamycin (50 μg/mL), chloramphenicol (50 μg/mL).

Site-directed mutagenesis of the chloramphenicol-acetyl transferase (*CAT*) gene in the Gateway cassettes

Inactivation of the type I *CAT* gene in the Gateway selection cassette of pDONR221 and pEF5/FRT/V5-DEST was achieved by site-directed mutagenesis. The oligonucleotides 5′-CCCCCGTTTTCACCTAAGGCA AATATTATAC-3′ (forward) and 5′-GTATAATATTT GCCTTAGGTGAAAACGGGGG-3′ (reverse) were used to introduce a nonsense mutation, replacing methionine 173 with a stop codon (underlined). This was designed to cause the premature interruption of the 219-residue type I CAT protein, resulting in the loss of the c-terminal α-Helix 5 [38]. This region allows the formation of the trimeric complex of identical subunits that constitutes the functional enzyme, having additional positive effects on the protein's solubility [39]. The mutation was expected to disrupt these properties and have profound deleterious effects on the enzyme's activity. The mutagenesis oligonucleotides were designed so that an *NcoI* site would be destroyed as a consequence of the mutation, which could then be used for diagnostic purposes. The oligonucleotides were used in combination with the QuikChange Lightning site-directed mutagenesis kit (Agilent), with the exception that transformation of the mutagenesis reactions was done into ccdB Survival™ 2 T1R competent cells to avoid the lethality of the *ccdB* gene in the Gateway selection cassette. After transformation, the cultures were spread on LB/agar dishes containing either ampicillin (pEF5/FRT/V5-DEST) or kanamycin (pDONR221) to allow the growth of clones containing the putatively mutant plasmids, since resistance to these antibiotics would be unaffected. 10–20 Colonies were picked and sequentially streaked first onto LB/agar Petri dishes supplemented with ampicillin or kanamycin, and secondly onto dishes additionally supplemented with chloramphenicol, in order to identify colonies that were sensitive to this antibiotic as a result of the mutation in the *CAT* gene. Colonies that failed to grow in the ampicillin/chloramphenicol (pEF5/FRT/V5-DEST) or kanamycin/chloramphenicol (pDONR221) dishes were identified, and their counterpart streak on the dish with a single antibiotic was used to rescue the clone and prepare plasmid. Since chloramphenicol sensitivity (Cam^S) could also result from spontaneous mutation of the *CAT* gene, the presence of the desired mutation was confirmed by digestion of the plasmids with *NcoI* and by sequencing. The resulting chloramphenicol-sensitive, mutant plasmids pDONR221_Cam^S and pEF5/FRT/V5-DEST_Cam^S were subsequently used in recombination reactions for the incorporation of the *attR4-attR3* Gateway cassette.

Construction of pDONR221-R4-R3

The standard Gateway cassette in pDONR221 was replaced by the *attR4-attR3* MultiSite Gateway cassette, resulting in vector pDONR221-R4-R3. This was done by using the methodology described by Magnani et al. [13], with modifications. Firstly, we PCR-amplified the *attR4-attR3*-flanked MultiSite Gateway cassette from vector pDEST-R4-R3 [13] in two sequential PCR reactions, by using oligonucleotide primers that added external *attB1* and *attB2* sites in juxtaposition to the existing *attR4* and *attR3* sites, respectively. The first PCR (10 cycles) was performed with the oligonucleotide primer pair: 5′-AAAGCAG GCTCAACTTTGTATAGAAAAGTTG-3 (*attR4fw*) and 5′-AAAGCTGGGTCAACTATGTATAATAAAGTTG-3′ (*attR3rv*), while the second PCR reaction (32 cycles) was carried out with the external primer pair: 5′-GGGGAC AAGTTTGTACAAAAAAGCAGGCTCA-3′ (*attB1-R4fw*) and 5′-GGGGACCACTTTGTACAAGAAAGCTGGGT CA-3′ (*attB2-R3rv*), by using a 1:10 dilution of the first PCR reaction in fresh Accuprime *Pfx* supermix as the source of template. The final PCR product, which contained an intact *CAT* gene, was used with plasmid pDONR221-Cam^S in a BP recombination reaction for 1h at 25 °C. After transformation of the reaction into ccdB Survival™ 2 T1R competent cells, the culture was plated onto an LB/agar dish containing kanamycin and chloramphenicol to select for colonies harbouring plasmids that had successfully recombined with the PCR product and were able to grow in the presence of chloramphenicol. Colonies were picked for preparation of plasmid and further analysis. The presence of the *attR4* and *attR3* sites was confirmed by sequencing of the vector. The integrity of the *ccdB* gene in pDONR221-R4-R3 was tested by transformation of the vector into Top10 competent cells, which are sensitive to the *ccdB* gene and failed to grow in the presence of the vector.

Construction of pEF5/FRT/V5-DEST-R4-R3

Adaptation of pEF5/FRT/V5-DEST for MultiSite Gateway cloning was carried out as described by Magnani et al. [13], with modifications. Briefly, an LR recombination reaction was carried out at 25°C for 16h between plasmids pEF5/FRT/V5-DEST_Cam^S (see above) and pDONR221-R4-R3. Importantly, the *attR4/attR3* sites in pDONR221-R4-R3 cannot recombine with the juxtaposed *attL1/attL2* sites [13]. After transformation of the reaction into ccdB Survival™ 2 T1R competent cells, the cultures were plated onto LB/agar dishes containing ampicillin and chloramphenicol to select for colonies harbouring the recombined pEF5/FRT/V5-DEST plasmid. In this way, the *attR1/attR2*-flanked standard Gateway cassette in pEF5/FRT/V5-DEST was substituted by the *attR4/attR3*-flanked MultiSite Gateway cassette, obtaining pEF5/FRT/V5-DEST-R4-R3, which was ready for a MultiSite LR recombination cloning reaction with the three modules that would make up the fusion protein of choice.

Entry clones containing the ORFs of interest

The entry clones for p65 and PAR-2 were generated by BP clonase-mediated recombination reactions (1 h, 25°C) between vector pDONR221 and purified PCR-products representing each of the full-length cDNAs. Nevertheless, purification of PCR products prior to BP cloning is not essential as the reactions also work well with non-purified PCR products. To make those entry clones apt for optional single-fragment recombination into a standard destination vector, part of a Kozak sequence (GCCGCC) was included in the forward primers, 5′to the ATG initiation codon, in order to optimise translation of the cDNA. Thus, when the same entry clone participated as a central element in the cloning of a fusion protein, the Kozak sequence translated into two alanine residues preceding the initial methionine. The stop codon was omitted from the cDNA-specific sequence of the reverse primers so as to allow the construction of fusions with a C-terminal module. The PCR products were generated by using a two-step PCR method similar to that used to amplify the *attR4-attR3*-flanked MultiSite Gateway cassette. In the first PCR reaction, a set of primers that contained part of the *attB1* (forward primer) or *attB2* (reverse primer) recombination sites followed by template-specific sequence (Additional file 3: Table S1), was used in combination with a template-containing plasmid, in a 10-cycle PCR reaction. A second, 32-cycle PCR reaction was set up by diluting 1/10 the first PCR reaction, and supplementing it with a new set of primers (Additional file 3: Table S1) that were complementary to the partial *attB1* (forward primer) or *attB2* (reverse primer) sequences incorporated into the PCR product by the first set of primers, and that extended the recombination sites to the full length recommended by Life Technologies for BP clonase-mediated cloning. This two-step method allowed the use of the second set of primers in multiple cloning projects since the specificity with regard to the cDNA relied on the sequence of the first set of primers, resulting in potential savings in the costs of oligonucleotide synthesis in the long term. It should be noted that this is not a technical requirement since one-step PCR reactions with primers providing full-length, flanking *attB* sites are widely used for Gateway cloning of PCR products. The cDNA fragment comprising positions 141 to 1793 of the human p65/RelA mRNA sequence (accession number NM_021975.3), excluding the natural stop codon, was amplified as described above from plasmid pCDNA3.1-p65 [40], with the primers indicated in Additional file 3: Table S1. The mouse thrombin receptor PAR-2 cDNA was amplified from the FANTOM Full Length cDNA clone number G8300117P07 (pFLCI-PAR2, The Institute of Physical and Chemical Research (RIKEN) [41]), with primers (Additional file 3: Table S1) that encompassed the sequence between positions from 114 to 1313 in the mouse mRNA (accession number NM_007974). For the fusion protein that contained the full-length mouse SIRT1 coding sequence, we used an IMAGE ORFeome collaboration clone (100066295) with the SIRT1 cDNA already flanked by *attL1/attL2* sites in vector pENTR223-SfiI [42]. Cultures of bacteria transformed with this plasmid were grown with LB medium containing 50 μg/ml spectinomycin. A mouse SIRT1 cDNA with a deletion that affected a region near the N-terminus of the protein (*ΔN-t-SIRT1)* was amplified from plasmid pUSEamp SIRα2 (Millipore). This internal deletion affected residues Leu7 to Ala123 but did not disrupt the rest of the ORF, causing the loss of a region encompassing the N-terminal nuclear location signal NLS1 [19]. The deletion was caused by off target annealing of the forward primer in the first PCR reaction (Additional file 3: Table S1), to a sequence lying 354 bp downstream of the target sequence. The full *attB1* and *attB2* sites were completed in a second PCR by using the universal set of primers, and the PCR product was cloned into pDONR221 through BP clonase-mediated recombination, as described above. All PCR primers used were designed so that the eventual *attB1/attB2*-flanked PCR products would be in frame in the context of three-fragment recombinations for the production of fusion proteins. The sequence of all clones was verified by sequencing.

Construction of the modules' collections

The collections of N-terminal and C-terminal modules were constructed as clones containing DNA inserts in vectors pDONR P4-P1R and pDONR P2R- P3, respectively. The inserts (italics) were PCR-amplified from the following plasmid templates: *mKate2*, pmKate2-C (Evrogen); *EGFP*, pLV-EGFP (kind gift from M. Perez-Pinzon, UM); *EYFP*, pEYFP-mito (Clontech); *ECFP* and *IRES_ECFP*, pYIC (Addgene plasmid 18673 [43]; *V5-6xHis epitope tag cassette*, pEF6/V5-His (Life Technologies). Construction of the clone containing the SV40 early polyadenylation signal has previously been described [44]. All inserts were amplified in a two-step PCR reaction, similarly to the way described in the section above, with the primers described in Additional file 3: Table S1. In this case, the primers attached flanking *attB4/attB1R* sites to the PCR products to be cloned into pDONRP4-P1R, while those to be cloned into pDONRP2R-P3 were furnished with *attB2R/attB3* sites. When the functional modules were cloned into pDONR-P4-P1R (N-terminal modules), the forward primer in the first pair contained a Kozak sequence in order to improve translation of the fusion protein. Furthermore, if the template sequence contained a stop codon, this was not included when designing the reverse primer of the first pair so as to avoid interrupting translation downstream of the module. Nevertheless, when the module was to be located at the C-terminal end of the fusion (a pDONRP2R-P3-based clone), the stop codon was allowed into the sequence of the reverse primer of the first pair. Cloning of the PCR products into their corresponding pDONR vector was carried out by using BP clonase, as described above. The sequence of all clones was verified by sequencing.

Cloning of fusion proteins into pEF5FRTV5DEST-R4-R3

Chimeric ORFs for the expression of fusion proteins were assembled by performing MultiSite LR recombination reactions between the three selected entry clones and the adapted pEF5FRTV5DEST-R4-R3 destination expression vector. Ten femtomoles of Maxiprep-quality DNA from each entry vector were mixed with 20 femtomoles of the destination expression vector and 2 μl of LR Clonase II plus enzyme mix (Life technologies) in a final volume of 10 μl and incubated for 16 h at 25°C, following instructions from the manufacturer. A 2-μl aliquot of the reaction was transformed into either Top10 or stbl3 *E. coli*, and the reaction was plated on LB-agar medium supplemented with ampicillin to select transformed bacteria. Colonies potentially containing recombined pEF5FRTV5DEST expression vectors with the chimeric ORF were picked with sterile pipette tips and streaked onto Petri dishes containing LB-agar medium plus ampicillin in order to amplify them. These cultures were subsequently streaked on Petri dishes containing LB-agar medium supplemented with ampicillin and chloramphenicol in order to check for the presence of colonies containing non-recombined destination expression vector that may have spontaneously mutated the *ccdB* gene in the Gateway cassette. The same colonies were also tested for growth on plates with LB-agar medium containing the antibiotic to which the entry clones that had been used in the LR reaction conferred resistance to (kanamycin, spectinomycin), since we observed an occasional phenomenon of cotransformation of the destination expression vector with the entry clones. This was observed even though both the pDONR series of plasmids and pEF5FRTV5DEST contain the same origin of replication and thus belong to the same incompatibility group. This is a phenomenon that has been thoroughly described elsewhere [45,46], and needs to be taken into account as it could be a confounding factor in plasmid preparations that are destined to be transfected. Future refinements of this method will have to be devised in order to avoid such cotransformation events in high-throughput applications. Only the colonies that grew exclusively on LB plus ampicillin were used for further tests. In our hands, screening of about 20–25 colonies per transformation was sufficient to find colonies harbouring the expression vector in the absence of "piggybacking" entry clones. The integrity of the chimeric ORFs in the recombined expression vectors was checked both by restriction digest and sequencing. As a guide to the expected colony yield, colony counting after transformation of half the reaction volume in a dedicated series of reactions resulted in 258 ± 54 (mean ± SD, n = 5) colonies per Petri dish. This number is lower than the range suggested by Life technologies in the Multisite Gateway cloning kit user's manual (1000–5000 colonies, when the whole transformation is plated), which could be caused by procedural differences introduced as a result of our toolkit's design. In any case, colony numbers expected from Multisite Gateway reactions are still below those produced in single-fragment LR recombinations (see Life Technologies' Gateway user manuals) because of the participation of more DNA fragments, which should be taken into account when planning Multisite Gateway LR recombination reactions.

Expression of fusion proteins

Expression vectors encoding chimeric ORFs were introduced into HeLa cells or Raw264.7 murine monocytes, in transient transfection experiments. Both cell lines were kindly provided by Dr Andy Clark (University of Birmingham, UK). In the case of HeLa cells, 1.5×10^4 cells per well were seeded on top of glass coverslips in 24-well plates. Cells were transfected with expression vectors for mKate2 fused to wt SIRT1 or ΔN-t-SIRT1 (both with N-terminal mKate2), or PAR-2 (with mKate2 fused to the C-terminus of PAR2), as well as for PAR-2 fused to EGFP (also in a C-terminal fusion to PAR-2). One μg of plasmid DNA was transfected with 3 μl of Lipofectamine reagent (Life Technologies), following the manufacturer's instructions. Forty-eight hours after transfection, cells transfected with the PAR-2 fluorescent fusions were either left untreated (control) or treated for 1 h with the PAR-2 specific agonist AC55541 (Tocris) at a 5 μM final concentration. At the end of the treatments, cells were fixed with 4% paraformaldehyde, rinsed in cold methanol followed by a brief wash in H_2O and mounted in Fluoromount G (Southern Biotech). Cells were observed under a Nikon Eclipse E1000 fluorescence microscope. For the Raw264.7 cell line, 10^5 cells per well were seeded on 24-well plates. Cells were transfected with 1 μg of maxiprep-quality plasmid DNA and 3.2 μl of jetPEI-Macrophage reagent (Polyplus transfection) per well, according to the manufacturer's instructions. The transfected expression vectors encoded for fusions of p65 with the red fluorescent protein mKate2. Forty-eight hours after transfection, cells were either left in control conditions or treated with LPS, and were observed live under an Olympus IX70 fluorescence inverted- microscope. For the time-lapse experiment, Raw 264.7 cells seeded on a 25 mm Ø glass coverslip in a well of a six-well plate were transfected with the expression vector for the PAR2: mKate2 fusion protein and mounted in a chamber equipped with an incubation system with temperature control, in a Leica TCS-SL confocal inverted microscope (63X objective lens). A transfected cell was imaged live, with micrographs taken at 20 sec intervals, both under control conditions or after treating with 5 μM AC55541.

Additional files

Additional file 1: Figure S1. Single fragment and MultiSite Gateway recombinational cloning. In the Gateway system, DNA fragments (such as a cDNA) can be PCR-amplified with primers that attach flanking *attB1/attB2* sites (B1, B2), and cloned into a compatible vector by carrying out a BP recombination (BP rec.). This generates a so-called Entry clone where the DNA fragment is flanked by *attL1/attL2* sites (L1, L2), and that can be subsequently used to shuttle the DNA fragment into destination vectors that provide specific functions. In standard single-fragment Gateway cloning, an *attL1/attL2*-flanked cDNA in the example is transferred to a destination vector that contains compatible *attR1/attR2* sites through an LR-recombination reaction (LR rec.). On the other hand, in the MultiSite Gateway cloning system, three different entry clones with DNA fragments flanked by sequence variants of the *attL* and *attR* sites (L3, L4, R3, R4) participate in a multi-fragment LR-recombination reaction with the promoter-less destination vector pDEST-R4-R3. This vector contains a Gateway cassette that is flanked by *attR4/attR3* sites, which conditions the order of recombination of the three fragments in the resulting destination vector owing to the nature of their respective flanking *att* sites, as indicated.

Additional file 2: Figure S2. Mutation of the *CAT* gene in the Gateway cassette of pEF5FRT-DEST abolishes resistance to chloramphenicol. Cultures of ccdB-resistant *E. coli* transformed with pEF5FRT-DEST encoding a wild type (a,c), or a mutant version of the *CAT* gene (b, d), were streaked on LB-agar dishes containing ampicillin (a,b) or ampicillin plus chloramphenicol (c,d). While bacteria transformed with either of the plasmids were able to grow in the presence of ampicillin, further supplementation of the medium with chloramphenicol specifically prevented the growth of bacteria transformed with the plasmid containing the mutation of the *CAT* gene (d).

Additional file 3: Table S1. Sequence of the oligonucleotides used for PCR amplification. Two-step PCRs were set up with the primers indicated on the table (fw: forward, rv: reverse) in order to attach the appropriate *attB* sites to the functional modules to be cloned by BP clonase-mediated recombination. The module-specific primers were used in the first PCR and contain part of the *att* sequence. The universal external primers were used in the second PCR to complete the *att* sites. In the module-specific primers, sequence in capitals corresponds to the oligonucleotide segment that anneals to the template, while the sequence in bold type is annealed by the universal external primer that will complete the corresponding *att* site. The same forward and reverse primers were used for the PCR amplification of EGFP, ECFP and EYFP, since the mutations dictating the fluorescence wavelength lie beyond the sequence annealed by the primers. The N-terminal V5-6xHis module was PCR-amplified with a three-step PCR. The first forward module-specific primer (a) attached a Kozak sequence and an initiation methionine codon to the cassette containing the epitope tags, but no *att*-related sequence (Ø), while the second PCR was carried out with a second forward primer (b) that provided the seed for the *attB4* site. This site was completed in the last PCR, which was carried out with the corresponding external universal primers. Only one reverse module-specific primer was used in the first and second PCRs for this module.

Additional file 4: Figure S3. Versatility of the cloning toolkit. *A.* Simultaneous construction of vectors expressing versions of the same fusion protein coupled to different fluorescent modules, offering a choice of optical properties in experiments where individual or multiple fusion proteins are expressed. *B.* A library of mutations can be generated in the vector encoding the ORF of interest so recombination of the library with intact functional modules would allow the generation of a homogeneous range of expression vector mutation libraries to be screened on different model systems. *C.* Fusion proteins can be constructed so that the functional modules flanking the ORF of interest are in either of the two possible orders, to evaluate putative effects on protein function.

Additional file 5: Movie 1. Complete time-lapse sequence recorded on the PAR2:mKate2-expressing cell shown in Figure 4C. The sequence spans an almost one-hour period, with the AC55541 agonist added right at the start.

Additional file 6: Movie 2. Time-lapse recording of the same cell as in Additional file 5: Movie 1, recorded under control conditions for 11 min before adding the AC55541 agonist.

Competing interests
The authors declare that they have no competing interests.

Authors' contributions
TS and AMP conceived the idea. TS designed the toolkit. RB, NI and TS carried out the cloning and transfection experiments and analyzed the data. TS and AMP wrote the paper. All authors read and approved the final manuscript.

Acknowledgements
We thank Alejandro Vaquero (IDIBELL, Spain) for sharing plasmid pUSEamp SIRα2. We are grateful to Valérie Petegnief (IIBB, Spain) for technical help and encouragement, as well as for critically reviewing the manuscript. We thank Maria Calvo and her team at the Confocal Microscopy Unit from the "Centres Científics i Tecnològics" (University of Barcelona, Spain) for their help in the time-lapse experiments. This work was supported by grants from the Spanish Ministry of Innovation and Science through FEDER funds (SAF2008-04515-CO2-02 and SAF2011-30492). T. S. is a participant of the "Programa d'estabilització d'investigadors de la Direcció d'Estratègia i Coordinació del Departament de Salut" from the "Generalitat de Catalunya", Spain. We thank the Unit of Scientific Information Resources for Research (URICI) at CSIC for their support on the publication fee through their Open Access Publication Support Initiative.

Author details
[1]Department of Brain Ischemia and Neurodegeneration, Institut d'Investigacions Biomèdiques de Barcelona (IIBB)-Consejo Superior de Investigaciones Científicas (CSIC), Barcelona, Spain. [2]Institut d'Investigacions Biomèdiques August Pi i Sunyer (IDIBAPS), Barcelona, Spain. [3]Institut de Medicina Predictiva i Personalitzada del Càncer (IMPPC), Badalona, Barcelona, Spain.

References

1. Kultima JR, Sunagawa S, Li J, Chen W, Chen H, Mende DR, Arumugam M, Pan Q, Liu B, Qin J, *et al*: **MOCAT: a metagenomics assembly and gene prediction toolkit.** *PLoS One* 2012, **7:**e47656.
2. Uchiyama T, Miyazaki K: **Functional metagenomics for enzyme discovery: challenges to efficient screening.** *Curr Opin Biotechnol* 2009, **20:**616–622.
3. Godzik A: **Metagenomics and the protein universe.** *Curr Opin Struct Biol* 2011, **21:**398–403.
4. Craig JW, Chang FY, Kim JH, Obiajulu SC, Brady SF: **Expanding small-molecule functional metagenomics through parallel screening of broad-host-range cosmid environmental DNA libraries in diverse proteobacteria.** *Appl Environ Microbiol* 2010, **76:**1633–1641.
5. Rual JF, Hill DE, Vidal M: **ORFeome projects: gateway between genomics and omics.** *Curr Opin Chem Biol* 2004, **8:**20–25.
6. Simpson JC, Wellenreuther R, Poustka A, Pepperkok R, Wiemann S: **Systematic subcellular localization of novel proteins identified by large-scale cDNA sequencing.** *EMBO Rep* 2000, **1:**287–292.
7. Crivat G, Taraska JW: **Imaging proteins inside cells with fluorescent tags.** *Trends Biotechnol* 2012, **30:**8–16.
8. Hartley JL, Temple GF, Brasch MA: **DNA cloning using in vitro site-specific recombination.** *Genome Res* 2000, **10:**1788–1795.
9. Cheo DL, Titus SA, Byrd DR, Hartley JL, Temple GF, Brasch MA: **Concerted assembly and cloning of multiple DNA segments using in vitro site-specific recombination: functional analysis of multi-segment expression clones.** *Genome Res* 2004, **14:**2111–2120.
10. Brasch MA, Hartley JL, Vidal M: **ORFeome cloning and systems biology: standardized mass production of the parts from the parts-list.** *Genome Res* 2004, **14:**2001–2009.
11. Lamesch P, Li N, Milstein S, Fan C, Hao T, Szabo G, Hu Z, Venkatesan K, Bethel G, Martin P, *et al*: **hORFeome v3.1: a resource of human open reading frames representing over 10,000 human genes.** *Genomics* 2007, **89:**307–315.
12. Rolfs A, Hu Y, Ebert L, Hoffmann D, Zuo D, Ramachandran N, Raphael J, Kelley F, McCarron S, Jepson DA, *et al*: **A biomedically enriched collection of 7000 human ORF clones.** *PLoS One* 2008, **3:**e1528.
13. Magnani E, Bartling L, Hake S: **From Gateway to MultiSite Gateway in one recombination event.** *BMC Mol Biol* 2006, **7:**46.
14. von Stetten D, Noirclerc-Savoye M, Goedhart J, Gadella TW Jr, Royant A: **Structure of a fluorescent protein from Aequorea victoria bearing the obligate-monomer mutation A206K.** *Acta Crystallogr Sect F Struct Biol Cryst Commun* 2012, **68:**878–882.
15. Espagne A, Erard M, Madiona K, Derrien V, Jonasson G, Levy B, Pasquier H, Melki R, Merola F: **Cyan fluorescent protein carries a constitutive mutation that prevents its dimerization.** *Biochemistry* 2011, **50:**437–439.
16. Roure A, Rothbacher U, Robin F, Kalmar E, Ferone G, Lamy C, Missero C, Mueller F, Lemaire P: **A multicassette Gateway vector set for high throughput and comparative analyses in ciona and vertebrate embryos.** *PLoS One* 2007, **2:**e916.
17. Chen Y, Qiu S, Luan CH, Luo M: **Domain selection combined with improved cloning strategy for high throughput expression of higher eukaryotic proteins.** *BMC Biotechnol* 2007, **7:**45.
18. Shaner NC, Steinbach PA, Tsien RY: **A guide to choosing fluorescent proteins.** *Nat Methods* 2005, **2:**905–909.
19. Tanno M, Sakamoto J, Miura T, Shimamoto K, Horio Y: **Nucleocytoplasmic shuttling of the NAD+ –dependent histone deacetylase SIRT1.** *J Biol Chem* 2007, **282:**6823–6832.
20. Shcherbo D, Murphy CS, Ermakova GV, Solovieva EA, Chepurnykh TV, Shcheglov AS, Verkhusha VV, Pletnev VZ, Hazelwood KL, Roche PM, *et al*: **Far-red fluorescent tags for protein imaging in living tissues.** *Biochem J* 2009, **418:**567–574.
21. Dery O, Thoma MS, Wong H, Grady EF, Bunnett NW: **Trafficking of proteinase-activated receptor-2 and beta-arrestin-1 tagged with green fluorescent protein. beta-Arrestin-dependent endocytosis of a proteinase receptor.** *J Biol Chem* 1999, **274:**18524–18535.
22. Carlotti F, Chapman R, Dower SK, Qwarnstrom EE: **Activation of nuclear factor kappaB in single living cells. Dependence of nuclear translocation and anti-apoptotic function on EGFPRELA concentration.** *J Biol Chem* 1999, **274:**37941–37949.
23. Johannessen CM, Boehm JS, Kim SY, Thomas SR, Wardwell L, Johnson LA, Emery CM, Stransky N, Cogdill AP, Barretina J, *et al*: **COT drives resistance to RAF inhibition through MAP kinase pathway reactivation.** *Nature* 2010, **468:**968–972.
24. Kwan KM, Fujimoto E, Grabher C, Mangum BD, Hardy ME, Campbell DS, Parant JM, Yost HJ, Kanki JP, Chien CB: **The Tol2kit: a multisite gateway-based construction kit for Tol2 transposon transgenesis constructs.** *Dev Dyn* 2007, **236:**3088–3099.
25. Love NR, Thuret R, Chen Y, Ishibashi S, Sabherwal N, Paredes R, Alves-Silva J, Dorey K, Noble AM, Guille MJ, *et al*: **pTransgenesis: a cross-species, modular transgenesis resource.** *Development* 2011, **138:**5451–5458.
26. Nagels Durand A, Moses T, De Clercq R, Goossens A, Pauwels L: **A MultiSite Gateway vector set for the functional analysis of genes in the model Saccharomyces cerevisiae.** *BMC Mol Biol* 2012, **13:**30.
27. Dupuy D, Li QR, Deplancke B, Boxem M, Hao T, Lamesch P, Sequerra R, Bosak S, Doucette-Stamm L, Hope IA, *et al*: **A first version of the Caenorhabditis elegans Promoterome.** *Genome Res* 2004, **14:**2169–2175.
28. Dupuy D, Bertin N, Hidalgo CA, Venkatesan K, Tu D, Lee D, Rosenberg J, Svrzikapa N, Blanc A, Carnec A, *et al*: **Genome-scale analysis of in vivo spatiotemporal promoter activity in Caenorhabditis elegans.** *Nat Biotechnol* 2007, **25:**663–668.
29. Sasaki Y, Sone T, Yahata K, Kishine H, Hotta J, Chesnut JD, Honda T, Imamoto F: **Multi-gene gateway clone design for expression of multiple heterologous genes in living cells: eukaryotic clones containing two and three ORF multi-gene cassettes expressed from a single promoter.** *J Biotechnol* 2008, **136:**103–112.

30. Petersen LK, Stowers RS: **A Gateway MultiSite recombination cloning toolkit.** *PLoS One* 2011, **6:**e24531.
31. *The Drosophila Gateway™ Vector Collection.* http://emb.carnegiescience.edu/labs/murphy/Gateway%20vectors.html#_Overview.
32. Alberti S, Gitler AD, Lindquist S: **A suite of Gateway cloning vectors for high-throughput genetic analysis in Saccharomyces cerevisiae.** *Yeast* 2007, **24:**913–919.
33. Engler C, Kandzia R, Marillonnet S: **A one pot, one step, precision cloning method with high throughput capability.** *PLoS One* 2008, **3:**e3647.
34. Lu Q: **Seamless cloning and gene fusion.** *Trends Biotechnol* 2005, **23:**199–207.
35. Tsvetanova B, Peng L, Liang X, Li K, Hammond L, Peterson TC, Katzen F: **Advanced DNA assembly technologies in drug discovery.** *Expert Opin Drug Discov* 2012, **7:**371–374.
36. Sarrion-Perdigones A, Falconi EE, Zandalinas SI, Juarez P, Fernandez-del -Carmen A, Granell A, Orzaez D: **GoldenBraid: an iterative cloning system for standardized assembly of reusable genetic modules.** *PLoS One* 2011, **6:**e21622.
37. Weber E, Engler C, Gruetzner R, Werner S, Marillonnet S: **A modular cloning system for standardized assembly of multigene constructs.** *PLoS One* 2011, **6:**e16765.
38. Leslie AG: **Refined crystal structure of type III chloramphenicol acetyltransferase at 1.75 A resolution.** *J Mol Biol* 1990, **213:**167–186.
39. Van der Schueren J, Robben J, Volckaert G: **Misfolding of chloramphenicol acetyltransferase due to carboxy-terminal truncation can be corrected by second-site mutations.** *Protein Eng* 1998, **11:**1211–1217.
40. Gutierrez H, O'Keeffe GW, Gavalda N, Gallagher D, Davies AM: **Nuclear factor kappa B signaling either stimulates or inhibits neurite growth depending on the phosphorylation status of p65/RelA.** *J Neurosci* 2008, **28:**8246–8256.
41. Carninci P, Kasukawa T, Katayama S, Gough J, Frith MC, Maeda N, Oyama R, Ravasi T, Lenhard B, Wells C, *et al*: **The transcriptional landscape of the mammalian genome.** *Science* 2005, **309:**1559–1563.
42. Strausberg RL, Feingold EA, Grouse LH, Derge JG, Klausner RD, Collins FS, Wagner L, Shenmen CM, Schuler GD, Altschul SF, *et al*: **Generation and initial analysis of more than 15,000 full-length human and mouse cDNA sequences.** *Proc Natl Acad Sci U S A* 2002, **99:**16899–16903.
43. Nie M, Htun H: **Different modes and potencies of translational repression by sequence-specific RNA-protein interaction at the 5'-UTR.** *Nucleic Acids Res* 2006, **34:**5528–5540.
44. de la Rosa X, Santalucia T, Fortin PY, Purroy J, Calvo M, Salas-Perdomo A, Justicia C, Couillaud F, Planas AM: **In vivo imaging of induction of heat-shock protein-70 gene expression with fluorescence reflectance imaging and intravital confocal microscopy following brain ischaemia in reporter mice.** *Eur J Nucl Med Mol Imaging* 2013, **40:**426–438.
45. Goldsmith M, Kiss C, Bradbury AR, Tawfik DS: **Avoiding and controlling double transformation artifacts.** *Protein Eng Des Sel* 2007, **20:**315–318.
46. Velappan N, Sblattero D, Chasteen L, Pavlik P, Bradbury AR: **Plasmid incompatibility: more compatible than previously thought?** *Protein Eng Des Sel* 2007, **20:**309–313.

15

Exploring the transcription activator-like effectors scaffold versatility to expand the toolbox of designer nucleases

Alexandre Juillerat*†, Marine Beurdeley†, Julien Valton†, Séverine Thomas, Gwendoline Dubois, Mikhail Zaslavskiy, Jérome Mikolajczak, Fabian Bietz, George H Silva, Aymeric Duclert, Fayza Daboussi and Philippe Duchateau*

Abstract

Background: The past decade has seen the emergence of several molecular tools that render possible modification of cellular functions through accurate and easy addition, removal, or exchange of genomic DNA sequences. Among these technologies, transcription activator-like effectors (TALE) has turned out to be one of the most versatile and incredibly robust platform for generating targeted molecular tools as demonstrated by fusion to various domains such as transcription activator, repressor and nucleases.

Results: In this study, we generated a novel nuclease architecture based on the transcription activator-like effector scaffold. In contrast to the existing Tail to Tail (TtT) and head to Head (HtH) nuclease architectures based on the symmetrical association of two TALE DNA binding domains fused to the C-terminal (TtT) or N-terminal (HtH) end of FokI, this novel architecture consists of the asymmetrical association of two different engineered TALE DNA binding domains fused to the N- and C-terminal ends of FokI (TALE::FokI and FokI::TALE scaffolds respectively). The characterization of this novel Tail to Head (TtH) architecture in yeast enabled us to demonstrate its nuclease activity and define its optimal target configuration. We further showed that this architecture was able to promote substantial level of targeted mutagenesis at three endogenous loci present in two different mammalian cell lines.

Conclusion: Our results demonstrated that this novel functional TtH architecture which requires binding to only one DNA strand of a given endogenous locus has the potential to extend the targeting possibility of FokI-based TALE nucleases.

Keywords: Transcription activator-like effectors, TALE, TALEN, Protein engineering, Genome editing

Background

Transcription activator-like effectors (TALEs), a group of bacterial plant pathogen proteins, have recently emerged as new engineerable scaffold for production of engineered DNA binding domains with chosen specificities [reviewed in [1]]. The targeting specificity of this family of proteins is driven by a central core composed of multiple repeated units. These 33 to 35 amino acids repeated units are nearly identical to one another except for two polymorphic amino acids called RVDs (repeat variable di residue), responsible for the specific recognition of a unique nucleotide [2,3]. In addition to this central core domain, the N-terminal domain of TALE has been reported to play a key role in TALEs specificity and binding mechanism. This domain displays a strong specificity bias toward a thymine nucleotide, the so called "T0", systematically located at the 5'end of the TALE target [2]. These different biochemical features were confirmed by the high resolution structure of TALE/DNA complexes, illustrating how the TALE protein wraps around its DNA target, from the 5' T0 to the 3' last nucleotide, in an N- to C-terminal orientation [4-6]. The particular DNA binding properties of TALE DNA binding domain, their exquisite specificity as well as their modularity have been used to develop engineered TALE nucleases named TALEN with tailored DNA specificity. The original TALEN architecture developed by Christian *et al.* [7],

* Correspondence: alexandre.juillerat@cellectis.com; philippe.duchateau@cellectis.com
†Equal contributors
CELLECTIS S.A, 8 Rue de la Croix Jarry, Paris 75013, France

consisted of a custom TALE DNA binding domain linked to the N-terminal end of the non-specific FokI nuclease domain (TALE::FokI scaffold, Figure 1A). Because FokI needs to dimerize to catalyze a double strand break (DSB), TALEN work by pairs. Each pair unit binds in a Tail to Tail (TtT) orientation to adjacent binding sites, starting by a 5' T0 and respectively located on the sense and antisense strand of the DNA. Such symmetrical architecture, originally designed to respect (i), the natural organization of the endonuclease and DNA binding domains of FokI (ii), the orientation of TALE DNA binding and (iii), the requirement of a T0, led to an incredibly robust TALE-based nuclease platform. Considered as the gold standard TALEN architecture, it has been extensively optimized and used for different gene editing applications [8,9]. Furthermore, the versatility of

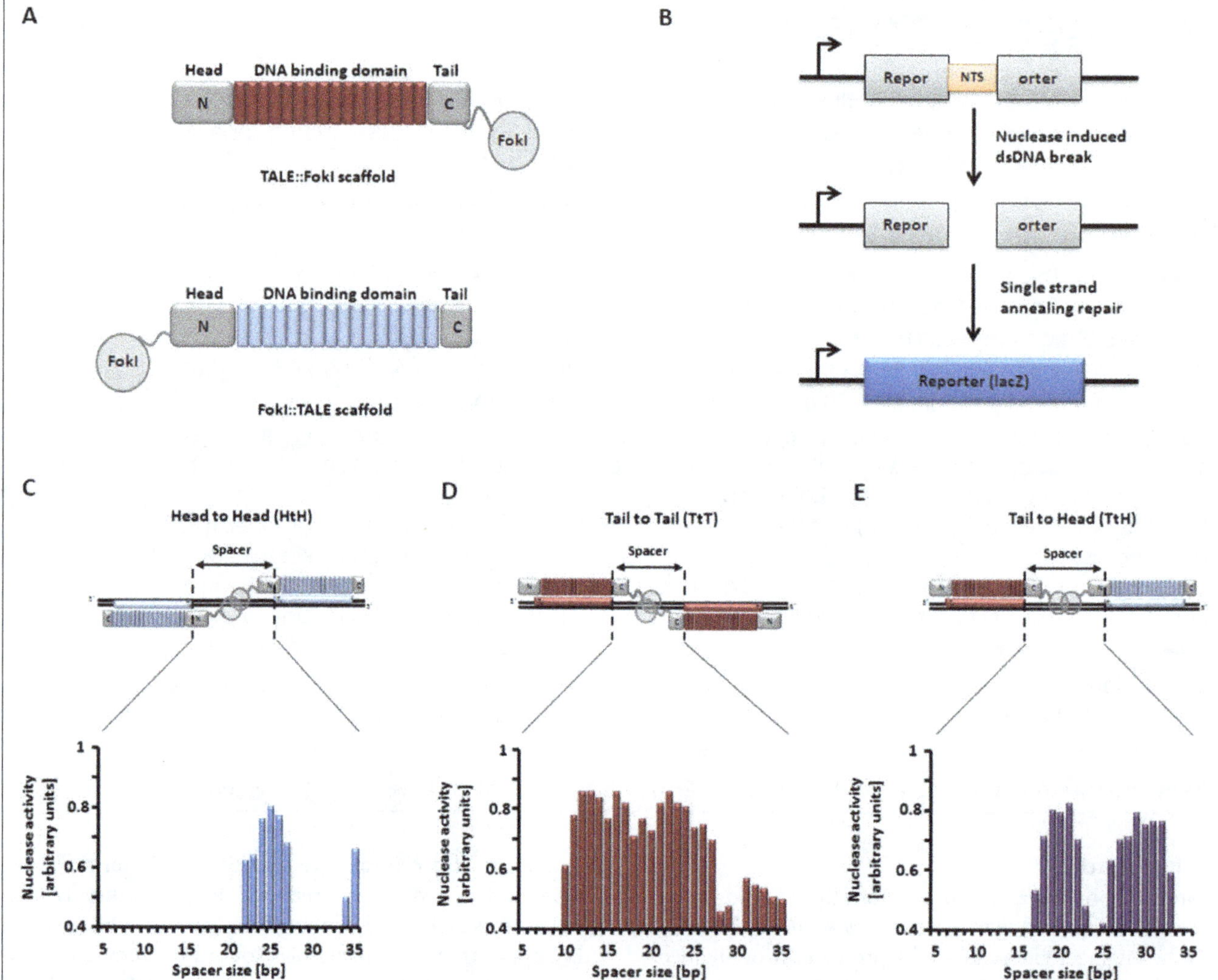

Figure 1 Design, creation and *In vivo* characterization of the three nuclease architectures based on the FokI catalytic domain in yeast. **(A)** Schematic representation of the two different scaffolds used in this study including the positions of N and C-terminal domains, DNA binding domain as well as the Tail and Head positions. **(B)** Schematic representation of the yeast extrachromosomal single strand annealing (SSA) assay. The reporter plasmid containing a Nuclease Target Sequence (NTS) is flanked by overlapping truncated LacZ genes sequences. Cleavage of the target sequence in yeast leads to the restoration of the LacZ marker through the single strand annealing (SSA) pathway of recombination. The restoration of the functional LacZ gene is quantified by a β-galactosidase activity assay and related to the nuclease efficiency. **(C, D and E)** Representative examples of activity measurements from the yeast SSA assay for the three architectures obtained on the same filter. **(C)** HtH architecture where two FokI:: TALE scaffolds are facing each other on the two DNA strands and in a head to head orientation. **(D)** Classical TtT architecture where two TALE::FokI scaffolds are facing each other on the two DNA strands and in a Tail to Tail orientation. **(E)** TtH architecture where a TALE::FokI and a FokI:: TALE scaffolds are facing each other on the same DNA strand and in a Tail to Head orientation. The nuclease activity measured for the three architectures in yeast using the single strand annealing assay (SSA) as a function of target spacer length (5-35 bp) is displayed at the bottom of each figure panels. For each filter, three controls (negative control, weak nuclease and strong nuclease) were measured multiple times (n > 100). Standard deviation on these activity measurements were typically of 0.05.

the TALE scaffold was demonstrated by the accumulation of studies reporting fusion (N-terminal as well as C-terminal) of the TALE core to various catalytic domains [10-21]. In particular, we [22] and others [23] have reported the development of TALE-based nuclease with an N-terminal fusion FokI catalytic domain.

In this study, we further explored and exploited the versatility of the TALE scaffold to develop a novel asymmetrical hybrid TALE and FokI-based nuclease architectures, the Tail to Head (TtH) architecture referred herein as TtH. We demonstrated the potential of this architecture to generate targeted mutagenesis at different endogenous loci in mammalian cells. In contrast to the conventional TALEN architecture, the TtH architecture only required one DNA strand of a given locus to efficiently bind and process it. Thus, our work presents a new advance in the development of TALE nucleases and further extends their targeting possibilities.

Results

Design and evaluation of the Head to Head (HtH) symmetrical architecture in a yeast SSA assay

To investigate the versatility of the TALE scaffold and generate alternative nuclease architectures, we used a yeast-based nuclease activity assay [24]. This assay was previously demonstrated to be suitable to assess the intrinsic nuclease activity of TALE nuclease without being biased by epigenetic modifications or chromatin context. In addition, we have previously found a good correlation between data obtained with the yeast SSA assay, an extrachromosomal SSA assay in CHO-K1 and chromosomal disruption experiments in CHO-KI [25,26]. We thus believed that the yeast model system could serve as an appropriate and representative assay to compare characteristics of different nuclease architectures. This assay relies on two yeast strains, one expressing the nuclease of interest and the other the target sequence flanked by overlapping truncated LacZ genes (Figure 1B). After mating of the two strains, the restoration, upon target cleavage, of the LacZ marker though the SSA pathway of recombination recontitute a functional LacZ gene. The resulting β-galactosidase can further be quantified and related to the nuclease efficiency. The experimental conditions were optimized to avoid saturation of the signal, thus allowing a direct comparison on the whole range of activities.

In order to evaluate alternative configurations to the standard TtT TALEN (symmetrical association of two TALE::FokI scaffolds, Figure 1A), we designed a construction harboring a FokI catalytic domain fused to the N-terminal domain of a TALE (FokI::TALE scaffold), leading to the symmetrical head-to-head nuclease architecture referred herein as HtH (Figure 1C). Throughout this study we used the avirulence protein AvrBs3 as scaffold (accession number P14727). The first step for developing the alternative HtH and TtH architectures was to re-engineer a fusion protein to create an efficient FokI::TALE scaffold. Early works performed on TALE protein showed that the first 152 amino acids of the N-terminal domain could be deleted (Δ152 variant) without affecting the protein activity [27]. We used an approach previously described to create an active I-TevI based TALE nuclease [22] and fused the FokI catalytic domain (amino acids 388 to 583, accession number P14870) to the N-terminal end of the Δ152 TALE variant via a 4 aminoacids (-GSSG-) flexible linker. In addition, we removed most of the C-terminal end of the TALE domain (keeping only the first 11 amino acids, amino acids 887 to 897, accession number P14727) to minimize the global size of the final protein. To allow the specific targeting of the desired DNA sequences, we used the canonical RVD/nucleotide association code (NI: A, HD:C, NN:G and NG:T) [2,3].

A RVD array targeting an 18 base pairs sequence of interest (ATATAAACCTAACCCTCT, Additional file 1: Table S2) was cloned into the FokI::TALE backbone. The nuclease activity of the resulting construction was tested in yeast using the extrachromosomal single strand annealing (SSA) assay (Figure 1B) and homodimeric DNA targets containing respectively two identical recognition sequences juxtaposed with the 5' ends proximal (Figure 1C). To determine the optimal distance for cleavage activity between the two recognition domains, a series of homodimeric targets were designed with spacers ranging from 5 to 35 bp (Figure 1C, Additional file 1: Table S2). Furthermore, we also prepared the classical TtT TALEN targeting the same DNA sequence to serve as a reference for the currently used architecture (Figure 1D). In this study, we used a + C40 (amino acids 887 to 926, accession number P14727, followed by a 4 aminoacids –ISRS- linker) TALEN scaffold as, in our hands, this truncation presented a good balance to obtain a high activity associated with a good specificity (narrow spacer window). We additionally designed series of homodimeric targets containing respectively two identical recognition sequences juxtaposed with the 3' ends proximal (Figure 1D, Additional file 1: Table S3) with the same spacing described above. The results obtained from this SSA assay showed that, despite not preserving the natural N-terminus (DNA binding domain) to C-terminus (catalytic domain) layout of the wild-type FokI, similar levels of activity for both architectures can be obtained (Figure 1C and 1D). Interestingly, one major difference between the two configurations was the spacing pattern reached by the HtH architecture, with a much narrower window of cleaved spacers compared to the ones obtained for the classical architecture (22 to 27 bp for the HtH nuclease versus 10 to 27 bp for the TtT nuclease).

Design and evaluation of the Tail to Head (TtH) asymmetrical architecture in a yeast SSA assay

Having demonstrated that the FokI::TALE scaffold display a high nuclease activity in a symmetrical HtH configuration, we next evaluated its ability to pair up with the TALE::FokI fusion scaffold and produce an active nuclease (Additional file 1: Figure S1E). The nuclease activity of the resulting asymmetric tail-to-head architecture, referred herein as TtH, was assessed in yeast on a collection of hybrid asymmetric targets. These targets contained two different recognition sequences juxtaposed with the 3′-5′ ends proximal and separated by a DNA spacer ranging from 5 to 35 bp (Figure 1E, Additional file 1: Table S4). The yeast SSA assay results showed activity levels comparable to the ones observed for TtT and HtH architectures, with two distinct windows of cleaved DNA spacers (18 to 22 bp and 27 to 33 bp; Figure 1E). This result suggested an optimal cleavage every one helix turn of DNA. Interestingly, the optimal cleavage distance of TtH architecture was 5 bp shorter than the one obtained for HtH architecture (20 and 25 bp respectively).

Nuclease activities of TtH asymmetrical architecture in mammalian cells and molecular characterization of nuclease-induced events

Once the activity of the new TtH architecture was demonstrated in yeast, we next investigated its activity in mammalian cells. Two loci of interest for potential therapeutic applications, previously chosen to investigate other tailored-made nucleases [25], DMD (gene involved in the Duchenne Muscular Dystrophies) and RAG1 (V(D)J recombination-activating protein 1), were selected in the human genome and an additional locus of interest for bioproduction, the fucosyltransferase 8 (FUT8) gene, was chosen in the Chinese hamster genome.

Because the nuclease activity of HtH architecture towards endogenous loci had never been reported, we also characterized this architecture in the following experiments. Targets were selected for both nuclease architectures according to the spacer profile (25 bp and 20 bp for HtH and TtH respectively) determined previously with the yeast SSA assay (Figure 1C and 1E). The nucleases were then assembled using the optimal scaffolds containing an additional N-terminal SV40 nuclear localization sequence to improve their *in vivo* nuclear targeting (Additional file 1: Table S5). These nucleases were then assayed for their ability to promote targeted insertions or deletions of nucleotides (indels) via error prone non-homologous end joining (NHEJ, Figure 2A), in the adequate cell-line (293H or CHO-KI, Table 1).

Three days post transfection, genomic DNAs were recovered and amplified by locus specific PCRs (370 to 630 bp). PCR amplicons were then analyzed by deep sequencing to determine the amount of Indels promoted by the different nucleases at their respective target site. Deep sequencing analysis demonstrated that two out of the three HtH nucleases displayed significant levels of targeted mutagenesis (2% and 35% of mutagenesis frequencies, Table 1). For the TtH conformation, all three nucleases showed activity on their respective target sequence, with Indel frequencies ranging from 2 to 9% (Table 1). However, the level of targeted mutagenesis generated by these two architectures (HtH and TtH displayed respectively 12% and 5% mean Indels frequencies) was lower than that reported for the classical TtT in two large scale studies (22% and 16% mean Indel frequencies) [28,29].

We next compared the NHEJ-dependent molecular events promoted by the nuclease activity of the two different architectures versus the classical TtT. Toward this goal, we generated 10 TtT TALEN, performed targeted mutagenesis experiments in 293H cells and recovered the resulting deep-sequencing dataset (Additional file 1: Table S6 and S7). We first compared the deletion length induced by the three different architectures and found similar patterns (p-value = 0.3852) with a large proportion of deletions smaller than 20 bp (Figure 2B-D), a feature previously described for the conventional TtT architecture [30]. However the important error bars obtained for some deletion sizes indicated a variability of DNA repair outcomes from one locus to another. Such variability could be due to several parameters including the RVD composition (DNA binding affinities) and the presence of micro-homologies in the targeted locus.

We then compared the position of mutagenic events within the spacer of each architecture target. Considering the fact that the optimal distance of cleavage was different for the two TtT and HtH symmetrical architectures, we hypothesized that the position of mutagenic events within the target spacer of TtH asymmetrical architecture would be eccentric. Interestingly, a statistical analysis of the deletion profiles revealed that its activity led to a significant shift of the deletion pattern (*t-test*, p-value = 0.00155 with respect to the TtT architecture) towards the TALE::FokI binding site (Figure 2E, Figure 3A and B). Due to their symmetrical configuration, the TtT and HtH architectures were expected to cleave right in the middle of their optimal target (Figure 1B) and thus, 7 to 8 bp and 12 to 13 bp away from the 3' end of their respective FokI::TALE and TALE::FokI binding sites. However, due to its asymmetrical configuration, the TtH architecture was rather expected to cleave 2 to 3 bp away from the middle of its optimal target. The consistency between our experimental data and theoretical expectations indicated that the position of cleavage catalyzed by the TtH nuclease is constrained by its FokI::TALE scaffold component.

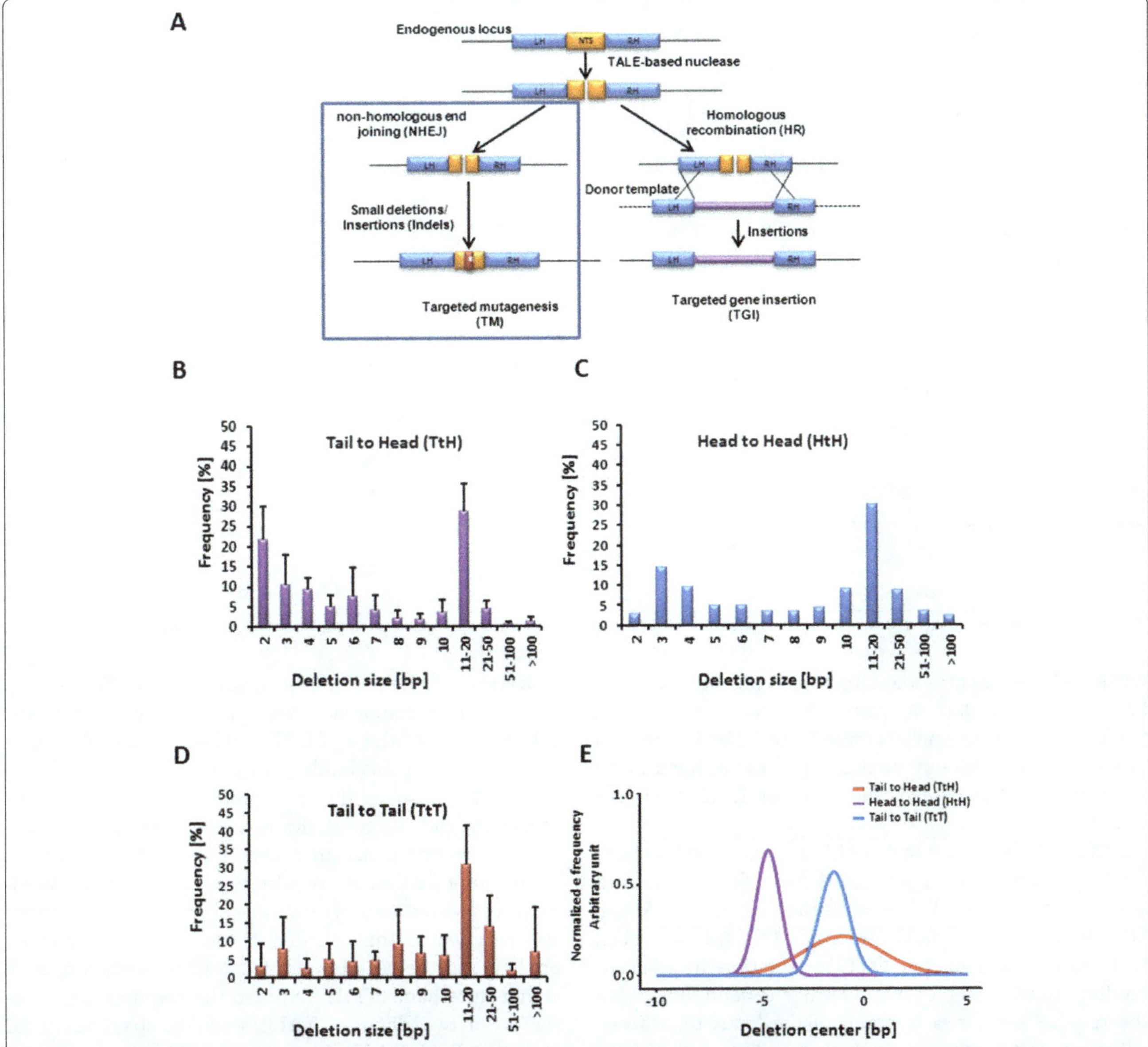

Figure 2 ***In vivo*** **nuclease activity of the three nuclease architectures in mammalian cells. (A)** Schematic representation of nuclease-mediated gene inactivation via the error-prone NHEJ pathway. **(B)** Size distribution of the deletion events induced at the endogenous locus by the TtH architecture. Loci presenting at least 20 events were taken into account to generate the figure. Error bars denote s.d. Student t test performed to compare deletion patterns induced by TtT and TtH architectures showed no statistical difference (p-value = 0.3852). **(C)** Same as for **(A)** but for the HtH nuclease architecture. One locus presenting at least 20 events was used. **(D)** Same as for **(A)** but for the TtT nuclease architecture. **(E)** Representation of the localization of the deletion center for the three architectures. The Gaussian curves having the same mean and variance of deletion centers for each of the three TtH, HtH, and TtT architectures are represented. The areas under the curves have been normalized to 1. *t-test*, p-value = 0.00155 with respect to the TtT architecture. For the TtT and HtH architectures, due to the odd number of nucleotides present in their spacer (15 or 25 respectively), we arbitrary chose to place the center of the spacer at 8 or 13 bp, explaining the shift of the deletion center close to -1. Data from 3 loci (DMD, FUT8 and RAG) were used for the HtH and TtH architectures. Data from 10 loci (APC, MLH, CD52, NR3C3, LIG4, BBC3, NR3C2, M2K, PPARD, ERBB2) were used for the TtT architectures.

Discussion

In this paper, we explored the versatility of the TALE scaffold and exploited it to generate a novel asymmetric FokI-based TALE nuclease architecture. This architecture called TtH TALE nuclease consisted of the asymmetrical association of TALE::FokI and FokI::TALE scaffolds (Figure 1A), two different engineered TALE DNA binding domains fused to the N- and C-terminal ends of FokI nuclease domain. Its nuclease activity was characterized in yeast toward extrachromosomal surrogate targets as well as in mammalian cells at different endogenous loci. Our results showed that this architecture was active in yeast and

Table 1 Activities of the TtH and HtH nuclease architectures at their endogenous cognate targets

Architectures and endogenous loci			Total events [%]	Total events [nb]	Insertion [nb]	Deletion [nb]	Wt [nb]	Reads [nb]
TtH	DMD	NC_000023.10: 32,364,567-32,364,620	1.8	193	58	141	10538	10731
					(16)	(48)		
ctrl	DMD	NC_000023.10: 32,364,567-32,364,620	0	0	0	0	7837	7837
TtH	RAG1	NC_000011.9: 36,594,622-36,594,675	2.6	59	18	41	2221	2280
					(7)	(24)		
ctrl	RAG1	NC_000011.9: 36,594,622-36,594,675	0	0	0	0	5494	5494
TtH	FUT8	NW_003613860.1 673,480-673,533	8.7	459	134	330	4817	5276
					(60)	(80)		
ctrl	FUT8	NW_003613860.1 673,480-673,533	0.033	2	1	1	6129	6131
HtH	DMD	NC_000023.10: 32,364,534-32,364,592	0.01	1	1	0	9147	9148
ctrl	DMD	NC_000023.10: 32,364,534-32,364,592	0.01	1	1	0	9540	9541
HtH	RAG1	NC_000011.9: 36,594,571-36,594,629	1.2	10	5	6	861	871
					(3)	(6)		
ctrl	RAG1	NC_000011.9: 36,594,571-36,594,629	0.051	2	1	1	3913	3915
HtH	FUT8	NW_003613860.1: 673,442-673,500	30.5	1791	1191	613	4088	5879
					(64)	(159)		
ctrl	FUT8	NW_003613860.1: 673,442-673,500	0	0	0	0	7223	7223

The number of unique events (considering the size and position) is indicated in brackets. Control (ctrl) indicates a tranfection with an empty vector plasmid.

further allowed to promote targeted mutagenesis at multiple endogenous loci in mammalian cells. Importantly, the TtH architecture only required one DNA strand of a given locus to efficiently bind and process it. Such novel configuration thus extends the range of TALE nucleases applications.

Today, the conventional TALE nuclease architecture used by most investigators results from the symmetrical association of two TALE::FokI scaffolds (Figure 1A and 1D). To be active as nuclease entity, this scaffolds need to bind in a Tail to Tail (TtT) orientation to adjacent binding sites bearing a T0 at their 5' ends and located on the reverse and forward strand of the locus to process. Although highly efficient, such architecture is unable to process loci devoided of thymidine residues either on one or two strands. To overcome such requirement for T0 and at the same time, to extend the range of TALE-based applications, two different groups employed rational design and directed evolution to modify the TALE N-terminal domain, a region reported to play a key role in the T0 specificity of TALE [31,32]. In another study, such requirement for T0 was reported to be overcome by using a new modular base-per-base binding domains (M3BD) scaffold from *Burkholderia rhizoxinica* [33].

Relaxing the T0 specificity of TALE binding domain might not be the only strategy to extend the range of TALE-based applications. Alternative approaches, exploiting the versatility of the TALE scaffold, could also be considered to abrogate the requirement of two T0 located on adjacent and anti-parallel TALE binding sites while still delivering efficient TALE nuclease activity. To develop such alternative approach, we first ruled out altering the T0 specificity of the TALE N-terminal domain, considering that keeping this binding anchor at the 5' end of each TALE DNA binding domain would be beneficial for the specificity and safety of the resulting designer nuclease. We thus sought to set up a novel dimeric TALE nuclease architecture that would require two adjacent TALE binding sites located on only one strand of the locus of interest. Because of the oriented fashion of TALE domain binding and the requirement for FokI dimerization, such architecture named TtH, required the asymmetrical association of two different TALE scaffolds, the TALE::FokI and the FokI::TALE in a Tail to Head orientation (Figure 1E).

We developed and tested combinations of the FokI::TALE and TALE::FokI scaffolds and found that the symmetrical HtH and TtT architectures as well as the asymmetrical TtH architecture could efficiently catalyze targeted DSBs in a yeast-based nuclease assay. Although the different architectures displayed similar activity levels, they showed marked differences regarding the properties of their targets. Interestingly, the narrower spacer range cleaved by the two new HtH and TtH architectures could represent an advantage by reducing the number of potential off-site targets in a genome of interest. Indeed, off-site targets are usually determined as sequences (combinations of left + right, left + left, right + right sequences) diverging from the intended target site by a few base pairs. In addition, to be considered as a potential

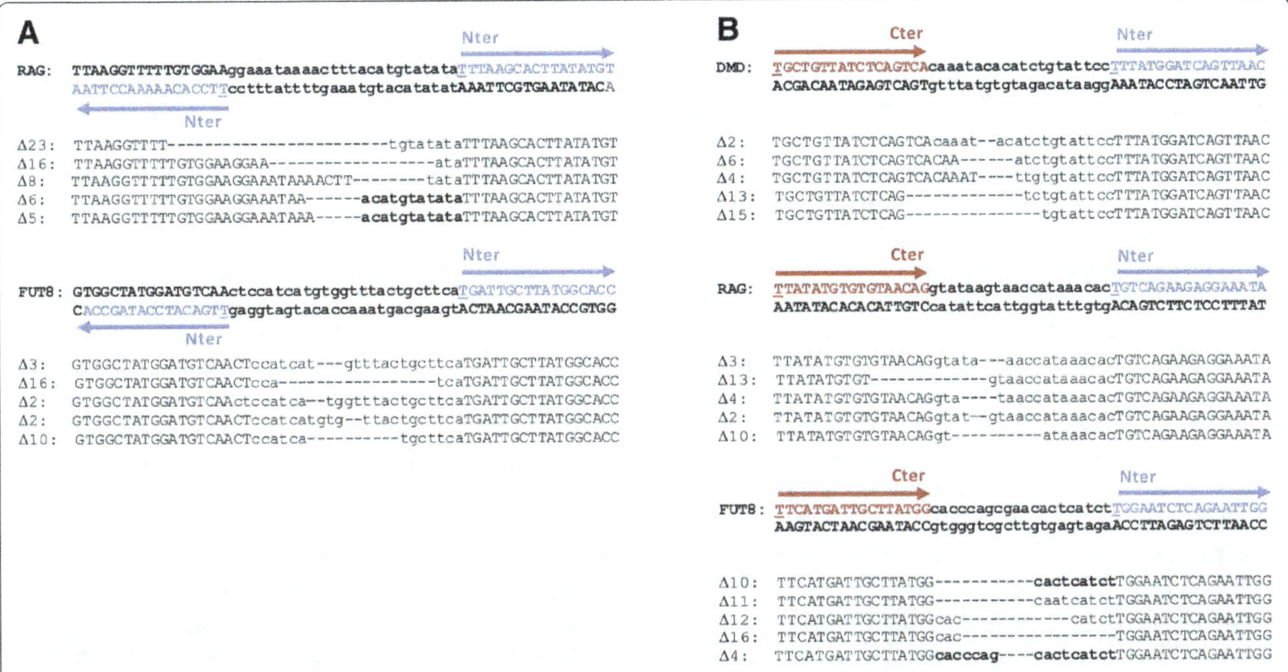

Figure 3 ***Characterization of mutagenic events promoted by*** **the HtH and TtH nuclease architectures in mammalian cells. (A)** Alignment of the WT genomic sequence and predominant deletion events induced by the HtH architecture at two different endogenous loci. FokI::TALE scaffold binding sites are represented by capital letters. **(B)** Same as **(A)** but for the TtH architecture at the three different endogenous loci. The targeted sequence is colored with respect to the scaffold binding site considered (blue: FokI::TALE, red: TALE::FokI). The position 0 is underlined.

off-site target, the two binding sites have to be separated by a spacer compatible with a nuclease activity. Based on the results presented in this study (Figure 1D), 25 different spacers have to be taken in account for the "classical" TtT architecture. Regarding the HtH and TtH architectures only 8 or 16 different spacer lengths need to be considered respectively (Figures 1C and E). These differences resulted in an approximate 3 and 2 fold higher numbers of potential off-site targets for the TtT compared to the HtH and TtH architectures respectively. In addition, we believed that the global nuclease activity and specificity could still be improved by optimizing the flexibility and/or rigidity of the aminoacid linker between the FokI and TALE domain. Indeed, Mercer and colleagues have recently shown that fusion of recombinase catalytic domain to alternative truncations of TALE N-terminal domain (Δ120 or Δ128) could enhance the efficiency of their chimeric TALE recombinase system [12].

As the levels of activities obtained in the yeast SSA assay were fully satisfactory, we evaluated the performance of the two HtH and TtH architectures in a chromosomal context. We thus generated three pairs of nucleases for both architectures following the guidelines (spacer length) obtained previously in yeast, to target a total of six loci in two mammalian cell types (CHO-KI and 293H). The efficiency of these nucleases to induce DSB events was monitored 3 days post transfection by measuring Indels generated by NHEJ at their cleavage sites. We cannot exclude that variation in nuclease activity is dependent on multiple parameters and is thus not exclusively resulting from the difference in architecture. Indeed, the RVD composition of the nuclease, the targeted DNA sequence and the presence of microhomologies within the targeted loci are likely to influence several biochemical parameters such as overall DNA binding, cleavage efficiency and global DNA repair outcome. Nevertheless, we found that both architectures induced a substantial level of targeted mutagenesis (Table 1). Additionally, the molecular characterization of deletion events allowed us to observe that the DSB occurred in the middle of the spacer region for the TtT and HtH architecture while being shifted toward the TALE::FokI scaffold binding site of the new TtH architecture.

While TtH architecture does not significantly increase the number of targetable loci, it could be endowed with a new technological advantage for the field of gene therapy. Indeed, when considering the human chromosome 1 as a model, and using standard criteria for the array (15.5 repeats) and spacer (10-16 and 20-25 bp) length [26], we estimated that the classical TtT architecture could target about 99.8% of this chromosome. The remaining 0.2% comprised a total of 2945 regions of 68 to 1983 bp (representing a total of 346422 nucleotides) that were devoided of any TtT nuclease target. We noted that more than half of these sequences could be potentially targeted using the TtH configuration (array of 15.5 repeats and spacer of 17-23 and 26-33 bp, Figure 1E) and that an important proportion of these sequences was composed of highly repetitive motives. Expansion of triplets or quadruplets is commonly linked to several genetic disorders and different neurological syndromes [34]. Their expansion induces aberrant protein expressions and subsequent aggregation, as well as formation and persistence of RNA:DNA hybrids responsible for genomic instability and inhibition of replication. Processing such pathogenic sequences represents important therapeutic potentials to cure their related genetic disease and, in that matter, the advantages of the new TtH architecture are twofold with respect to conventional TALEN architecture. First, it could process and thus stimulate contraction of expanded sequences harboring thymidine on one unique strand (CAG, GAA, CTG and CCTG). Second, through the well known ability of FokI to cleave DNA:RNA hybrids along with the capacity of TALE domain to interact with such molecule [35], it could also reduce the deleterious downstream effects of DNA:RNA duplex via their targeted processing. Noteworthy, such approach could also be used in general to process DNA: RNA hybrids.

Finally, besides its potential for specific gene therapy application, the TtH architecture could be beneficial for the field of genome editing by allowing for more precise positioning of TALE nuclease without affecting their T0 specificity, one of the hallmark of their DNA specificity.

Conclusions

Overall, although a larger dataset would be desirable, our results demonstrate that the level of genome modifications that can be obtained in mammalian cells with the new TtH architecture is compatible with most, if not all, genome editing applications. An additional benefit of this particular asymmetrical architecture is the availability of both N- and C-termini that can also expand the "cargo" possibility of TALE-based nucleases. We believe that this particular TtH architecture will further expand the possibilities of the TALE-based nuclease technology for targeting sequences with biased nucleotide composition (e.g. highly repetitive motives) or DNA-RNA hybrids.

Methods

TALE arrays

All TALE arrays were obtained from Cellectis Bioresearch (Paris, France). TALEN™ is a trademark owned by Cellectis Bioresearch. Sequences of TALE-nuclease backbones, TALE RVD array composition and/or relevant targets are presented in the Additional file 1. For experiments in mammalian cell lines, the TALE-based nucleases were expressed under the control of either an EF1a promoter (HtH and TtH architectures) or a CMV promoter (TtT architecture).

Extrachromosomal SSA assay in yeast

TALE-based nuclease containing yeast strain were gridded at high gridding density (~20 spots/cm^2) on nylon filters placed on solid agar containing YP-glycerol plates, using a colony gridder (QpixII, Genetix). A second layer, consisting of reporter-harboring yeast strains, was gridded on the same filter for each target. Membranes were incubated overnight at 30°C to allow mating. To select for diploids, filters were then places and incubated for 2 days at 30°C on medium containing glucose (2%) as the carbon source but lacking leucine (for the TALE nuclease left arm), tryptophan (for the target) and supplemented with G418 (for the TALE nuclease right arm, if required). To induce the expression of the TALE-based nuclease, filters were transferred onto YP-galactose-rich medium for 24-48 hours at 30°C or 37°C. To monitor nuclease activity, through the β-galactosidase activity, filters were finally placed on solid agarose medium containing 0.02% X-Gal in 0.5 M sodium phosphate buffer, pH 7.0, 0.1% SDS, 6% dimethyl formamide (DMF), 7 mM β-mercaptoethanol, 1% agarose and incubated at 37°C for up to 48 h.

Filters were scanned and each spot was quantified using the median values of the pixels constituting the spot. We attribute the arbitrary values 0 and 1 to white and dark pixels, respectively. β-Galactosidase activity is directly associated with the efficiency of homologous recombination, thus with the cleavage efficiency of the TALE-based nuclease. Any value >0 is considered as the consequence of cleavage.

Nuclease transfection in 293H cells

Human 293H cells (Life Technologies) were cultured at 37°C with 5% CO_2 in DMEM complete medium supplemented with 2 mM L-glutamine, penicillin (100 IU/ml), streptomycin (100 μg/ml), amphotericin B (Fongizone: 0.25 μg/ml, Life Technologies,) and 10% FBS. Adherent 293H cells were seeded at 1.2 10^6 cells in 10 cm Petri dishes one day before transfection. Cell transfection was performed using the Lipofectamine 2000 reagent according to the manufacturer's instructions (Invitrogen). In brief, 2.5 μg (for the HtH and TtH architectures) or 12 μg (for the TtT architecture) of each of the two nuclease expression vector pairs, and 10 ng of GFP expression vector (to monitor transfection efficiencies) were mixed with 0.3 ml of DMEM without FBS (5 μg final DNA amount). In another tube 25 μL of Lipofectamine were mixed with 0.3 ml of DMEM without FBS. After 5 minutes incubation, both DNA and Lipofectamine mixes were combined and incubated for 20 min at RT. The mixture was transferred to a Petri dish containing the 293H cells in 9 ml of complete medium and then cultured at 37°C under 5% CO_2. Three days post-transfection, the cells were washed with phosphate-buffered saline

(PBS), trypsinized, resuspended in 5 ml complete medium and the percentage of GFP positive cells was measured by flow cytometry (Guava EasyCyte) in order to monitor transfection efficacy.

Nuclease transfection in CHO-KI cells

CHO-K1 cells (ATCC) were cultured at 37°C with 5% CO_2 in F-12 K complete medium (Gibco) supplemented with 2 mM L-glutamine, penicillin (100 IU/ml), streptomycin (100 μg/ml), amphotericin B (Fongizone: 0.25 μg/ml, Life Technologies,) and 10% FBS. Cell transfection was performed by electroporation with the Nucleofector Kit T for CHO-K1 cells (Lonza) according to the manufacturer's protocol. Cells (1 x 10^6 cells) were transfected with 5 μg of the two nuclease expression vector pairs and 10 ng of GFP expression vector (to monitor transfection efficiencies) (10 μg final DNA amount), then plated in a 10 cm dish in complete medium (F-12 K medium, Gibco) supplemented with 2 mM L-glutamine, penicillin (100 IU/ml), streptomycin (100 μg/ml), Fongizone (0.25 μg/ml) and 10% FBS. Three days post-transfection, the cells were washed with phosphate-buffered saline (PBS), trypsinized, resuspended in 5 ml complete medium and the percentage of GFP positive cells was measured by flow cytometry (Guava EasyCyte) in order to monitor transfection efficacy.

Targeted mutagenesis

Cells were pelleted by centrifugation and genomic DNA was extracted using DNeasy Blood & Tissue Kit (Qiagen) according to the manufacturer's instructions. PCR of the endogenous locus (370-630 bp final product size) were performed using the oligonucleotide sequences presented in the Additional file 1 and purified using the AMPure kit (Invitrogen). Amplicons were further analyzed by deep sequencing using the 454 system (Roche) [36].

Competing interests

All co-authors are present or former Cellectis employees.

Authors' contributions

AJ, MB, JV and PD conceived the study and designed experiments. AJ, MB, JV, ST and GD performed experiments. FB, GHS, FD and AD provided conceptual and technical advices. AJ, MB, JV, MZ, and JM analyzed experiments. AJ, MB, JV and PD wrote the manuscript with support from all authors. All authors read and approved the final manuscript.

Acknowledgements

The authors acknowledge the contribution of Frederic Cedrone and the Cellectis Nuclease Production Platform.

Funding

Cellectis.

References

1. Doyle EL, Stoddard BL, Voytas DF, Bogdanove AJ: **TAL effectors: highly adaptable phytobacterial virulence factors and readily engineered DNA-targeting proteins.** *Trends Cell Biol* 2013, **23**(8):390–398.
2. Boch J, Scholze H, Schornack S, Landgraf A, Hahn S, Kay S, Lahaye T, Nickstadt A, Bonas U: **Breaking the code of DNA binding specificity of TAL-type III effectors.** *Science* 2009, **326**(5959):1509–1512.
3. Moscou MJ, Bogdanove AJ: **A simple cipher governs DNA recognition by TAL effectors.** *Science* 2009, **326**(5959):1501.
4. Deng D, Yan C, Pan X, Mahfouz M, Wang J, Zhu JK, Shi Y, Yan N: **Structural basis for sequence-specific recognition of DNA by TAL effectors.** *Science* 2012, **335**(6069):720–723.
5. Mak AN, Bradley P, Cernadas RA, Bogdanove AJ, Stoddard BL: **The crystal structure of TAL effector PthXo1 bound to its DNA target.** *Science* 2012, **335**(6069):716–719.
6. Stella S, Molina R, Yefimenko I, Prieto J, Silva G, Bertonati C, Juillerat A, Duchateau P, Montoya G: **Structure of the AvrBs3-DNA complex provides new insights into the initial thymine-recognition mechanism.** *Acta Crystallogr D Biol Crystallogr* 2013, **69**(Pt 9):1707–1716.
7. Christian M, Cermak T, Doyle EL, Schmidt C, Zhang F, Hummel A, Bogdanove AJ, Voytas DF: **Targeting DNA double-strand breaks with TAL effector nucleases.** *Genetics* 2010, **186**(2):757–761.
8. Sun N, Zhao H: **Transcription activator-like effector nucleases (TALENs): a highly efficient and versatile tool for genome editing.** *Biotechnol Bioeng* 2013, **110**(7):1811–1821.
9. Scharenberg AM, Duchateau P, Smith J: **Genome engineering with TAL-effector nucleases and alternative modular nuclease technologies.** *Curr Gene Ther* 2013, **13**(4):291–303.
10. Zhang F, Cong L, Lodato S, Kosuri S, Church GM, Arlotta P: **Efficient construction of sequence-specific TAL effectors for modulating mammalian transcription.** *Nat Biotechnol* 2011, **29**(2):149–153.
11. Mercer AC, Gaj T, Sirk SJ, Lamb BM, Barbas CF: **Regulation of Endogenous Human Gene Expression by Ligand-Inducible TALE Transcription Factors.** In *ACS Synth Biol.* 3rd edition; 2013.
12. Mercer AC, Gaj T, Fuller RP, Barbas CF 3rd: **Chimeric TALE recombinases with programmable DNA sequence specificity.** *Nucleic Acids Res* 2012, **40**(21):11163–11172.
13. Cong L, Zhou R, Kuo YC, Cunniff M, Zhang F: **Comprehensive interrogation of natural TALE DNA-binding modules and transcriptional repressor domains.** *Nat Commun* 2012, **3**:968.
14. Yanik M, Alzubi J, Lahaye T, Cathomen T, Pingoud A, Wende W: **TALE-PvuII fusion proteins - novel tools for gene targeting.** *PLoS One* 2013, **8**(12):e82539.
15. Maeder ML, Angstman JF, Richardson ME, Linder SJ, Cascio VM, Tsai SQ, Ho QH, Ser JD, Reyon D, Bernstein BE, Costello JF, Wilkinson MF, Joung JK: **Targeted DNA demethylation and activation of endogenous genes using programmable TALE-TET1 fusion proteins.** *Nat Biotechnol* 2013, **31**(12):1137–1142.
16. Miyanari Y, Ziegler-Birling C, Torres-Padilla ME: **Live visualization of chromatin dynamics with fluorescent TALEs.** *Nat Struct Mol Biol* 2013, **20**(11):1321–1324.
17. Boissel S, Jarjour J, Astrakhan A, Adey A, Gouble A, Duchateau P, Shendure J, Stoddard BL, Certo MT, Baker D, Scharenberg AM: **megaTALs: a rare-cleaving nuclease architecture for therapeutic genome engineering.** *Nucleic Acids Res* 2014, **42**(4):2591–2601.
18. Mendenhall EM, Williamson KE, Reyon D, Zou JY, Ram O, Joung JK, Bernstein BE: **Locus-specific editing of histone modifications at endogenous enhancers.** *Nat Biotechnol* 2013, **31**(12):1133–1136.
19. Owens JB, Mauro D, Stoytchev I, Bhakta MS, Kim MS, Segal DJ, Moisyadi S: **Transcription activator like effector (TALE)-directed piggyBac transposition in human cells.** *Nucleic Acids Res* 2013, **41**(19):9197–9207.
20. Konermann S, Brigham MD, Trevino AE, Hsu PD, Heidenreich M, Cong L, Platt RJ, Scott DA, Church GM, Zhang F: **Optical control of mammalian endogenous transcription and epigenetic states.** *Nature* 2013, **500**(7463):472–476.
21. Sun N, Zhao H: **A single-chain TALEN architecture for genome engineering.** *Mol Biosyst* 2014, **10**(3):446–453.
22. Beurdeley M, Bietz F, Li J, Thomas S, Stoddard T, Juillerat A, Zhang F, Voytas DF, Duchateau P, Silva GH: **Compact designer TALENs for efficient genome engineering.** *Nat Commun* 2013, **4**:1762.
23. Li T, Huang S, Jiang WZ, Wright D, Spalding MH, Weeks DP, Yang B: **TAL nucleases (TALNs): hybrid proteins composed of TAL effectors and FokI DNA-cleavage domain.** *Nucleic Acids Res* 2010, **39**(1):359–372.

24. Arnould S, Chames P, Perez C, Lacroix E, Duclert A, Epinat JC, Stricher F, Petit AS, Patin A, Guillier S, Roll S, Prieto J, Blanco FJ, Bravo J, Montoya G, Serrano L, Duchateau P, Paques F: **Engineering of large numbers of highly specific homing endonucleases that induce recombination on novel DNA targets.** *J Mol Biol* 2006, **355**(3):443–458.
25. Daboussi F, Zaslavskiy M, Poirot L, Loperfido M, Gouble A, Guyot V, Leduc S, Galetto R, Grizot S, Oficjalska D, Perez C, Delacote F, Dupuy A, Chion-Sotinel I, Le Clerre D, Lebuhotel C, Danos O, Lemaire F, Oussedik K, Cedrone F, Epinat JC, Smith J, Yanez-Munoz RJ, Dickson G, Popplewell L, Koo T, VenDriessche T, Chuah MK, Duclert A, Duchateau P, Paques F: **Chromosomal context and epigenetic mechanisms control the efficacy of genome editing by rare-cutting designer endonucleases.** *Nucleic Acids Res* 2012, **40**(13):6367–6379.
26. Juillerat A, Dubois G, Valton J, Thomas S, Stella S, Marechal A, Langevin S, Benomari N, Bertonati C, Silva GH, Daboussi F, Epinat JC, Montoya G, Duclert A, Duchateau P: **Comprehensive analysis of the specificity of transcription activator-like effector nucleases.** *Nucleic Acids Res* 2014, **42**(8):5390–5402.
27. Gurlebeck D, Szurek B, Bonas U: **Dimerization of the bacterial effector protein AvrBs3 in the plant cell cytoplasm prior to nuclear import.** *Plant J* 2005, **42**(2):175–187.
28. Kim Y, Kweon J, Kim A, Chon JK, Yoo JY, Kim HJ, Kim S, Lee C, Jeong E, Chung E, Kim D, Lee MS, Go EM, Song HJ, Kim H, Cho N, Bang D, Kim JS: **A library of TAL effector nucleases spanning the human genome.** *Nat Biotechnol* 2013, **31**(3):251–258.
29. Reyon D, Tsai SQ, Khayter C, Foden JA, Sander JD, Joung JK: **FLASH assembly of TALENs for high-throughput genome editing.** *Nat Biotechnol* 2012, **30**(5):460–465.
30. Chen S, Oikonomou G, Chiu CN, Niles BJ, Liu J, Lee DA, Antoshechkin I, Prober DA: **A large-scale in vivo analysis reveals that TALENs are significantly more mutagenic than ZFNs generated using context-dependent assembly.** *Nucleic Acids Res* 2013, **41**(4):2769–2778.
31. Tsuji S, Futaki S, Imanishi M: **Creating a TALE protein with unbiased 5'-T binding.** *Biochem Biophys Res Commun* 2013, **441**(1):262–265.
32. Lamb BM, Mercer AC, Barbas CF 3rd: **Directed evolution of the TALE N-terminal domain for recognition of all 5' bases.** *Nucleic Acids Res* 2013, **41**(21):9779–9785.
33. Juillerat A, Bertonati C, Dubois G, Guyot V, Thomas S, Valton J, Beurdeley M, Silva GH, Daboussi F, Duchateau P: **BurrH: a new modular DNA binding protein for genome engineering.** *Sci Rep* 2014, **4**:3831.
34. Mirkin SM: **Expandable DNA repeats and human disease.** *Nature* 2007, **447**(7147):932–940.
35. Yin P, Deng D, Yan C, Pan X, Xi JJ, Yan N, Shi Y: **Specific DNA-RNA hybrid recognition by TAL effectors.** *Cell Rep* 2012, **2**(4):707–713.
36. Valton J, Dupuy A, Daboussi F, Thomas S, Marechal A, Macmaster R, Melliand K, Juillerat A, Duchateau P: **Overcoming transcription activator-like effector (TALE) DNA binding domain sensitivity to cytosine methylation.** *J Biol Chem* 2012, **287**(46):38427–38432.

16

Molecular cloning and RNA interference-mediated functional characterization of a Halloween gene *spook* in the white-backed planthopper *Sogatella furcifera*

Shuang Jia, Pin-Jun Wan, Li-Tao Zhou, Li-Li Mu and Guo-Qing Li*

Abstract

Background: Ecdysteroid hormones ecdysone and 20-hydroxyecdysone play fundamental roles in insect postembryonic development and reproduction. Five cytochrome P450 monooxygenases (CYPs), encoded by Halloween genes, have been documented to be involved in the ecdysteroidogenesis in insect species of diverse orders such as Diptera, Lepidoptera and Orthoptera. Up to now, however, the involvement of the Halloween genes in ecdysteroid synthesis has not been confirmed in hemipteran insect species.

Results: In the present paper, a Halloween gene *spook* (*Sfspo, Sfcyp307a1*) was cloned in the hemipteran *Sogatella furcifera*. SfSPO has three insect conserved P450 motifs, i.e., Helix-K, PERF and heme-binding motifs. Temporal and spatial expression patterns of *Sfspo* were evaluated by qPCR. *Sfspo* showed three expression peaks in late second-, third- and fourth-instar stages. In contrast, the expression levels were lower and formed three troughs in the newly-molted second-, third- and fourth-instar nymphs. On day 3 of the fourth-instar nymphs, *Sfspo* clearly had a high transcript level in the thorax where PGs were located. Dietary introduction of double-stranded RNA (dsRNA) of *Sfspo* into the second instars successfully knocked down the target gene, and greatly reduced expression level of *ecdysone receptor* (*EcR*) gene. Moreover, knockdown of *Sfspo* caused lethality and delayed development during nymphal stages. Furthermore, application of 20-hydroxyecdysone on *Sfspo*-dsRNA-exposed nymphs did not increase *Sfspo* expression, but could almost completely rescue *SfEcR* expression, and relieved the negative effects on nymphal survival and development.

Conclusion: In *S. furcifera*, *Sfspo* was cloned and the conservation of SfSPO is valid. Thus, SfSPO is probably also involved in ecdysteroidogenesis for hemiptera.

Keywords: *Sogatella furcifera*, Halloween gene, Ecdysteroidogenesis, RNA interference, Lethality, Development

Background

20-Hydroxyecdyone (20E), an active form of ecdysteroid, regulates insect postembryonic development and reproduction. Because of the absence of the enzymes involving in squalene synthesis, insects cannot synthesize 20E *de novo,* and must obtain precursor sterols from their food [1], or their associated yeasts or fungi [2]. Rice planthoppers reportedly harbored yeast-like symbionts (YLSs), mainly in mycetocytes formed by abdominal fat body cells [3-8]. The YLSs synthesize ergosta-5,7,24(28)-trienol [9-12]. Ergosta-5,7,24(28)-trienol is then converted into cholesterol in planthoppers [10,11].

The ecdysteroid biosynthesis in the prothoracic glands (PGs) begins from conversion of cholesterol into 7-dehydrocholesterol (7dC), mediated by a Rieske oxygenase *Neverland* [13,14]. The conversion of 7dC into 2,22,25-trideoxyecdysone (ketodiol) is a series of hypothetical and unproven reactions, and is called 'Black Box' [15]. In *Drosophila melanogaster* and *Bombyx mori*, CYP307A1/A2 (SPOOK/SPOOKIER, SPO/SPOK) [16,17] and CYP6T3

* Correspondence: liguoqing001234@yahoo.com.cn
Education Ministry Key Laboratory of Integrated Management of Crop Diseases and Pests, College of Plant Protection, Nanjing Agricultural University, Nanjing 210095, China

[18] have been proven to be involved in the 'Black Box'. Moreover, a paralog SPOOKIEST (SPOT, CYP307B1) was also found in CYP307 family [16,17]. RNAi mediated knockdown of *spok* in the PGs results in arrest of molting in *D. melanogaster*. Feeding two 3-oxo steroids, cholesta-4,7-diene-3,6-dione-14α-ol (Δ4-diketol) and 5β [H]cholesta-7-ene-3,6-dione-14a-ol (diketol), in the RNAi-treated larvae triggered molting, enhanced amounts of ecdysteroids and induced 20E inducible genes [19]. These results indicate that Δ4-diketol and diketol are components of the ecdysteroid biosynthetic pathway and lie downstream of a step catalyzed by SPOK/SPO. SPO- and/or SPOK-like proteins had found in other insect species in Diptera such as *Bemisia tabaci* [20], in Coleoptera such as the red flour beetle *Tribolium castaneum* [21], in Hymenoptera such as *Apis mellifera* [22], in Lepidoptera such as *Spodoptera littoralis* [23], *Manduca sexta* [17] and *Holcocerus hippophaecolus* [24], in Orthoptera such as *Schistocerca gregaria* [25], and in Hemiptera such as *Acyrthosiphon pisum* [26]. Up to now, however, involvement of SPO in ecdysteroidogenesis has not been confirmed in other insect species except *D. melanogaster* and *B. mori*.

Most actions of 20E are mediated through their nuclear receptor, the ecdysone receptor (EcR) and its heterodimer partner ultraspiracle. Mutations in and RNA interference (RNAi) against *EcR* cause phenotypic defects and lethality in *T. castaneum* [27], and in *Laodelphgax striatellus* and *Nilaparvata lugens* [28]. Moreover, *EcR* expression is regulated by ecdysteroids through a positive feedback loop directly [29] or indirectly in *D. melanogaster* [30].

The white-backed planthopper, *Sogatella furcifera*, was a secondary pest of rice before 1980s. However, since the mid-1980s, its population dramatically increased following a nationwide adoption of hybrid rice in China [31]. *S. furcifera* causes serious damage to rice plants by sucking the phloem sap and blocking the phloem vessels, and by acting as a virus vector to transmit Southern rice black-streaked dwarf virus [32-34]. Even though the complete genome sequence of *S. furcifera* is still unavailable, the transcriptome data have been published [35]. These data prompt us to identify and characterize the Halloween genes. Since dietary ingestion of double-stranded RNA (dsRNA) can effectively knock down target genes in planthoppers [36-39], our second goal in the present paper is to study the influence of the Halloween gene dsRNAs on the performance of *S. furcifera* nymphs, and the rescuing effects of 20E application on the negative influences of *spo*-dsRNA in the nymphs. Our results suggest that SfSPO play a critical role in ecdysteroidogenesis in *S. furcifera*.

Results

Molecular cloning and sequence analysis

Complete coding sequence of *S. furcifera* Halloween gene *Sfspo* (*spo, cyp307a1*) was obtained. Its open reading frame (ORF) encoded a putative protein with the length of 510 amino acid residue (Figure 1).

SPO sequence is similar to those from other insects. Insect CYPs have five insect conserved P450 motifs, i.e., WxxxR (Helix-C), GxE/DTT/S (Helix-I), ExxR (Helix-K), PxxFxPE/DRF (PERF motif) and PFxxGxRxCxG/A (heme-binding domain), where 'x' means any amino acid [40]. For SfSPO, Helix-C and Helix-I are not conserved. Helix-C had the amino acid sequence of H/YxxPR, and the amino acid sequence of Helix-I was GGHSA/V (Figure 1).

In insects, SPO belongs to CYP2 family. The N-terminus of SfSPO has one of the common characters in microsomal P450s, consisting many hydrophobic residues followed by a proline/glycine (P/G) rich region (Figure 1).

Temporal and spatial transcript profiles

At our experiment temperature, *S. furcifera* second-, third- and fourth-instar nymphs lasted an average of 2.0, 2.0 and 3.0 days. *Sfspo* showed three expression peaks in day 2 of second-instar, day 2 of third-instar and day 3 of fourth-instar nymphs. In contrast, the expression levels were lower and formed three troughs in the newly-molted second-, third- and fourth-instar nymphs (Figure 2A).

The spatial distribution of *Sfspo* on day 3 of the fourth-instar nymphs was also tested using qPCR. *Sfspo* clearly had a high transcript level in the thorax where PGs were located. Moreover, trace amounts of transcripts were found in the head and abdomen (Figure 2B).

Dietary ingestion of dsRNA on expression of *Sfspo* and *EcR* genes

During 6 days of continuous exposure to dsRNA-contained diet and 1 day after experiment, mRNA abundance of *Sfspo* in the surviving nymphs was examined by q-PCR. The mRNA level of *Sfspo* in treated nymphs respectively reduced by 63.0%, 87.8%, 76.2%, 93.9%, 81.8%, 92.2% and 94.5%, respectively, comparing to that in ds*egfp*-exposed controls (Figure 3A). This indicated that the RNAi-mediated knockdown of *Sfspo* was successful.

Since SfSPO is expected to act in other genes in the same signaling pathway, the possible effect of *Sfspo* knockdown was examined on the transcript level of *SfEcR*, which was one of 20E heterodimeric nuclear receptors and was regulated by 20E through a positive feedback loop directly [29] or indirectly in *D. melanogaster* [30]. As expected, during 6 days of continuous exposure to dsRNA-contained diet and 1 day after experiment, *SfEcR* expression levels in nymphs decreased by 76.0%, 88.3%, 71.7%, 88.8%, 82.1%, 85.9% and 89.2% respectively, when compared with that in ds*egfp*-ingested planthoppers (Figure 3B).

Effect of dsRNA on nymph survival

Six day ingestion of dsRNA-contained diet caused nymphal lethality. The mortality reached up to 20% in nymphs that

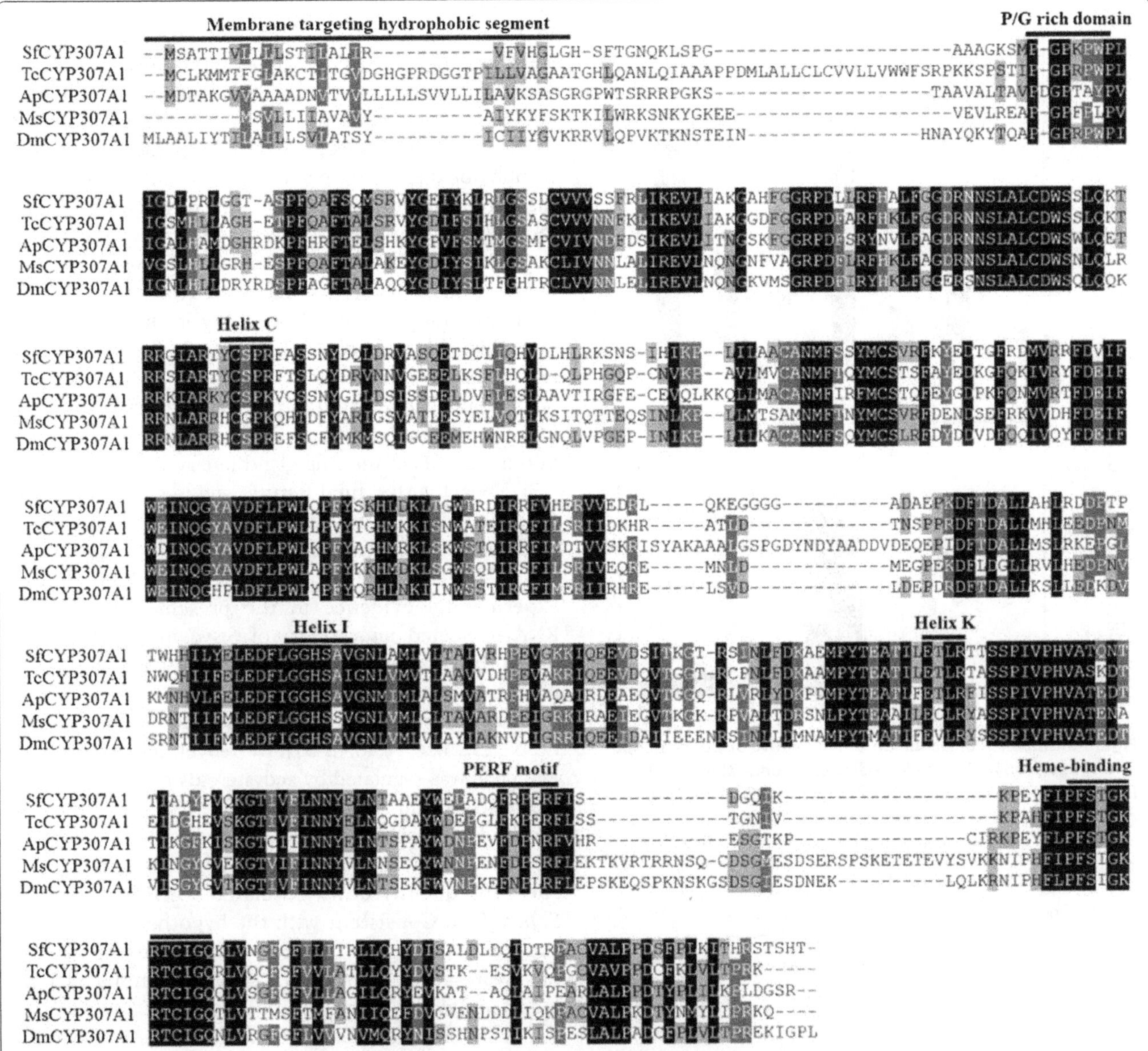

Figure 1 Alignment of CYP307A1 (SPOOK, SPO) sequences from five insect species. SPO originates from *Manduca sexta* (Ms) (ABI74778), *Drosophila melanogaster* (Dm) (NP_647975), *Tribolium castaneum* (Tc) (XP_969587), *Acyrthosiphon pisum* (Ap) (XP_001946295) and *Sogatella furcifera* (Sf), respectively. Amino acids with 100%, 80%, and 60% conservation are shaded in black, dark grey and light grey. The characteristic P450 structure, membrane targeting hydrophobic segment, P/G rich domain, Helix C, Helix I, Helix K, PERF motif and Heme-binding domain are shown in the figure.

had ingested ds*Sfspo*. In most cases, nymphs died during the period of ecdysis. In contrast, less than 5% of the planthoppers on normal or *egfp*-dsRNA-contained diets died (Figure 3C).

Effects of dsRNA on nymph development

Six day period of continuous exposure to dsRNA-contained diet significantly delayed nymphal development. 100% of the nymphs on normal and *egfp*-dsRNA-contained diets became the fourth instars after experiment. In contrast, 24% of the individuals on *Sfspo*-dsRNA-contained diets remained in the third-instar (Figure 3D).

Rescue experiment

Application of 300 pg of 20E did not affect the expression level of *Sfspo*. In contrast, 20E application almost completely rescued *SfEcR* expression at mRNA level. Moreover, 20E application to *Sfspo*-dsRNA-exposed nymphs almost completely overcame the negative effects on the survival and the development (Figure 4).

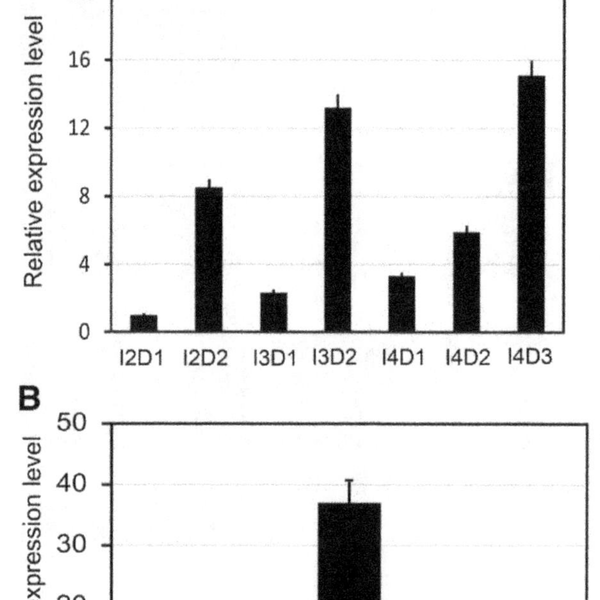

Figure 2 Graphic representation of the relative *Sfspo* transcript levels measured in the whole bodies of second-, third- and fourth-instar (I2D1, I2D2, I3D1, I3D2, I4D1, I4D2 and I4D3) nymphs at 24 h intervals (A) and the head, thorax and abdomen of day 3 fourth-instar nymphs (B). For each sample, 3 independent pools of 5–10 nymphs were measured in technical triplicate using qRT-PCR. The values were calculated using the $2^{-\Delta\Delta Ct}$ method. The relative expression levels were the ratios of relative copy numbers in individuals of specific developmental stage or specific body part to that in I2D1 or head. The columns represent averages with vertical bars indicating SE.

Discussion

Since the fundamental phenomena such as molting and metamorphosis are conserved during arthropod evolution, the Halloween genes are expected to be well conserved in insects [23,25,26,41-44], and in other arthropods [1,45]. In the present paper, the presence of *Sfspo* was demonstrated in *S. furcifera*. The primary structure of SfSPO has three insect conserved P450 motifs, i.e., Helix-K, PERF and heme-binding motifs. Similar structural characters have been documented in SPO- and SPOK-like proteins from other insect species of diverse orders such as Diptera [16,17,20], Coleoptera [21], Hymenoptera [22], Lepidoptera [17,23], Orthoptera [25], and Hemiptera [26]. The N-terminus of SfSPO has one of the common characters in microsomal P450s, consisting many hydrophobic residues followed by a proline/glycine (P/G) rich region. Consistent with the structural features, SPO is detected in endoplasmic reticulum (ER) when the corresponding gene is transfected to *Drosophila* S2 cells [17,46]. Moreover, *Sfspo* showed three expression peaks in late second-, third- and fourth-instar stages. In contrast, the expression levels were lower and formed three troughs in the newly-molted second-, third- and fourth-instar nymphs. In the fourth-instar nymphs of the brown planthopper *N. lugens* [47] and in the sixth-instar larvae of a lepidopteran species *S. littoralis*, the level of ecdysteroid showed a peak in the later instar stage. In *D. melanogaster* larval stage, expression patterns of *Dmspo* gene undergoes dramatic fluctuations, consistent with circulating ecdysteroid quantity in the haemolymph: being high in late seconds, low in early third and high in late thirds [17]. Furthermore, we found in this study that *Sfspo* clearly had a high transcript level in the thorax where PGs were located. Similarly, *Dmspo* is expressed primarily in the PG cells of the ring gland in larval and adult stages [17]. Thus, the structural features and temporal and spatial expression patterns suggest that SfSPO might be involved in the ecdysteroidogenesis in *S. furcifera*.

The suggestion is further confirmed by three lines of experimental evidence in the present paper. Firstly, RNAi-mediated knockdown of *Sfspo* in *S. furcifera* reduced the expression level of *SfEcR* at the mRNA level. In other insect species, mutations in or RNAi against the Halloween enzymes caused a decrease in ecdysteroid titers [23,25,26,44,46,48-51]. Moreover, the expression of *EcR* gene was regulated by ecdysteroids through a positive feedback loop in *D. melanogaster* [29,52]. Accordingly, it can be hypothesized that RNAi-mediated knockdown of *Sfspo* negatively affects ecdysteroidogenesis in *S. furcifera*, and subsequently down-regulated *SfEcR* expression in *S. furcifera*. Consistent with the hypothesis, our rescue experiment revealed that 20E application almost completely rescued *SfEcR* expression in nymphs that had ingested ds*Sfcyp307a1*.

The second line of experimental evidence is that RNAi-mediated knockdown of *Sfspo* in *S. furcifera* caused phenotypic defects similar to insects whose ecdysteroid synthesis was disturbed or whose ecdysteroid-mediated signaling had been inhibited [53,54]. In the present paper, we found that ingestion of ds*Sfspo* caused nymphal lethality and developmental delay. Since the average second- and third-instar periods of the nymphs in our experimental conditions was respectively about 2 days and the deaths mainly occurred in the sixth day after dsRNA exposure, it means that the nymphs died during the third ecdysis. In fact, we also observed many abnormal and lethal ecdysis individuals on *Sfspo*-dsRNA contained diet, whereas most of the larvae on control normally molted. Similar phenomena have been observed in other two rice planthoppers, *L. striatellus* and *N. lugens*, in which silencing of *EcR* expression by *in vivo* RNAi to inhibit ecdysteroid-mediated signaling generated phenotypic defects in molting and

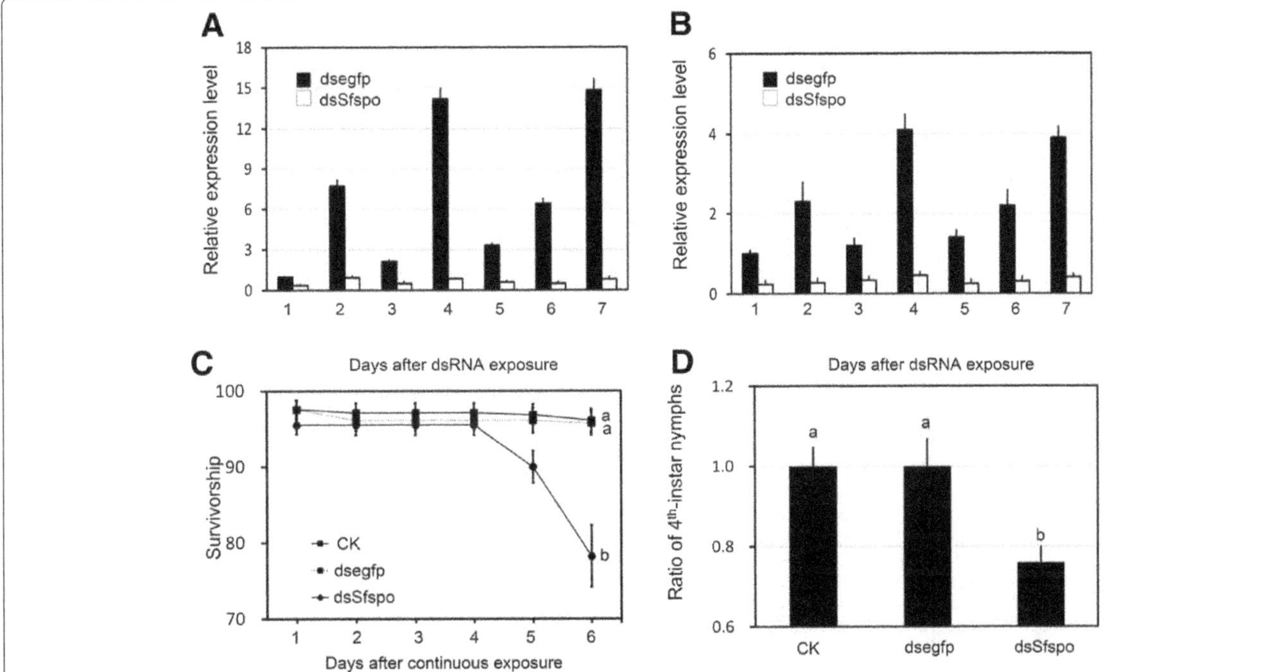

Figure 3 Effects of dietary ingestion of ds*Sfspo* on the relative *Sfspo* (A) and *SfEcR* (B) transcript levels, survival (C) and development (D) of *L. striatellus* nymphs. The nymphs were continuously ingested dsRNA from the second- through the fourth-instar stage. The relative transcript level for each sample was measured daily from 3 independent pools of 5–10 nymphs. The survival was calculated daily from 10 biological replicates, with each replicate of 10 individuals. The percentage of the fourth-instar was estimated from those in 10 biological replicates that survived through all experimental period. The values represent averages with vertical bars indicating SE, which topped with the same letters are not statistically significantly different at P = 0.05.

resulted in lethality in most of the treated nymphs. Intriguingly, apparent wing defects in morphogenesis and melanization occurred in *L. striatellus* nymphs subjected to ds*EcR* microinjection [54].

It has long been known that topical application of 20E could trigger physiological response such as regulation of diapause in the fourth-instar planthopper nymphs [55]. In the present paper, we tested whether 20E could rescue the negative effects of *Sfspo*-dsRNA ingestion on nymphs. Our results revealed that 20E application to *Sfspo*-dsRNA-exposed nymphs almost completely relieved the negative effects on the survival and the development. Thus, we provided the third line of evidence to support the suggestion that SfSPO plays critical roles in ecdysteroidogenesis in *S. furcifera*.

Conclusions

In the present paper, we cloned *Sfspo* and found that the conservation of SfSPO is valid in *S. furcifera*. Thus, SfSPO is probably also involved in ecdysteroidogenesis for hemiptera.

Methods

Insect culture and chemicals

S. furcifera adults were collected from Nanjing (32.0° N, 118.5° E), Jiangsu Province in China in 2010. The strain has been reared routinely on rice (*Oryza sativa*), in an insectary under controlled temperature (28 ± 1°C), photoperiod (16 h light/8 h dark) and relative humidity (more than 80%) since then, with wild stock injections every summer. Rice variety (Taichung Native 1) was grown in soil at 30–35°C under a long day photoperiod (16 h light/8 h dark) in a growth incubator. The planthoppers were transferred to fresh seedlings every 10–14 days to assure sufficient nutrition.

At laboratory reared by above protocol, *S. furcifera* eggs hatched into nymphs within 7 days. Nymphs went through 5 instars, with the average periods of the first-, second-, third-, fourth- and fifth-instar stages of 2.5, 2.0, 2.0, 3.0 and 3.0 days, respectively. Upon reaching full size, the fifth-instar nymphs emerged as adults.

20E was purchased from Sigma, and was purified by reverse-phase HPLC before experiments.

Sequence assembly and homology searches

Raw nucleotide reads of *S. furcifera* were downloaded from the NCBI Sequence Read Archive (SRA) database with its accession number SRP009194, and assembled into unigenes using Trinity software [56]. The annotated SPO from 4 representative insect species *A. pisum*, *T. castaneum*, *M. sexta* and *D. melanogaster* were downloaded from NCBI reference sequences (RefSeq) database. These protein

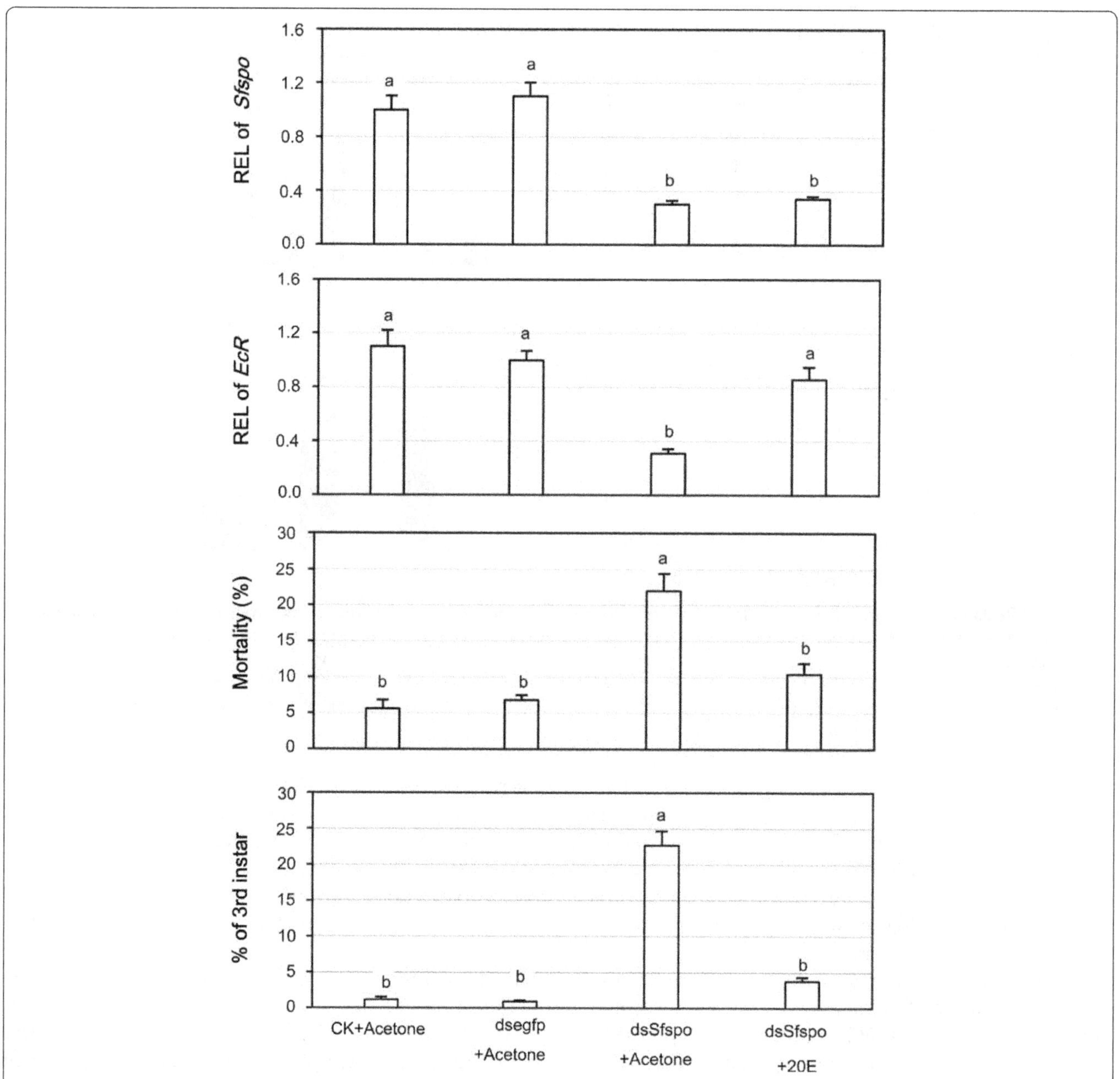

Figure 4 Relative expression level (REL) of *spo* and *EcR* gene, mortality and percentage of third-instar nymphs in *S. furcifera* nymphs subjected to both ds*Sfspo* exposure and 20E application. The nymphs were continuously ingested dsRNA from the second-instar through the early fourth-instar stage. Two-days after dsRNA exposure, the nymphs received 0.03 μL acetone or 300 pg of 20E in 0.03 μL acetone. The values represent averages with vertical bars indicating SE, which topped with the same letters are not statistically different at P = 0.05.

sequences were used for TBLASTN searches of *S. furcifera* transcriptome data to identify hits at a cutoff E-value of 1.0^{-5}. The nucleotide sequences of hits resulting from initial searches were annotated by blasting (BLASTX, e-values < 10^{-5}) against a local protein database containing NCBI non-redundant proteins.

Molecular cloning

Total RNA was extracted from the fourth-instar nymphs using TRIzol reagent according to the manufacturer's instructions (Invitrogen), and was treated for 30 min at 37°C with RNase free DNase I (Ambion, Austin, TX) to eliminate traces of chromosomal DNA. The purity and amount of RNA were determined by NanoDrop ND-1000 spectrophotometer (Nanodrop Technologies, Rockland, DE, USA). First-strand cDNA was synthesized from the total RNA using the reverse transcriptase (M-MLV RT) (Takara Bio., Dalian, China) and an oligo $(dT)_{18}$ primer, and was used as a template for polymerase chain reaction (PCR) to authenticate the sequences of the selected unigenes. The primers based on the sequences were designed using Primer3 software [57]. Once initial *Sfspo* unigenes were authenticated, they were aligned to the full cDNA sequence of the gene from the 4 representative insect species mentioned above. Some short sequence gaps between two aligned unigenes were found. Specific primers were designed based on the two unigenes between each gap, and the gaps were filled by PCR. The final cDNA sequence was authenticated using the primers listed in Table 1. Thermal cycling conditions were 94°C for 5 min, followed by 35 cycles of 94°C for 30 sec, 55°C for 45 sec and 72°C for 3 min. The last cycle was followed by final extension at 72°C for 10 min. Each 50 μL PCR reaction contained 2 μL of cDNA template, 5 μL of 10× LA Taq buffer (Mg^{2+} Free), 4 μL of $MgCl_2$ (25 mM), 4 μL of dNTP mixture (2.5 mM/each), 1 μL of forward and 1 μL of reverse primers (10 μM), 0.5 μL of LA Taq polymerase (Takara Bio.) (5 U/μL) and 32.5 μL of double distilled H_2O.

The 5'- and 3'-RACE Ready cDNA were synthesized following the manufacturer's instructions, primed by oligo (dT) primer and the SMART II A oligonucleotide using the SMARTer RACE cDNA amplification kit (Takara Bio.). Antisense and sense gene-specific primers (Table 1) corresponding to the 5'- and 3'-end of the sequence obtained above, and the universal primers in the SMARTer RACE kit (Takara Bio.) were used to amplify the 5'-end and the 3'-end. The components of reaction have been described above. Thermal cycling conditions were 94°C for 3 min; followed by 5 cycles of 94°C for 30 sec, 72°C for 5 min; and another 5 cycles of 94°C for 30 sec, 70°C for 30 sec, 72°C for 5 min; and followed by 25 cycles of 94°C for 30 sec, 68°C for 30 sec, 72°C for 5 min. The last cycle was followed by final extension at 72°C for 10 min.

The amplified product was separated by 1.2% agarose gel and purified with Wizard DNA Gel Extraction Kit (Promega, Madison, Wis., USA), and then cloned into pGEM-T easy vector (Promega). Several independent subclones were sequenced on an Applied Biosystems 3730 automated sequencer (Applied Biosystems, Foster City, Calif., USA) from both directions.

After full-length cDNA was obtained, we designed primers (Table 1) to verify the complete ORF with the same PCR conditions outlined above. ORF was predicted using the editseq program of DNAStar (http://www.dnastar.com) and the features of the protein were determined by TargetP. The resulting sequence was submitted to GenBank (KC579454). The annotated SPO-like proteins from the 4 representative insect species mentioned above were aligned with the predicted LsSPOK using ClustalW2.1 [58].

Table 1 Primers used in RT-PCR, 5' and 3' RACE, synthesizing dsRNA, and performing qRT-PCR

Primer	Sequence (5' to 3')	Amplicon size (bp)
Primers used in RT-PCR		
spoFp	ACGGCCAGTCCATTTCAG	344
spoRp	TGTTGGATGAGGCAGTCG	
Primers used in 5'-RACE		
spoGSP	GATGCCAAGTGAGGGTGGGATC	
spoNGSP	TCGGTGAAGTCTTTGGGCTCGG	
Primers used in 3'RACE		
spoGSP	ACGCGACATTCGCCGCTTTGTG	
spoNGSP	CGATGCGCTGCTCGCTCACCTT	
Primers used in PCR for End to End		
spoFp	CGTCGTGAACACCCTTAT	2090
spoRp	GCCCGGTACACTATTATCTT	
Synthesizing the dsRNAs		
spoFd	CCCTCACTTGGCATCAC	415
spoRd	TCGGGTTTCTTTATTTGTC	
egfpup	AAGTTCAGCGTGTCCG	414
egfpdown	CTTGCCGTAGTTCCAC	
Performing the qPCR		
spoFq	CAACTCCATACACATCAAGCCACTG	119
spoRq	ACCGACGCACCATATCTCTGAAC	
EcRFq	AATGAGTTCGAGCACCCTAGCGAA	129
EcRRq	AATGGTGATTTCGGTGATGTGGCG	
RPL9Fq	TGTGTGACCACCGAGAACAACTCA	131
RPL9Rq	ACGATGAGCTCGTCCTTCTGCTTT	
ARFFq	CACAATATCACCGACTTTGGGATTC	141
ARFRq	CAGATCAGACCGTCCGTACTCTC	

Preparation of dsRNA

A 415 bp cDNA sequence of *Sfspo* and a 414 bp fragment of enhanced green fluorescent protein gene *egfp* (control) were individually subcloned into pEASY-T3 vector (TransGen Biotech, Beijing, China), and the diluted plasmids were used as templates for amplification of these target sequences by PCR, using specific primers (Table 1) conjugated with the T7 RNA polymerase promoter (5'-taatacgactcactataggg-3') and the PCR conditions described above. The PCR products were purified with Wizard H SV Gel (Promega) and used as templates for dsRNA synthesis with the T7 Ribomax TM Express RNAi System, according to the manufacturer's instructions (Promega). The reaction products were treated with RNase and DNase I to degrade single-strand RNA and DNA template, respectively, at 37°C for one hour, following manufacturer's directions. The synthesized dsRNA was isopropanol precipitated, resuspended in Nuclease-free water, and quantified by a spectrophotometer (NanoDrop TM 1000) at 260 nm. The purity and integrity were determined by agarose gel electrophoresis. The dsRNA stocks can be stored for several weeks at −80°C until use.

Bioassay

Previously reported dietary dsRNA-introducing procedure [38,39] was used, with small modifications. Briefly, glass cylinders, 12 cm in length and 2.8 cm in internal diameter, were used as feeding chambers. Twenty first-instar nymphs were carefully transferred into each chamber and pre-reared for one day to the second-instar stage, on liquid artificial diet (according to Dr. Fu et al. [59]) between two layers of stretched Parafim M (Pechiney Plastic Packaging Company, Chicago, IL, USA) that was placed at both ends of the chamber. The artificial diet containing one of the dsRNAs at the concentration of 0.5 mg/ml [38,39] were then used to feed the second-instar nymphs. The diet was changed and dead nymphs were removed daily.

Two experiments were carried out. The first had three treatments including non-dsRNA diet (blank control), ds*egfp* diet (negative control) and ds*Sfspo* diet. The experiment lasted for 6 days. The second bioassay was a rescue experiment. Since topical application of 300 pg of 20E was enough to trigger physiological response in the fourth-instar planthopper nymphs [55], 300 pg of 20E was used in the second bioassay. After exposed to dsRNA for 2 days, the nymphs were anesthetized with carbon dioxide. A 0.03 μL aliquot of acetone with or without 300 pg of 20E was topically applied to the dorsal thoracic surface of the nymphs with a 10-μL microsyringe connected to a microapplicator (Hamilton Company, Reno, NV). And then, the nymphs were continuously exposed to dsRNA for another 4 days. There were four treatments including: (1) nymphs on non-dsRNA diet and applied acetone (blank control); (2) nymphs on ds*egfp* diet and applied acetone (negative control); (3) nymphs on ds*Sfspo* diet and applied acetone; (4) nymphs on ds*Sfspo* diet and applied 20E. All treatments in both experiments were replicated 25 times (25 chambers), and a total of 250 nymphs in each treatment were used (100 nymphs for bioassays and 150 nymphs for q-PCR).

Mortality was recorded daily. The surviving nymphs after bioassay were collected and frozen. The instars of the surviving nymphs were identified by head capsule width and the number of rhinaira (sensilla clusters) bearing on the pedicle of the antennae [60,61].

Real-time quantitative PCR

Total RNA samples were prepared from the whole bodies of the of second-, third- and fourth-instar (I2D1, I2D2, I3D1, I3D2, I4D1, I4D2 and I4D3) nymphs, from the head, thorax and abdomen of I4D3 nymphs on rice, and from nymphs subjected to 6-day's bioassays, using SV Total RNA Isolation System Kit (Promega). Each sample contained 10 nymphs and repeated in biological triplicate. Purified RNA was subjected to DNase I to remove any residual DNA according to the manufacturer's instructions. In a preliminary experiment, we estimated the expression stability of four house-keeping genes (*Actin*; *ADP-ribosylation factor*, *ARF*; *ribosomal protein RPL9*; *translation elongation factor 1α EF1α*), and found that *ARF* and *RPL9* were the most stable house-keeping genes and selected as internal controls. The primers of the Halloween genes *Sfspo*, *EcR* gene, *ARF* and *RPL9* were designed with Beacon Designer 7 (Table 1). Putative mRNA abundance of *Sfspo* the Halloween and *EcR* genes in each nymphal sample was estimated by qPCR using SYBR Premix Ex Taq™ (Perfect Real Time) (Takara Bio.) and ABI 7500 Real-Time PCR System (Applied Biosystems) according to the manufacturer's instruction. The reaction mixture consisted of 2 μL of cDNA template (corresponding to 50 ng of the starting amount of RNA), 10 μL of SYBR Premix Ex Taq (Takara Bio.), 1 μL of forward primer (10 μM), 1 μL of reverse primer (10 μM), 0.4 μL of Rox Reference Dye (50×) in a final reaction volume of 20 μL. A reverse transcription negative control (without reverse transcriptase) and a non-template negative control were included for each primer set to confirm the absence of genomic DNA and to check for primer-dimer or contamination in the reactions, respectively. The following standard qPCR protocol was used: denaturing at 95°C for 30 sec, followed by 40 cycles of 95°C for 5 sec and 60°C for 34 sec. After amplification, the melting curves were determined by heating the sample up to 95°C for 15 sec, followed by cooling down to 60°C for 1 min, and heating the samples to 95°C for 15 sec.

The generation of specific PCR products was confirmed by sequencing and gel electrophoresis. Each primer pair was tested with a 10-fold logarithmic dilution of a cDNA mixture to generate a linear standard curve (crossing point CP plotted vs. log of template concentration), which was used to calculate the primer pair efficiency. All experiments were repeated in technical triplicate. Data were analyzed by the $2^{-\Delta\Delta Ct}$ method [62], using the geometric mean of *ARF1* and *RP18* for normalization according to the strategy described previously [62,63].

Data analysis

The data were given as means ± SE, and were analyzed by ANOVAs or a repeated measures ANOVA followed by the Tukey-Kramer test, using SPSS for Windows (SPSS, Chicago, IL, USA).

Abbreviations

PCR: Polymerase chain reaction; RT-PCR: Reverse transcriptase PCR; qRT-PCR: Quantitative real-time PCR; cDNA: Complementary DNA; CYP: Cytochrome P450 monooxygenase; dsRNA: Double-stranded RNA; EcR: Ecdysone receptor; E: Ecdysone; 20E: 20-Hydroxyecdysone; YLS: Yeast-like symbionts; RNAi: RNA interference; ORF: Open reading frame; ML: Maximum-likelihood; SE: Standard error; ANOVA: Analysis of variance.

Competing interests

The authors declare that they have no competing interests.

Authors' contributions

SJ and PJW performed most of the experimental procedures, and data analysis. LTZ and LLM performed partial experiments, assisted in manuscript revising and provided helpful discussions. GQL wrote the manuscript, conceived and supervised the research. All authors read and approved the final manuscript.

Acknowledgments

This research was supported by the National Basic Research Program of China (973 Program, No. 2010CB126200). We thank Drs Z. Han, and S. Dong of our laboratory for useful discussions during the course of this research.

References

1. Iga M, Kataoka H: **Recent studies on insect hormone metabolic pathways mediated by cytochrome P450 enzymes.** *Biol Pharm Bull* 2012, **35**(6):838–843.
2. Behmer ST, David Nes W: **Insect sterol nutrition and physiology: a global overview.** *Advances in insect physiology* 2003, **31**:1–72.
3. Chen CC, Cheng LL, Hou RF: **Studies on the intracellular yeast-like symbiote in the brown planthopper, *Nilaparvata lugens* Stal.** *J Appl Entomol* 1981, **92**:440–449.
4. Pang K, Dong S-Z, Hou Y, Bian Y-L, Yang K, Yu X-P: **Cultivation, identification and quantification of one species of yeast-like symbiotes, Candida, in the rice brown planthopper, Nilaparvata lugens.** *Insect Sci* 2012, **19**(4):477–484.
5. Dong SZ, Pang K, Bai X, Yu XP, Hao PY: **Identification of two species of yeast-like symbiotes in the brown planthopper, Nilaparvata lugens.** *Curr Microbiol* 2011, **62**(4):1133–1138.
6. Noda H, Saito T: **The role of intracelular yeastlike symbiotes in the development of *Laodelphax striatellus* (Homoptera: Delphacidae).** *Appl Environ Microbiol* 1979, **14**:453–458.
7. Noda H, Saito T: **Histological and histochemical observation of intracellular yeastlike symbiotes in the fat body of the smaller brown planthopper, *Laodelphax striatellus* (Homoptera: Delphacidae).** *Applied Entomology and Zoology* 1977, **12**:134–141.
8. Noda H: **Preliminary histological observation and population dynamics of intracellular yeast-like symbiotes in the smaller brown planthopper, *Laodelphax striatellus* (Homoptera: Delphacidae).** *Applied Entomology and Zoology* 1974, **9**:275–277.
9. Noda H, Wada K, Saito T: **Sterols in *Laodelphax striatellus* with special reference to the intracellular yeastlike symbiotes as a sterol source.** *J Insect Physiol* 1979, **25**(5):443–447.
10. Eya BK, Kenny PT, Tamura SY, Ohnishi M, Naya Y, Nakanishi K, Sugiura M: **Chemical association in symbiosis, Sterol donors in planthoppers.** *J Chem Ecol* 1989, **15**(1):373–380.
11. Wetzel JM, Ohnishi M, Fujita T, Nakanishi K, Naya Y, Noda H, Sugiura M: **Diversity in steroidogenesis of symbiotic microorganisms from planthoppers.** *J Chem Ecol* 1992, **18**(11):2083–2094.
12. Noda H, Koizumi Y: **Sterol biosynthesis by symbiotes: cytochrome P450 sterol C-22 desaturase genes from yeastlike symbiotes of rice planthoppers and anobiid beetles.** *Insect Biochem Mol Biol* 2003, **33**(6):649–658.
13. Yoshiyama T, Namiki T, Mita K, Kataoka H, Niwa R: ***Neverland* is an evolutionally conserved Rieske-domain protein that is essential for ecdysone synthesis and insect growth.** *Development* 2006, **133**:2565–2574.
14. Yoshiyama-Yanagawa T, Enya S, Shimada-Niwa Y, Yaguchi S, Haramoto Y, Matsuya T, Shiomi K, Sasakura Y, Takahashi S, Asashima M: **The conserved rieske oxygenase DAF-36/neverland is a novel cholesterol-metabolizing enzyme.** *J Biol Chem* 2011, **286**(29):25756–25762.
15. Gilbert LI, Warren JT: **A molecular genetic approach to the biosynthesis of the insect steroid molting hormone.** *Vitam Horm* 2005, **73**:31–57.
16. Namiki T, Niwa R, Sakudoh T, Shirai K, Takeuchi H, Kataoka H: **Cytochrome P450 CYP307A1/Spook: a regulator for ecdysone synthesis in insects.** *Biochemical and Biophysical Research Communication* 2005, **337**:367–374.
17. Ono H, Rewitz KF, Shinoda T, Itoyama K, Petryk A, Rybczynski R, Jarcho M, Warren JT, Marqués G, Shimell MJ: ***Spook* and *Spookier* code for stage-specific components of the ecdysone biosynthetic pathway in Diptera.** *Dev Biol* 2006, **298**(2):555–570.
18. Ou Q, Magico A, King-Jones K: **Nuclear receptor DHR4 controls the timing of steroid hormone pulses during *Drosophila* development.** *PLoS Biol* 2011, **9**(9):e1001160.
19. Ono H, Morita S, Asakura I, Nishida R: **Conversion of 3-oxo steroids into ecdysteroids triggers molting and expression of 20E-inducible genes in *Drosophila melanogaster*.** *Biochem Biophys Res Commun* 2012, **421**(3):561–566.
20. Luan J-B, Ghanim M, Liu S-S, Czosnek H: **Silencing the ecdysone synthesis and signaling pathway genes disrupts nymphal development in the whitefly.** *Insect Biochem Mol Biol* 2013, **43**(8):740–746.
21. Hentze JL, Moeller ME, Jørgensen AF, Bengtsson MS, Bordoy AM, Warren JT, Gilbert LI, Andersen O, Rewitz KF: **Accessory gland as a site for prothoracicotropic hormone controlled ecdysone synthesis in adult male insects.** *PLoS One* 2013, **8**(2):e55131.
22. Yamazaki Y, Kiuchi M, Takeuchi H, Kubo T: **Ecdysteroid biosynthesis in workers of the European honeybee *Apis mellifera* L.** *Insect Biochem Mol Biol* 2011, **41**(5):283–293.
23. Iga M, Smagghe G: **Identification and expression profile of *Halloween* genes involved in ecdysteroid biosynthesis in *Spodoptera littoralis*.** *Peptides* 2010, **31**(3):456–467.
24. Zhou J, Zhang H, Li J, Sheng X, Zong S, Luo Y, Weng Q: **Molecular cloning and expression profile of a Halloween gene encoding CYP307A1 from the seabuckthorn carpenterworm, Holcocerus hippophaecolus.** *J Insect Sci* 2012, **13**:56.
25. Marchal E, Badisco L, Verlinden H, Vandersmissen T, Van Soest S, Van Wielendaele P, Vanden Broeck J: **Role of the *Halloween* genes, *Spook* and *Phantom* in ecdysteroidogenesis in the desert locust, Schistocerca gregaria.** *J Insect Physiol* 2011, **57**(9):1240–1248.
26. Christiaens O, Iga M, Velarde R, Rougé P, Smagghe G: ***Halloween* genes and nuclear receptors in ecdysteroid biosynthesis and signalling in the pea aphid.** *Insect Mol Biol* 2010, **19**:187–200.
27. Tan A, Palli SR: **Edysone receptor isoforms play distinct roles in controlling molting and metamorphosis in the red flour beetle, Tribolium castaneum.** *Mol Cell Endocrinol* 2008, **291**(1):42–49.
28. Wu W-J, Wang Y, Huang H-J, Bao Y-Y, Zhang C-X: **Ecdysone receptor controls wing morphogenesis and melanization during rice planthopper metamorphosis.** *J Insect Physiol* 2012, **58**(3):420–426.
29. Karim FD, Thummel C: **Temporal coordination of regulatory gene expression by the steroid hormone ecdysone.** *EMBO J* 1992, **11**(11):4083–4093.
30. Varghese J, Cohen SM: **microRNA miR-14 acts to modulate a positive autoregulatory loop controlling steroid hormone signaling in *Drosophila*.** *Science* 2007, **21**(18):2277–2282.
31. Matusmura M, Sanada-Morimura S: **Recent status of insecticide resistance in Asian rice planthoppers.** *Japan Agricultural Research Quarterly* 2010, **44**(3):225–230.
32. Zhang P, Mar TT, Liu W, Li L, Wang X: **Simultaneous detection and differentiation of Rice black streaked dwarf virus (RBSDV) and Southern rice black streaked dwarf virus (SRBSDV) by duplex real time RT-PCR.** *Virol J* 2013, **10**(1):24.
33. Matsukura K, Towata T, Sakai J, Onuki M, Okuda M, Matsumura M: **Dynamics of Southern rice black-streaked dwarf virus in rice and implication for virus acquisition.** *Phytopathology* 2013. doi:10.1094/PHYTO-10-12-0261-R(ja).
34. Fujita D, Kohli A, Horgan FG: **Rice resistance to planthoppers and leafhoppers.** *Crit Rev Plant Sci* 2013, **32**(3):162–191.
35. Xu Y, Zhou W, Zhou Y, Wu J, Zhou X: **Transcriptome and comparative gene expression analysis of *Sogatella furcifera* (Horváth) in response to southern rice black-streaked dwarf virus.** *PLoS One* 2012, **7**(4):e36238.
36. Zha W, Peng X, Chen R, Du B, Zhu L, He G: **Knockdown of midgut genes by dsRNA-transgenic plant-mediated RNA interference in the hemipteran insect *Nilaparvata lugens*.** *PLoS One* 2011, **6**(5):e20504.
37. Li J, Chen QH, Lin YJ, Jiang TR, Wu G, Hua HX: **RNA interference in *Nilaparvata lugens* (Homoptera: Delphacidae) based on dsRNA ingestion.** *Pest Manag Sci* 2011, **67**(7):852–859.
38. Chen J, Zhang D, Yao Q, Zhang J, Dong X, Tian H, Zhang W: **Feeding-based RNA interference of a trehalose phosphate synthase gene in the brown planthopper, Nilaparvata lugens.** *Insect Mol Biol* 2010, **19**(6):777–786.
39. He P, Zhang J, Liu NY, Zhang YN, Yang K, Dong SL: **Distinct expression profiles and different functions of odorant binding proteins in *Nilaparvata lugens* Stål.** *PLoS One* 2011, **6**(12):e28921.

40. Werck-Reichhart D, Feyereisen R: **Cytochromes P450: a success story.** *Genome Biol* 2000, **1**(6):reviews 3003.3001–3009.
41. Petryk A, Warren JT, Marqués G, Jarcho MP, Gilbert LI, Kahler J, Parvy JP, Li Y, Dauphin-Villemant C, O'Connor MB: ***Shade* is the *Drosophila* P450 enzyme that mediates the hydroxylation of ecdysone to the steroid insect molting hormone 20-hydroxyecdysone.** *Proc Natl Acad Sci USA* 2003, **100**(24):13773–13778.
42. Rewitz KF, Rybczynski R, Warren JT, Gilbert LI: **Developmental expression of *Manduca shade*, the P450 mediating the final step in molting hormone synthesis.** *Mol Cell Endocrinol* 2006, **247**(1–2):166–174.
43. Rewitz KF, Rybczynski R, Warren JT, Gilbert LI: **The *Halloween* genes code for cytochrome P450 enzymes mediating synthesis of the insect moulting hormone.** *Biochem Soc Trans* 2006, **34:**1256–1260.
44. Yamazaki Y, Kiuchi M, Takeuchi H, Kubo T: **Ecdysteroid biosynthesis in workers of the European honeybee *Apis mellifera* L.** *Insect Biochem Mol Biol* 2011, **41:**283–293.
45. Rewitz K, Gilbert L: ***Daphnia Halloween* genes that encode cytochrome P450s mediating the synthesis of the arthropod molting hormone: Evolutionary implications.** *BMC Evol Biol* 2008, **8**(1):60.
46. Warren JT, Petryk A, Marqués G, Parvy JP, Shinoda T, Itoyama K, Kobayashi J, Jarcho M, Li Y, O'Connor MB: ***Phantom* encodes the 25-hydroxylase of *Drosophila melanogaster* and *Bombyx mori*: a P450 enzyme critical in ecdysone biosynthesis.** *Insect Biochem Mol Biol* 2004, **34**(9):991–1010.
47. Kobayashi M, Uchida M, Kuriyama K: **Elevation of 20-hydroxyecdysone level by buprofezin in *Nilaparvata lugens* Stål nymphs.** *Pestic Biochem Physiol* 1989, **34**(1):9–16.
48. Niwa R, Matsuda T, Yoshiyama T, Namiki T, Mita K, Fujimoto Y, Kataoka H: **CYP306A1, a cytochrome P450 enzyme, is essential for ecdysteroid biosynthesis in the prothoracic glands of *Bombyx* and *Drosophila*.** *J Biol Chem* 2004, **279**(34):35942–35949.
49. Chávez VM, Marqués G, Delbecque JP, Kobayashi K, Hollingsworth M, Burr J, Natzle JE, O'Connor MB: **The *Drosophila disembodied* gene controls late embryonic morphogenesis and codes for a cytochrome P450 enzyme that regulates embryonic ecdysone levels.** *Development* 2000, **127**(19):4115–4126.
50. Niwa R, Sakudoh T, Namiki T, Saida K, Fujimoto Y, Kataoka H: **The ecdysteroidogenic P450 *Cyp302a1/disembodied* from the silkworm, *Bombyx mori*, is transcriptionally regulated by prothoracicotropic hormone.** *Insect Mol Biol* 2005, **14**(5):563–571.
51. Warren JT, Petryk A, Marqués G, Jarcho M, Parvy JP, Dauphin-Villemant C, O'Connor MB, Gilbert LI: **Molecular and biochemical characterization of two P450 enzymes in the ecdysteroidogenic pathway of *Drosophila melanogaster*.** *Proc Natl Acad Sci USA* 2002, **99**(17):11043–11048.
52. Varghese J, Cohen SM: **microRNA miR-14 acts to modulate a positive autoregulatory loop controlling steroid hormone signaling in *Drosophila*.** *Science Signalling* 2007, **21**(18):2277–2282.
53. Tan AJ, Palli SR: **Edysone receptor isoforms play distinct roles in controlling molting and metamorphosis in the red flour beetle, Tribolium castaneum.** *Mol Cell Endocrinol* 2008, **291:**42–49.
54. Wu WJ, Wang Y, Huang HJ, Bao YY, Zhang CX: **Ecdysone receptor controls wing morphogenesis and melanization during rice planthopper metamorphosis.** *J Insect Physiol* 2012, **58:**420–426.
55. Miyake T, Haruyama H, Mitsui T, Sakurai A: **Effects of a new juvenile hormone mimic, NC-170, on metamorphosis and diapause of the small brown planthopper, Laodelphax striatellus.** *J Pestic Sci* 1992, **17**(1):75–82.
56. Grabherr MG, Haas BJ, Yassour M, Levin JZ, Thompson DA, Amit I, Adiconis X, Fan L, Raychowdhury R, Zeng Q, *et al*: **Full-length transcriptome assembly from RNA-Seq data without a reference genome.** *Nat Biotechnol* 2011, **29**(7):644–652.
57. Rozen S, Skaletsky H: **Primer3 on the WWW for general users and for biologist programmers.** *Methods Mol Biol* 2000, **132**(3):365–386.
58. Larkin MA, Blackshields G, Brown NP, Chenna R, McGettigan PA, McWilliam H, Valentin F, Wallace IM, Wilm A, Lopez R, *et al*: **Clustal W and Clustal X version 2.0.** *Bioinformatics* 2007, **23**(21):2947–2948.
59. Fu Q, Zhang Z, Hu C, Lai F, Sun Z: **A chemically defined diet enables continuous rearing of the brown planthopper, Nilaparvata lugens (Stål) (Homoptera: Delphacidae).** *Applied Entomology and Zoology* 2001, **36**(1):111–116.
60. Sun HX, Hu XJ, Shu YH, Zhang GR: **Observation on the antennal sensilla of *Sogatella furcifera* (Horváth) (Homoptera: Delphacidae) with scanning electron microscope.** *Acta Entomologica Sinica* 2006, **49**(2):349–354.
61. Ding JH, Hu CL, Fu Q, He JC, Xie MC: *A colour atlas of commonly encountered delphacidae in china rice regions.* HangZhou, China: HangZhou Science and Technology Press; 2012:10–26.
62. Pfaffl MW: **A new mathematical model for relative quantification in real-time RT-PCR.** *Nucleic Acids Res* 2001, **29:**e45.
63. Vandesompele J, De Preter K, Pattyn F, Poppe B, Van Roy N, De Paepe A, Speleman F: **Accurate normalization of real-time quantitative RT-PCR data by geometric averaging of multiple internal control genes.** *Genome Biol* 2002, **3:**RESEARCH0034.

17

Lowering the quantification limit of the Qubit™ RNA HS Assay using RNA spike-in

Xin Li[1,2], Iddo Z Ben-Dov[3], Maurizio Mauro[1,2] and Zev Williams[1,2*]

Abstract

Background: RNA quantification is often a prerequisite for most RNA analyses such as RNA sequencing. However, the relatively low sensitivity and large sample consumption of traditional RNA quantification methods such as UV spectrophotometry and even the much more sensitive fluorescence-based RNA quantification assays, such as the Qubit™ RNA HS Assay, are often inadequate for measuring minute levels of RNA isolated from limited cell and tissue samples and biofluids. Thus, there is a pressing need for a more sensitive method to reliably and robustly detect trace levels of RNA without interference from DNA.

Methods: To improve the quantification limit of the Qubit™ RNA HS Assay, we spiked-in a known quantity of RNA to achieve the minimum reading required by the assay. Samples containing trace amounts of RNA were then added to the spike-in and measured as a reading increase over RNA spike-in baseline. We determined the accuracy and precision of reading increases between 1 and 20 pg/μL as well as RNA-specificity in this range, and compared to those of RiboGreen®, another sensitive fluorescence-based RNA quantification assay. We then applied Qubit™ Assay with RNA spike-in to quantify plasma RNA samples.

Results: RNA spike-in improved the quantification limit of the Qubit™ RNA HS Assay 5-fold, from 25 pg/μL down to 5 pg/μL while maintaining high specificity to RNA. This enabled quantification of RNA with original concentration as low as 55.6 pg/μL compared to 250 pg/μL for the standard assay and decreased sample consumption from 5 to 1 ng. Plasma RNA samples that were not measurable by the Qubit™ RNA HS Assay were measurable by our modified method.

Conclusions: The Qubit™ RNA HS Assay with RNA spike-in is able to quantify RNA with high specificity at 5-fold lower concentration and uses 5-fold less sample quantity than the standard Qubit™ Assay.

Keywords: Lower quantification limit, Minimum RNA concentration, Plasma RNA, Qubit™ RNA HS Assay, RNA quantification, RNA spike-in

Background

Recent studies utilizing trace amounts of RNA present in biospecimens such as biofluids, single cells and minute clinical samples have revealed their novel functions and biomedical potentials [1-14]. RNA quantification is an important and necessary step prior to most RNA analyses. However, it can be very challenging to quantify RNA present in the pg/μL ranges found in biofluids and minute cell and tissue samples [6]. After purification using most commercial RNA isolation kits, the concentrations of purified plasma RNA samples are often less than 200 pg/μL. UV spectrophotometry commonly used for nucleic acid quantification has a lower quantification limit around 4 ng/μL, and is therefore not suitable for measuring RNA samples with such low concentrations [15-17].

An alternative approach is fluorescence-based RNA quantification that utilizes the fluorescent property of nucleic acid binding dyes. Unbound dyes are nearly non-fluorescent, but upon binding to nucleic acid, the complex exhibits a large increase in fluorescence, thereby greatly amplifying nucleic acid signal for detection at concentrations much lower than that required

* Correspondence: zev.williams@einstein.yu.edu
[1]Department of Obstetrics & Gynecology and Women's Health, Albert Einstein College of Medicine, 10461 Bronx, NY, USA
[2]Department of Genetics, Albert Einstein College of Medicine, 10461 Bronx, NY, USA
Full list of author information is available at the end of the article

by UV spectrophotometry [15,16,18-21]. An example of fluorescence-based RNA quantification methods is the Qubit™ RNA HS Assay (Life Technologies, Thermo Fisher Scientific Inc.).

The Qubit™ RNA HS Assay is highly selective for RNA over DNA [22] and provides a minimum "reading" (RNA concentration in the Qubit™ working solution) of 25 pg/μL with high confidence (deviation from ideal < 20%). Up to 20 μL of RNA sample can be added in a 200 μL Qubit™ Assay, and therefore RNA samples with a minimum starting concentration of 250 pg/μL can be accurately quantified. However, this minimum concentration is still relatively high compared to levels of RNA found in certain biological specimens. Moreover, the assay consumes a minimum of 5 ng of RNA sample, which may leave insufficient RNA for downstream applications. Thus, these detection limitations to the Qubit™ Assay can hinder the analysis and application of some biological samples with extremely low RNA quantities.

Here we used an RNA spike-in to set a baseline reading of the Qubit™ Assay and measured RNA sample as an increase over RNA spike-in. This method was validated to accurately measure RNA at lower concentrations and require less sample compared to standard Qubit™. We tested the utility of this spike-in approach by measuring plasma RNA samples that fell below the detection limit of the standard Qubit™ Assay. We named the modified assay the Spike-in Qubit™ RNA HS Assay because this optimization takes advantage of an RNA spike-in.

Methods

Validation of the Spike-in Qubit™ RNA HS assay

The Qubit™ RNA HS Assay Kit (Life Technologies, Thermo Fisher Scientific Inc.), Qubit™ 2.0 Fluorometer (Life Technologies, Thermo Fisher Scientific Inc.) and Axygen PCR-05-C tubes (Axygen) were used for all measurements. The Qubit™ working solution was made according to manufacturer's instructions. We added 180 μL of working solution to each assay tube, up to 20 μL of RNA, and water to bring the final volume to 200 μL. 10 μL of the Qubit™ RNA Standard solutions were used for standard tubes. "RNA spike-in" was made by diluting the Qubit™ RNA Standard #2 (10 ng/μL rRNA) included in the Qubit™ RNA HS Assay kit to 2.5 ng/μL and 2 μL was added into each tube. 18 μL of water was added into one tube for RNA spike-in alone reading. "RNA sample" was made by diluting the Qubit™ RNA Standard #2 to 250 pg/μL and increasing volumes of RNA sample were added into remaining tubes to create expected reading increases of 1, 2, 3, 4, 5, 7.5, 15 and 20 pg/μL over RNA spike-in alone. Assay tubes were vortexed for 2–3 s, centrifuged briefly (~5 s), and then incubated at room temperature for 2 min to allow the assay to reach optimal fluorescence before measuring with the Qubit™ Fluorometer. Each tube was measured three times to obtain the average reading. Reading increase was calculated by subtracting RNA spike-in reading from that of spike-in plus sample RNA reading. The experiment was repeated four times using independently prepared RNA spike-in, RNA sample, and working solution. In addition, the total RNA from human trophoblast cells was used as "RNA sample" and tested as described above for a total of four independent experiments.

Comparison between the Spike-in Qubit™ RNA HS Assay and Quant-iT™ RiboGreen® RNA Assay

Based on the original concentration measured by NanoDrop® ND-1000 (Thermo Fisher Scientific Inc.), the trophoblast total RNA was diluted to 60 pg/μL in water. A mixture of RNA and DNA (60 pg/μL each) was prepared by diluting in water the trophoblast total RNA and the Qubit™ DNA Standard #2 (10 ng/μL DNA) included in the Qubit™ DNA HS Assay kit. 18 μL of each sample was measured by the Spike-in Qubit™ as described above and by the Quant-iT™ RiboGreen® RNA Assay (Life Technologies, Thermo Fisher Scientific Inc.) following manufacturer's instructions using a BioTek Synergy™ 4 Multi-Mode Microplate Reader (BioTek). Four independent repeats were performed with each method.

Quantification of plasma RNA samples

The standard Qubit™ RNA HS Assay was performed according to manufacturer's instructions. The Spike-in Qubit™ RNA HS Assay was performed as described above. For the Spike-in Qubit™ Assay, RNA sample concentration was calculated as: [Sample] = reading increase (pg/μL) x assay volume (μL) ÷ sample volume for Spike-in Qubit™ (μL). Three independent measurements using separately prepared RNA spike-in and Qubit™ reagents were made for each sample. A detailed step-by-step protocol for the Spike-in Qubit™ RNA HS Assay is provided in the Supplementary Methods.

Sample preparation and RNA isolation

After obtaining informed consent, 8.5 mL of peripheral blood samples were collected into Vacutainer™ tubes containing acid citrate dextrose solution A (BD) and immediately inverted eight times to mix anticoagulant additive with blood. Blood samples were centrifuged at 1,900 g for 10 min at 4°C in a swinging-bucket centrifuge (Eppendorf) and plasma was aspirated using disposable transfer pipets (VWR), aliquoted into 2 mL polypropylene microcentrifuge tubes (Sarstedt) and centrifuged at 16,050 g for 5 min at 4°C in a benchtop centrifuge (Eppendorf). The final plasma was pooled and mixed in 15 mL tubes (Falcon), then aliquoted into 1.5 mL DNA LoBind tubes (Eppendorf), snap frozen in liquid nitrogen and stored at −80°C. Plasma RNA isolation was performed using three commercial RNA isolation kits: miRCURY™ RNA Isolation Kit – Biofluids (Exiqon), mirVana™ PARIS™ Kit (Life Technologies, Thermo Fisher Scientific Inc.) and miRNeasy Micro Kit (Qiagen) following manufacturer recommended protocols. The miRCURY™ kit is designed to isolate RNA shorter than 1000 nucleotides (nt) and the mirVana™ and miRNeasy kits isolate total RNA.

After informed consent was obtained, tissues were collected from manual vacuum aspiration. To isolate trophoblast cells, chorionic villi were identified and floated in DPBS supplemented with 10 mg/ml Gentamycin, minced with sterile scalpel blades, placed in 35 ×10 mm dishes and incubated with 3.3 mg/ml Collagenase (Sigma) for about 2 hours at 37°C in the cell incubator. Cells were further washed twice with Amniomax complete medium (Amionax basal medium plus F100 supplement; GIBCO) and cultured in Amniomax complete medium for 7 days. RNA from trophoblast cells was extracted with All Prep® DNA/RNA/Protein mini kit (Qiagen) according to the manufacturer's protocol. All aspects of these studies were reviewed and approved by the Institutional Review Board at Albert Einstein College of Medicine.

Data analyses

Data analyses were performed using Excel® for Mac 2011 (Microsoft) and GraphPad Prism 6 (GraphPad Software). Relative error (RE) was calculated as RE = (average measured reading – expected reading) ÷ expected reading. Coefficient of variation (CV) was calculated as CV = standard deviation ÷ average measured reading. Deviation from ideal was calculated as the sum of the absolute value of RE and CV. A two-way ANOVA interaction analysis was performed to determine whether the differences between the measurements of the 60 pg/μL RNA sample and those of the mixed RNA and DNA sample (60 pg/μL each) were consistent for the Spike-in Qubit™ RNA HS Assay and the Quant-iT™ RiboGreen® RNA Assay.

Results and discussion

The lower reading limit of the Qubit™ RNA HS assay for accurate quantification is 25 pg/μL (deviation from ideal < 20%). In order to quantify samples that fall below this limit, prior to adding the RNA sample that was to be measured, we added 5 ng of RNA spike-in (2.5 ng/μL Qubit™ RNA standard #2) into a 200 μL Qubit™ assay to generate an expected baseline reading of 25 pg/μL. Then, the RNA sample was added to the baseline and the increase over baseline would correspond to the reading of RNA sample (Figure 1).

To evaluate the ability of our Spike-in Qubit™ Assay to measure increases in the range of 1 to 20 pg/μL, an RNA sample (250 pg/μL Qubit™ RNA standard #2) was added at increasing volumes into Qubit™ assay tubes containing RNA spike-in to create expected reading increases ranging from 1 to 20 pg/μL. As shown in Figure 2A, the Spike-in Qubit™ Assay achieved optimal linear regression with slope of 1.0324, and R^2 of 0.99837 after four independent experiments (Additional file 2: Table S1). To represent a typical RNA sample that might introduce additional influences on fluorometric

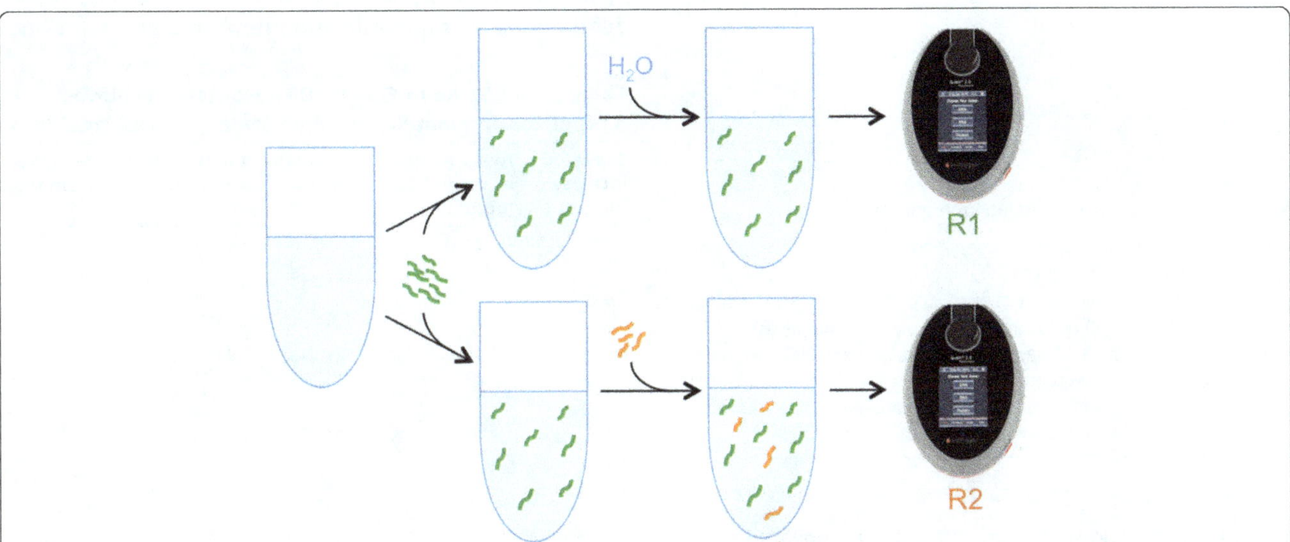

Figure 1 Schematic diagram of the Spike-in Qubit™ RNA HS Assay. RNA spike-in (green) was added to reach the lower quantification limit of Qubit™. Then nuclease-free water or RNA sample (orange) was added for Qubit™ measurement. R1 is the reading for RNA spike-in alone and R2 for RNA spike-in plus RNA sample. The reading for RNA sample is (R2 - R1) and RNA sample concentration is calculated as [sample] = (R2 – R1) (pg/μL) × assay volume (μL) ÷ sample volume for the assay (μL).

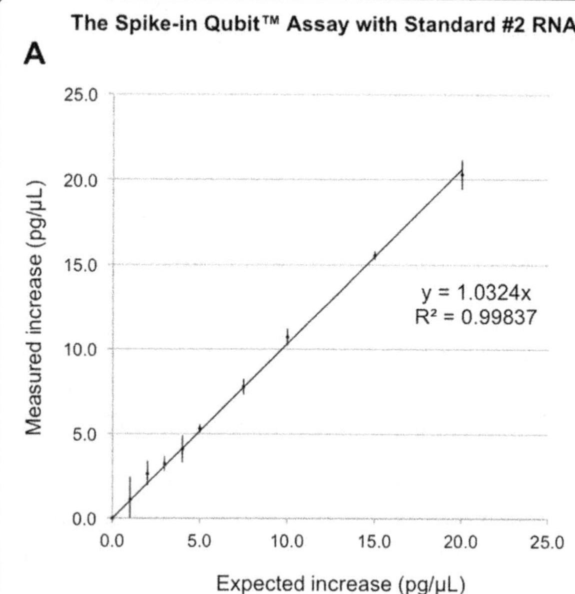

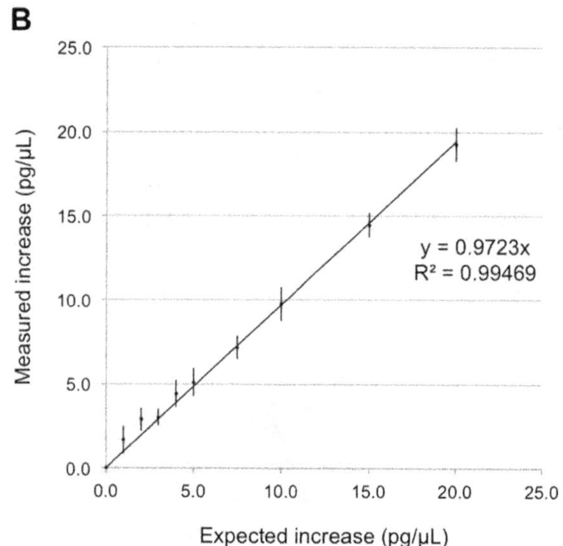

Figure 2 Reading increases between 1 and 20 pg/µL show a strong linear correlation in the Spike-in Qubit™ RNA Assay. RNA spike-in alone or with increasing amounts of a 250 pg/µL Qubit™ RNA Standard #2 sample **(A)** or a 250 pg/µL trophoblast total RNA sample **(B)** was measured by the Qubit™ Assay. Reading increases over RNA spike-in were plotted against expected reading increases. Regression line equation, coefficient of determination (R^2) and error bars indicating standard deviation are shown. N = 4 independent repeats.

measurement, the same experiment was repeated with a 250 pg/µL human trophoblast total RNA sample. Strong linear correlation was also evident with slope of 0.9723 and R^2 of 0.99469 (Figure 2B) after four independent experiments (Additional file 2: Table S2).

Table 1 The Spike-in Qubit™ RNA HS Assay achieves 5 pg/µL lower quantification limit for the Qubit™ RNA standard #2

Expected increase (pg/µL)	Ave. measured increase ± SD (pg/µL)	Relative error	Coefficient of variation	Deviation from ideal
1	1.1 ± 1.3	12.5%	114.6%	127.1%
2	2.7 ± 0.7	32.5%	27.6%	60.1%
3	3.2 ± 0.4	7.5%	13.6%	21.1%
4	4.1 ± 0.8	2.5%	19.9%	22.4%
5	5.3 ± 0.2	6.3%	4.6%	10.9%
7.5	7.8 ± 0.5	3.6%	5.8%	9.4%
10	10.7 ± 0.5	6.9%	4.7%	11.6%
15	15.5 ± 0.3	3.4%	1.7%	5.1%
20	20.3 ± 0.9	1.6%	4.4%	6.0%

A 250 pg/µL Qubit™ RNA Standard #2 RNA sample was used to assess the accuracy and precision of reading increases between 1 and 20 pg/µL using the Spike-in Qubit™. Average measured reading increase ± standard deviation (SD), relative error, coefficient of variation, and deviation from ideal expressed in percent are listed in the table. N = 4 independent repeats.

We assessed the accuracy and precision of the Spike-in Qubit™ Assay for all tested reading increases in order to determine the lower quantification limit that passed accuracy and precision requirements. Accuracy is inversely correlated to the relative error (RE) and precision is inversely correlated to the coefficient of variation (CV). The Qubit™ RNA HS Assay used "deviation from ideal", the combined value of the absolute value of RE and CV, to evaluate the precision and accuracy and set deviation from ideal of <20% for its lower quantification limit in the core quantification range. Using the same criterion, we determined that Spike-in Qubit™ Assay achieved a lower quantification limit of 5 pg/µL for both

Table 2 The Spike-in Qubit™ RNA HS Assay achieves 5 pg/µL lower quantification limit for trophoblast total RNA

Expected increase (pg/µL)	Ave. measured increase ± SD (pg/µL)	Relative error	Coefficient of variation	Deviation from ideal
1	1.7 ± 0.8	65.8%	50.6%	116.5%
2	2.9 ± 0.7	44.2%	23.9%	68.0%
3	3.0 ± 0.5	0.6%	16.7%	17.3%
4	4.4 ± 0.8	10.6%	18.4%	29.1%
5	5.1 ± 0.8	2.2%	16.3%	18.5%
7.5	7.2 ± 0.7	−4.1%	9.6%	13.7%
10	9.7 ± 1.0	−2.6%	10.5%	13.0%
15	14.5 ± 0.8	−3.6%	5.2%	8.8%
20	19.3 ± 1.0	−3.7%	5.3%	9.0%

A 250 pg/µL trophoblast total RNA sample was used to assess the accuracy and precision of reading increases between 1 and 20 pg/µL using the Spike-in Qubit™. Average measured reading increase ± standard deviation (SD), relative error, coefficient of variation, and deviation from ideal expressed in percent are listed in the table. N = 4 independent repeats.

Table 3 The spike-in Qubit™ RNA HS Assay reduces minimal RNA concentration and sample consumption

	Qubit™	Spike-in Qubit™
Lower Quantification Limit (pg/μL)	25	5
Max. Sample Volume (μL)	20	18*
Min. Sample Concentration (pg/μL)	250	55.6
Min. Sample Quantity (ng)	5	1

The lower quantification limit and maximum sample volume for the Qubit™ Assay and Spike-in Qubit™ Assay and their corresponding minimal sample concentration and quantity are listed in the table. *2 μL RNA Spike-in is added into each assay tube, leaving maximally 18 μL for RNA sample in a 200 μL assay.

Qubit™ Standard #2 RNA and trophoblast total RNA, which is 80% less than that of the standard Qubit™ Assay (25 pg/μL) (Tables 1 and 2). This new lower quantification limit allowed quantification of RNA samples with original concentrations as low as 55.6 pg/μL and reduced minimal sample consumption from 5 ng to 1 ng (Table 3).

We also cross-compared the precision and accuracy of the Spike-in Qubit™ Assay with those of the Quant-iT™ RiboGreen® RNA Assay that has a lower quantification limit of 1 pg/μL. A 60 pg/μL trophoblast total RNA sample was measured in four independent repeats with both methods (Additional file 2: Table S3 and S4). As shown in Table 4, the RNA concentration measured by the Spike-in Qubit™ was 63.6 ± 3.4 pg/μL and 53.4 ± 1.3 pg/μL by the Quant-iT™. Both methods achieved good precision with CVs of 5.4% and 2.5% for the Spike-in Qubit™ and the Quant-iT™, respectively. In contrast to the Quant-iT™ RiboGreen® RNA Assay and most other methods including UV spectrometer, the Qubit™ RNA Assay is reported to be selective to RNA over DNA [15,16,18,22]. DNA at 8 times higher concentration of the lower quantification limit (25 pg/μL) is not detectable in the Qubit™ RNA Assay and an equal mixture of DNA and RNA up to 200 pg/μL each in the assay does not affect the reading of RNA [22]. To test if the Spike-in Qubit™ maintains RNA specificity at low readings, we mixed the 60 pg/μL RNA with 60 pg/μL DNA. The reading for the mixture was 64.2 ± 6.2 pg/μL, similar to the 60 pg/μL RNA concentration measured by the Spike-in Qubit™. In contrast, the presence of DNA significantly increased the quantification value by the Quant-iT™ from 53.4 ± 1.3 pg/μL for the RNA sample to 190.1 ± 6.1 pg/μL for the RNA plus DNA sample. A two-way ANOVA analysis determined that the differences in the measurements of the RNA sample and those of the mixed RNA and DNA sample were significantly different between the Spike-in Qubit™ RNA HS Assay and the Quant-iT™ RiboGreen® RNA Assay ($P < 0.0001$).

Table 4 The Spike-in Qubit™ RNA HS Assay maintains high precision and RNA specificity in the extended lower reading range

Sample	Spike-in Qubit™		Quant-it™	
	Reading ± SD (pg/μL)	[RNA] ± SD (pg/μL)	Reading ± SD (pg/μL)	[RNA] ± SD (pg/μL)
60 pg/μL RNA	5.7 ± 0.3	63.6 ± 3.4	4.8 ± 0.1	53.4 ± 1.3
60 pg/μL RNA + 60 pg/μL DNA	5.8 ± 0.6	64.2 ± 6.2	17.1 ± 0.6	190.1 ± 6.1

A 60 pg/μL RNA sample and a mixture of 60 pg/μL RNA and 60 pg/μL DNA were measured by the Spike-in Qubit™ and the Quant-iT™ RiboGreen® Assays. Average reading ± standard deviation (SD) and corresponding RNA concentration ± SD are listed in the table. N = 4 independent repeats.

RNA samples purified from serum or plasma are present at low concentration (~30 ng per 1 mL of plasma), making their quantification challenging [6,23,24]. To determine whether our newly validated Spike-in Qubit™ Assay would enable accurate measurements from these samples which would normally be at too low of a concentration to be measured with standard Qubit™, we used both assays to measure plasma RNA samples purified using three commercial RNA isolation kits. All three samples fell below the detection limit of the standard Qubit™ and therefore their concentrations could not be determined (Table 5). In contrast, The Spike-in Qubit™ Assay achieved quantification for all samples while consuming 25-50% less samples than the standard Qubit™ (Table 5).

We speculate that the Spike-in Qubit™ approach may work because the readings of RNA spike-in alone and with additional RNA samples fall into the linear and high-precision quantification range of the Qubit™. It ensures that the reading increase over RNA spike-in baseline is of high precision and linear therefore of high accuracy. There are limitations to the Spike-in Qubit™ RNA HS Assay. First, it requires extra steps to prepare and add RNA spike-in. However, these steps only consist of diluting the Qubit™ RNA Standard #2 (that comes pre-made in the kit) and adding it into a master mix. Therefore, the time and risk of introducing error are minimal. In addition, because an excess amount of Qubit™ RNA Standard #2 is provided in the Qubit™ Assay kit, there is no need to purchase additional reagents. As the validation for reading increases between 1 and 5 pg/μL was performed by adding RNA samples in 0.8 to 4 μL volumes, Pipetting of small volumes could have contributed to variation in measurements due to pipetting error.

Conclusions

The Spike-in Qubit™ RNA HS Assay reported here achieved accurate and precise RNA quantification at a new lower quantification limit of 5 pg/μL, 5-fold lower than that of the standard Qubit™, while maintaining the RNA specificity of the original assay. This improvement lowers minimal RNA concentration measurable from

Table 5 The spike-in Qubit™ RNA HS Assay enables quantification of RNA samples purified from plasma

Kit	Qubit™			Spike-in Qubit™			
	Sample vol. (μL)	Reading (pg/μL)	[RNA] (pg/μL)	Sample Vol. (μL)	Reading ± SD (pg/μL)	[RNA] ± SD (pg/μL)	CV
1	20	<20	N.D.	10	5.3 ± 0.7	106.0 ± 14.0	13.2%
2	20	< 20	N.D.	15	13.2 ± 1.8	176.4 ± 23.4	13.3%
3	20	< 20	N.D.	15	6.0 ± 0.3	80.0 ± 3.5	4.4%

Plasma RNA samples purified using the three kits listed in the Methods were quantified by the Qubit™ and the Spike-in Qubit™ Assays. Sample volume used for quantification, Qubit™ reading or Spike-in Qubit™ reading increase ± standard deviation (SD) and corresponding RNA concentration ± SD and coefficient of variation (CV) are listed in the table. " < 20" indicates the reading is below the Qubit™ detection limit and therefore sample RNA concentration could not be determined (N.D.) For the Spike-in Qubit™, RNA sample concentration was calculated as described in the Methods. N = 3 independent repeats.

250 pg/μL to 55.6 pg/μL and reduces minimal RNA consumption from 5 ng to 1 ng. As demonstrated in the successful quantification of plasma RNA samples, the Spike-in Qubit™ RNA HS can be readily used to quantify RNA samples having low concentrations and limited quantities.

Additional files

Additional file 1: The Spike-in Qubit™ RNA HS Assay Protocol.

Additional file 2: Table S1. Four independent Spike-in Qubit™ measurements of reading increases between 1 and 20 pg/μL using a 250 pg/μL Qubit™ Standard #2 RNA sample. **Table S2.** Four independent Spike-in Qubit™ measurements of reading increases between 1 and 20 pg/μL using a 250 pg/μL trophoblast total RNA sample. **Table S3.** Four independent Spike-in Qubit™ measurements of a 60 pg/μL trophoblast total RNA sample or a mixure of trophoblast total RNA and Qubit™ Standard #2 DNA (60 pg/μL each). **Table S4.** Four independent Quant-iT™ RiboGreen® measurements of a 60 pg/μL trophoblast total RNA sample or a mixure of trophoblast total RNA and Qubit™ Standard #2 DNA (60 pg/μL each). **Table S5.** Three independent Spike-in Qubit™ measurements of plasma RNA samples.

Abbreviations

CV: Coefficient of variation; RE: Relative error; SD: Standard deviation.

Competing interests

The authors declare that they have no competing interests.

Authors' contributions

XL, IZB-D and ZW designed research; XL and MM performed research; XL, IZB-D, MM and ZW analyzed data and wrote the paper. All authors read and approved the final manuscript.

Acknowledgements

We thank Drs. Ryung Kim, Shan Wei, Andrzej Breborowicz, Thomas Tuschl, and the Vijg and Suh Laboratories for their assistance. This work was supported by NIH grant 1U19CA179564-01 and HD068546. IZB-D is supported by the I-CORE Program of the Planning and Budgeting Committee and The Israel Science Foundation (Grant No. 41/11).

Author details

[1]Department of Obstetrics & Gynecology and Women's Health, Albert Einstein College of Medicine, 10461 Bronx, NY, USA. [2]Department of Genetics, Albert Einstein College of Medicine, 10461 Bronx, NY, USA. [3]Nephrology and Hypertension, Hadassah – Hebrew University Medical Center, 91120 Jerusalem, Israel.

References

1. Koh W, Pan W, Gawad C, Fan HC, Kerchner GA, Wyss-Coray T, et al. Noninvasive in vivo monitoring of tissue-specific global gene expression in humans. Proc Natl Acad Sci U S A. 2014;111(20):7361–6.
2. Kosaka N, Izumi H, Sekine K, Ochiya T. microRNA as a new immune-regulatory agent in breast milk. Silence. 2010;1:7.
3. Menke TB, Warnecke JM. Improved conditions for isolation and quantification of RNA in urine specimens. Ann N Y Acad Sci. 2004;1022:185–9.
4. Michael A, Bajracharya SD, Yuen PS, Zhou H, Star RA, Illei GG, et al. Exosomes from human saliva as a source of microRNA biomarkers. Oral Dis. 2010;16(1):34–8.
5. Mitchell PS, Parkin RK, Kroh EM, Fritz BR, Wyman SK, Pogosova-Agadjanyan EL, et al. Circulating microRNAs as stable blood-based markers for cancer detection. Proc Natl Acad Sci U S A. 2008;105(30):10513–8.
6. Williams Z, Ben-Dov IZ, Elias R, Mihailovic A, Brown M, Rosenwaks Z, et al. Comprehensive profiling of circulating microRNA via small RNA sequencing of cDNA libraries reveals biomarker potential and limitations. Proc Natl Acad Sci U S A. 2013;110(11):4255–60.
7. Ben-Dov IZ, Tan YC, Morozov P, Wilson PD, Rennert H, Blumenfeld JD, et al. Urine microRNA as potential biomarkers of autosomal dominant polycystic kidney disease progression: description of miRNA profiles at baseline. PLoS One. 2014;9(1):e86856.
8. Deng Q, Ramskold D, Reinius B, Sandberg R. Single-cell RNA-seq reveals dynamic, random monoallelic gene expression in mammalian cells. Science. 2014;343(6167):193–6.
9. Hedegaard J, Thorsen K, Lund MK, Hein AM, Hamilton-Dutoit SJ, Vang S, et al. Next-generation sequencing of RNA and DNA isolated from paired fresh-frozen and formalin-fixed paraffin-embedded samples of human cancer and normal tissue. PLoS One. 2014;9(5):e98187.
10. Patel AP, Tirosh I, Trombetta JJ, Shalek AK, Gillespie SM, Wakimoto H, et al. Single-cell RNA-seq highlights intratumoral heterogeneity in primary glioblastoma. Science. 2014;344(6190):1396–401.
11. Ramskold D, Luo S, Wang YC, Li R, Deng Q, Faridani OR, et al. Full-length mRNA-Seq from single-cell levels of RNA and individual circulating tumor cells. Nat Biotechnol. 2012;30(8):777–82.
12. Shalek AK, Satija R, Adiconis X, Gertner RS, Gaublomme JT, Raychowdhury R, et al. Single-cell transcriptomics reveals bimodality in expression and splicing in immune cells. Nature. 2013;498(7453):236–40.
13. Trapnell C, Cacchiarelli D, Grimsby J, Pokharel P, Li S, Morse M, et al. The dynamics and regulators of cell fate decisions are revealed by pseudotemporal ordering of single cells. Nat Biotechnol. 2014;32(4):381–6.
14. Yick CY, Zwinderman AH, Kunst PW, Grunberg K, Mauad T, Chowdhury S, et al. Gene expression profiling of laser microdissected airway smooth muscle tissue in asthma and atopy. Allergy. 2014;69(9):1233–40.
15. Glasel JA. Validity of nucleic acid purities monitored by 260 nm/280nm absorbance ratios. Biotechniques. 1995;18(1):62–3.
16. Ingle JD, Crouch SR. Spectrochemical Analysis. Englewood Cliffs, NJ: Prentice Hall; 1988.
17. Desjardins PR, Conklin DS. Microvolume quantitation of nucleic acids. In: Ausubel FM, editor. Current protocols in molecular biology, vol. Appendix 3. 2011. p. 3J.
18. Jones LJ, Yue ST, Cheung CY, Singer VL. RNA quantitation by fluorescence-based solution assay: RiboGreen reagent characterization. Anal Biochem. 1998;265(2):368–74.

19. Le Pecq JB, Paoletti C. A new fluorometric method for RNA and DNA determination. Anal Biochem. 1966;17(1):100–7.
20. Labarca C, Paigen K. A simple, rapid, and sensitive DNA assay procedure. Anal Biochem. 1980;102(2):344–52.
21. Rye HS, Dabora JM, Quesada MA, Mathies RA, Glazer AN. Fluorometric assay using dimeric dyes for double- and single-stranded DNA and RNA with picogram sensitivity. Anal Biochem. 1993;208(1):144–50.
22. Dallwig J, Hagen D, Cheung C-y, Thomas G, Yue S: Methine-substituted cyanine dye compounds. In. US Patent US 7,776,529 B2; 2010.
23. Burgos KL, Javaherian A, Bomprezzi R, Ghaffari L, Rhodes S, Courtright A, et al. Identification of extracellular miRNA in human cerebrospinal fluid by next-generation sequencing. RNA. 2013;19(5):712–22.
24. Rykova EY, Wunsche W, Brizgunova OE, Skvortsova TE, Tamkovich SN, Senin IS, et al. Concentrations of circulating RNA from healthy donors and cancer patients estimated by different methods. Ann N Y Acad Sci. 2006;1075:328–33.

PERMISSIONS

All chapters in this book were first published in MB, by BioMed Central; hereby published with permission under the Creative Commons Attribution License or equivalent. Every chapter published in this book has been scrutinized by our experts. Their significance has been extensively debated. The topics covered herein carry significant findings which will fuel the growth of the discipline. They may even be implemented as practical applications or may be referred to as a beginning point for another development.

The contributors of this book come from diverse backgrounds, making this book a truly international effort. This book will bring forth new frontiers with its revolutionizing research information and detailed analysis of the nascent developments around the world.

We would like to thank all the contributing authors for lending their expertise to make the book truly unique. They have played a crucial role in the development of this book. Without their invaluable contributions this book wouldn't have been possible. They have made vital efforts to compile up to date information on the varied aspects of this subject to make this book a valuable addition to the collection of many professionals and students.

This book was conceptualized with the vision of imparting up-to-date information and advanced data in this field. To ensure the same, a matchless editorial board was set up. Every individual on the board went through rigorous rounds of assessment to prove their worth. After which they invested a large part of their time researching and compiling the most relevant data for our readers.

The editorial board has been involved in producing this book since its inception. They have spent rigorous hours researching and exploring the diverse topics which have resulted in the successful publishing of this book. They have passed on their knowledge of decades through this book. To expedite this challenging task, the publisher supported the team at every step. A small team of assistant editors was also appointed to further simplify the editing procedure and attain best results for the readers.

Apart from the editorial board, the designing team has also invested a significant amount of their time in understanding the subject and creating the most relevant covers. They scrutinized every image to scout for the most suitable representation of the subject and create an appropriate cover for the book.

The publishing team has been an ardent support to the editorial, designing and production team. Their endless efforts to recruit the best for this project, has resulted in the accomplishment of this book. They are a veteran in the field of academics and their pool of knowledge is as vast as their experience in printing. Their expertise and guidance has proved useful at every step. Their uncompromising quality standards have made this book an exceptional effort. Their encouragement from time to time has been an inspiration for everyone.

The publisher and the editorial board hope that this book will prove to be a valuable piece of knowledge for researchers, students, practitioners and scholars across the globe.

LIST OF CONTRIBUTORS

Ellen De Keyser, Laurence Desmet and Jan De Riek
Institute for Agricultural and Fisheries Research (ILVO)-Plant Sciences Unit, Caritasstraat 21, 9090, Melle, Belgium

Erik Van Bockstaele
Institute for Agricultural and Fisheries Research (ILVO)-Plant Sciences Unit, Caritasstraat 21, 9090, Melle, Belgium
Department for Plant Production, Ghent University, Coupure links 653, 9000, Ghent, Belgium

Lisette Quaade Sørensen, Jesper Erup Larsen, Paiman Khorsand-Jamal and Rasmus John Normand Frandsen
Eukaryotic Molecular Cell Biology Group, Department of Systems Biology, The Technical University of Denmark, Søltofts Plads building 223, DK-2800 Kgs., Lyngby, Denmark

Kristian Fog Nielsen
Metabolic Signaling and Regulation group,Department of Systems Biology, The Technical University of Denmark,Søltofts Plads building 221, DK-2800 Kgs., Lyngby, Denmark

Erik Lysøe
Bioforsk–Norwegian Institute of Agricultural and Environmental Research, Høgskoleveien 7, Ås 1430, Norway

Tadashi Nakagawa
Department of Cell Proliferation, United Center for Advanced Research and Translational Medicine, Graduate School of Medicine, Tohoku University, Sendai 900-8575, Japan

Koushik Mondal and Patrick C Swanson
Department of Medical Microbiology and Immunology, Creighton University, 2500 California Plaza, Omaha, NE 68178, USA

Christopher Edge, Clare Gooding and Christopher WJ Smith
Department of Biochemistry, University of Cambridge, Tennis Court Road, Cambridge CB2 1QW, UK

Abhijit Rath and Arrigo De Benedetti
Department of Biochemistry and Molecular Biology, Louisiana State University Health Sciences Center, 1501 Kings Highway, Shreveport, LA 71130, USA

Robert Hromas
Department of Medicine, College of Medicine, University of Florida & Shands, Gainesville, FL 32610-0277, USA

Piotr M Skowron, Joanna Jezewska-Frackowiak, Joanna Zebrowska and Agnieszka Zylicz-Stachula
Division of Molecular Biotechnology, Department of Chemistry, Institute for Environmental and Human Health Protection, University of Gdansk, Wita Stwosza 63, 80-952, Gdansk, Poland

Jolanta Vitkute and Goda Mitkaite
Thermo Fisher Scientific, V.A. Graiciuno 8, LT-02241, Vilnius, Lithuania

Danute Ramanauskaite
Department of Botany and Genetics, Vilnius University, M.K. Ciurlionio 21/27, LT-03101, Vilnius, Lithuania

Arvydas Lubys
Thermo Fisher Scientific, V.A. Graiciuno 8, LT-02241, Vilnius, Lithuania
Department of Botany and Genetics, Vilnius University, M.K. Ciurlionio 21/27, LT-03101, Vilnius, Lithuania

Mohea Couturier
Université Versailles St-Quentin, 45 avenue des Etats-Unis, Versailles 78035, France
Institut de Génétique et Microbiologie, UMR 8621 CNRS, Université Paris-Sud, Bât. 409, Orsay Cedex 91405, France
Université d'Evry-Val d'Essonne, Boulevard François Mitterrand, Evry 91025, France.
Department of Molecular Biosciences, The Wenner-Gren Institute,
Stockholm University, Stockholm, Sweden

Florence Garnier
Université Versailles St-Quentin, 45 avenue des Etats-Unis, Versailles 78035, France

Institut de Génétique et Microbiologie, UMR 8621 CNRS, Université Paris-Sud, Bât. 409, Orsay Cedex 91405, France

Anna H Bizard
Université Versailles St-Quentin, 45 avenue des Etats-Unis, Versailles 78035, France
Institut de Génétique et Microbiologie, UMR 8621 CNRS, Université Paris-Sud, Bât. 409, Orsay Cedex 91405, France
Institute of Cellular and Molecular Medicine (ICMM), Center for Healthy Ageing (CEHA), University of Copenhagen, Blegdamsvej 3B, København N DK-2200, Denmark

Marc Nadal
Université Versailles St-Quentin, 45 avenue des Etats-Unis, Versailles 78035, France
Institut de Génétique et Microbiologie, UMR 8621 CNRS, Université Paris-Sud, Bât. 409, Orsay Cedex 91405, France
Université Paris Diderot, 5 rue Thomas Mann, Paris 75013, France

Sarah E Reks, Vera McIlvain, Xinming Zhuo and Barry E Knox
Departments of Neuroscience & Physiology, Ophthalmology and Biochemistry & Molecular Biology, State University of New York Upstate Medical University, Syracuse, NY 13210, USA

Guang-Hua Luo, Xiao-Huan Li, Hui-Fang Guo, Qiong Yang, Zhi-Chun Zhang, Bao-Sheng Liu and Ji-Chao Fang
Institute of Plant Protection, Jiangsu Academy of Agricultural Sciences, Nanjing 210014, China

Zhao-Jun Han and Min Wu
Education Ministry Key Laboratory of Integrated Management of Crop Diseases and Pests, College of Plant Protection, Nanjing Agricultural University, Nanjing 210095, China

Lu Qian
Jiangsu Entry-Exit Inspection and Quarantine Bureau, Nanjing 210001, China

Katie Rose Boissonneault
Department of Biological Sciences, Plymouth State University, MSC 64, 17 High St., Plymouth, NH 03264, USA
Koch Institute, Massachusetts Institute of Technology, 76-553, 77 Massachusetts Avenue, Cambridge, MA 02139, USA

David E Housman
Koch Institute, Massachusetts Institute of Technology, 76-553, 77 Massachusetts Avenue, Cambridge, MA 02139, USA

Brooks M Henningsen
Department of Biological Sciences, Plymouth State University, MSC 64, 17 High St., Plymouth, NH 03264, USA
Mascoma Corporation, 67 Etna Road Suite 300, Lebanon, NH 03766, USA

Stephen S Bates
Fisheries and Oceans Canada, Gulf Fisheries Centre, P.O. Box 5030, Moncton, New Brunswick E1C 9B6, Canada

Deborah L Robertson
Biology Department,Clark University, 950 Main Street, Worcester, MA 01610, USA

Sean Milton
Koch Institute, Massachusetts Institute of Technology, 76-553, 77 Massachusetts Avenue, Cambridge, MA 02139, USA
Vertex Pharmaceuticals, 130 Waverly Street, Cambridge, MA 02139, USA

Jerry Pelletier
Department of Biochemistry, McGill University, 3655 Promenade Sir William Osler, Montreal, Quebec H3G 1Y6, Canada

Deborah A Hogan
Department of Microbiology and Immunology, Vail Building Room 208, Dartmouth Medical School, Hanover, NH 03755, USA

Sara Ali, Chitralekha Bhattacharya and Rui Zhu and Angabin Matin
Department of Genetics, University of Texas, MD Anderson Cancer Center, 1515 Holcombe Blvd, Houston, TX 77030, USA

Donna A MacDuff, Mark D Stenglein, April J Schumacher, Zachary L Demorest and Reuben S Harris
Department of Biochemistry, Molecular Biology, and Biophysics, University of Minnesota, 321 Church
Street SE, Minneapolis, MN 55455, USA

Namrata Karki and Sita Aggarwal
Pennington Biomedical Research Center, 6400 Perkins Road, Baton Rouge, LA 70808, USA

Renate Schloemer and Elke Deuerling
Molecular Microbiology, University of Konstanz, Constance 78457, Germany

Silke Mueller
Screening Center Konstanz, University of Konstanz, Constance 78457, Germany

Rainer Nikolay
Molecular Microbiology, University of Konstanz, Constance 78457, Germany
Institute of Medical Physics and Biophysics, Charité-Universitaetsmedizin Berlin, Berlin 10117, Germany

Carina Modig, Huthayfa Mujahed, Hazem Khalaf and Per-Erik Olsson
Örebro Life Science Center, School of Science and Technology, Örebro University, Örebro SE-701 82, Sweden

Peter Kling
Department of Zoology, Göteborg University, Göteborg SE-405 30, Sweden

Jonas von Hofsten
Department of Molecular Biology, Umeå University, Umeå SE-901 87, Sweden

Noa Iglesias
Department of Brain Ischemia and Neurodegeneration, Institut d'Investigacions Biomèdiques de Barcelona (IIBB)-Consejo Superior de Investigaciones Científicas (CSIC), Barcelona, Spain

Raquel Buj
Department of Brain Ischemia and Neurodegeneration, Institut d'Investigacions Biomèdiques de Barcelona (IIBB)-Consejo Superior de Investigaciones Científicas (CSIC), Barcelona, Spain
Institut de Medicina Predictiva i Personalitzada del Càncer (IMPPC), Badalona, Barcelona, Spain

Anna M Planas and Tomàs Santalucía
Department of Brain Ischemia and Neurodegeneration, Institut d'Investigacions Biomèdiques de Barcelona (IIBB)-Consejo Superior de Investigaciones Científicas (CSIC), Barcelona, Spain
Institut d'Investigacions Biomèdiques August Pi i Sunyer (IDIBAPS), Barcelona, Spain

Alexandre Juillerat, Marine Beurdeley, Julien Valton, Séverine Thomas, Gwendoline Dubois, Mikhail Zaslavskiy, Jérome Mikolajczak, Fabian Bietz, George H Silva, Aymeric Duclert, Fayza Daboussi and Philippe Duchateau
CELLECTIS S.A, 8 Rue de la Croix Jarry, Paris 75013, France

Shuang Jia, Pin-Jun Wan, Li-Tao Zhou, Li-Li Mu and Guo-Qing Li
Education Ministry Key Laboratory of Integrated Management of Crop Diseases and Pests, College of Plant Protection, Nanjing Agricultural University, Nanjing 210095, China

Xin Li, Maurizio Mauro and Zev Williams
Department of Obstetrics & Gynecology and Women's Health, Albert Einstein College of Medicine, 10461 Bronx, NY, USA
Department of Genetics, Albert Einstein College of Medicine, 10461 Bronx, NY, USA

Iddo Z Ben-Dov
Nephrology and Hypertension, Hadassah - Hebrew University Medical Center, 91120 Jerusalem, Israel

Index

www.ingramcontent.com/pod-product-compliance
Lightning Source LLC
Chambersburg PA
CBHW082050190326
41458CB00010B/3499

* 9 7 8 1 6 8 2 8 6 6 5 1 1 *